Soil Quality and Agricultural Sustainability

Edited by

Rattan Lal

Ann Arbor Press
Chelsea, Michigan

Cataloging-in-Publication Data

Soil quality and agricultural sustainability / R. Lal (editor).
p. cm.
Includes bibliographical references and index.
ISBN 1-57504-082-4
1. Soils—Quality—Congresses. 2. Soil management—Congresses.
3. Sustainable agriculture—Congresses. I. Lal, R.
S590.2.S6286 1998
631.4′22—dc21 98-9659
CIP

ISBN 1-57504-082-4

Ann Arbor Press
121 South Main Street
Chelsea, MI 48118
Ann Arbor Press is an imprint of Sleeping Bear Press

Printed in the United States of America

10 9 8 7 6 5 4 3 2 1

PREFACE

Soil resources of the world are finite, easily degraded by misuse and mismanagement, nonrenewable over the human time frame, and shrinking because of degradation and conversion to nonagricultural uses. Soil degradation is a severe global issue, and predominant degradative processes are accelerated soil erosion, depletion of soil organic matter and plant nutrients, decline of soil structure, and salinization. Most of these degradative processes are more severe in the tropics than in the temperate climate, in marginal rather than prime agricultural lands, and in resource-based and subsistence agriculture rather than science-based or commercial farming. Soil degradation and poverty go hand-in-hand because soil degradation is driven by the strong interaction between biophysical and socioeconomic factors.

Soil degradation is a double-edged sword because of its adverse impact on agronomic productivity and also on environmental quality. Soil degradation jeopardizes soil's capacity to perform its functions, e.g., agronomic, environmental, industrial, engineering, athletic, cultural, and aesthetical. Soil's capacity to function, or soil quality is undermined by various degradative processes. The concept of soil quality has been in use since the 1950s, especially by pedologists who developed methodology and criteria for land evaluation and soil capacity assessment. The use of the specific term "soil quality" since the late 1980s is related to the issues of sustainability, particularly with regard to agricultural sustainability.

The concept of agricultural sustainability addresses the issue of food security and environmental quality. The issue of food security is more important now than ever before because of the shrinking land resources, persistence of food deficit and hunger in Africa and South Asia, and the inability of the scientific community to break the agrarian stagnation and bring about the green revolution in Africa.

The green revolution in South Asia was ushered in by the introduction of improved varieties responsive to inputs of fertilizers and other managerial inputs under irrigated conditions. However, the same strategy has thus far evaded Africa and other parts of the tropics probably because of the extremely severe soil-related constraints to crop production. It is the soil quality, its productivity and environmental regulatory capacity, that is being questioned in the regions with a perpetual food deficit and severe degradation of soil and other natural resources. In other words, the green revolution has evaded Africa despite the availability of new germ plasm and improved varieties because of the inherently poor quality of its soil resources and their high susceptibility to degradation.

It was with this background, and in view of the strong link between soil quality and agricultural sustainability, that a workshop was organized in July 1996. The workshop was sponsored by The Ohio State University, USAID, and International Agricultural Research Centers. In addition to these sponsoring organizations, workshop participants were drawn from agricultural research institutions in North America, Europe, Asia, Africa, and South America.

This volume is primarily based on the papers presented at the workshop. Several additional papers were invited to fill in the obvious regional or thematic gaps. A total of 26 chapters are presented in six sections dealing with : (i) basic concepts and global issues, (ii) nutrient and water inputs for soil quality management, (iii) soil quality management in Asia, (iv) soil quality management in Africa, (v) soil quality management in tropical America, and (vi) future priorities. These chapters were written by highly professional and knowledgeable researchers representing a wide range of disciplines including soil science, agronomy, ecology, economics, and planning and managing agricultural research and development.

The publication of the proceedings was delayed due to some unforeseen circumstances. However, all authors were extremely cooperative in revising the manuscripts and providing the state-of-the-art information. The logistic support was provided by Ms. Lynn Everett in organizing the workshop. Ms. Brenda Swank did an excellent job in typing, communication, and day-to-day operation of putting the volume together. The support received from Dr. Christine Bergmark and Dr. Rob Bertram of USAID is highly appreciated. The importance of the workshop and its organizational quality were greatly enhanced by active participation of Dr. Bobby Moser, Dr. David Hansen, and Dr. Fred Miller of The Ohio State University.

The subject matter of this volume will be of interest to graduate students, researchers, policy makers, and those interested in agricultural development in the tropics. Themes discussed are of equal interest to biophysical and social scientists.

Rattan Lal
Editor

Columbus, Ohio

List of Contributors

Mohamed Abdel Monem
Farm Resource Management Program
International Center for Agricultural Research in the Dry Areas
P.O. Box 5466
Aleppo, Syria

G.D. Anderson
Edgerston Tofts
Camptown, Jedburgh
Scotland TD8 6NG
United Kingdom

John M. Antle
Department of Agricultural Economics and Economics
Montana State University
Bozeman, MT 59717

Jens B. Aune
Centre for International Environment and Development Studies (Noragric)
P.O. Box 5001
1432 Aas, Norway

Mateete A. Bekunda
Department of Soil Science
Makerere University
P.O. Box 7062
Kampala, Uganda

Kevin F. Bronson
Soil and Water Sciences Division
International Rice Research Institute
P.O. Box 933
Manila, Philippines

Bernard H. Byrnes
International Fertilizer Development Center
P.O. Box 2040
Muscle Shoals, AL 35662-2040

Robert J. Carsky
International Institute of Tropical Agriculture
PMB 5320
Ibadan, Nigeria

Eric T. Craswell
International Board for Soil Research and Management
P.O. Box 9-109
Bangkhen, Bangkok 10900
Thailand

Charles C. Crissman
International Potato Center
Box 17-21-1977
Quito, Ecuador

Charles R. Curtis
Department of Plant Pathology
The Ohio State University
Columbus, OH 43210-1087

Olaf Erenstein
CIMMYT
Lisboa 27, Apdo. Postal 6-641
06600 Mexico, D.F.
Mexico

Dennis P. Garrity
International Center for Research in Agroforestry
Jalan Gunung Batu No. 5
P.O. Box 161
Bogor 16001
Indonesia

B. S. Ghuman
Department of Soils
Punjab Agricultural University
Ludhiana – 141 004
Punjab, India

Larry Harrington
CIMMYT
Lisboa 27, Apdo. Postal 6-641
06600 Mexico, D.F., Mexico

Deborah T. Hellums
International Fertilizer Development Center
P.O. Box 2040
Muscle Shoals, AL 35662-2040

Peter R. Hobbs
International Maize and Wheat Improvement Center
P.O. Box 5186
Kathmandu, Nepal

S. S. Jagtap
Resource and Crop Management Division
International Institute of Tropical Agriculture
PMB 5320, Ibadan, Nigeria

Fedilis S. B. Kaihura
National Soil Service
P.O. Box 5088
Tanga, Tanzania

B. Kayombo
Department of Agricultural Engineering and Land Planning
Sokoine University of Agriculture
P.O. Box 3003
Morogoro, Tanzania

Jacob W. Kijne
International Irrigation Management Institute
P.O. Box 2075
Colombo, Sri Lanka

Method Kilasara
Department of Soil Science
P.O. Box 3008
Sokoine University of Agriculture
Morogoro

John M. Kimble
National Soil Survey Center
USDA-NRCS
Federal Building, Room 152
100 Centennial Mall North
Lincoln, NE 68508-3866

Jean Kotto-Same
Institut Recherche de la Agricole pour le Developpement
P.O. Box 2067, Knolbisson
Yaounde, Cameroon

Ignance K. Kullaya
ARI Lyamungu
P.O. Box 3004
Moshi, Tanzania

Marcel Kuper
International Irrigation Management Institute
P.O. Box 2075
Colombo, Sri Lanka

Rattan Lal
The Ohio State University
School of Natural Resources
2021 Coffey Road
Columbus, OH 43210

Maurice J. Mausbach
USDA-NRCS
Washington, D.C. 20013

Fred P. Miller
The Ohio State University
School of Natural Resources
2021 Coffey Road
Columbus, OH 43210

G.C. Mrema
Association for Strengthening Agricultural Research in Eastern and Central Africa
P.O. Box 765
Entebbe, Uganda

A. Naazie
International Livestock Research Institute
IITA Campus
Ibadan, Nigeria

R. S. Narang
Department of Agronomy
Punjab Agricultural University
Ludhiana – 141 004
Punjab, India

J. Robert Okalebo
Moi University
Chepkoilel Campus
P.O. Box 1125
Eldoret, Kenya

Dimas Vital Siqueira Resck
EMBRAPA-CPAC
C. Postal 08223
73301-970 Planaltina, DF
Brazil

John Ryan
Farm Resource Management Program
International Center for Agricultural Research in the Dry Areas
P.O. Box 5466
Aleppo, Syria

N. Sanginga
Resource and Crop Management Division
International Institute of Tropical Agriculture
PMB 5320, Ibadan
Nigeria

Ken Sayre
CIMMYT
Lisboa 27, Apdo. Postal 6-641
06600 Mexico, D.F.
Mexico

Cathy A. Seybold
Soil Quality Institute, NRCS
Oregon State University
Corvallis, OR 97331-7306

Bal Ram Singh
Department of Soil Science
P.O. Box 5028
Agricultural University of Norway
1432 Aas, Norway

J. W. Smith
International Livestock Research Institute
IITA Campus
PMB 5320
Ibadan, Nigeria

Carlos R. Spehar
EMBRAPA-CPAC
C. Postal 08223
73301-970 Planaltina, DF
Brazil

G. Tian
Resouce and Crop Management Division
International Institute of Tropical Agriculture
PMB 5320, Ibadan
Nigeria

Goro Uehara
Department of Agronomy and Soil Science
University of Hawaii
1910 East-West Road, Room 101
Honolulu, HI 96822-2319

B. Vanlauwe
Resource and Crop Management Division
International Institute of Tropical Agriculture
PMB 5320, Ibadan
Nigeria

Raúl R. Vera
Lowland Savanna Program
Centro Internacional de Agricultura Tropical
A.A. 6713
Cali, Colombia

Jeffrey W. White
CIMMYT
Lisboa #27
Apdo. Postal 6-641
06600 Mexico, D.F.
Mexico

Paul L. Woomer
Tropical Soil Biology and Fertility Programme
P.O. Box 30592
Nairobi, Kenya

Contents

Part 1

Basic Concepts and Global Issues

Part II

Nutrient and Water Inputs for Soil Quality Management

Part III

Soil Quality Management in Asia

Part IV

Soil Quality Management in Africa

Part V

Soil Quality Management in Tropical America

Part VI

Future Priorities

One

Basic Concepts and Global Issues

ONE

Soil Quality and Agricultural Sustainability

R. Lal

INTRODUCTION

The growing concern of agricultural sustainability expressed during the last two decades of the twentieth century is being carried into the twenty-first century because of the limited arable land resources, rapidly increasing world population especially in developing countries of the tropics and subtropics, conversion of agricultural land to other uses, and persistence of hunger and malnutrition in several regions of the world. World grain harvested area increased steadily for three decades from 587 million hectares (ha) in 1950 to 732 million ha in 1981. The grain harvested area has progressively declined from a high in 1981 to about 680 million ha in 1995 (Brown et al., 1997). The per capita grain harvested area has steadily declined from 0.23 ha in 1950 to only 0.12 ha in 1995, and is expected to continue to decline with the projected increase in population. World irrigated area also increased substantially from 1960 to 1980 but the rate of expansion has been low. World total irrigated land area was 139 million ha in 1961, 167 million ha in 1970, 209 million ha in 1980, 240 million ha in 1990, and 249 million ha in 1994. Per capita irrigated land area peaked at 0.047 ha in 1979 and decreased to 0.044 ha in 1994. Therefore, the concern is whether the basic necessities of life can be met at an acceptable standard of living and quality of life, with per capita grain harvested area of about 0.10 ha and including irrigated land area of only 0.04 ha.

The problem is confounded by shrinkage of the arable land area due to soil degradation and conversion to nonagricultural uses. Soil degradation is a major global issue. Although reliable statistics are not available, the degraded area increased from 10% of the world's arable land in early 1970s (Biswas and Biswas, 1974) to 40% in early 1990s (Oldeman, 1994). Furthermore, 6% of world arable land had been severely degraded by early 1990s. The problem of soil degradation is accentuated by increasing demographic pressure, subsistence agriculture that depends on the fertility-mining and resource-exploitative practices, and a combination of harsh climate and fragile soils.

In addition to natural and anthropogenic soil degradative processes, arable land resources are also shrinking due to conversion to other nonagricultural uses. The world population more than doubled from around 2.5 billion in 1950 to over 5.3 billion in 1990, and is expected to at least double again before stabilizing. About 90% of the current increase in

Figure 1.1. Desurfacing for brick making in India.

world population is due to that in the developing countries. Therefore, urbanization is a major issue in most developing countries where prime agricultural land is rapidly being converted to urban centers for expansion of human habitats, industrial use, and to development of infrastructure. The land area being desurfaced for brick making in Asia (especially India and China) is phenomenal (Figure 1.1). In the predominantly agricultural state of Haryana in northwestern India, percent of net cultivated land area desurfaced for brick making increased 10 times during the 5-year period from 1989 to 1993 (Table 1.1).

Consequently, there is a widespread problem of hunger and malnutrition throughout the developing world. The absolute number of chronically underfed people in the world was 942 million in 1970, 976 million in 1975, 846 million in 1980 and 786 million in 1990 (Uvin, 1994; Young, 1997). The absolute number of chronically underfed population is rapidly increasing in sub-Saharan Africa, which was 94 million in 1970, 1120 million in 1975, 128 million in 1980, and 175 million or 37% of the total population in 1990. Despite the Green Revolution, the absolute number of chronically underfed population in South Asia was 277 million in 1990.

The issue of chronic malnutrition and hunger is confounded by increasing world pop-

Table 1.1. Land Area Desurfaced for Brick Making in Haryana, India[a]

Year	% of the Net Cultivated Area[b]
1989	0.055
1990	0.121
1991	0.210
1992	0.385
1993	0.472

Crop stand of *Brassica* Sp on desurfaced soil is 38 to 84% of the normal.

[a]Recalculated from Agrawal et al., 1994.
[b]Net cultivated area = 3.63 x 106 ha.

ulation and decreasing per capita cereal production. Steady increases in per capita cereal production occurred from about 280 kg in 1950 to 370 kg in 1985 (Dyson, 1996). However, the per capita cereal production has declined since mid-1980s and was only 350 kg in 1993. Therefore, the question of agricultural sustainability is extremely relevant. Are the world's soil resources adequate to feed its growing population? Can soil productivity be enhanced to increase agricultural output? What are the indicators of soil's capacity to increase and maintain productivity? These questions have led to a considerable interest in data on productivity trends observed in long-term experiments conducted around the world (Leigh and Johnston, 1994; Barnett et al., 1995; Lal and Stewart, 1995).

AGRICULTURAL SUSTAINABILITY

It refers to the ability of agricultural systems to remain productive, efficiently and indefinitely. Quantitatively, it implies trends in agricultural production over time. A non-negative trend in production of a system over time implies that the system is sustainable. Herdt and Steiner (1995) outlined three aspects of sustainability of a system: space, time, and dimension.

The spatial aspect of sustainability refers to the scale of its assessment or quantification. Sustainability may be assessed at the scale of a cell, plant, crop, farming system, regional system, and a global system. At a farm level, the sustainability is appropriately assessed at a cropping or farming system level. In addition to scale, sustainability is dynamic and a time-dependent entity because agricultural production systems are constantly changing due to changing demands of the population that develops these for meeting its needs. Therefore, sustainability is assessed over time, and a logical time span is a human generation or 25 years. There are three dimensions of assessing the sustainability, e.g., biophysical, economic, and social (Herdt and Steiner, 1995). The biophysical dimension is related to the quantity of output (Mg of yield/ha), the economic dimension to the value of the output (gross or net), and the social dimension to the capacity of the system to support the farming community.

Sustainability must be assessed in relation to all three dimensions. The biophysical output (biomass or grain yield per ha) may change due to change in soil properties over time (erosion, compaction, salinization, waterlogging, etc.), introduction of new cultivars, and change in the input used. The economic output may change over time independent of the biophysical output, and the social carrying capacity may change due to change in food habits and standard of living.

Lal (1994) proposed several indices of measuring sustainability. These indices include the following:

Productivity (P)

It refers to production per unit of resources used (Eq. 1).

$$P = p/R \tag{1}$$

where P is productivity, p is total biophysical production (biomass, grain yield), and R is resource input. It is relevant to assess temporal changes in productivity. There are several indications that crop yields of the rice-wheat system in South Asia per unit input of fertilizer input is declining. In that event, the productivity of the system is not sustainable. There are several reasons for the unsustainability of the rice-wheat system. Important among these is the decline in soil organic matter content. These soils have low soil or-

Figure 1.2. Crop residue removal from farmland in the Indo-Gangetic plains is a major factor responsible for low and declining soil organic matter content.

ganic matter content and all of the wheat straw is removed for feeding cattle (Figure 1.2) and most rice straw is also either removed or burned (Figure 1.3). Further, more soil puddling (Figure 1.4) for rice cultivation leads to destruction of soil structure, and the latter is exacerbated by depletion of soil organic matter content.

Total Factor Productivity (TFP)

Rather than considering a single input (e.g., fertilizer input in Fig. 1.5), the total factor productivity is computed by considering the cost of all factors of production (Eq. 2).

$$TFP = p / \sum_{i=1}^{n} (R_i \cdot C_i) \tag{2}$$

where p is total production, R_i is the resource used, C_i is the cost of the specific input, and n is the number of input (resources) used in achieving total production. The TFP index is

Figure 1.3. Crop residue and other biomass is burned in Africa and elsewhere in the tropics.

Figure 1.4. Puddling of soil for planting rice destroys soil structure, and may have a negative impact on the following wheat crop grown on the same land throughout the rice-wheat belt in South Asia.

a measure of total output relative to an index of total managed inputs (e.g., seeds, water, traction, agricultural chemicals, labor, land resources, etc.).

Total Natural Resource Productivity (TNRP)

In addition to managed inputs, there are some indirect costs associated with specific output which may be due to the degradation of natural resources, e.g., decrease in topsoil depth due to erosion, increase in soluble salt content due to salinization, fall in the groundwater table due to irrigation, eutrophication of natural water due to losses of fertilizer, etc. Therefore, TNRP can be computed from TFP by using Equation 3.

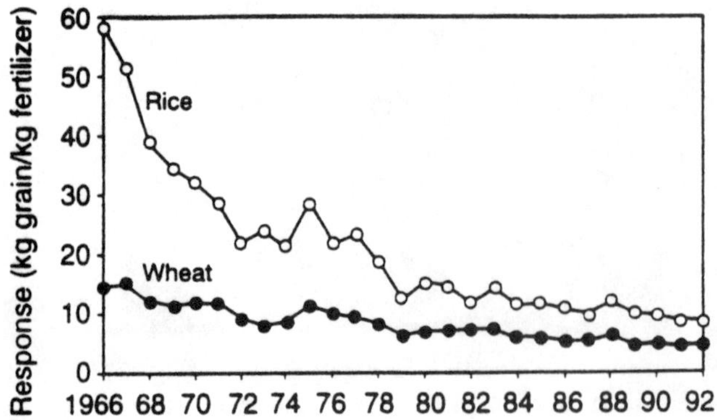

Figure 1.5. Average response of rice and wheat to fertilizer, all India, 1966–1992. From: Fertilizer Association of India (various).

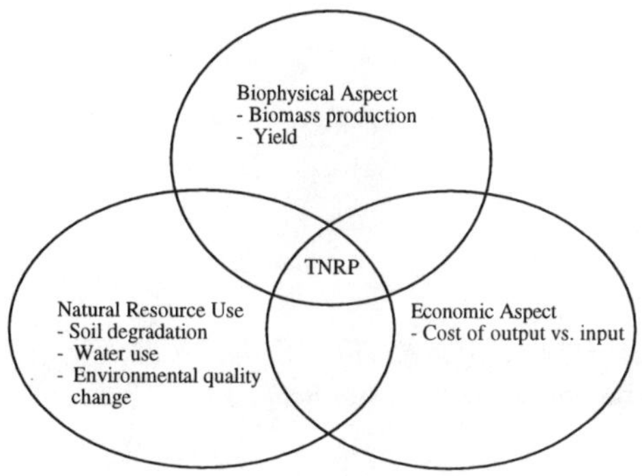

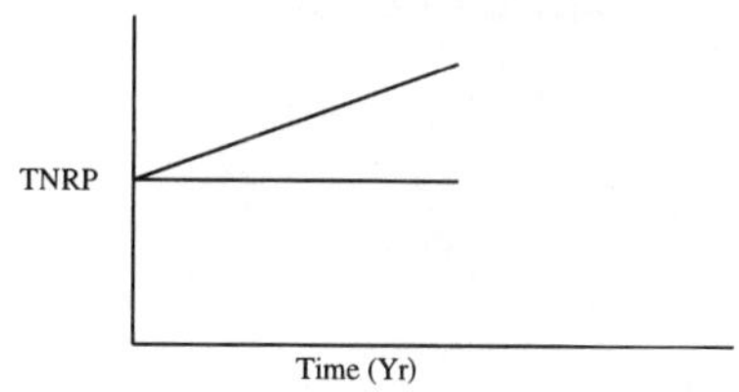

Figure 1.6. Sustainability implies a non-negative trend in TNRP, output per unit input of managed and natural resources used.

$$\text{TNRP} = (\text{TFP})/\Delta S \qquad (3)$$

where ΔS is the change in critical soil properties. If soil degradation is due to soil erosion, then TFP is divided by the annual loss of topsoil. The denominator may be soil organic carbon content, nitrate concentration in the groundwater, fall in the groundwater level, etc. Lal (1994) referred to the TNRP as the coefficient of sustainability or index of sustainability, and Herdt and Steiner (1995) refered to TNRP as "total social factor productivity." The TNRP is also an index of the health of the natural resource, and combines biophysical, economic, and social dimensions of assessment of sustainability. Agricultural sustainability, therefore, is a non-negative trend in TNRP over time (Fig.1.6).

SOIL QUALITY

Since the late 1980s, soil scientists have denoted the soil's capacity to perform specific functions by the term "Soil Quality" (SSSA, 1987; Larson and Pierce, 1991; Papendick and Parr, 1992; Parr et al., 1992; Lal, 1993; Pierce and Larson, 1993). Doran and Parkin (1994) defined soil quality as "the capacity of a soil to function within ecosystem boundaries to sustain biological productivity, maintain environmental quality, and promote plant and animal health." Johnson et al. (1997) proposed that "soil quality is a measure of the condition of the soil relative to the requirements of one or more species and/or to any human needs or purposes."

Most of these definitions refer "soil quality" to its capacity to perform specific functions. In terms of agricultural production, therefore, soil quality refers to its ability to sustain productivity. There exists a strong link between soil quality and agricultural sustainability (Fig. 1.7). As soil quality increases, TFP and TNRP also increase. If an agricultural system is unsustainable, it may partly be due to the fact that soil quality is declining over time. In other words, there is a negative trend in TRP or TNRP due to decline in soil quality caused by reduction in topsoil depth by accelerated erosion, increase in soil salin-

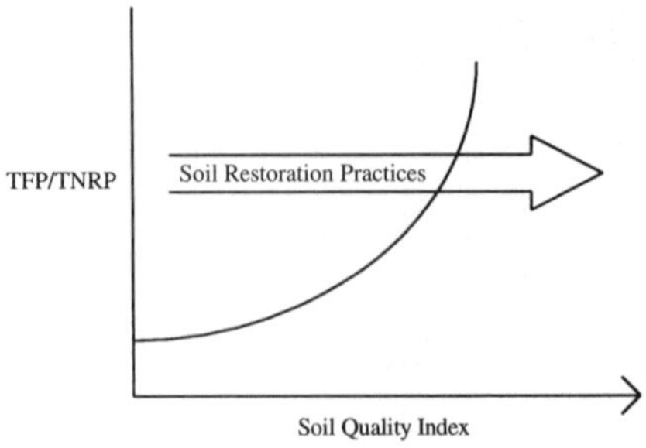

Figure 1.7. Relationship between soil quality and agricultural productivity or sustainability.

Table 1.2. Soil Physical Constraints to High Crop Yield in Haryana, India[a]

Soil Physical Constraint	% of Net Cultivated Area[b]
Crusting	43.2
Wind erosion	26.0
Waterlogging	17.2
Hard-setting (hardening)	17.2
High infiltration rate (permeability)	14.8
Shallow topsoil	5.5
High subsoil mechanical impedance	1.9
Water erosion	1.9

[a]Modified from Agrawal et al., 1994.
[b]Net cultivated area = 3.63×10^6 ha; total land area = 4.42×10^6 ha.

ity due to waterlogging, drop in the groundwater table due to excessive irrigation, depletion of soil organic matter content due to residue removal or burning, or depletion of soil fertility due to inadequate application of fertilizers and manures. The negative trend in TNRP may also be due to other factors; e.g., eutrophication of natural waters, emissions of radiatively-active gases from soil to the atmosphere, etc. Adoption of soil restorative measures can improve soil quality and transform an unsustainable agricultural or land use system into a sustainable system (Fig. 1.7). An important step in improving soil quality and achieving agricultural sustainability, therefore, is to identify soil-related constraints to crop production. An example of such a constraint evaluation is shown by the data in Table 1.2. Adoption of improved restorative technologies can alleviate constraints, improve TFP and TNRP, and help achieve sustainability.

INTERACTIVE EFFECTS OF SOIL QUALITY ON AGRICULTURAL SUSTAINABILITY AND ENVIRONMENT QUALITY

There is a close link between soil quality, economic progress, and environmental quality. Soil is the most basic of all natural resources. In addition to biomass production, it regulates the environment. There is indeed a close link between soil quality and environment quality on the one hand, and soil quality and economic productivity on the other (Fig. 1.8). Decline in soil quality can set in motion the degradative trends that lead to reduction in agricultural production, decrease in industrial (agri-based industries) output, increase in rural and urban unemployment, and reduction in GAP and GDP. What begins as soil degradation eventually translates into environmental degradation and political instability (Brown, 1997). On the other hand, decline in soil quality has a strong adverse impact on soil's environmental/ecological functions that maintain water quality and en-

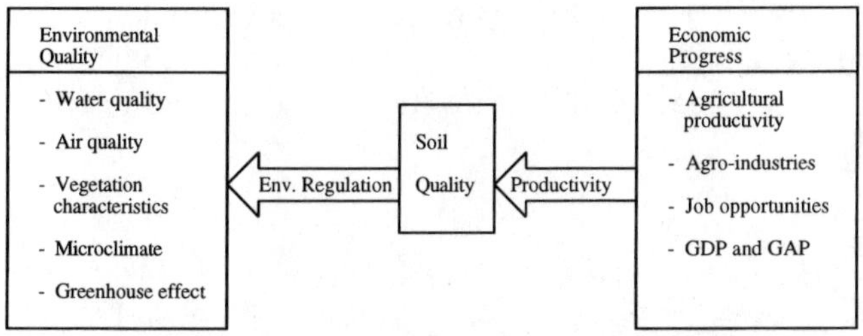

Figure 1.8. Interactive effects of soil quality on economic progress and environmental quality.

hance air quality. Soil degradation influences water balance, biomass production and the vegetation cover, and the microclimate. Soil degradation leads to desertification.

Therefore, restoring and maintaining high soil quality is the principal strategy for attaining economic progress and improving environmental quality. Choice of land use and soil management practices should be made with due consideration to restoration and maintenance of high soil quality.

NEED FOR A POSITIVE AND A HOLISTIC APPROACH TO SOIL QUALITY ENHANCEMENT AND NATURAL RESOURCE MANAGEMENT

There are numerous examples of success stories and bright spots regarding the adoption of improved technologies for restoration of soil quality and improvements in agricultural productivity. The Green Revolution in South Asia is an example of what can be achieved by using improved technology. However, there is a need to apply similar strategies in sub-Saharan Africa and elsewhere in the densely populated regions of the tropics to restore degraded soils and improve productivity.

It is important that soil scientists and agronomists involved in natural resources management adopt a holistic approach. They cannot afford to work in isolation, and must work in close cooperation with scientists from other disciplines; e.g., geography, geology, ecology, biology, climatology, and hydrology. Soil degradative trends are set in motion by numerous interactive processes, and these processes must be studied by an interdisciplinary approach. Soil scientists around the world are seeking a new paradigm of reaching out to other disciplines in addressing global problems of broader interest. The narrow disciplinary approach can be counterproductive and miss an opportunity to effectively address practical issues that are appropriately addressed through a multidisciplinary approach.

Soil scientists must also adopt a positive approach to natural resources management (Table 1.3). They must think about soil conservation rather than erosion, soil fertility enhancement rather than nutrient depletion and imbalance, soil restoration rather than

Table 1.3. Positive Approach to Soil Natural Resources Management

Positive		Negative
Soil conservation		Soil erosion
Fertility enhancement	rather	Nutrient depletion and imbalance
Soil restoration	than	Degradation and desertification
Judicious management		Low input systems

degradation and desertification, and judicious use of input rather than low input systems. There are technologies available to restore degraded soils, and improve their productivity. A positive approach, based on interdisciplinary and holistic strategies, can lead to widespread adoption of these technologies.

REFERENCES

Agrawal, R.P., V.K. Phogat, T. Chand, and M.S. Grewal. Improvements of Soil Physical Conditions in Haryana. Dept. of Soil Sci., CCS Haryana Agric. Univ., Hisar, India, 1994.

Barnett, V., R. Payne and R. Steiner, Eds. *Agricultural Sustainability: Economic, Environmental and Statistical Considerations.* J. Wiley & Sons, Chichester, U.K., 1995.

Biswas, A.K. and M.R. Biswas. 1974. Environmental Considerations of Increasing World Food Production. UNEP, Nairobi, Kenya.

Brown, L.R. The Agricultural Link: How Environmental Deterioration Could Disrupt Economic Progress. Worldwatch Paper 136, Washington, D.C., 1997.

Brown, L.R., M. Renner, and C. Flavin. 1997. Vital Signs: The Environmental Trends That Are Shaping Our Future. Worldwatch Institute.

Doran, J.W. and T.B. Parkin. Defining and Assessing Soil Quality, in *Defining Soil Quality for a Sustainable Environment,* J.W. Doran, D.C. Coleman, D.F. Bezdicek and B.A. Stewart, Eds. Soil Sci. Soc. Am. Special Publication No. 35, Madison, WI, pp. 3–21, 1994.

Dyson, T. *Population and Food,* Routledge, London, U.K., 1996.

Herdt, R.W. and R.A. Steiner. 1995. Agricultural Sustainability: Concepts and Conundrums, in *Agricultural Sustainability: Economic, Environmental and Statistical Considerations,* Barnett, V., R. Payne, and R. Steiner, Eds. J. Wiley & Sons, Chichester, U.K., 1995, pp. 1–13.

Johnson, D.L., S.H. Ambrose, T.J. Bassett, M.L. Bowen, D.E. Crummey, J.S. Isaacson, D.N. Johnson, P. Lamb, M. Saul and A.E. Winter-Nelson. Meaning of Environmental Terms. *J. Environ. Qual.* 26: 581–589, 1997.

Lal, R. Tillage Effects on Soil Degradation, Soil Resilience, Soil Quality and Sustainability. *Soil Tillage Res.* 27: 1–8, 1993.

Lal, R. Methods and Guidelines for Assessing Sustainable Use of Soil and Water Resources in the Tropics. Soil Management Support Services, USDA-NRCS, Washington, D.C., 1994.

Lal, R. and B.A. Stewart, Eds. *Soil Management: Experimental Basis for Sustainability and Environmental Quality.* CRC/Lewis Publishers, Boca Raton, FL. 1995.

Larson, W.E. and F.J. Pierce. Conservation and Enhancement of Soil Quality, in *Evaluation for Sustainable Land Management in the Developing World.* Vol. 2. IBSRAM Proc. 12 (2). Bangkok, Thailand. Int. Board for Soil Res. and Management, 1991.

Leigh, R.A. and A.E. Johnston, Eds. *Long-Term Experiments in Agricultural and Ecological Sciences.* CAB International, Wallingford, U.K., 1994.

Oldeman, L.R. The Global Extent of Soil Degradation, in D.J. Greenland and I. Szabolcs, Eds., *Soil Resilience and Sustainable Land Use,* CAB International, Wallingford, U.K., pp. 99–118, 1994.

Papendick, R.I. and J.F. Parr. Soil Quality—The Key to a Sustainable Agriculture. *Am. J. Altern. Agric.* 7: 2–3, 1992.

Parr, J.F., R.I. Papendick, S.B. Hornick, and R.E. Meyer. Soil Quality: Attributes and Relationship to Alternative and Sustainable Agriculture. *Am. J. Altern. Agric.* 7: 5–11, 1992.

Pierce, F.J. and W.E. Larson. Developing Criteria to Evaluate Sustainable Land Management, in *Proc. of the 8th Int'l Soil Management Workshop; Utilization of Soil Survey Information for Sustainable Land Use,* J.M. Kimble, Ed. May 1993. USDA-SCS, National Soil Surv. Center, Lincoln, NE, pp. 7–14, 1993.

SSSA (Soil Science Society of America). *Glossary of Soil Science Terms.* SSSA, Madison, WI, 1987.

Uvin, P. *The Hunger Report: 1993.* Brown University, Providence, RI, 1994.

Young, E.M. *World Hunger.* Routledge, London, U.K., 1997.

Two

Soil-Land Degradation: Interventions for Sustaining Civilization

Fred P. Miller

INTRODUCTION

Soil scientists and other natural scientists tend to view land and soil degradation from a physical-biogeochemical paradigm. It is a relatively straightforward, albeit daunting, task to observe, measure, and spatially display at various scales land-soil degradation as manifested by accelerated erosion, compaction, hard-setting, loss of organic matter and microbial diversity, drawdown of nutrient stocks, salinization, and other manifestations of degradation. The physical-biochemical dimensions of the processes responsible for land-soil degradation are relatively well understood. While it is beyond the scope of this chapter to review the nature and extent of soil and land degradation, various soil-land degradation databases are available at local, regional, and global scales such as those found in Oldeman et al., 1990; Dregne and Chou, 1992; World Resources Institute, 1992; and USDA 1992 National Resources Inventory, 1994.

If soil scientists view soil and land degradation in physical-biogeochemical terms, economists, at least until recently, traditionally have viewed such degradation in very limited terms, if at all. Economists tend to measure only that part of the natural resource base utilization equation which sees natural resources as providers and producers of products easily valued and measured in an economic marketplace. Economists usually ignore any deterioration of natural resources or functional loss of ecosystem processes. Degraded soil and land, therefore, are not recognized as depreciated assets in calculations of a country's gross domestic product (GDP), which is simply a gross measure of market activity and money changing hands. GDP makes no distinction as to whether such activity is desirable or undesirable (e.g., soil-land degradation).

Social scientists view soil and land degradation from the perspective of individual, community, and social-political institutions. This hierarchy of social systems imposes different kinds of demands and pressures on soil and land resources and brings a variety of values and incentives to bear on natural resources. Social scientists bring to the natural resources management table a variety of possible strategies to intervene in the social processes that are important drivers of soil and land degradation.

Soil scientists, therefore, must not think of themselves as the sole curators of the soil and land or as the principal source of wisdom in managing soil and landscapes. There are many more factors and disciplines involved in both the processes responsible for soil and land degradation and the control interventions than what soil science brings to the table. To minimize soil and land degradation requires both technical and sociopolitical interventions. Soil scientists, therefore, must address the problems and solutions of soil and land degradation as members of a team of disciplines, not as a solo act.

ROOT CAUSES OF SOIL-LAND DEGRADATION

If the physical, chemical, and biological causes and consequences of soil and land degradation are relatively well known, then the next logical question is, "What are the root causes that provoke and drive these environmental perturbations"? Then we must ask "What strategies and interventions are necessary to address these root causes and minimize these environmental impacts and abuses to those ecosystem processes that undergird the sustainability of natural resources and ultimately the sustainability of civilization?"

At the apex of the root causes and drivers of soil and land degradation is the demand by people for sustenance and well-being coupled with economic frameworks which essentially ignore the degradation of natural resources in the equations used to calculate socioeconomic "progress"; e.g., GDP. Cohen (1995) has stated that basic ecology coupled with human purpose begets applied ecology which includes agriculture, forest management, grassland-range management, and wildlife-fisheries management. Applied ecology, therefore, aims first and foremost at satisfying human wants. Human, and by extension, civilization's sustenance and well-being are inextricably linked to the earth's natural resources base and those ecological processes and functions (Table 2.1) that sustain this natural resource base.

Table 2.1. Ecosystem Goods and Services.[a] Healthy ecosystems carry out a diverse array of processes that provide both goods and services to humanity. Here, goods refer to items given monetary value in the marketplace, whereas the services from ecosystems are valued, but rarely bought or sold

Ecosystem processes include:
- Hydrologic flux and storage
- Biological productivity
- Biogeochemical cycling and storage
- Decomposition
- Maintenance of biological diversity

Ecosystem "goods" include:
- Food
- Construction materials
- Medicinal plants
- Wild genes for domestic plants and animals
- Tourism and recreation

Ecosystem "services" include:
- Maintaining hydrological cycles
- Regulating climate
- Cleansing water and air
- Maintaining the gaseous composition of the atmosphere
- Pollinating crops and other important plants
- Generating and maintaining soils
- Storing and cycling essential nutrients
- Absorbing and detoxifying pollutants
- Providing beauty, inspiration, and recreation

[a]Modified from Ehrlich and Ehrlich, 1991, Lubchenco et al., 1993, and Richardson, 1994.

At the global scale, much environmental degradation is caused by peoples' need to counter the specter of hunger and poverty. Another cause of environmental degradation is the lack of appreciation and understanding of ecological processes (Table 2.1) that sustain ecosystems and their natural resources. The slow, relentless, and usually imperceptible but insidious processes that induce soil erosion and other kinds of soil-land degradation occur at such rates and scales and with such ubiquitousness that human perception of their occurrence and consequences are either nonexistent, minimal, or trivialized as not being important enough to be concerned about. Exacerbating this situation is the fact that such degradation and ecosystem deterioration fall outside traditional economic marketing-pricing systems, thereby becoming invisible to national accounting systems and political processes.

CURTAILING DEGRADATION: TECHNICAL FIXES—SOCIAL INTERVENTIONS

Whether driven by poverty and hunger or production-profit motives, curtailing soil and land degradation and maintaining ecological integrity provide very challenging tasks for natural resources managers. The application of technical and cultural "fixes" for controlling soil-land degradation is not limited by people's understanding of degradation-controlling practices nor necessarily by their ability to implement these fixes. The major limitations as to why people allow or tolerate soil-land degradation have much to do with their perception of the problem and the value they attach to its solution. Thus, curtailing soil and land degradation rests as much, if not more, in the realm of the social sciences as in the natural or physical sciences.

There is much evidence and experience, both scientifically derived (Halcrow et al., 1982; Hooks et al., 1983; Lovejoy and Napier, 1986; Napier, 1990) and anecdotal, that clearly tells us that the availability of sound information and cultural-technical solutions to land-natural resources user-owners does not necessarily result in sustainable natural resources management. Therefore, other incentives or socioeconomic-political strategies and interventions must be considered if meaningful and long-term solutions to soil-land degradation are to be implemented. Technical solutions for soil and land degradation must be coupled to socioeconomic and political strategies that provide incentives for intervening in what drives soil and land degradation.

CAPTURING THE MOMENTUM OF SOCIETY'S INTEREST IN SUSTAINABILITY

To address the complex issue of soil and land degradation, the time seems ripe for the social and natural scientists to join ranks and take advantage of the public's growing interest in and understanding of the concepts of sustainable natural resources management and sustainable development. These concepts, along with the concept of ecology, are becoming well entrenched as part of the lexicon of the global environmental and political vocabulary. Gallagher et al. (1995) recently stated, "*one of the defining social themes of the decade has been ecological awareness. Consequently, ecology now stands at the interface between science and public policy.*" Sustainability, therefore, seems to have become to the public the conceptual equivalent of maintaining ecological integrity.

While the concept of sustainability has attained both legitimacy and respect as an ecological and economic development paradigm, Viederman (1993) suggests that as the concept of sustainability "*gains currency, it seems to be losing whatever precision it may have had.*" The difficulty in trying to develop a precise and long-term operational definition of sustainability lies in the dynamic nature and breadth of its fundamental components and dimensions; namely, ecological, economic, and social.

Ecological sustainability, while based on scientific principles, has both spatial and temporal dimensions and rhythms, the character of which is related to biological diversity, stability, productivity, and climatic anomalies, as well as resilience from naturally or anthropically-imposed stresses. The economic dimension of sustainability is neither static nor stable as demands for natural resources and their use vary with the dynamics of human populations and value systems these populations attach to natural resources and ecological goods and services. Reluctant as many natural scientists are to consider or affirm the social dimensions of sustainability, this third component of the sustainability dimension trinity plays a major, if not primary, role in sustainable natural resources management and precludes defining sustainability with both stability and precision as well as confining natural resources management to the physical and natural sciences.

The social dimensions of sustainability involve complex issues such as equity, access, stewardship, cultural differences, lifestyle choices, sociopolitical institutions, and amenity values, all of which are subject to change over time. Peoples' demand for natural resources stem from their needs for raw materials to enhance their well-being, however defined, and to satisfy their amenity values. Amenity values include both use and existence values such as aesthetics, recreation value, or the satisfaction that a desired amenity (scenic vistas, old growth forest, herd of elephants) exists. Furthermore, it is becoming clear that more people and their institutions recognize the value of ecosystem goods and support services (Randall, 1994) as indicated in Table 2.1 and the vulnerability of various ecosystems and land uses when inappropriately utilized.

INSTITUTIONAL RECOGNITION OF AN ECOLOGICAL TENET

Institutions such as the insurance and banking-lending industries are among those on the "front lines" of financial vulnerability to global change and ecological disasters caused by improper land uses. Global change-induced events such as increased incidences of floods, droughts, fires, and heat outbreaks as well as damage caused by high probability events such as hurricanes/cyclones, earthquakes, landslides, volcanic eruptions, and expansive soils, are beginning to cause these financial institutions to understand and pay attention to one of the fundamental tenets of ecology, namely, "you cannot do everything equally everywhere."

Damage from expansive soils in the U.S. is at least $6 billion annually, ranking as the nation's most costly natural hazard (U.S. Acad. Sciences, 1987). Economic losses from weather-related natural disasters alone have averaged over 25 billion dollars worldwide since 1990, about 20-25% of which are insured (Flavin, 1996). To reduce their financial exposure to these risks, insurance companies are curtailing insurance coverage in hurricane-vulnerable coastal areas and other natural disaster-prone areas. A single hurricane (Hurricane Andrew in 1992) in southern Florida cost the insurance industry $15.5 billion in just a few hours.

While banks and lending institutions have profited handsomely from financially underwriting urban and suburban growth and development, the underbelly of this growth and development is manifested in poorly planned land uses, inefficient infrastructure development, dissatisfaction with the resulting quality of life as manifested through congestion, ethnic and economic segregation, crime, air pollution, and the uglification of much of the developed and developing world. One of the largest financial institutions in the U.S. recently issued its own manifesto (Pierce, 1995) on this underbelly and found itself guilty of exacerbating these problems even though it profited greatly from the process. As will be discussed subsequently, the relationship of quality of life to profit and gross domestic product (GDP) is not necessarily highly correlated.

Robert Malthus (1798) recognized two centuries ago the decreased "power of the land" resulting from land degradation. And farmers worldwide have learned through experience that not all soils are created equal—some are more productive than others while still others are more vulnerable to erosion and degradation. Therefore, one cannot farm or manage natural resources equally everywhere.

Soil and land degradation offer particularly difficult ecological impacts to quantify economically. Pimentel et al. (1995) calculated the annual direct and indirect damage from soil erosion alone at $44 billion in the U.S. and $400 billion worldwide, although these figures have been challenged as too high by Crosson (1995). While soil scientists and agronomists have quantified lost productivity due to soil erosion (Pierce et al., 1984; Larson et al., 1983; Follett and Stewart, 1985; Jennings et al., 1990; Weesies et al., 1994; Pimentel et al., 1995), there has not been a parallel effort by economists to economically quantify the depreciated soil resource degraded by erosion and other causes. Nevertheless, agriculturalists, natural resources managers, and various institutions are beginning to recognize the necessity to account for this depreciation as forfeited or lost value of ecosystem goods and services.

SUSTAINABILITY AND PRODUCTION AGRICULTURE NEXUS

Natural scientists and natural resources managers must sell society on the necessity for maintaining ecological integrity in order to sustain the natural resource base that undergirds human well-being and civilization. This concept, while scientifically sound, must be marketed to society as an extension of the existing paradigm of agricultural production and profitability, not as a new paradigm devoid of the necessity for high yielding agriculture and profitability. Otherwise, the tendency is for society to assume that they are expected to go, in the parlance of one extreme, "totally green." Thus, natural resources managers must make a linkage between sustainability and production agriculture in its broadest sense.[1]

To sustain agricultural production and natural resources goods and services, therefore, farmers and other natural resources managers must take as their starting point the care and maintenance of the natural resource base and ecological processes. It is this natural resource base and these ecological processes that provide the long-term capability to sustain biological productivity and profitability and sustain the capacity to serve human needs. If biological productivity becomes the primary goal of natural resources and ecosystem management, then sustainability is at risk of being compromised through retrofitted management strategies that commonly exacerbate the deterioration of ecological processes and the environment.

While more nutrients, more pesticides, narrowed crop-biological diversity, and bigger, more powerful equipment may result in higher yields and/or profits in the short term, such investments and production systems usually are not sustainable over long periods without continually ratcheting up inputs at the expense of compromising those ecosystem processes that ultimately undergird the capacity for sustained production. Therefore, the

[1] The U.S. National Academy of Science has defined agriculture to encompass "*the entire system that grows, processes, and provides food and fiber for the globe. It includes: the management of natural resources such as surface and groundwater, forests, and other lands for commercial or recreational uses, and wildlife; the social, physical, and biological environments; the public policy issues that relate to the system. All activities, practices, and processes of the public and private sectors involved in agriculture and forestry are contained within the system.*"

"health" or ecological integrity of an ecosystem is measured, not by its short-term biological productivity, but by its ability to sustain biological productivity over time.

The key to sustaining long-term biological production, therefore, is the maintenance of those ecological processes (Table 2.1) that undergird the ecosystem's capability for production. Applied ecology, as manifested through agriculture, forestry, and fisheries-wildlife management, must have as its foundation the maintenance of ecological processes. It is these processes that are the machinery analogue in the agricultural factory (Vandermeer, 1992). Attributes such as soil quality, balanced soil fertility, and other indicators of sustainability are but manifestations of "healthy" ecological processes and natural resources that have not had their physical, chemical, and/or biological integrity compromised.

THE "CURRENCIES" OF SUSTAINABILITY

The most common and understood "currencies" of ecosystems are their measures of biological productivity and monetary worth. A given hectare, farm field, or ecosystem (e.g., pond) has the capacity to yield so much grain, cassava, beef, forest products, or fish over a period of time. Biological productivity indicators can be converted to monetary value and expressed in terms of net monetary yield per unit of land or ecosystem. But these currencies provide no measure of ecosystem or natural resource depreciation, energy investment, "health"/quality, or social benefit. Other "currencies" have been devised to measure ecosystem output in relation to resource and energy investments as well as social benefits and costs.

One of the challenges facing natural resource and ecosystem managers, be they farmers, foresters, wildlife or fisheries managers, is the task of monitoring the "health" or quality of those ecosystem processes that are requisite for sustaining biological productivity. Farmers must manage their soil resources such that their nutrient stocks are maintained and such that their organic matter and microbial processes are not jeopardized through erosion, compaction, or excessive tillage and tilth reduction. Similarly, fisheries managers must not overharvest the aquatic ecosystem and must be alert to changes or reductions in biodiversity and other indicators of deteriorating ecosystem processes.

Simply "pushing" an ecosystem for higher production through greater resource and energy investments may generate more biological and monetary output "currency," but may jeopardize the ecosystem's capacity for sustaining such levels of output "currency." Thus, different "currencies" must be recognized that reflect the sustainability of ecosystems as well as provide measures of economic and social value to ecosystem and natural resources management. Several of these currencies will be briefly reviewed.

Monetary Currencies; When Prices Miss the Mark

We are all familiar with simple monetary pricing and at least the rudimentary principles undergirding the economics of supply and demand. Farmers produce and sell products for a price and generate income. But such transactions may provide no measure of how well or how poorly farmers managed their production resources. Similarly, a nation could exhaust its mineral reserves, cut its forests, erode its soils, pollute its aquifers, and harvest its fish and hunt its wildlife to extinction—all without affecting measured income and even show an increase in national wealth (Repetto, 1989).

Differences in management of natural resources can send false information and signals to policymakers and those interested in the monetary output and carrying capacity of a unit or country. As Repetto (1989) points out, the consequences of such exploitative natural resources management can be illusory gains in income and permanent losses in wealth. Such monetary "currency" measures do not reflect the loss of valuable assets such as the natural resource base and damage to those ecosystem processes that sustain this base. As the nineteenth-century writer John Ruskin noted, an economy produces "illth" as well as wealth, yet the conventional measures of economic well-being lump the two together, thereby masking the negative "costs" of wealth (Cobb et al., 1995).

The fundamental definition of income encompasses the notion of sustainability, such as when businesses account for depreciation of assets (e.g., machinery) and extend dividends without reducing net worth. Therefore, companies reinvest portions of their profits in research and development and in new equipment in order to sustain their capacity to produce, compete, and remain profitable. But as Repetto (1989) argues, failure to extend this concept to natural resource assets is a major inconsistency of economic accounting and monetary policies in that such accounting portrays the depletion of capital (e.g., soil) as current income. Such "currency" accounting inconsistencies distort the sustainability of economic performance, whether for a farm or a country.

Economists have recognized this inconsistency. The UN-sponsored World Commission on Environment and Development (1987) in its study entitled "Our Common Future," stated that "in all countries, rich or poor, economic development must take full account in its measurement of growth of the improvement or deterioration in the stock of natural resources." Several countries (e.g., Norway, France, former West Germany) have begun to establish environmental accounting systems (Repetto, 1989). Without taking into account the environmental and ecological "costs" (e.g., soil-land degradation, impaired natural resources) of development and economic output, a nation can be blind-sided by environmental decay (Hoehn and Walker, 1993). Without such environmental accounting, the exploitation of soils that results in soil and land deterioration and events such as earthquakes and hurricanes/cyclones can end up contributing to a country's GDP because there is no minus sign on the "national adding machine" to subtract these expenses or document liabilities to natural resource sustainability (Cobb et al., 1995).

Off-site benefits from soil erosion control in the U.S. may account for as much as two-thirds of total soil erosion control benefits, although the downstream costs and benefits of soil loss and its control can vary dramatically from site to site (Hitzhusen, 1991). Aside from soil erosion's impact on productivity and income reduction, external or off-site costs include such elements as increased costs for drinking water treatment, damage to recreational uses, dredging and delays to commercial shipping, and water quality-biological impacts. Because of the great variability of land degradation impacts downstream and benefits of control strategies, targeting limited resources for soil erosion and land degradation control intervention measures is becoming increasingly important with both economic and ecological benefits.

It is encouraging, therefore, that economists are beginning to account, in monetary currencies, for the costs of depreciated natural resource bases and compromised ecosystem processes, thereby accounting for the real costs of production. Heretofore, soil and land degradation impacts have, for the most part, not been accounted for in the calculus of what constitutes true profit and wealth generated from using and managing natural resources and ecosystems. The costs of "mining" or sustaining the natural resource base and ecosystem processes have, for the most part, been ignored through the traditional "currencies" of economic markets and economic indicators such as GDP. The world community, as Herman Daly (Cobb et al., 1995) noted, can no longer treat the earth as a business in liquidation.

Energy "Currency"—Thermodynamic Basis of Sustainability

Thermodynamics govern the processes of energy capture within and transmission through biological systems. These thermodynamically-governed processes result in both matter being produced by and energy exiting the biological system (Ulanowicz, 1986). Biological ecosystems, in combination with the atmosphere and hydrosphere (climate), therefore, are the conductors and transmitters of energy and material that drive both soil formation and biological sustenance. Thus, ecosystems can be described and analyzed in terms of material and energy flows which are absolute requisites for the sustenance of life.

Energy, its capture and flow through biological systems and its function in driving ecological processes, is another of the "currencies" of natural resources and ecosystem sustainability that must be recognized. The quantity and duration of energy transport manifests itself through biological energy capture and climate which, in turn, determine the character of the biosphere that imparts a stamp of unique morphology on the resulting soil (Targulian et al., 1990).

The application of thermodynamic principles and models to soil science dates back at least to 1935 when the energy relationship of water in soil was examined (Schofield, 1935). Later, soil genesis models based on thermodynamics were developed (Runge, 1973). Runge (1973) used water flow as the energy currency for transforming soil parent material into a more ordered soil profile, i.e., reducing entropy. This model, however, was based on thermodynamic principles more appropriate to *closed systems* (Hoosebeek and Bryant, 1992).

More recently, Addiscott (1995) reviewed these aforementioned thermodynamic approaches and argued that a form of thermodynamics that deals with flows of matter and energy in *open systems* is necessary. This approach leads to nonequilibrium thermodynamics which describes open systems, such as soil-plant ecosystems, that tend toward steady states characterized by minimum production of entropy. Ecosystem sustainability can be defined on the basis of this thermodynamic approach; i.e., the balance between ordering and dissipative processes.

Solar energy flow provides the energy currency that sustains the biosphere, including soil ecosystems, and drives the self-organization (i.e., order or the reduction of entropy) of the biosphere (Morowitz, 1970). This solar energy flow can result in effective work or products, such as biological production, soil ecosystems, ecosystem processes, and their sustenance, only if infinite isothermal reservoirs at high (energy source) and low (thermal sink) temperatures exist (Addiscott, 1995). For the biosphere (and soil ecosystems), these isothermal reservoirs are provided by the sun and outer space, respectively. These two isothermal reservoirs allow for solar energy to drive the earth's climatic engine and flow through the biosphere, resulting in work as manifested in net primary production, soil development, and the sustenance of ecosystem processes and renewable natural resources. The resulting energy that is not used or captured by the biosphere is "exhausted" or flows to the outer space "sink," thereby maintaining a constant temperature on earth while sustaining the biosphere and its requisite processes such as ecological cycles and maintaining a balance between ordering and dissipative processes.

Biology provides the most obvious model of these ordering and dissipative processes (Addiscott, 1995). Ordering processes such as photosynthesis and its complementary processes result in high molecular weight compounds (e.g., carbohydrates, cellulose, proteins) that have complex, ordered structures derived from smaller molecules of $C0_2$, H_20, NH_3, and various elements and minerals. Thus, agriculture can be defined thermodynamically as the building of large molecules (e.g., carbohydrates and proteins in grains,

Table 2.2. Ordering and Dissipative Processes in Soil-Plant Systems Categorized as Biological or Physical[a] (The pairs are not necessarily exact opposite)

Ordering Processes/Entropy Decreases	Dissipative Processes/Entropy Increases
Biological	
Photosynthesis	Respiration
Growth	Senescence
Formation of humus	Decomposition of humus
Physical	
Water flow (profile development)	Water flow (erosion, leaching)
Flocculation	Dispersion
Aggregation	Disaggregation
Structure development	Breakdown of structure
General Characteristics	
Larger units	Smaller units
Fewer of them	More of them
More ordered	Less ordered

[a]From Addiscott, 1995.

meat, milk, etc.) from small molecules and elements. Similarly, soil development, soil ecosystems, and other ecosystems are driven and sustained by ordering processes, giving rise to unique soil profiles and stable land ecosystems derived from various parent materials and biological processes and biological components.

Dissipative processes are also at work within the biosphere, even under conditions of a steady state of minimum entropy production. These dissipative processes are exemplified by such processes as respiration and senescence that degrade photosynthesized molecules and structures back into smaller molecules such as CO_2, H_2O and less complex mineral compounds and structures. Ordering and dissipative processes in the soil-plant ecosystem can be categorized (Table 2.2). Soil and land degradation occur when dissipative processes become excessive; e.g., soil profile destruction by erosion resulting from tillage and poor residue management or overgrazing.

Sustainable agriculture and sustainable ecosystem management can be defined using this thermodynamic-ordering-dissipative model with solar energy being the driving currency. The goal of sustaining agriculture and ecosystems is to strive for minimal production of entropy; i.e., minimization of dissipative processes caused by soil and land perturbations such as excessive erosion, nutrient losses, soil compaction, soil structure deterioration, and salinization. This model can be used also to understand the concept of ecosystem resilience when man-induced ecological perturbations push an ecosystem away from its steady state condition.

The cutting of the eastern U.S. deciduous forests and the cultivation of these forest soils and subsequently the prairie soils were major perturbations of these soil ecosystems. But the biological capacity of these soil ecosystems was not destroyed. Under proper management, these soil ecosystems have resulted in a virtual agricultural cornucopia while maintaining at least some semblance of a suitable balance of ordering and dissipative processes. Other ecosystems; e.g., in parts of North America, the Sahel of Africa, the Middle East, and elsewhere, have been deteriorated to the point where their natural resource base and ecosystem processes have been destroyed or severely to moderately compromised. One criterion of sustainability advocated by Addiscott (1995) is that there should be the maintenance of a proper balance between ordering and dissipative processes; i.e., the maintenance of minimum entropy production by managing agriculture and ecosystems that permit the establishment of steady states.

> Continuous agriculture, or even monoculture, may therefore be one of the more sustainable options available, but with four provisos:
>
> 1. The system must achieve a steady state.
> 2. The flows within the system, particularly that of nitrogen, must be of appropriate magnitude.
> 3. The capacity for self-organization must be retained in the system; this means making sure that the biological potential is maintained.
> 4. The maintenance of the steady state must not involve excessive expenditure of energy, and entropy production, outside the system, in fertilizer manufacture, for example (Addiscott, 1995).

Addiscott quotes the distinguished quantum physicist, Schrödinger (1967), who commented that the living organism has the "astonishing gift of concentrating a 'stream of order' on itself and thus escaping the decay into atomic chaos." The sustainable management of soil-land ecosystems and biological systems such as agriculture and forests involves the maintenance of those soil-land biological processes (Schrödinger's 'gift') such that these ecosystems attain steady states characterized by minimum production of entropy. Such a state can be monitored by how much dissipative by-products flow out of the system. Therefore, water quality (as measured by turbidity-sediment, nitrate, phosphorus, etc.) can be a "flow indices" of ecosystem health as measured by dissipative product exhausts to show how well a biological or soil-land ecosystem is doing (Ulanowicz, 1986).

If monetary "currency" can be used to measure the true wealth of nations as both product output and natural resource-ecosystem depreciation are accounted for, then so too can the "currency" of energy be used to measure true biological-ecosystem output versus resource-energy investments. Table 2.3 provides an example of the energy "currency" equivalent of agricultural yield (output) to production investments. Most world grain production systems capture more solar energy than is invested in their production, at least as measured at the farm gate before processing, packaging, storage, and other value-added processes result in far more energy invested compared to the biologically-captured solar energy in the grain. For many vegetable and fruit crops as well as animal product (e.g., meat, milk, eggs) production systems, biological energy capture is less than the production energy equivalent invested, thereby resulting in a negative energy output/input ratio.

Such production systems are sustainable only if the natural resource base and ecosystem processes undergirding these production systems are sustained. Thus, it behooves soil and natural scientists to develop soil-land quality indicators and tools that will provide measures of soil-land "health," quality, or sustainability and strategies and practices that soil-land managers can use to maintain ecosystem "health" or sustainability. It is relatively simple to assess a soil-land ecosystem once "the wheels have fallen off"; i.e., once it has been degraded. But diagnostic tools and procedures must be available to assess soil-land ecosystem "health," quality, or sustainability *before* visible signs are obvious and interventions become too costly or are too late. Thus, different "currencies" are needed to determine the "value" of ecosystems and their functional worth toward the enhancement of human well-being.

Table 2.3. Energy Inputs per Hectare for Average (mid-1980s) Corn Production in the U.S.[a]

Inputs and Output	Kcal/ha
Inputs	
Labor	5,000
Machinery	1,018,000
Fuel	1,255,000
Irrigation	2,250,000
Electricity	100,000
Nitrogen	3,192,000
Phosphorus	473,000
Potassium	240,000
Lime	134,000
Seed	520,000
Insecticides	300,000
Herbicides	800,000
Drying	660,000
Transportation	90,000
Total Inputs	11,037,000
Output	
Yield (7000 Kg corn)	24,500,000
Ratio: Kcal output/Kcal input	
Irrigated	2.22
Nonirrigated	2.60 approx.

[a]From Pimentel and Wen Dazhong, 1990; Pimental et al., 1989. Data from Pimentel and Wen Dazhong (see complete citation in References section) in *Agroecology*, McGraw-Hill Publishing Co., N.Y. copyrighted, 1990, reproduced with permission of McGraw-Hill Companies.

The "Currency" of "Emergy"

Odum (1996) developed another "currency" concept for environmental accounting; namely, "emergy." This concept measures both the work of nature and that of humans in generating products and services and, ultimately, real wealth. The common sense notion that it takes more energy (resources, effort) to build an automobile than a bicycle extends to the understanding that the automobile has greater value than the bicycle. The concept of "emergy" is designed to give a quantitative measure to this common sense principle, but recognizes the different levels of prior effort, embodied energy, and energy transformations involved in generating different kinds of products (Odum, 1996).

Even though a calorie of energy derived from sunlight, fossil fuel, electricity, nuclear fission, and human thinking contains the same amount of energy, Odum's (1996) "emergy" concept differentiates these energy sources with respect to the different levels of prior effort and train of transformations that were involved in yielding the same calorie. Unlike a monetary valuation of "pricing" as an index of wealth where there is no common carrier of value, the concept of "emergy" uses a component of measuring value that is common to all resources and processes; namely, energy, and tracks the magnitude of embodied energy and energy transformation pathways to arrive at a scientifically defined measure of resource value which is a more realistic indicator of wealth.

The fallacy of using simple monetary indicators of value, particularly as applied to natural resources management and the environment, already has been reviewed. Such fallacious accounting for major environmental impacts such as the *Exxon Valdez* oil spill or Chernobyl accident can actually translate to increased monetary indicators such as GDP that do not discriminate between "illth" and true wealth since environmental depreciation is not taken into account. Such depreciation must also accrue to soil and land degradation.

Odum's (1996) "emergy" concept is one means of accounting for environmental depreciation and calculating real wealth in terms of a "currency" common to all resources; namely, embodied energy and those energy transformations expended to yield this embodied energy. The "emergy" concept extends the concept of thermodynamic energy conversions in the biosphere and economic development to a common "currency" that recognizes net energy and prior energy transformations in products and processes and can relate such "emergy" to monetary value and real wealth. This concept also can be applied to soil and land degradation.

The "Currency" of Soil Quality

Until recently, soil quality was described or discussed primarily in terms of desirable soil and landscape properties such as tilth, bulk density, organic carbon or organic matter, pH, adequate rooting depth or volume, texture, landscape character (e.g., topographic complexity, aspect), spatial distribution and uniformity, nutrient balance, and cation exchange capacity. While these properties and attributes of soils and landscapes are important characteristics and indicators of soil and land quality, they are not in themselves useful indicators of the quality (health) or sustainability of the processes that are responsible for sustaining these attributes of soil quality. Because soils are biologically dynamic systems under the constant influence of the dynamics of climatic and hydrologic activity which drive soil formation (and soil degradation under improper management), the sustenance, definition, and quantification of soil quality must be transacted in "currencies" related to processes rather than properties. Soil scientists have only recently begun to shift their attention to this need (Doran and Parkin, 1994).

If ecosystem processes and services identified in Table 2.1 are requisite to soil and land quality, then the definition and sustenance of soil-land quality must focus on those processes that result in desirable soil-land attributes and goods. Therefore, the "currencies" of soil quality must be defined in terms of processes that result in and sustain desired soil-land properties and attributes. Defining and measuring soil-land quality indicators based on soil-land properties and attributes results in measuring only the output or end product of processes that lead to desired (quality) soils and landscapes. Such focus on properties and attributes rather than the processes responsible for sustaining these desired properties and attributes is to miss the point of sustainable soil quality.

For example, many soils that were immediately cultivated after clearing forests or breaking the prairie had very desirable soil properties. Eventually, many of these soils became degraded through erosion, nutrient drawdown, salinization, and other degradation phenomena because the processes that sustained these soils were compromised. Often, the processes sustaining soil quality become compromised long before obvious degradation occurs as manifested by erosion, loss of organic matter and biodiversity, altered hydrologic behavior, nutrient imbalance, tilth deterioration and other attributes of degradation and reductions in quality.

Process indicators of soil and land quality can be defined in terms of alterations in hydrologic flux, changes in surface and subsurface water quality, reduced soil biodiversity, changes in carbon dioxide flux from soils, changes in biogeochemical cycling, and other process indicators. Soils may still generate very high biological output (yield) under high intensity management. But such output can temporarily mask the deterioration of those processes responsible for sustained productivity as more and more production resources and cultural energy are invested. Thus, soil and ecosystem quality can deteriorate before biological indicators or management investment indicators reflect obvious signs of soil-ecosystem deterioration.

Several analogies can illustrate the point of assessing properties vs. processes as indicators of quality. Human beings may exhibit many characteristics of perfect physical specimens even though some of their many metabolic and vital organ processes may become slowly compromised (diseased) before any outward manifestation of deteriorated health is detected. A physician does not rely on outward appearance and physical observations of a patient to determine suitable health. Indicators of human physiological processes are assessed through blood analysis, analyses of metabolic exhausts such as basal metabolism (CO_2/O_2 ratios), urine and stool samples, neurological functions, and other measures of physiological processes. Similarly, maintenance of finely machined and tuned engines such as jet aircraft engines relies on analysis of lubricating oils for excessive wear rather than relying simply on routine lubricant changes at prescribed intervals. Thus, the maintenance of human health and machines is turning more toward process assessment to sustain performance quality rather than relying on other indicators of quality that may mask process deterioration and result in corrective strategies being too late or too costly.

Doran and Parkin (1994), Harris and Bezdicek (1994), Larson and Pierce (1994) and others have proposed various schemes, indices, and data sets for describing and assessing soil quality. These scientists recognize the importance of processes as the "currency" for sustaining soil and landscape quality. Should solar energy be instantaneously shut off from the earth, soils and landscapes would still exhibit many properties and attributes that are deemed necessary to the definition of quality. However, such soils and landscapes would be mere terrestrial component cadavers, incapable of supporting those processes that are requisite to the sustenance of soil quality. Thus, soil scientists must make sure they are dealing in the proper "currency" when defining and measuring quality with respect to soil and landscape utility, behavior, and sustainability.

The "Currency" of Social Benefits and Costs

As Cobb et al. (1995) point out, not only do traditional indicators of economic activity, such as GDP, mask the impaired functions of ecosystems and the deterioration of natural resources, but so too do such economic indicators mask the breakdown of social structure, even portraying such breakdown as economic gain. Social structure, whether measured in terms of community and political stability or family unity, contributes greatly to human well-being as well as positively to economies, but is itself ignored or undervalued because it is invisible to economic indicators or indexes of "progress."

Dr. Ralph Cummings, Sr., considered by many to be the father of the 'Green Revolution' in southeast Asia, once remarked (personal communication), when asked by a group of students what he considered to be the key to the success of this agricultural accomplishment, quickly responded that political and social stability were absolute requisites for such success. Otherwise, the necessary institution-building and institutional support for such a program were doomed to failure and enhanced human well-being and true gains in national wealth would have been forfeited. But economic indicators such as GDP ignore the contributions of these social benefits.

Natural resources managers have begun to appreciate, if not understand, amenity values that people and society bring to the table of resources management decisions. No longer are natural resources management decisions unidimensional or driven by the simple "currencies" of biological output or resource harvests. For example, forests are valued by a variety of constituencies, not only for their lumber and fiber, but because they provide unique habitats for certain wildlife, watershed qualities, or attributes that contribute to human spiritual well-being and other valued purposes. Other amenity values include "existence" values, knowing that a natural resource (e.g., herd of elephants, community of gorillas, a pristine vista, old growth forest) is a part of the earth's biological community and endowment of natural resources. Such amenity values do not have a common "currency," but play major roles in the business of determining how natural resources are to be used.

A society's as well as an individual's degree of stewardship toward the environment is couched in both culture as well as circumstances. The basic necessities of human needs coupled with the common goal of enhanced human well-being, inexorably linked as they are to natural resources and the environment, become the primary drivers of how and to what degree natural resources and the environmental are used, managed, and cared for. Hungry and desperate people cannot and do not spend emotional and philosophical capital on their responsibility and an ethic as to how best to sustainably manage natural resources and ecosystems. To curtail soil and land degradation under the juggernaut of poverty, ignorance, and desperation requires the building of social capital, measured in "currencies" of political stability, building institutions to serve society, education, population control, and other attributes.

To attempt technical "fixes" for controlling soil and land degradation without deflecting or slowing this aforementioned juggernaut is simply an exercise in futility. While "currencies" such as GDP ignore the depreciated landscape as soils and other national resources are liquidated for short-term gains, it is obvious that other "currencies" depicting social benefits and costs must be "minted" to build the necessary social capital to address issues such as soil and land degradation, poverty, and unsupportable population growth. Again, soil scientists and other natural resources managers cannot "go it alone" in attempting to "fix" these problems without understanding the "currencies" that are involved in driving the problems and are required as part of the solutions.

The "Currency" of Human Capital

Investing in human capital through education and supportive institutions and infrastructure is analogous to companies reinvesting capital into research, development, and equipment replacement. Companies make these investments in order to sustain their future. Companies can then declare "dividends" after such reinvestments as the "currency" of added value which builds real wealth. Investing in human capital also adds value to people, making them better global citizens, reducing the chance of their making decisions out of desperation on the use and management of natural resources.

Educated people will understand and use the tools of sustainable natural resources management. Such tools have resulted in much land saved from human encroachment that can sustain other uses such as wildlife habitats and protect fragile or unique ecosystems (Waggoner, 1994). Attempting to curtail soil and land degradation without investing in the "currencies" of human capital and social benefits is to ignore the primary drivers of ecosystem deterioration.

Epilogue on the "Currencies" of Conservation

Having argued that the natural resource sustainability dimension trinity can be categorized into the physical-biological (natural-ecological), economic, and social dimensions, the social and economic dimensions are the most difficult to define common "currencies" that measure real sustainable wealth. Progress in curtailing soil and land degradation on a sustained basis seems futile unless such "currencies" are identified and accepted in global markets.

Cobb et al. (1995) argue that "economists are demanding that their profession start to take into account of the larger economy in which the market is grounded—the natural and social spheres, which they have in the past dismissed as the netherworlds of externality. . . . the pursuit of GDP has been undermining traditional values and social cohesion, much as it has been destroying the natural habitat." These authors go on to suggest that society, especially in the developed world, has begun to understand that "the pronouncements from economic experts are fundamentally out of sync with the experience of their own lives; that economics must be about more than just the production and consumption of stuff; and that we need larger goals and better ways to measure our achievements as a nation."

People are beginning to recognize the difference between real wealth and John Ruskin's concept of "illth." There is a teachable moment for natural resources managers to begin to put the world, or at least certain parts of it, on a sustainable natural resources management course. But to try to address this large issue through reductionist means or by single disciplines is tantamount to putting a finger in the dike. Soil scientists have no business addressing soil and land degradation intervention strategies apart from other disciplines whose contributions are absolutely necessary if the root causes or drivers of degradation are to be blunted.

CONCLUSIONS

Soil and land degradation are driven by combinations of forces, such as poverty; excessive population; low productivity; lack of knowledge, ability, and desire, or disincentives to adopt technology; and poorly defined or inadequate land tenure systems. These elements must be addressed in any interventions or policies initiated to retard land degradation. If sustainable agricultural systems are to be implemented, then systems components, such as resources, technology, and institutions, must be accommodated (Crosson, 1986).

Miller and Wali (1995) set forth several premises of sustainability (as modified from Edwards and Wali, 1993), as guidelines for developing objectives for, and approaches to sustainability.

The Premises of Sustainability

The following list summarizes the premises of sustainability:

- Human history is a saga of people adapting to their natural environments as modified into cultural landscapes (Worster, 1990), through inherent human predisposition for population growth.
- Populations of people and domestic animals are high and increasingly concentrated.
- Land remains the paramount resource base for all societies (Cook, 1992) and the cultural landscape societies create strongly reflects the character and quality of this land.
- Land and good-quality soils are a finite resource and will continue to provide the primary basis for human sustenance.
- Prime agricultural lands are already developed in virtually all countries.
- The development of additional lands for agricultural purposes requires substantial investments to improve soil fertility, availability of water, drainage, irrigation, and erosion control.

- Changes in climate could alter the extent and distribution of land that is suitable for agriculture.
- Agriculture practices, coupled with poor management, have been responsible for considerable environmental degradation.
- Traditional agricultural systems, some of which are sustainable, are disappearing. They are being replaced by farming systems that are more intensive and (or) dependent on finite fossil fuels and off-farm resources.
- Nonfossil fuel inputs (e.g., water and phosphorus) for industrial agriculture are finite.
- The trend of economic development is toward increasing global interdependence in food and energy.
- Lack of hard currency in developing countries, coupled with high population density and poverty, encourages exploitation of land and increases the pressure to produce exports, thereby creating unsustainable conditions.
- Global indebtedness in the developed countries reduces capital investments in, and development assistance and technical aid to, developing countries.
- Simultaneous breakdown of large political structures increases transboundary problems; thus, approaches to global sustainable development will have to be multinational.
- The trend toward specialization and fragmentation of knowledge is a deterrent to the development of holistic agricultural systems and of policies for sustainability.

Sustainable Development Objectives

To satisfy the conditions for sustainability, the following sustainable-development objectives are imperative:

- Implement technologies and incentives for sustained production of food, fodder, fiber, and biomass fuels on a per capita basis.
- Develop strategies for improved management and use of the indigenous biological system as a production resource and the reduction of dependency on petrochemical and other finite resources.
- Develop protocols for improved environmental quality.
- Strive for improved per capita income and greater equity in its distribution.

Approaches to Achieve Sustainability

In response to Burton and Kates' (1986) challenge to the international community to "formulate goals for where we wish and expect the world to be . . . and develop ways of getting there," soil scientists have a responsibility of contributing to this agenda. Both the probability of, and capacity for, a sustainable future rest largely on our ability to tap the Earth's natural resources with sustainable management strategies. We must adopt the following approaches to achieve sustainability:

- Inventory the world's natural resources and characterize them for behavior prediction under various management regimes. Data must be standardized, periodically updated for trend analysis, and disseminated.
- Investigate soil processes to determine why (rather than simply how) soils respond to various treatments. Agricultural production must be viewed as feedback loops from ecological processes (e.g., energy flows, biogeochemical cycles, population dynamics).
- Develop measures or indices of soil quality that can be used to determine present soil health, assess degrees of soil deterioration, and predict threshold soil conditions before a soil or ecosystem "goes over the brink" of economic or ecological deterioration.
- Characterize soils to predict their response to water management strategies, since water scarcity looms large on the horizon for many countries and regions, within the next quarter century.

- Assess the reservoir of existing technologies that are available but unused, either in total or in their optimal potential; such an assessment would be helpful in determining future research directions and policy interventions to provide incentives for technology adoption.
- Characterize degraded lands and formulate best use-least cost measures for their rehabilitation and management.
- Establish a series of landscape-level agroecosystem model sites representing points along a continuum around the world.
- Establish a rigorous and systematic understanding of alternative property rights and entitlement, their origins and implications for sustainability.
- Integrate ecological and economic policy.
- Develop and use synthetic integrative tools, such as simulation modeling, to examine options for, and outcomes of, various management scenarios.

SUMMARY

The combination of the earth's quality land and soil resources and the resiliency of many ecosystems, coupled with the application of science and technology, gives both hope and time for intervening in land degradation processes. The biological capacity of the earth may be able to sustain a global population of 10-12 billion people at an acceptable level of well-being, but only if the earth's ecological systems are well managed and, thus, sustained.

Common experience indicates that the availability of information as well as education does not usually result in people making informed or ecologically-sensitive decisions about how to manage ecosystems sustainably. Technical advances and scientific information have always outpaced society's ability or willingness to embrace and adopt it. Soil loss beyond soil replacement levels and land degradation still occur in developed nations where databases and technical solutions are available. This situation prevails because the drivers of soil and land degradation and the incentives for controlling degradation are couched in economic and sociopolitical "currencies," not in technical or scientific "currencies." Until society, through its governments and other institutions, decides to provide incentives for controlling soil and land degradation, available technological "fixes" voluntarily implemented will usually not be adequate. Socioeconomic interventions and incentives are usually required.

Hayes (1993) pointed out that, with the latest version of the global model, we are at a singular moment in the history of the world: "the one and only transition from abundance to scarcity, from growth to stasis." Our global ecosystem has withstood much human abuse, usually bending without breaking. However, its capacity to endure even greater demands from its inhabitants may not be far from the point of irreversible damage. We cannot be driven to global conclusions interpreted from anecdotal evidence and limited case studies (Hudson, 1989). Neither can we acquiesce to timidity for action.

Nearly half a century ago, Ortega y Gasset (1944) stated: "Life cannot wait until the sciences have explained the universe scientifically. We cannot put off living until we are ready." But we can proceed with what we know and an understanding of what we must do. For the sustainability of civilization itself is at risk if we do not act at least with prudence if not aggressiveness.

REFERENCES

Addiscott, T.M. Entropy and Sustainability. *Eur. J. Soil Sci.* 46:161–168, 1995.
Burton, I. and R.W. Kates. The Great Climateric, 1798–2048: The Transition to a Just and Sus-

tainable Environment, in *Geography, Resources, and Environment.* Vol. II. R.W. Kates and I. Burton, Eds., University of Chicago Press, Chicago, IL. pp. 339–360, 1986.

Cobb, C., T. Halstead, and J. Rowe. If the GDP is Up, Why is America Down? *The Atlantic Monthly.* pp. 59–78, 1995.

Cohen, J.E. *How Many People Can the Earth Support?* W.W. Norton & Co. New York, 1995.

Cook, M.G. Sustainable Agriculture in a Survival Setting, pages 90–97 in *Soil Conservation for Survival,* K. Tato and H. Hurni, Eds. Soil Water Conserv. Soc., Ankeny, IA, 1992.

Crosson, P.R. Agricultural Development—Looking to the Future, in *Sustainable Development of the Biosphere,* W.C. Clark and R.E. Munn, Eds. Cambridge University Press, New York, 1986, pp. 104–136.

Crosson, P.R. Soil Erosion Estimates and Costs, Letters to Science, *Science.* 269:461–464, 1995.

Doran, J.W. and T.B. Parkin. Defining and Assessing Soil Quality, in *Defining Soil Quality for a Sustainable Environment,* J.W. Doran, D.C. Coleman, D.F. Bezdicek, and B.A. Stewart, Eds. Special Pub. No. 35. Soil Science Society of America. Madison, WI, 1994.

Dregne, H.E. and N.-T. Chou. Global Desertification, Dimensions, and Costs, in *Degradation and Restoration of Arid Lands,* H.E. Dregne, Ed. Texas Tech University Press, Lubbock, 1992, pp. 249–282.

Edwards, C.E. and M.K. Wali. The Global Need for Sustainability in Agriculture and Natural Resources, in *Agriculture and the Environment.* C.E. Edwards, M.K. Wali, D.J. Horn and F. Miller, Eds. Elsevier Science Publishers, Amsterdam, The Netherlands. 1993, pp. vii–xxv.

Ehrlich, P.R. and A.H. Ehrlich. *Healing the Planet: Strategies for Resolving the Environmental Crisis.* Addison-Wesley Publishing Co., Reading, MA, 1991.

Flavin, C. Facing Up to the Risks of Climate Change. *State of the World.* L. Brown et al., Eds. Worldwatch Institute. 1996, pp. 21–39.

Follett, R.F. and B.A. Stewart, Eds. *Soil Erosion and Crop Productivity.* Amer. Society of Agronomy, Crop Science Society of America, and Soil Science Society of America. Madison, WI, 1985.

Gallagher, R.B., J. Fischman, and P.J. Hines. Big Questions for a Small Planet. *Science.* 269:283, 1995.

Halcrow, H.G., E.O. Heady, and M.L. Cotner, Eds. *Soil Conservation Policies, Institutions, and Incentives.* Soil and Water Conservation Society of America, Ankeny, IA, 1982.

Harris, R.F. and D.F. Bezdicek. Descriptive Aspects of Soil Quality/Health, in *Defining Soil Quality for a Sustainable Environment.* J.W. Doran, D.C. Coleman, D.F. Bezdicek, and B.A. Stewart, Eds. Special Pub. No. 35. Soil Science Society of America. Madison, WI, 1994.

Hayes, B. Balanced on a Pencil Point. *Am. Sci.* 81(6):510–516, 1993.

Hitzhusen, F. Soil Erosion Policy—A Downstream Economic Perspective. *Ohio's Challenge,* Vol. 4, Issue 2. 1991, pp. 3–5.

Hoehn, J.P. and D.R. Walker. When Prices Miss the Mark: Methods for Valuing Environmental Change. Policy Brief. The Environmental and Natural Resources Policy and Training Project. EPAT/MUCIA, Arlington, VA. No. 3: 1–6, 1993.

Hooks, G.M., T.L. Napier, and M.V. Carter. Correlates of Adoption Behaviors: The Case of Farm Technologies. *Rural Sociol.* 48(2):309–324, 1983.

Hoosbeek, M.R. and R.B. Bryant. Towards the Quantitative Modeling of Pedogenesis—A Review. *Geoderma.* 55:183–210, 1992.

Hudson, W.J. Population, Food, and the Economy of Nations, in *Food and Natural Resources,* D. Pimentel and C.W. Hall, Eds. Academic Press, San Diego, CA, 1989 pp. 275–299.

Jennings, M.D., B.C. Miller, D.F. Bezdicek, and D. Granatstein. Sustainability of Dryland Copping in the Palouse: An Historical View. *J. Soil Water Conserv.* 45(1):75–80, 1990.

Larson, W.E. and F.J. Pierce. The Dynamics of Soil Quality as a Measure of Sustainable Management, in *Defining Soil Quality for a Sustainable Environment.* J.W. Doran, D.C. Coleman, D.F. Bezdicek, and B.A. Stewart, Eds. Special Pub. No. 35. Soil Science Society of America. Madison, WI, 1994.

Larson, W.E., F.J. Pierce, and R.H. Dowdy. The Threat of Soil Erosion to Long-Term Crop Production. *Science.* 219:458–465, 1983.

Lovejoy, S.B. and T.L. Napier, Eds. *Conserving Soil: Insights from Socioeconomic Research.* Soil and Water Conservation Society of America. Ankeny, IA, 1986.

Lubchenco, J., P.G. Risser, A.C. Janetos, J.R. Gosz, B.D. Gold, and M.M. Holland. Priorities for an Environmental Science Agenda for the Clinton-Gore Administration: Recommendations for Transition Planning. *Bull. Ecolog. Soc. Am.* 74:4–8, 1993.

Malthus, T.R. *An Essay on the Principle of Population, as it Affects the Future Improvement of Society. With Remarks on the Speculations of Mr. Godwin, M. Condorcet, and Other Writers.* Complete 1st ed. and partial 7th ed. (1872) reprinted in *On Population,* G. Himmelfarb, Ed. Modern Library, New York, 1798 (1960).

Miller, F.P. and M. Wali. Soils, Land Use and Sustainable Agriculture: A Review. *Can. J. Soil Sci.* 75(4):413–422, 1995.

Morowitz, H.J. *Entropy for Biologists.* Academic Press, New York, 1970.

Napier, T.L. The Evolution of U.S. Soil Conservation Policy: From Voluntary Adoption to Coercion, in *Soil Erosion on Agricultural Land,* J. Boardman, I. Foster, and J. Dearing, Eds. Wiley Publishers, London, 1990, pp. 627–644.

Odum, H.T. *Environmental Accounting: Emergy and Environmental Decision Making.* John Wiley and Sons, Inc., New York, 1996.

Oldeman, L.R., V.W. P. van Engelen, and J.H.M. Pulles. The Extent of Human-Induced Soil Degradation. Annex 5 of L.R. Oldeman, R.T.A. Hakkeling, and W.G. Sombroek. World Map of the Status of Human-Induced Soil Degradation: An Exploratory Note. 2nd Rev. ed. International Soil Reference and Information Centre, Wageningen, The Netherlands, 1990, Tables 1–7.

Ortega y Gasset, J. *Mission of the University.* Princeton University Press, Princeton, NJ, 1944.

Peirce, N.R. A Megabank Speaks Out on Sprawl. *Nat. J.* 27(11):699, 1995.

Pierce, F.J., R.H. Dowdy, W.E. Larson, and W.A.P. Graham. Soil Productivity in the Corn Belt: An Assessment of Erosion's Long-Term Effects. *J. Soil Water Conserv.* 39:131–136, 1984.

Pimentel, D., L.E. Armstrong, C.A. Flass, F.W. Hopf, R.B. Landy, and M.H. Pimentel. Interdependence of Food and Natural Resources in *Food and Natural Resources.* D. Pimentel and C.W. Hall, Eds. Academic Press, Inc., San Diego, pp. 31–48, 1989.

Pimentel, D. and Wen Dazhong. Technological Changes in Energy Use in U.S. Agricultural Production, in *Agroecology.* Carroll, C.R., J.H. Vandermeer, and P. Rosset, Eds. McGraw-Hill Publishing Co., N.Y. pp. 147–164, 1990.

Pimentel, D., C. Harvey, P. Resosudarmo, K. Sinclair, D. Kurz, M. McNair, S. Crist, L. Shpritz, L. Fitton, R. Saffouri, and R. Blair. Environmental and Economic Costs of Soil Erosion and Conservation Benefits. *Science.* 267: 1117–1123, 1995.

Randall, A. Biodiversity and a Safe Minimum Standard of Conservation. Gordon H. Barker Memorial Paper Series in Natural Resources Policy. School of Natural Resources, University of Wisconsin-Madison. 1994, p. 28.

Repetto, R. Wasting Assets. *Technol. Rev.* 1989, pp. 39–44.

Richardson, C.J. Ecological Functions and Human Values in Wetlands: A Framework for Assessing Forestry Impacts. *Wetlands.* 14:1–9, 1994.

Runge, E.C.A. Soil Development Sequences and Energy Models. *Soil Sci.* 115:183–193, 1973.

Schofield, R.K. The pF of the Water in Soil. *Trans. Third International Congress of Soil Science.* 2:37–48, 1935.

Schrödinger, E. *What is Life*? Cambridge University Press. Cambridge, U.K., 1967.

Targulian, V.O., R.W. Arnold, and B.G. Rozanov. Pedosphere. Chapter II in *Global Soil Change.* R.W. Arnold, I. Szabolcs, and V.O. Targulian, Eds. Report on IIASA-ISSS-UNEP Task Force on the Role of Soil in Global Change. Intl. Institute for Applied Systems Analysis. Laxenburg, Austria, 1990, pp. 21–29.

Ulanowicz, R.E. *Growth and Development—Ecosystem Phenomenology,* Springer-Verlag, New York, 1986.

U.S. Department of Agriculture. Summary Report, 1992 National Resources Inventory. Washington, D.C, 1994.

U.S. National Academy of Sciences and U.S. National Academy of Engineering. Confronting

Natural Disasters. Commission on Engineering and Technical Systems. National Research Council. National Academy Press, Washington, D.C, 1987.
Vandermeer, J.H. Thoughts on Agriculture and the Environment in a Post-Modern World. Symposium on Enhancing the Future of the Land Grant System. National Academy of Sciences. Board on Agriculture. Irving, CA, April 3–4, 1992. p. 14.
Viederman, S. A Dream of Sustainability. *Renewable Resour. J.* 11(2):14–15, 1993.
Waggoner, P.E. How Much Land Can Ten Billion People Spare for Nature? Council on Agricultural Science and Technology, Ames, IA. Task Force Rep. 121, 1994.
Weesies, G.A., S.J. Livingston, W.D. Hosteter, and D.L. Schertz. Effect of Soil Erosion on Crop Yield in Indiana: Results of a 10 Year Study. *J. Soil Water Conserv.* 49(6):597–600, 1994.
World Commission on Environment and Development. *Our Common Future.* Oxford University Press, New York, p. 383, 1987.
World Resources Institute. World Resources, 1992–1993. A report by the World Resources Institute. Oxford University Press, New York, 1992.
Worster, D. Seeing Beyond Culture. *J. Am. History* 76: 1142–1147, 1990.

THREE

Assessment of Soil Quality

Maurice J. Mausbach and Cathy A. Seybold

INTRODUCTION

Assessment of soil quality presumes procedures exist to measure it and standards have been developed to determine the relative quality of soil under various land uses and management systems. Assessments of soil quality can be, in the simplest form, visual observations of soil quality or may be very complex involving many laboratory analyses and the calculation of soil quality indices.

In this chapter we discuss the assessment of soil quality at various levels of scale ranging from the farm or field level to the regional or national level. We will first review the definition of soil quality; then discuss soil quality indicators and reference values; and finish with a discussion of soil quality assessments.

SOIL QUALITY DEFINITION

Soil quality has different meanings according to the perspective of the person: for the land manager or crop advisor, it may mean sustaining or enhancing productivity and maximizing profits, while maintaining the soil resource for future generations; for the conservationist, it may mean protecting the environment and sustaining the soil resource for future generations; for consumers of agriculture products, it may mean plentiful, healthful, and inexpensive food for present and future generations; and for the environmentalist, it may mean a holistic, interconnected view of soil in an ecosystem with respect to maintenance or enhancement of biodiversity, water quality, nutrient cycling, and biomass production. How we view soil quality largely depends on the scale of concern. For the farmer, the field or points within the field are important, for the citizens of a community the impact of soil quality within their watershed is important, and for national policy makers, a national assessment of the nation's soil quality and its trends are important. Each perception of soil quality and the scale of concern impacts how it is measured and assessed. The key is to establish commonalties of assessment procedures to facilitate communications and actions related to soil quality.

Soil quality is a composite picture of the condition of a specific soil to function for a specific use. Simply put, soil quality is *the capacity of a soil to function* (Pierce and Larson, 1993; Karlen et al., 1996). A more complete definition of soil quality as proposed by Karlen et al. (1996) is:

> The capacity of a specific kind of soil to function, within natural or managed ecosystem boundaries, to sustain plant and animal productivity, maintain or enhance water and air quality, and support human health and habitation.

Soil quality is a holistic view of soil in its landscape setting and how it functions within the ecosystem. Soil quality encompasses two general points of view: (1) as inherent properties of a soil; and (2) the dynamic nature of soils as influenced by human use and management decisions (Larson and Pierce, 1991; Pierce and Larson, 1993). With respect to inherent properties, a soil is a result of the factors of soil formation—climate, topography, vegetation, parent material, and time (Jenny, 1941). Each soil, therefore, has an innate capacity to function; e.g., some soils then are able to function at higher levels than other soils. This view of the definition is useful for comparing the abilities of one soil against another and is often used to evaluate the worth or suitability of soils for specific uses.

The second view of quality relates to the dynamic nature of soil as influenced by human use and management and is an integration of the kind of soil, its natural ability to function, and its use and management. This view is often referred to as soil health and is the focal point for the assessment of soil quality (National Research Council, 1993; Doran and Parkin, 1994; Larson and Pierce, 1991, Larson and Pierce, 1994; Harris and Bezdicek, 1994; Acton and Gregorich, 1995; Karlen et al., 1996; Howard, 1993; de Haan et al., 1993). Assessing the dynamic nature of soil quality requires a reference or baseline value for assessment of soil quality. This reference condition and subsequent soil health assessments should be made within areas of a specific soil series and land use.

SOIL QUALITY INDICATORS AND REFERENCE VALUES

Proposed measures of soil quality employ the use of soil quality indicators that reflect the capacity of the soil to function (Larson and Pierce, 1991; Pierce and Larson, 1993; Doran and Parkin, 1994; Karlen et al., 1994b). The functions of soil as proposed by Karlen et al. (1996) are:

1. sustaining biological activity, diversity, and productivity;
2. regulating and partitioning water and solute flow;
3. filtering, buffering, degrading, immobilizing, and detoxifying organic and inorganic materials, including industrial and municipal by-products and atmospheric deposition;
4. storing and cycling nutrients and other elements within the earth's biosphere; and
5. providing support of socioeconomic structures and protection for archeological treasures associated with human habitation.

Doran and Parkin (1994) suggested that soil quality indicators should (1) encompass ecosystem processes and relate to process oriented modeling, (2) integrate soil physical, chemical, and biological properties and processes, (3) be accessible to many users and applicable to field conditions, (4) be sensitive to variations in management and climate, and (5) where possible, be components of existing soil databases. Indicators must also be easily measured and measurements must be reproducible (Gregorich et al., 1994). Arshad and Coen (1992) suggested indicators should be sensitive enough to detect changes in soil as a result of degradation.

Seybold et al. (1996) discuss the importance of scale in assessing soil quality. The effects of scale are usually addressed in the selection of indicators (Karlen et al., 1996), or by generalization of point scale data through the use of statistical sampling frames. Temporal effects, such as moisture content, crop, or tillage on soil quality indicators are usually avoided by selecting indicators that are not affected by these short-term factors or timing the measurement of the indicator to alleviate the effect.

Several minimum data sets of indicators have been proposed (Larson and Pierce, 1991; Arshad and Coen, 1992; Doran and Parkin, 1994; Gregorich et al., 1994; Larson and Pierce, 1994; Papendick et al., 1995). An example of a minimum data set is presented in Table 3.1. The concept of how indicators are related to soil function and quality is illustrated in Figure 3.1. One or two indicators can sufficiently represent each function; however, indicators may be related to more than one soil function (Fig. 3.1).

For indicators to be useful in assessing soil quality, standards or reference values must be established as a baseline from which to assess the current state of soil quality. Larson and Pierce (1991) suggested that the current state of soil must be compared to some standard to assess the degree of excellence of soil quality. Pierce and Larson (1993) proposed defining limits for the indicators that represent a desired level of quality and then show when the soil is at risk or 'out of control.' They used statistical quality control concepts developed for manufactured goods and services to establish upper and lower limits of changes that are based on natural variation. An example of this approach is shown in Figure 3.2.

Table 3.1. Proposed Minimum Data Set of Physical, Chemical, and Biological Indicators for Screening the Condition, Quality, and Health of Soil[a]

Indicators of Soil Condition	Relationship to Soil Condition and Function; Rationale as a Priority Measurement
	Physical
Texture	Retention and transport of water and chemicals; modeling use, soil erosion and variability estimate
Depth of soil, topsoil, and rooting	Estimate of productivity potential and erosion; normalizes landscape and geographic variability
Infiltration and bulk density	Potential for leaching, productivity, and erosivity; SBD needed to adjust analyses to volumetric basis
Water holding capacity	Related to water retention, transport, and erosivity; available H_2O; calculate from SBD, texture, and OM
	Chemical
Soil organic matter (OM)	Defines soil fertility, stability, and erosion extent; use in process models and for site normalization
pH	Defines biological and chemical activity thresholds; essential to process modeling
Electrical conductivity	Defines plant and microbial activity thresholds; presently lacking in most process models
Extractable N, P, and K	Plant available nutrients and potential for N loss; productivity and environmental quality indicators
	Biological
Microbial biomass C and N	Microbial catalytic potential and repository for C and N; modeling: early warning of manag. effects on OM
Potentially mineralizable N	Soil productivity and N supplying potential; process modeling (surrogate indicator of biomass)
Soil respiration, water content, and temperature	Microbial activity measure (in some cases plants) process modeling; estimate of biomass activity

[a]From Seybold, et al., 1996. SBD = soil bulk density.

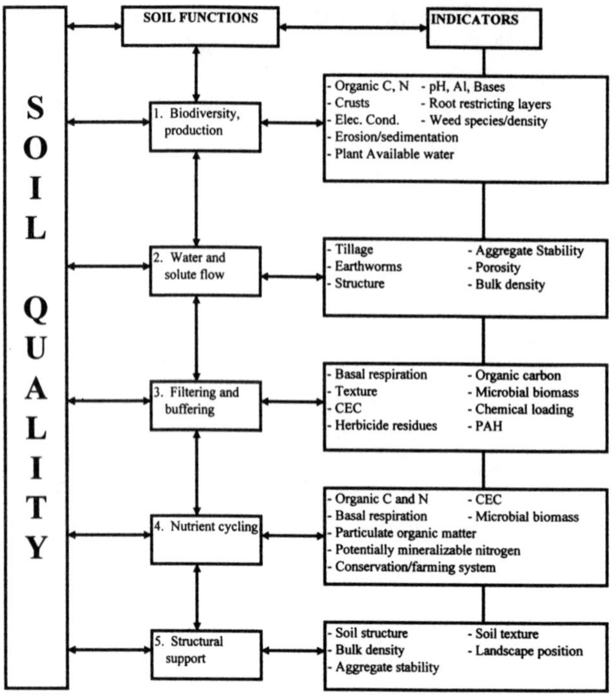

Figure 3.1. Graphical representation of the concept of soil quality using soil function and indicators of soil quality (from Seybold et al., 1996).

Reference values can be given as specific limits or ranges for each indicator for a particular soil or groups of similar soils. These ranges or limits are based on the values for indicators which define a condition representative of a soil functioning at full potential (Seybold et al., 1996). Full potential is defined as the maximum capacity of a soil to function under a particular use. We envision three soil conditions for developing indicator reference values: (1) the natural condition; (2) specific human use; and (3) altered soils. For soils in their natural condition, reference values represent the inherent ability of a soil to function as defined by the soil forming factors and processes in its native or natural state. Reference values for soils that are intensively managed, as in production

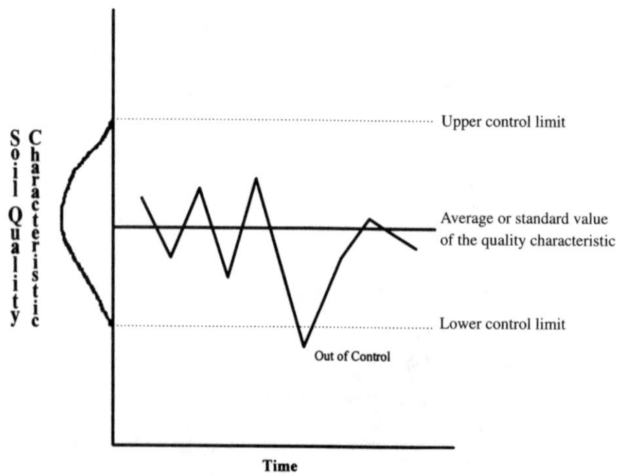

Figure 3.2. Concept of a control chart (from Pierce and Larson, 1993).

agriculture, should be established based on what is sustainable under a specific land use as defined by Best Management Practices (BMPs). Sustainability in this context means sustained yields and profitability as well as a healthy environment including healthy air, water, plant and animal resources. For the third case, reference values on altered soils are based on present conditions of the soil in conjunction with BMPs. For example, soils that have been drained will have reference values that reflected the properties of the soil after drainage.

Reference values can be expressed in many ways. The simplest is to establish an acceptable range for each indicator that represents a soil functioning at full potential. In this form a soil is either healthy (with the acceptable range) or at risk (outside of the acceptable range). Another way to express reference values is to develop equations or scoring functions (Table 3.2) (Karlen et al., 1994b). Scoring functions show gradations from very poor quality to very good or excellent soil quality for a particular indicator. These functions are usually normalized to a scale from 0 to 1, with 1 being excellent quality and 0 being very poor quality. Examples of scoring functions are given in Table 3.2. The examples represent three scenarios: (1) more is better, (2) less is better, and (3) an optimum value.

ASSESSMENT OF SOIL QUALITY

Regardless of scale, measuring soil quality requires a sampling procedure that takes into account temporal and spatial variability of an area. There are several approaches for sampling an area; selection of an approach depends on the required precision and resources available to do the assessment. Approaches for field scale assessments range from observational, where observations are grouped according to indicators (Romig et al., 1995); to an intensive grid sampling and use of geostatistics to summarize the data (Smith et al., 1993, 1994). For assessments of soil quality on a regional scale, randomized sampling of points in a longitudinal type of surveys has been used (Nusser and Goebel, 1996; Hellkamp et al., 1995).

The approach of Romig et al. (1995) is very useful for the observational or qualitative measure of soil quality. Their approach involves (1) interviewing producers on their concepts, perceptions and measures of a healthy soil; (2) summarizing and ranking the results of the interviews; (3) ranking the measures of soil health into three classes—healthy, impaired, unhealthy; and (4) developing a score card with criteria for each class of soil health (Table 3.3). The scores for each indicator are totaled and used as a baseline for assessing soil quality for a field.

Gomez et al. (1996) present a method for measuring sustainability of agricultural systems at the farm level. Their approach can be adapted to measure soil quality because many of their indicators relate to the soil. These indicators include yield, soil depth, organic matter, and permanent ground cover. Threshold levels for the indicators are set by what is average in the community or at a value 20% more than the average. An index for each indicator is then calculated by dividing the value for the indicator by the threshold value and averaging over all indicators to give an index for the farm. If the index is less than one, it is determined that the specific farming system is not sustainable.

For regional or national measurements of soil quality, we favor sampling points over time using a randomized statistical sampling framework similar to the longitudinal survey employed by the National Resources Inventory (NRI) of the Natural Resources Conservation Service (NRCS). The NRI consists of a two-stage sampling design that accommodates varying intensities for a national survey as well as more detailed designs for a re-

Table 3.2. Soil Quality Indicators and Example Scoring Functions[a]

Indicators	Soil Functions	Range/Limits for Scoring Function	Scoring Function
Physical			
% Stable aggregate (*more is better*)	Regulating and partitioning water and nutrient flow	30–60 %	
Porosity—surface 75 mm (*optimum*)	Regulating and partitioning water and nutrient flow	20–80 %	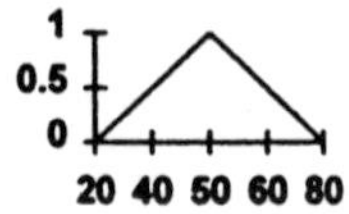
Bulk density (*less is better*)	Regulating and partitioning water and nutrient flow	1.3–2.1 mg/m^3	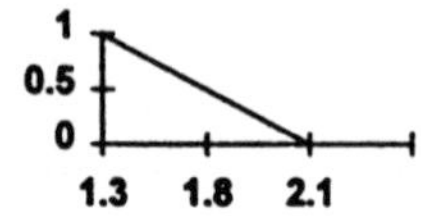
Rooting depth (*more is better*)	Biological activity	60–250 cm	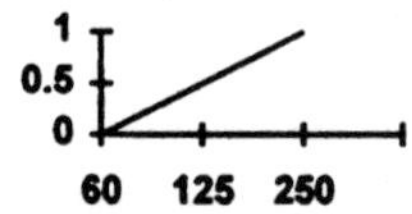
Biological			
Microbial biomass (*more is better*)	Biological activity, nutrient cycling, and filtering and buffering	75-700 mg C/kg	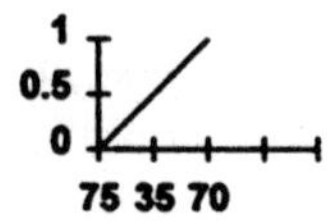
Respiration (*more is better*)	Biological activity, nutrient cycling	0.5-8.0 mg C/kg	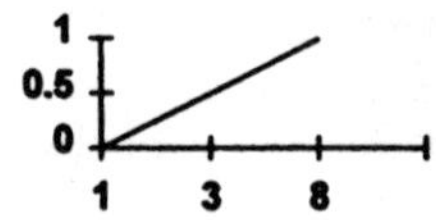
Total nitrogen (*more is better*)	Biological activity, nutrient cycling	1.5 - 5.0 mg/cm^3	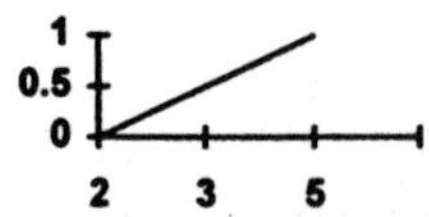
Total C in surface 75 mm (mg/cc) (*more is better*)	Storing and cycling nutrients and other elements, regulating and partitioning water and nutrient flow	15-50	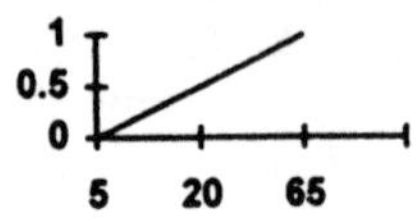
Chemical			
Plant available P (*optimum*)	Biological activity, nutrient cycling	7.5 - 150 mg/kg	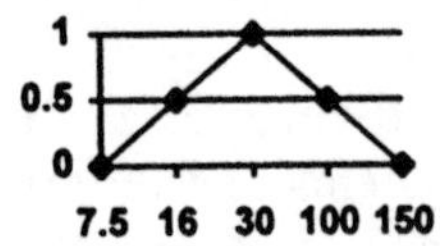
Nitrate nitrogen (*optimum*)	Biological activity, nutrient cycling	3 - 50 mg/kg	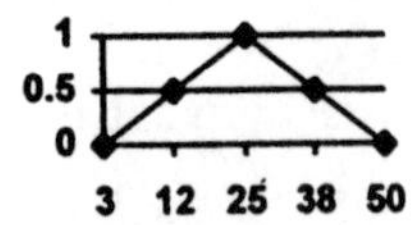
Exchangeable K (*optimum*)	Biological activity, nutrient cycling	45 - 525 mg/kg	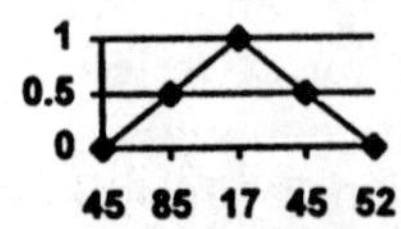
pH Surface 250 mm (*optimum*)	Sustaining biological activity, diversity, and productivity, storing and cycling nutrients and other elements	5.5 - 8.2	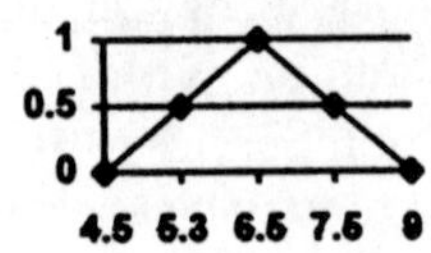

[a]From Karlen et al., 1994b.

Table 3.3. Example Section of the Observational Approach to Measuring Soil Health[a]

Descriptive Properties	Score
1. Earthworms 0 Little sign of worm activity 2 Few worm holes or castings 3 Worm holes and castings numerous	
2. Erosion 0 Severe erosion, considerable topsoil moved, gullies formed 2 Moderate erosion, signs of sheet and rill erosion, some topsoil blows 4 Little erosion evident, topsoil resists erosion by water or wind	
3. Tillage Ease 0 Plow scours hard, soil never works down 2 Soil grabs plow, difficult to work, needs extra passes 4 Plow field in higher gear, soil flows and falls apart, mellow	
4. Soil Structure 0 Soil is cloddy with big chunks, or dusty and powdery 2 Soil is lumpy or will not hold together 4 Soil is crumbly, granular	
5. Color (moist) 0 Soil color is tan, light yellow, orange, or light gray 2. Soil color is brown, gray, or reddish 4 Soil color is black, dark brown, or dark gray	
6. Compaction 0 Soil is tight, compacted, cannot get into it, thick hardpan 2 Soil packs down, thin hardpan or plow layer 4 Soil stays loose, does not pack, no hardpan	
7. Infiltration 0 Water does not soak in, sits on top or runs off 2 Water soaks in slowly, some runoff or puddling after a heavy rain 4 Water soaks right in, soil is spongy, no ponding	

[a]From Romig et al., 1995.

gional or watershed survey (Nusser and Goebel, 1996). Expansion factors enable the generalization of the point data to the geographic area of interest. The point data can also be tied to the soil resource inventories to extend the information from the point to larger areas.

Quantification of Soil Quality Assessments

Monitoring of trends is a way of quantifying change in soil quality over time and requires establishing baseline values for the various indicators and measuring change in the indicators over time. Changes in the indicators reflect the combined effects of land use and climate. Soil quality is then evaluated by the trend lines as described by Pierce and Larson (1993) and Larson and Pierce (1994). If the change in a soil quality indicator is positive and more is of better quality, then the soil can be regarded as improving or aggrading in quality. Conversely if the trend line is negative, then soil quality is degrading. Calculating the slope of the trend lines is a way of quantifying change in soil quality. A disadvantage of this approach is that it requires measurement of the indicators for at least two points in time and thus does not provide immediate assessment of soil quality. In addition, this approach is somewhat misleading if a soil is functioning at the highest quality attainable and cannot improve or conversely, if it is functioning at its lowest quality and cannot go lower. Both cases show a static or no change trend but are completely different. It is most desirable to show state of soil quality on a normalized scale in addition to trends.

In quantifying the current state of soil quality and as a base from which to monitor change with time, an index of soil quality can be calculated using the concepts of Karlen and Stott (1994). For each indicator, a scoring function and realistic baseline and threshold values are established using procedures described in the previous section. Examples of baseline values and scoring functions are shown in Table 3.2. (Values for the scoring functions in Table 3.2 are summaries from the literature and thus are not soil specific nor specific for land use. We show the values for demonstration purposes only and urge readers to develop scoring functions based on soil specific properties and research for specific land uses.) After these values have been established for a soil (or group of soils that function similarly for a particular use), weights can be assigned based on relative importance of each soil function and indicator as determined by the land use (Table 3.4). Karlen et al. (1994a) describe a nested scheme for assignment of weights to soil functions and their respected soil quality indicators as shown in Table 3.4. The scores for each indicator are multiplied by the appropriate weight. An overall soil quality rating is calculated by summing the weighted score for each soil function.

Examples of Assessment Using a Soil Quality Index

An example calculation of a soil quality index is given in Table 3.4 using data from Harris et al. (1996). The treatments are undisturbed Conservation Reserve Program (CRP) land that had been in grass for 9 years, plowed CRP land, and no-till CRP land. Data were collected after the first year of returning the CRP land to production. Soil quality indices were calculated by multiplying the weighted index value by the scoring function value for each soil quality indicator and soil function (Table 3.4). The individual indices for the soil quality indicators were summed within the treatments to give an overall soil quality index for each treatment. The soil quality index values are 0.53, 0.40, and 0.44 for the CRP, plowed CRP and no-till CRP, respectively. These values are different than those presented by Harris et al. (1996) but include two additional soil functions (regulating and partitioning water and filtering and buffering) in addition to the production function in the calculations. In addition, for simplicity, we used linear scoring functions rather than the more representative curvilinear functions used by Harris et al. (1996).

The soil quality index values seem low for the CRP treatment and may reflect the need to adjust the scoring function and reference values for the indicators given in Table 3.2. The values in Table 3.2 are very general and have not been adjusted for specific soil or land uses. However, the relative values among the treatment reflect the soil quality difference, with the CRP treatment showing the highest quality or health.

In a separate study Karlen et al. (1994) calculated soil quality indices for two long-term (10 yr) studies that attempt to show the effects of crop residue on soil quality ratings. The results showed soil quality indices of 0.45, 0.68, and 0.86, respectively, for the removal, normal, and double residue no-till treatments. In a follow-up to this study, Harris et al. (1996) report that 3 years after the residue management experiment was terminated, yields averaged 11.7, 12.0, and 13.0 mg/ha for the removal, normal, and

Table 3.4. Assignment of Weights to Soil Function and Indicators for Calculating Soil Quality Indices for Three Treatments: CRP, Plowed, and No-Till[a]

Soil Function		Soil Indicator		Weighted	Soil Indicator Value (Score)			Soil Quality Index[c]		
	Weight-1		Weight-2[d]	Index[b]	CRP	Plowed	No-Till	CRP	Plowed	No-Till
Sustaining biological activity	0.40	Rooting medium	0.30							
		Rooting depth cm	(0.60)	0.072	150 (0.6)	100 (0.3)	125 (0.5)	0.043	0.022	0.036
		Bulk density g/cm^3	(0.40)	0.048	1.07 (1.0)	1.39 (0.9)	1.31 (1.0)	0.048	0.043	0.048
		Water relations	0.30							
		Porosity %	(0.60)	0.072	60 (0.5)	48 (0.9)	51 (1.0)	0.036	0.065	0.072
		Total carbon mg/kg	(0.40)	0.048	23.9 (0.6)	16.6 (0.3)	16.0 (0.3)	0.029	0.014	0.014
		Nutrient relations	0.30							
		pH	(0.30)	0.048	6.3 (0.9)	6.4 (0.9)	6.4 (0.9)	0.043	0.043	0.043
		Available P mg/kg	(0.10)	0.016	38 (1.0)	6 (0)	13 (0.5)	0.016	0	0.008
		Exch. K mg/kg	(0.10)	0.016	267 (0.9)	137 (0.9)	167 (1.0)	0.014	0.014	0.016
		Total carbon mg/kg	(0.40)	0.064	23.9 (0.6)	16.6 (0.3)	16.0 (0.3)	0.038	0.019	0.019
		Nitrate nitrogen mg/kg	(0.10)	0.016	1.0 (0)	2.2 (0)	1.4 (0)	0	0	0
Regulating and partitioning water	0.30	Aggregate stability %	0.60	0.18	40 (0.3)	19 (0)	26 (0)	0.054	0	0
		Porosity %	0.20	0.06	60 (0.5)	48 (0.9)	51 (1.0)	0.03	0.054	0.06
		Bulk density g/cm^3	0.20	0.06	1.07 (1.0)	1.39 (0.9)	1.31 (1.0)	0.06	0.054	0.06
Filter and buffer	0.30	Aggregate stability	0.60	0.18	40 (0.3)	19 (0)	26 (0)	0.54	0	0
		Porosity	0.10	0.03	60 (0.5)	48 (0.9)	51 (1.0)	0.15	0.27	0.03
		Microbial processes	0.30							
		Total carbon mg/kg	(0.50)	0.045	23.9 (0.6)	16.6 (0.3)	16.0 (0.3)	0.27	0.014	0.014
		Total nitrogen mg/kg	(0.50)	0.045	2.9 (0.5)	3.6 (0.6)	2.9 (0.5)	0.23	0.027	0.023
Totals				**1.0**				**0.53**	**0.40**	**0.44**

[a]From Karlen et al., 1994b and Harris et al., 1996. Data for the three treatments is from Harris et al., 1996.
[b]The weighted index is calculated by multiplying the weight for the soil function (weight 1) by the weights for rooting medium (weight 2) by the weight for rooting depth (weight 3). For example, the weighted index for porosity is obtained by multiplying 0.40 (weight 1) times 0.30 (weight 2) times 0.60 (weight 3), which equals 0.072.
[c]The score is calculated by comparing the value of the indicator to the scoring function (graph in Table 3.2). For example, the score of the value of porosity for CRP is 0.5 as determined by the porosity of 60%.
[d]The weighted soil quality index is calculated by multiplying the weighted index and the score of the indicator. For example, the soil quality index for porosity and CRP treatment 0.072 times 0.5, which equals 0.036.

double residue treatments. Thus, yields correlate well with soil quality indices in this study.

SUMMARY

The assessment of soil quality is dependent on the development of soil quality indicators that relate to specific soil functions. The indicators can be qualitative, such as the use of observational features of the soil or plants growing on the soil to quantitative measure of physical, chemical, and biological indicators. Once a minimum data set of soil quality indicators has been established for an area, standards or reference values for soils functioning at full capacity must be established in order to assess the current state of soil quality. Standards or reference values should be established for specific kinds of soils or groups of soils that function similarly and for specific land uses based on best management practices. The standards represent sustainability of soil function with its ecosystem.

After reference values are established for groups of soils and specific land uses, assessments of soil quality are made by comparing the measured value to the reference value for an indicator. For the qualitative assessment, reference values are usually grouped into classes such as healthy, impaired, or unhealthy. For quantitative assessments, indices of the current state of soil quality can be calculated by representing the reference values or standards as scoring functions. These indices can be further refined by weighting the relative importance of each soil function and indicator for the soil function. Trends in the change of the indices are then monitored over time with respect to management systems.

REFERENCES

Acton, D.F. and L.J. Gregorich. Understanding Soil Health, in *The Health of Our Soils: Toward Sustainable Agriculture in Canada*: D.F. Acton and L.J. Gregorich, Eds. Centre for Land and Biological Resources Research, Research Branch, Agriculture and Agri-Food Canada, Ottawa, Ont., 1995, pp. 5–10.

Arshad, M.A. and G.M. Coen. Characterization of Soil Quality: Physical and Chemical Criteria. *Am. J. Altern. Agric.* 7:25–31, 1992.

de Haan, F.A.M., W.H. van Riemsdijk, and S.E.A.T.M. van der Zee. General Concepts of Soil Quality, in *Integrated Soil and Sediment Research: A Basis for Proper Protection.* H. J.P. Eijsackers and T. Hamers, Eds., Proceedings of the First European Conference on Integrated Research for Soil and Sediment Protection and Remediation (EUROSOL), Maastricht, The Netherlands, 1–12 Sept. 1992. Kluwer Academic Publishers, Dordrecht, 1993, pp. 155–170.

Doran, J.W. and T.B. Parkin. Defining and Assessing Soil Quality, in *Defining Soil Quality for a Sustainable Environment.* J.W. Doran, D.C. Coleman, D.F. Bezdicek, and B.A. Stewart, Eds. SSSA Spec. Pub. No. 35, Soil Sci. Soc. Am., Am. Soc. Agron., Madison, WI, 1994, pp. 3–21.

Gomez, A.A., D.E. Swete Kelly, and J.K. Syers. Measuring the Sustainability of Agricultural Systems at the Farm Level. Proceedings, Workshop on Advances in Soil Quality for Land Management, Ballarat, Australia, 17–19 April, 1996.

Gregorich, E.G., M.R. Carter, D.A. Angers, C.M. Monreal, and B.H. Ellert. Towards a Minimum Data Set to Assess Soil Organic Matter Quality in Agricultural Soils. *Can. J. Soil Sci.* 74:367–85, 1994.

Harris, R.F. and D.F. Bezdicek. Descriptive Aspects of Soil Quality/Health, in *Defining Soil Quality for a Sustainable Environment.* J. W. Doran, D. C. Coleman, D. F. Bezdicek, and B. A.

Stewart, Eds. SSSA Spec. Pub. No. 35, Soil Sci. Soc. Am., Am. Soc. Agron., Madison, WI, 1994, pp. 23–35.

Harris, R.F., D.L. Karlen, and D.J. Mulla. A Conceptual Framework for Assessment and Management of Soil Quality and Health, in *Methods for Assessing Soil Quality*. J.W. Doran and A.J. Jones, Eds. Soil Science Society of America, Madison, WI. (in press), 1996.

Hellkamp, A.S., J.M. Bay, K.N. Easterling, G.R. Hess, B.F. McQuaid. 1995. Mid-Atlantic Study, interim report, 1994.

Howard, P.J.A. Soil Protection and Soil Quality Assessment in the EC. *Sci. Total Environ.* 129:219–239, 1993.

Jenny, H. *Factors of Soil Formation*. McGraw-Hill, New York, 1941.

Karlen, D.L., M. J. Mausbach, J.W. Doran, R.G. Cline, R.F. Harris, and G.E. Schuman. Soil Quality: Concept, Rationale, and Research Needs. *Soil Sci. Soc. Am. J.* 60 (in press), 1996.

Karlen, D.L., N.C. Wollenhaupt, D.C. Erbach, E.C. Berry, J.B. Swan, N.S. Eash, and J.L. Jordahl. Crop Residue Effects on Soil Quality Following 10-Years of No-till Corn. *Soil Tillage Res.* 31:149–167, 1994a.

Karlen, D.L., N.C. Wollenhaupt, D.C. Erbach, E.C. Berry, J.B. Swan, et al. Long-Term Tillage Effects on Soil Quality. *Soil Tillage Res.* 32:313–27, 1994b.

Karlen, D.L. and D.E. Stott. A Framework for Evaluating Physical and Chemical Indicators of Soil Quality, in *Defining Soil Quality for a Sustainable Environment*. J.W. Doran, D.C. Coleman, D.F. Bezdicek, and B.A. Stewart, Eds. SSSA Spec. Pub. No. 35, Soil Sci. Soc. Am., Am. Soc. Agron., Madison, WI, 1994, pp. 53–72.

Larson, W.E. and F.J. Pierce. The Dynamics of Soil Quality as a Measure of Sustainable Management, in *Defining Soil Quality for a Sustainable Environment*. J.W. Doran, D.C. Coleman, D.F. Bezdicek, and B.A. Stewart, Eds. SSSA Spec. Pub. No. 35, Soil Sci. Soc. Am., Am. Soc. Agron., Madison, WI, 1994, pp. 37–51.

Larson, W.E. and F.J. Pierce. Conservation and Enhancement of Soil Quality, in *Evaluation for Sustainable Land Management in the Developing World*, Vol. 2: Technical papers. Bangkok, Thailand: International Board for Research and Management, 1991. IBSRAM Proceedings No. 12(2), 1991, pp. 175–203.

National Research Council. Soil and Water Quality: An Agenda for Agriculture. National Academy Press, Washington, D.C., 1993.

Nusser, S.M. and J.J. Goebel. The National Resources Inventory: A Long-Term Multi-Resource Monitoring Program. *Environ. Ecol. Stat.* (in review), 1996.

Papendick, R.I., J.F. Parr, and J. Van Schilfgaarde. 1995. Soil Quality: New Perspective for a Sustainable Agriculture, in *Proceedings for International Soil Conservation Organization*. New Delhi, India, December 4–8, 1994.

Pierce, F.J. and W.E. Larson. Developing Criteria to Evaluate Sustainable Land Management, in *Proceedings of the Eighth International Soil Management Workshop: Utilization of Soil Survey Information for Sustainable Land Use, May 3, 1993*. J. M. Kimble, Ed. USDA Soil Conservation Service, National Soil Survey Center, Lincoln, NE, 1993, pp. 7–14.

Romig, D.E., M.J. Garlynd, R.F. Harris, and K. McSweeney. How Farmers Assess Soil Health and Quality. *J. Soil Water Conserv.* 50:229–236.,

Seybold, C. A., M. J. Mausbach, D. L. Karlen, and H. H. Rogers. Quantification of Soil Quality. *Adv. Soil Sci.* (in review), 1996.

Smith, J. L., J.J. Halvorson, and R.I. Papendick. Multiple Variable Indicator Kringing: A Procedure for Integrating Soil Quality Indicators, in *Defining Soil Quality for a Sustainable Environment*. J.W. Doran, D.C. Coleman, D.F. Bezdicek, and B.A. Stewart, Eds. SSSA Spec. Pub. No. 35, Soil Sci. Soc. Am., Am. Soc. Agron., Madison, WI, 1994, pp. 149–157.

Smith, J.L., J.J. Halvorson, and R.I. Papendick. Using Multiple-Variable Indicator Kriging for Evaluating Soil Quality. *Soil Sci. Soc. Am. J.* 57:743–749, 1993.

FOUR

Minimum Data Sets Needed for Soil and Climatic Characterization

John M. Kimble

WHAT IS A MINIMUM DATA SET

There is no such thing as a minimum data set or a magic pill (data set) that we can or should collect to solve all problems. We always seem to be striving to make long lists, saying "This is the minimum." The problem is, researchers do not know all the questions which will be asked in the future, so how can we say this is the minimum needed? Did we know 50 years ago about acid rain? Global climate change? Sea level rise? Salinization? Heavy metal pollution? The answer may be both "yes" and "no." We now measure the levels of Pb, Cd, Cr, and even carbon levels in soils, yet we have no baseline to judge these against. How much change is recent? What is the real background level? We know, or think we know, that levels of nitrates are increasing in soils, but are we really sure? Where is the minimum data needed to relate present values to? We need to consider these questions when we talk/define minimum data sets.

Different kinds of research do not all require the same amount of data. Data needed for fertility are different from work on sustainable agriculture, erosion studies, soil quality/degradation, land use planning, and so on. But all studies share some basic needs, requiring the same basic information, and that could be considered the starting point for a minimum data set. However, there is no agreement about what this so-called minimum data set should actually contain.

The concept of a minimum data set is also discussed by Uehara (in this volume); he points out, in general, the needed data for experimental work and lists in generic terms a minimum data set for a specific use (but not for all uses). As time passes, we will have an ever-expanding need for data. Other books and papers on data sets (see "References") are useful but should not be considered a panacea for all needs, as they were written by a specific group for a specific topic at hand. Instead of listing properties or discussing other lists, this chapter discusses issues important in creating data sets.

If we look at present and future customer needs, we can, in the long run, develop data sets which service many purposes. "Serving a purpose" or "having use" is the key here. Many times, large data files for a project become the ultimate minimum data set; i.e., *nothing*. Computer archived files may be large, and the person who developed the file/database may well know what it contains, who obtained the data, how it was measured, and what the significance of the contained values are, but if there is no documentation as to the methodology used to generate the data, no one else will be able to use them. Data collected without any geo-reference also are of limited use. Saying it is from Jose's Farm is not the real world. It may be a very extensive, well documented data set but is still of limited use without a location identifier attached to it.

The book on site characterization (IBSRAM, 1988) included a chapter on chemical properties to be measured and monitored for soil management experiments (Kimble and Holzhey, 1988), a chapter on physical properties (Babaloa, 1988), and another on climatic measurements (Shing et al., 1988). Such lists can serve a useful purpose. However, what we really need to ask about such lists is: what is a minimum data set now and what will it be in the future? Does it go beyond my specific interest? Even though I have written chapters about what I consider to be the minimum data needed for soil management experiments, what I wrote then was for a specific use at a specific time. Nor can data be collected at only a set point in time, particularly climatic data. You can have an extensive climatic data set for one growing season, but is that a typical or nontypical season? Will it happen again? Climatic data need to be collected while experiments are underway because changes in solar radiation, rainfall, and so on, can vary greatly. Data also need to be collected before an experiment is started and after the experiment is completed. The bottom line is, without long-term data you cannot properly interpret short-term data sets.

If you are conducting a fertility experiment, NPK levels may be the only properties which are measured. Results, however, may have been affected by a lack of, or an excess of, some micronutrient, by soil variability, or by climatic abnormalities and if these constraints are not considered, your results may be hard to explain at times. Having more data about the site will allow others to use the data in relation to their data for that site or for other sites which may be similar. It will allow other researchers to use your data in combination with theirs, and everyone will come out ahead. Because of varying conditions in climatic and soil patterns, what is minimum in the tropics may have no relation to what would be considered minimum in cool temperate environments. Researchers, database system managers, system analysts, and database designers need to consider the expected outcomes of their work, i.e., how it may relate to other studies which may be planned and/or conducted at the same location and vice versa.

Another problem is what climatic information to store. Many models run on weekly or even monthly averages. Therefore, the only data set that may be available is one with "average" values, as it consumes less space. However, averages are just that, particularly with rainfall. You may have had the so-called normal rainfall for a given period (week or month), yet when it is averaged over time, the duration and intensity are not recorded. All of the rain could have come on one day, which is very ineffective for plant growth, yet when averages are used this is not known, so there actually may be little correlation to the average when you are looking at plant growth. Rainfall averages will not even show how nutrients move through a soil profile. If you record daily or even hourly climatic information, that needs to be stored. The researchers using the data can manipulate it any way they want, but if all that is stored is an average, no one knows what the data really were. If all you use is temperature in a model, how does that relate to cloud cover? It may be warm and moist, but crops can still do poorly if there is no sunshine. How you average data is also important. Do you just average the maximum and minimum temperature? This can lead to strange results. We live in a world of averages and they can really be misleading, or even meaningless.

As farming becomes a more precise science, the use of fertilizers is related more to soil tests and yields for specific areas within a field. You can no longer take samples and come up with an average value for a field. Specific samples are needed for specific areas of the field since averages are no longer considered a reliable way to maximize or enhance yields. Data storage is no longer the problem it once was, as computers have larger and larger storage capacities. In simple terms, if you measure a property, store it as measured. It can be averaged later, but averages cannot be unaveraged later!

One of the scenarios of global change is that night temperatures may increase. This can give a higher average daytime temperature but may not have as great an effect on crop growth as could be expected; again, the averages lie. A higher respiratory rate in a period without photosynthesis may have a greater effect than expected or than an average temperature would show.

The bottom line on a minimum data set is to record what you measure and enter it into a database. Others can then add to that database and, over time, the information and usefulness of the data set will grow. Also define how the data were obtained; do not simply list cation exchange capacity (CEC) or pH, as there are many methods for collecting each, and each method will give a valid and useful answer for a specific use but not for other uses. A data set with all kinds of information (numbers) but no details as to methodology is of less use than one that contains just a small amount of data where the methodology is well defined. Large quantities of poor quality data are not what we want.

The following is taken from Kimble and Holzhey (1988), "Pedon characterization data, or any soil survey data, are best used when the operations for collecting the data are well understood. Our mental pictures and conceptual definitions that aid in visualizing properties and processes are often different from the information supplied by an analysis. Also, results differ with different methods, even though the different methods may carry the same name or concept. It is difficult to compare one bit of data with another without knowing how both bits were gathered. This leads to operational definitions."

Also, as we move into the computer age and the collection of reams of data, we need to make sure the data collected are entered into a computer database. Long-term climatic information cannot be used if it is kept only as hard copies. Nor can data sets be alive and grow if they are just piles of paper. If it is just a pile of paper, a study can be done on the rate of growth of the paper pile (weight and size), but the usefulness of the actual data will be limited.

Data should not be collected just to build a database. However, we do need to spend money to collect needed data. Saving money by limiting data collection may seem wise at one point, but in the long run we often pay the piper when we find that little can be done with the limited data collected.

Data sets should be demand driven; i.e., what does the end-user want and need? Can you predict this? All too often the answer is "no." When studies are being developed, researchers need to consult their potential end-users, so that all needed data can be collected. Do not make these decisions alone, talk to potential end-users and collaborators, and work to develop a data set which is useable by many different researchers.

SITE DATA

All too often, research sites are not checked and documented before experimental work begins. In one instance a site in Africa was being used for agronomic trials but no one had ever looked at the soils or mapped the soils at the site. After a quick soils map was made, it was evident that the upper two-thirds of the area were deep soils (> 1.5 meters) and the lower third was only about 30 cm to rock. This would have a major influence on crop growth and compromise any fertility work or, at least, confound it greatly. What effect did the soil depth have on soil moisture? On crop growth? This was an unknown because it was not part of the experimental design.

We need site data before work begins. Site data requires both soil and climatic information. This will tell the researchers, if the site is uniform, if it is an extensive representative area, an odd (nonrepresentative) area where any information gathered would have no use to others or if there are any soil or climatic features which would affect the work to be done. For example, if a researcher is recording rainfall, and the area where the study is being conducted collects runoff or is subirrigated, the rainfall data will be useless. Another example of a compounding feature is the high level of boron at IRRI's site at Las Banos. This may influence crop selection, which is not at all related to the features being selected.

Soil maps and good climatic data are a must. They allow for agrotechnology transfer and for understanding confounding features. What are the rainfall patterns? Is there extensive runoff? Run-on onto the site? A climatic station which measures all kinds of parameters may not be sufficient, because how the pedosphere and atmospheric conditions are related is very important.

How much moisture can a profile store? You may have adequate rainfall, but if the profile is sand and the soil moisture storage is very low, the amount of rainfall may not be the controlling feature in your experimental work. The frequency and intensity may be the controlling feature and this may not be shown in averages.

When site characterization is not done, it may be that the researchers are not concerned about the soil and climatic data, or that they take soils for granted. Many experimental stations are on good soils and not related to other nearby areas or, conversely, are on very poor soils; without wider site data, the results may not be related at all to the rest of the world. Site data also may be neglected because it is expensive to collect and may not yield any short-term payoff, as far as a specific research topic. Many say site characterization is collection of data for the sake of data, but that is completely wrong. It can provide the basis for any potential future work, it will help explain variability, it allows the transfer of information to other sites. It can also allow that specific experiment to be compared to other similar experiments at other sites.

Background information allows for the measurement of change and a better understanding of processes. Soil maps are not just pretty pictures, they are an important part of a good data set. Part of any field work thus should be complete site characterization of chemical, physical, and climatic parameters and along with soil maps, the site characterization will include the background climatic information about the site, to provide a baseline for possible changes. Background monitoring and related information is not exciting research, but without it changes cannot be determined. Without baseline soils and climatic information, you will never know, for example, whether the systems being tested are good or bad for the environment.

WHO DECIDES WHAT IS NEEDED FOR A DATA SET?

The customer should decide, but all too often the ultimate customer is not even considered in research. Research is done to answer researcher-driven questions. Many of these are valid, and basic research is needed, but it also must relate to the real world. Most research is done by centers funded by governmental bodies and NGOs (federal, state, and international centers, UN agencies, etc.), yet many do not consider the needs of the smaller or even large landowner (ultimate customer) in deciding what is needed.

In a world of shrinking budgets, we need to get as much from each project as possible. We need to link projects together to share data and this is especially true of expensive background site information on climate and soils. Groups such as the NSF are now requesting that data be put into central files to which everyone has access. This makes individual researchers nervous and raises the question of who really owns the data. The researcher doing the collecting? Or the agency or individual providing the funding? In the future, the agency or individual funding the work may have a larger say. Some will say this will stop scientific work, but I do not feel this is the case. In fact, it may speed up the writing of papers and distribution of data, as researchers will become more motivated to use the data than to let it go unused to others. Individual researchers may not like this, but if donors say, "Give us your basic data so others can use it," and researchers refuse, then the eventual outcome may be the loss of future funding for that researcher.

I have been involved in recent meetings dealing with monitoring change in the Antarctic. There is a perceived need to see if changes are taking place. Many researchers have listed species of plants and animals to study, but there has been little discussion about gathering baseline data. The baseline data are what goes into the minimum data set. Basic research is the major interest, but without the mundane background data, of what use will be the data gathered? Many times NSF and other funding agencies do not consider background data collection to be real research and therefore do not fund it. But without it, we cannot interpret other research.

The time of going it alone is in the past. We need to work as teams and to make sure the collection and distribution of data takes a broad-based approach and that decisions about what is to be collected are likewise broadly based. If only one data element is added, it may make the data set much more useful to other researchers. With LTERs (Long Term Ecological Research Sites) and international centers we need to develop long-term plans to determine what data the overall scientific community believes are needed. We need to establish baselines for soils and climatic data which then can be used by all disciplines. What one individual thinks is necessary may not be what another one wants or needs. If researchers work as teams we can gather better information and spread the cost among different groups. The different groups need to feed their baseline data into a common database, and over time a very useful database will develop.

The answer to "Who decides what is needed?" should be "Everyone working together." It makes very little sense to collect data for just one experiment (in most cases). Many groups can use standard climatic data, soil maps, and supporting laboratory data. The amount and detail of such information should be determined by a broad group so that all will benefit in the long run.

WHAT ARE FUTURE NEEDS?

We face ever-changing environments. The Natural Resources Conservation Service (NRCS), formally the Soil Conservation Service (SCS), has been collecting soils data for over 60 years. This information is now used for many things which were not considered in the 1950s or even as recently as the 1980s. We are now concerned with issues including, but not limited to, acid rain, global climate change, carbon budgets, and wetlands. The data that have been collected can be used to answer questions that surround these issues. It was not planned when the data collection was started, but a standard data set was collected at many different sites. Also, samples were stored, which allows new tests and information to be developed and added to the existing data set. Future needs are not known, but even if you do not analyze for everything now, if samples are stored, determinations can be made in the future. Storage may seem like a waste of space, time, and money, yet it has allowed and will continue to allow for the expansion of what at one point in time was a minimum or complete data set and thus provides answers to new questions not thought of when an experiment began.

We live in a world with an ever-increasing population, and we have limited natural resources. If we are to develop sustainable systems, we need to know what our basic resources are and how to properly manage them. We need to continue to expand our data sets and storage of data. Data is a legacy which will allow future generations to go forward rather than repeat work.

In the future we are likely to need to reanalyze samples with new procedures. Most carbon determinations were once done by wet digestion but, because of environmental problems with this, more and more laboratories are now using dry combustion. Both procedures are valid but give different results. By rerunning some determinations on stored samples, the old and new data can be used within the same data set, with a built-in correction factor for the different levels of recovery. Laboratory methods will continue to change, but this does not mean stored data are not valid; it means we need to look at methods, document changes, and store the procedures used with the meta-data for each attribute. Some measurements will be dropped; others will be added. What is done must be well documented. Every piece of hard chemical, physical, and climatic data must be supported with viable and well defined meta-data.

If we show a temperature for a day, is it the average? The minimum? The maximum? How is the average figured? By adding the maximum and minimum and dividing by 2? During the Olympics it was easy to see how temperature data can be manipulated. Atlanta is very hot in the summer, but if you look at the arithmetic average (high plus low/2), it is not that bad. The average is not the true case if you are interested in daytime temperatures.. The daily average high is much greater than the average of the maximum and minimum. Data transformation such as this goes on all the time to present a view which someone chooses to present. We need to get beyond the averages and store all of the data so different groups with many different interests can use it. Data reduction should not be what is stored When data is reduced (averaged or otherwise transformed), the more need there is for explanation and documentation.

One of the biggest problems with existing data is the lack of meta-data to tell how chemical, physical, mineralogical, or even climatic information was gathered or determined. Without extensive supporting data, values called CEC are of little use. Even saying the CEC was measured with ammonium acetate is not that useful. It can be measured at pH 4.8 or 7, and each gives a correct number, but the results are different and need to be interpreted differently. Reams of data around the world will never be used because the meta-data on how the information was collected is not available. Soil temperature may be recorded, but the depth at which it was recorded is not given, so it is of little use. Climatic information may just be an average and, without details on how the averages were generated, is of little use.

In the end we may have a long list of properties to measure and store. And that may really not be that bad an idea.

CONCLUSIONS

There is no such thing as a minimum data set. There is only a minimum set of data for a specific research project, and that should be the minimum needed by your anticipated enduser. We can make all the soil and climatic maps we want at a scale of 1:250,000, but of what use are they in precision farming? To the farmer trying to make a living on one or two hectares? Maps and information can make pretty pictures, but what purpose do they serve? We need to consider the end uses in what we do and not just develop information to store in a database or to hang on the wall.

End-users, researchers, and funding agencies working as a team need to determine data set contents. The actual database structure should not be the primary focus, but the accessibility and usability of the data should be the main consideration when creating a data set. Data must be collected and stored in raw form so that a variety of end-users can process and use the data. We must remain flexible and have the ability to change what we collect and to store samples for future tests so we do not have to go back to the field to start the process over.

REFERENCES

Babaloa, B. Soil Physical Characterization for Soil-Based Research Networks, in *First Training Workshop on Site Selection and Characterization. IBSRAM Technical Notes No. 1.* IBSRAM, Bangkok, Thailand, 1988, pp. 165–180.

IBSNAT. *Experimental Design and Data Collection Procedures for IBSNAT: Minimum Data Set for Systems Analyses and Crop Simulation, 3rd Edition, Technical Report 1.* Department of Agronomy and Soil Science, University of Hawaii, Honolulu, 1988.

IBSNAT. Documentation for IBSNAT Crop Model and Output Files for the Decision Support System for Agrotechnology Transfer (DSSAT V 2.1). *Technical Report 5.* Department of Agronomy and Soil Science, University of Hawaii, Honolulu, 1990.

IBSRAM. First Training Workshop on Site Selection and Characterization. *IBSRAM Technical Notes No. 1.* IBSRAM, Bangkok, Thailand, pp. 1–298. 1988.

IBSRAM. Evaluation for Sustainable Land Management in the Developing World. Volume 2. Technical Papers. Bangkok, Thailand: International Board for Soil Research and Management, 1991. *IBSRAM Proceedings,* 12(2), 1991, pp. 1–631.

Kimble, J. M. and C.S. Holzhey. Chemical Properties to be Measured and Monitored for Soil Management Experiments, in *First Training Workshop on Site Selection and Characterization. IBSRAM Technical Notes No. 1.* IBSRAM, Bangkok, Thailand, 1988, pp. 181–194.

Lal, R. Methods and Guidelines for Assessing Sustainable Use of Soil and Water Resources in the Tropics, Soil Management Support Services, Technical Bulletin 21, Washington, D. C., 1994.

Michener, W.K., A.B. Miller, and R. Nottrott. Long-Term Ecological Research Network Core Data Set Catalog. Belle W. Baruch Institute for Marine Biology and Coastal Research, University of South Carolina, 1990, pp. 1–322.

Shing, P., A.K.S. Huda, and S.M. Virmani. In *First Training Workshop on Site Selection and Characterization. IBSRAM Technical Notes No. 1.* IBSRAM, Bangkok, Thailand, 1988, pp. 53–88.

SMSS. Soil Resource Inventories and Development Planning. *Technical Monograph No 1.* 1981. SCS-USDA, Washington, D.C., 1981, pp. 1–407.

Swindale, L.D. Ed. *Soil Resource Data for Agriculture Development.* Hawaii Agriculture Experiment Station, College of Tropical Agriculture, University of Hawaii. Honolulu, 1978, pp. 1–306.

Two

Nutrient and Water Inputs for Soil Quality Management

Five

STRATEGIES FOR FERTILIZER RESEARCH IN THE TROPICS

Bernard H. Byrnes and Deborah T. Hellums

Humanity is at a critical juncture. Food needs continue to increase with increasing population while the grain reserves are at all-time lows. The population continues to grow at an alarming rate, especially in the most food-deficit regions of the world. There is little land area that can be readily brought under economic agricultural production, and the amount of land suitable for additional irrigation capacity is very limited. The rate of irrigated land degradation through salinity and waterlogging is almost equal to the rate of new areas being developed for irrigation (Umali, 1993).

In the face of these physical limitations to meet food requirements, support for efforts to avoid the impending crisis has been declining. There are no longer the political pressures to maintain influence in poor countries through aid programs since the demise of Communism, and news reports of mass starvation have been losing their impact. Food aid, rather than efforts to increase indigenous production capacity, is often chosen by the developed countries with excess food production. It is, therefore, of paramount importance that the available scarce resources be put to the best possible use to increase food security in tropical countries. While this book addresses the broad issues of long-term soil fertility management in the tropics, this chapter deals with research on fertilizers that is needed to alleviate soil fertility[1] constraints to help meet world food requirements.

ADDRESSING LONG-TERM SOIL FERTILITY NEEDS

The movement of nutrients is an inexorable aspect of the natural world, but the rates of movement are greatly increased by agricultural uses of land as foodstuffs move from fields to points of consumption, generally the rapidly growing urban areas (Blackie, 1995). In addition, soil tillage causes more rapid degradation of organic material, loss of nutrients, and increased soil erosion. A basic goal of sustainability, to mimic natural systems as much as possible by maximizing the internal cycling of nutrients, is an impossible tenant to achieve because of the export of nutrients with harvested products, even with low ex-

[1]Soil fertility refers to the amounts of nutrients available for plant growth.

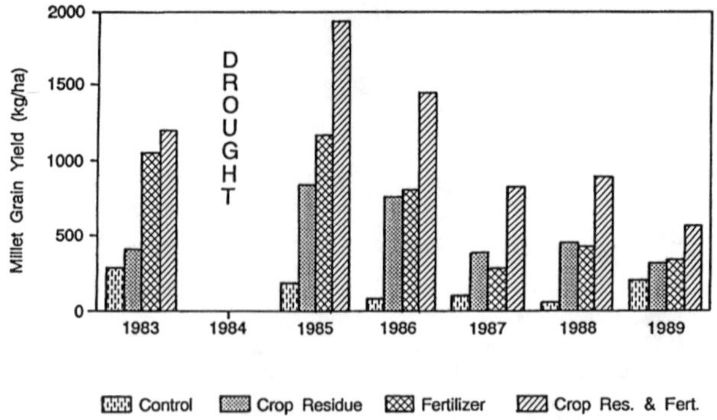

Figure 5.1. Effect of combined fertilizer and residue management on yield (Niger). From Vlek, 1990.

ternal input systems. The most economical way to replenish nutrients is normally with mineral fertilizers, but the utilization of locally available organic materials and recycling of nutrients in crop residues are very important to maintaining long-term fertility and productivity[2] of soils, particularly those of the tropics.

Organic wastes, crop residues, and green and animal manures are not only important sources of nutrients, but they also impart many additional benefits to soil productivity. These sources need to be utilized as much as possible, particularly in tropical environments. Studies have shown that productivity of many tropical soils cannot be maintained by the use of mineral fertilizers alone (Beets, 1990; Nambiar, 1995), and even the maintenance of crop residues can have very dramatic effects on tropical soils (Figure 5.1). Policy incentives will be needed to promote the effective use of organic wastes in both developed and developing countries. Poorly utilized materials are not only wasted resources, but become sources of pollution. As important as organic waste utilization is, the necessity of mineral fertilizers to improve the soil fertility of the weathered tropical soils is obvious, even with a rudimentary mass material balance, but this can be difficult for untrained people to understand. As with many other issues, however, the application of existing knowledge in less developed countries stands in stark contrast to the application in the industrialized countries (Breman, 1995).

In the United States, "precision agriculture" or site-specific management is the emerging crop management technology which many regard as revolutionary as the introduction of mechanization or hybrid varieties. It involves intensive mapping of soil characteristics and yield monitoring with computer data management and global positioning systems to vary fertilizer application rates, pesticide application rates, and seeding densities across fields. With these systems, 0.2 to 0.5 hectare units are sampled and characterized to provide automated management and record keeping, epitomizing site-specific management. In contrast, many tropical countries have no soil testing programs available nor promoted with farmers, even though fertilizers are widely used. There have been considerable efforts to improve fertilizer efficiencies, yet soil testing, the most basic method to achieve efficiency and promote good fertilizer management practices, is not utilized. Progress toward initiating soil testing in several Southeast Asian countries appears to be imminent, but support from scientists within the countries is often weak. Indonesia was spending $500 to $800 million (US) per year on fertilizer subsidies while providing no support for soil testing programs. This can only be explained by a lack of understanding of the impact that soil testing can have in improving efficiencies, protecting water resources, educating farmers about fertilizer use, and monitoring long-term trends in soil fertility. Scientists from developing countries often lament that they have no hope of reaching illiterate farmers with well-proven technologies such as soil testing. However, if information that benefits farmers is not presented to them, do they have any motivation to become literate?

[2]Soil productivity includes all factors which affect plant growth, including soil fertility, water availability, climate, etc.

While new technologies on how to manage fertilizers emerge periodically, the lack of application of existing knowledge about nutrients and soil management continues to be a major problem. The problem is not just with farmers. Research is often conducted with little understanding of well-established principles of soil fertility and fertilizer use, little appreciation for when nutrients are needed during crop growth, loss pathways, and how plant roots access nutrients. Soils researchers often appear to ignore soil testing as a method to explain results; are quite happy with blanket recommendations; and do little to promote the use of soil testing with administrators that need some good advice. There are many studies being published that measure crop response to fertilizers with no mention of initial soil fertility conditions, or with no description of how or when the fertilizers are applied. A recent article defines the most profitable NPK application rate for wheat for the whole of Pakistan (Shah et al., 1994). There is a tremendous need for the scientific and nonscientific community to realize that blanket recommendations cause inefficient use of fertilizers, pollution problems, and detrimental effects to soils. As the most yield-limiting macronutrients are applied and yields have increased, previously nonlimiting nutrients become deficient, but the deficiencies are not recognized because of the lack of adequate soil and plant tissue testing. Excessive buildup of soil fertility has not only occurred in developed countries, but also in countries like Indonesia. While this problem can and should be avoided, many other countries are apparently not applying enough P and K to maintain soil fertility. Without archives of soil test results and long-term experimentation, these trends are difficult to establish.

There are other examples of poor application of existing knowledge in soils research, extension, and in motivating farmers to use better techniques. Some that should be considered by international organizations concerned with long-term fertility in the tropics are:

1. Utilizing simple machines that band-apply P and N sources to improve efficiency through reducing P fixation and ammonia volatilization losses from urea. In some areas of Africa, where fertilizer use is already severely constrained, the "dollop" method of spooning a measured amount of a multinutrient fertilizer onto individual plants is used. Both the timing and the placement causes very inefficient utilization of the fertilizers.
2. Utilizing liming materials in humid areas, often in relatively low amounts, to avoid Al toxicities and maintain soil Ca and Mg levels. There is need for experimentation with coarser grinds of liming materials which will have long-term effects and have economic advantages.
3. The use of prophylactic application of soil ameliorants to soils that will develop salinity and alkalinity problems under irrigation. Vigorous policy intervention will likely be needed in many countries.
4. Research on minimum tillage systems to conserve organic material on the soil surface to improve soil structure and water infiltration, and reduce soil erosion. There may also be long-term benefits in nutrient efficiency through fertilizer placement in concentrated zones as is done in minimum tillage systems.
5. Research on the use of concentrated and efficient forms of fertilizers to achieve the greatest crop utilization in remote, disadvantaged areas. The use of other inputs and the development of output marketing potential are also needed to make the use of the fertilizers profitable in these areas. Without nutrient inputs the soil fertility declines and production needs will not be met.
6. Long-term research to establish cropping systems and methods to preserve crop residues are needed in all the tropical systems to preserve nutrients and reduce soil erosion losses.

7. Policy research and interventions to address risk in various agroecological situations, which often inhibits the use of needed inputs.

Unfortunately, financial support is difficult to find at an international level for research and promotion of accepted practices. Donors and researchers are often more interested in new solutions rather than to look for innovative ways to promote known technologies. Advances in soil fertility will be slow in a "pure research" context, mainly because most of the basic knowledge of soil fertility and fertilizer management has been pretty well established. There needs to be more effort in fostering the application of existing knowledge rather than conducting more research on fertilizers in the traditional sense. There are not likely to be any major breakthroughs in fertilizer products or management. We need to be more imaginative and conscientious in promoting adoption and application of what is already known, possibly by collaboration with NGOs, and by encouraging national extension programs to promote adoption.

Addressing Root Causes of Problems

Policy makers and soil fertility specialists frequently fail to identify the root causes of both physical and fertility decline of soils. These root causes need to be identified and can greatly impact the design of long-term experiments (Ofori, 1995). An example is the very low soil fertility status and low fertilizer use in sub-Saharan Africa (Table 5.1; Van Reuler, 1996; Vlek, 1993). Even though there have been numerous efforts to promote fertilizer use, the root cause of the lack of adoption is that farmers lack economic incentives to use fertilizers, due to poor input and output marketing systems, poor infrastructure for transportation and storage of produce, output taxation, and lack of land tenure. Local grain prices often fluctuate by a factor of 10 during the year, so it was only profitable to fertilize maize in Ghana, for instance, if the farmer stored the grain for 7.5 months (Pouzet, 1991). Food prices are kept artificially low to placate the large urban poor population and it is often cheaper to import foodstuffs than to buy locally produced foods. Fertilizer prices are extremely high in sub-Saharan Africa, for instance, due to low demand, lack of fertilizer marketing, high transport costs, and unfavorable monetary conditions (Table 5.2). Policy initiatives are needed to promote efficiency in input-output marketing if these countries are to achieve greater food sufficiency. The question needs to be reframed from how to improve soil fertility to how to produce market conditions under which it is profitable for farmers to adopt practices that increase soil productivity on a sustainable basis. In many cases, efforts are undertaken at national levels, when progress would be more likely attained if smaller areas with geographic and resource advantages were targeted.

Olsen (1992) stresses that in addition to multidisciplinary approaches,

Table 5.1. Regional Fertilizer Use per Hectare in 1994/1995[a]

Country	kg N-P_2O_5-K_2O/ha
Africa	19
Sub-Saharan Countries	9
Republic of South Africa	63
North African Countries	51
Asia	129
West Asia	60
South Asia	77
East Asia	216
South America	66
Central America	62
North America	92
Eurasia (former Soviet Countries)	20
Oceania	44
Western Europe	201
Eastern Europe	67
World	85

[a]From Bumb and Baanante (1996a), based on FAO data.

Table 5.2. Examples of Nitrogen Fertilizer Prices in Various Counries, 1985 vs. 1990[a]

Country	Price Per Ton N 1985 (US $)	Price Per Ton N 1990 (US $)	% Change In US $	% Change In Local Currency
Bangladesh	386	335	-15	+7
Turkey	216	260	+17	+389
Ghana	505	895	+44	+697
Zambia	1250	1294	+3	+1,384
Mexico	164	177	+7	+975
Venezuela	188	96	-95	+819
Poland	179	685	+74	+3,648
United States	275	284	+3	+3

[a]Based on Bumb and Baanante, 1996a.

more applied work is needed using existing data, and that greater emphasis should be placed on education and technology transfer by the international community. The publication of findings from research studies is not enough to accomplish any farmer adoption and frequently does not even achieve awareness by scientists in the poor target countries; we need to venture more into extension activities. Regarding policy development, Olsen points out that it also must take a more multidisciplinary approach, not regard certainties as requirements for implementation, consider the extremes in market and climatic fluctuations, and try to emphasize the multiple benefits achieved from good policy implementation. All of these considerations fit into the accepted concepts of sustainability and have important implications in long-term experimentation.

Few developing nations have long-term research, particularly those facing the most severe challenges in soil degradation due to poor land management (Ofori, 1995; Scherr and Yadav, 1996). Donors and politicians might be compelled to adopt policy regarding fertilizer management if long-term experiments comparing good land management practices with common farmer practices were available for their observation. The Auburn University long-term plots in a humid area on sandy soils are somewhat analogous to parts of West Africa, for instance. They clearly demonstrate the consequences of not replenishing soil nutrients in such environments. Many of the plots that have not received nutrients have difficulty supporting even weed growth, similar to the "white spots" of West Africa (Smaling, 1995).

Progress in Soil Fertility Research

The International Fertilizer Development Center (IFDC) intended to develop new fertilizers and new fertilizer management practices for tropical regions. The thought was that high P fixation, low soil pH, low cation exchange capacity, high temperatures, and high leaching rates in the tropics require new forms of fertilizers, or different management of existing fertilizers, than those used in temperate climates. There has also been emphasis on utilizing indigenous deposits of phosphate rocks in developing countries. The basic research has revealed the importance of ammonia volatilization losses from urea, greater understanding of direct application of phosphate rock, and greater understanding of the constraints for the development of phosphate rock deposits in developing countries. For the most part, these efforts have yielded few new fertilizers or new fertilizer production procedures. However, there still are many improvements in fertilizer use, production, marketing, and fertilizer-related policy in tropical countries that can be made, and could greatly contribute to better food security, alleviation of poverty, and resource conservation. The application of known knowledge can have a much larger impact than improved fertilizer forms.

Nitrogen Research

Research on coated fertilizers has generally shown that while the efficiency can be improved, coated materials are too expensive for use on commodity crops. Unless urea becomes extremely expensive, coating technologies have little application for field crop production. This premise is being tested by the Food and Agriculture Organization of the United Nations (FAO) in studies on rice production in Southeast Asia sponsored by the Chisso Corporation. There is likely more potential for their use on high-value vegetable crops in the tropics than on rice. Vegetables are heavily fertilized, nutrient use efficiencies are very low, and there is a large potential for nitrate leaching. Economic returns for coated fertilizer use are largely lacking relative to other uncoated fertilizers, and the possible effects on nitrate pollution are not considered to be of enough consequence to warrant policy implementation to bring about the use of coated fertilizers on such crops as vegetables.

A urease inhibitor has recently been commercially developed that reduces ammonia volatilization losses from surface-applied urea, particularly for no-till maize. The technology also includes a very effective spreading and carrier agent to facilitate the addition of the inhibitor on the urea. The product, marketed as Agrotain®[3] in the United States by IMC-Agrico, worked well on temperate soils planted to flooded rice, but its performance in numerous studies with flooded rice in the tropics has not been good (Byrnes and Freney, 1996). This is because the compound (N-[n-butyl] thiophosphoric triamide) is not a urease inhibitor itself, but must be converted to the actual inhibitor, its oxygen analogue, through yet unknown reactions in the soil. The oxygen analogues of the compound have shown considerably more potential for use on flooded rice, but more work is needed. Tropical crops, such as sugar cane and bananas, are likely crops on which this urease inhibitor may be successfully used, and some research is being conducted to assess this potential.

One conclusion of the urease inhibitor work with flooded rice has been that denitrification losses in flooded rice systems become more significant when the urea-N is spared from ammonia volatilization. Work on compounds to reduce denitrification losses has slowed greatly since research on the promising DowElanco product, XDE 474, was discontinued after toxilogical problems were found. Nitrapyrin (N-serve) and dicyandiamide (DCD) have not produced favorable results in tropical experiments, and apparently, use of these chemicals in the temperate countries is declining. An interesting development in denitrification research has been the work of Mosier and his coworkers to develop a coated (encapsulated) calcium carbide product that releases acetylene as water penetrates the coating. Acetylene gas in soils inhibits both nitrification and denitrification processes, and also methane production in flooded soils (Bronson and Mosier, 1991). Despite the apparent promise of this technology, support for the development of better products and testing of them has been limited by the lack of funding.

[3]Registered trademark of Freeport-McMoran Resource Partners, Limited Partnership, licensed to IMC-Agrico, Mundelein, Illinois.

Fertilizer production has been moving toward higher analysis products. This should benefit tropical countries with poor transportation infrastructure. Unfortunately, the education levels of the farmers and the oversight capabilities of the governments often reduce the benefits because the poorly educated farmers do not distinguish between types of fertilizers and the governments often do little quality assurance monitoring, which allows unscrupulous marketing of inferior goods.

Large tablets of urea with diammonium phosphate (2.8 g tablets with a 4:1 N:P ratio) to facilitate deep point-placement of the two nutrients together has performed extremely well in remote areas of India where farmers have little labor constraint but very high land and cash constraints (Daftardar et al., in press; Savant and Stangel, 1995). This work is an excellent example of application of the principles of sustainability, in which the researcher understands the entire context of the farming communities that the research is meant to assist (Blackie, 1995). Simple methods were developed to help farmers adopt line transplanting, and the transplanting geometry was modified to speed the fertilizer placement and to enhance nutrient uptake (Savant, 1995). The on-farm research in India on 127 sites over 3 years has produced an average of 1 additional ton of paddy yield from yields of about 3 tons with conventional fertilizer and crop management techniques (Table 5.3). The technologies are very attractive to small farmers, but adoption depends on sustained demonstration and extension work in carefully chosen regions. There are also efforts by the Indian Farmer's Fertilizer Corporation to promote the use of "supergranules," which contain only urea.

The government of Indonesia has brought about adoption of deep point-placement of large granules of urea on Java through its control of fertilizers and input packages. The government subsidized the fabrication of single-point applicators to assist the placement of 1 g urea supergranules produced by tableting machines, rather than briquetting machines as developed by the IFDC. The 1-g granules have a disadvantage compared to larger granules because three granules have to be placed at each placement site to achieve a reasonable N rate (73 kg N per ha), significantly increasing the labor requirement. The use of appropriately sized granules, NP granules where needed, and modified transplanting geometry would improve the likelihood of adoption of deep placement in transplanted rice in remote regions without labor constraints. As with many other cases, significant research results go largely unnoticed or unheeded without some efforts to help NARS become cognizant of the work.

Nitrogen demand will likely increase, particularly with the record low grain reserves, and continued population growth, keeping N prices relatively high. New production capacity in the Mideast and China which will be coming on line in the next 5 years is likely

Table 5.3. Average Rice Yields with Deep Point-Placement of Urea-DAP Briquettes Compared to the Same Rates of N and P Conventionally Applied (56 kg N and 14 kg P/ha)[a]

	Yield (tons/ha)		
Fertilizer Management	**1993**	**1994**	**1995**
Urea-DAP briquettes	4.55	3.94	4.09
Conventional application	2.99	3.11	2.99
(number of sites)	(26)	(51)	(40)

[a]From Daftardar et al., in press.

to cause some weakening of prices with the additional supply, assuming energy prices do not rise.

Phosphorus Research

Only 10% to 20% of water-soluble P in fertilizers is recovered by the initial crop, representing a considerably lower recovery rate than that associated with N fertilizers. However, since P is not appreciably lost from the system, the remaining P is converted to less soluble forms via a number of chemical reactions commonly referred to as P fixation. The less soluble forms not only furnish significant P to succeeding crops, but make succeeding application more effective by satisfying the P fixation capacity of soils.

Due to previous fertilization practices, many of the soils of North America and Western Europe have high P fertility levels so only a maintenance dose, which essentially offsets the P removed in the harvested crop, is required. Early growth response to applications of water-soluble P can often be observed, but significant yield responses to P application are unlikely. Increases in P fertilizer consumption in the developed countries are not expected, while it will increase rapidly in some developing countries (Engelstad and Hellums, 1993). By the year 2000, phosphate fertilizer consumption in North America and Western Europe will be less than 25% of global consumption (Schultz et al., 1993).

In contrast, to improve the P fertility status much of the tropics will require increased P inputs from mineral fertilizers. This is particularly true of the upland soils in Africa, Latin America, and Southeast Asia. Many of these soils are acidic, deficient in native P, and in some cases have a high P fixation capacity. Initially, these soils will require substantial P inputs to increase the soil P fertility and then lesser amounts to maintain fertility after production has been increased. The amount of P fertilizers required to increase soil P fertility will primarily depend on the P fixation capacity, off takes, and previous fertilization practices. Many soils of West Africa have low P fixation and application rates as low as 20 kg P/ha/yr can increase the P status as well as meet immediate crop demand. For highly weathered Oxisols, 400-800 kg P/ha may be needed to satisfy the P fixation capacity and obtain satisfactory crop growth, similar to the amounts needed in the Cerrados of Brazil. After the P fixation is satisfied, the situation is similar to that in developed countries except that erosional losses of P can be much higher without good soil conservation practices.

In the last decade, there have been significant increases in the use of P fertilizers in Asia, Latin America, and North Africa (Schultz et al., 1993; Bumb and Baanante, 1996b). However, the situation in sub-Saharan Africa (excluding the Republic of South Africa) remains bleak. All fertilizer inputs, including P, remain extremely inadequate (Smaling, 1995; Van Reuler, 1996) and the fertilizers are mostly used on export crops. Input/output marketing largely constrains use of the expensive fertilizers for food crops.

Presently, there is an increasing trend toward production and consumption of higher analysis fertilizers. In 1993/94, diammonium phosphate (18-46-0) and monoammonium phosphate (11-52-0) accounted for over 50% of the global phosphate consumption (Figure 5.2). This increase in market share by the ammonium phosphates has come primarily at the expense of triple superphosphate. Single superphosphate (0-20-0-6S) continues to be an important P source in countries with major SSP production capacity and in areas with known S deficiencies. While the ammoniated phosphates represent an economic means of addressing both P and N requirements, they do not contain nutrients (Ca and S) found in less concentrated P fertilizers (e.g., single superphosphate). As more ammoniated phosphates are used, the needs for Ca and S will have to be met by other products.

Considerable effort has been made to determine the economic feasibility of developing indigenous deposits of phosphate rock (PR) as a source for direct application of PR or to identify methods of beneficiation which produce more soluble P materials. While many of the tropical soils have characteristics that promote dissolution of PR (i.e., acidic pH and low levels of P and Ca), most of the PRs of the developing countries have low reactivity and cannot effectively supply sufficient P to support growth of annual crops. For the few PRs which have medium to high reactivity, numerous in-country studies have established that they can be appropriate alternatives to high analysis P fertilizers under certain agroclimatic conditions. However, on a P nutrient content basis, the cost of mining, processing, and transportation of the PR often results in unfavorable economics when compared to the present price of imported high analysis, water-soluble P fertilizers.

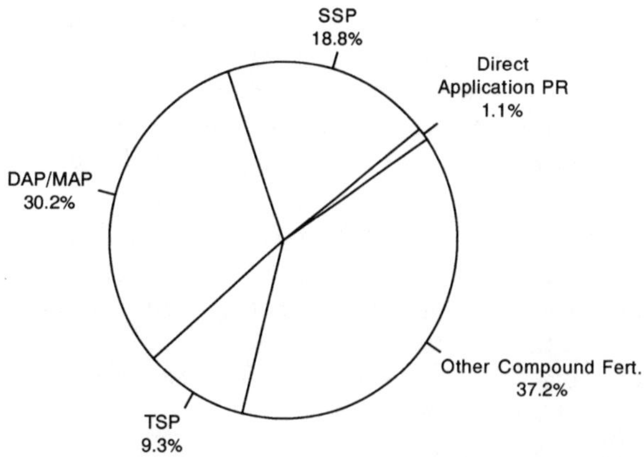

Figure 5.2. World phosphate fertilizer consumption—nutrient basis (1993/94). From International Fertilizer Association (IFA) World Fertilizer Consumption Statistics No. 27, Paris, December 1995.

An increase in the world market price for finished soluble P fertilizers may produce more favorable economics for direct application of ground PR sources, but this is not expected in the foreseeable future. While PR represents a finite, nonrenewable resource used to produce all P fertilizers, there are ample supplies of quality PR, enough for hundreds of years at current consumption rates. Presently, 75% of PR production is concentrated in the United States, the former Soviet Union, Morocco, and China, with Morocco expected to become the leading producer in the future (Van Kauwenbergh, 1995). Increased production is expected in Jordan, Iraq, and South Africa. With sufficient supplies of raw materials and high analysis products, P fertilizer prices are expected to remain fairly constant for the next few years, again assuming energy prices do not increase.

Research efforts should concentrate on maximizing the efficient use of soluble P fertilizers by promoting band application of P, planting in the same row location, placement with other nutrients, and more use of liming materials. These approaches provide P more efficiently in concentrated zones and provide an economic means to build P fertility. To foster adoption of these techniques, simple equipment is needed and this is feasible and adoptable.

Potassium Research

Potassium needs are increasing throughout the world as use of the other nutrients increases the off-take of K. For instance, combined rice and wheat production of South Asia have increased from about 57 million tons in 1950 to about 128 million tons in 1990. Under this increased production, soils become depleted of their K reserves more rapidly. With 30 kg K_2O per ton of grain removed, this means an addition of 2.1 million tons of K_2O is removed in the harvested crop each year. This figure represents twice the annual current use of K_2O for the region (Bumb and Baanante, 1996a).

In many tropical regions, little of the K-containing crop residue remains in the field. The low amount of K use is another illustration of the lack of application of known principles of soil fertility. China is importing large amounts of K, but much of the rest of the developing countries have not increased their importation of the nutrient to maintain or increase the needed K status of their soils. More soil testing, demonstration, and extension work is needed to help farmers know whether they need more K use in tropical areas with intensive cropping systems. Soil testing and data analysis of archival soil test information would help to document these changes and suggest research, extension, and policy initiatives that could help to ameliorate the decline in K status of tropical soils.

Research and demonstration work on improving the efficiency of K use, particularly the use of split applications in sandy, leached soils, and in saline areas, is needed. In rice and wheat systems, in which the straw is removed from the field, research is needed to determine if the straw can be left in the field until rain leaches the K from the straw, thereby contributing to the K fertility, while still maintaining the usefulness of the straw.

Other Nutrient Concerns

As described for the K situation, the off-take of other nutrients under intensive agricultural systems increases when the major nutrients are applied and yields increase. Following an axiom of Liebig's Law of the Minimum, ultimately other nutrients become limiting and the need for these nutrients will have to be diagnosed, and appropriate nutrients must be available and used by farmers to maintain crop yields. Even more use will be needed in many cases to increase yields. While farmers of developed countries are hiring crop consultants to help diagnose crop needs, farmers in developing countries often receive no advice at all. Yield stagnation (or decline) in many productive rice growing areas is likely explained by developed nutrient deficiencies caused by large off-takes of nutrients over the past 20 years, perhaps associated with declines in the organic matter content. A likely cause of the problem is S deficiencies, and the detection and correction of S fertility constraints should be of concern in many areas of the tropics.

Silicon off-take by rice is very high, and Si deficiencies of rice should be considered in rice-growing areas as a possible reason for yield stagnation and decline (Savant et al., 1996). Silicon often does not affect yields like other nutrients, but imparts many attributes, such as drought tolerance, and disease (blast) and insect (stem borer) resistance. Extraordinary results from applying small amounts of plant-available Si to rice nurseries in the form of burned rice hulls have been demonstrated (Savant et al., 1996; Sawant et al., 1994). This method should be tested by others to determine if there is a widespread need for more available Si in rice-growing areas.

OVERCOMING MISCONCEPTIONS OF SOIL FERTILITY

There are rampant misconceptions concerning soil fertility and fertilizer use that knowledgeable scientists need to help others understand. The first is that nutrient availability can be sustained without the use of external inputs (e.g., Repetto, 1989). But there are also critical misconceptions in the analysis of fertilizer use statistics. Many analysts of fertilizer use expect fertilizer use in specific areas to continue to increase indefinitely. This is not reasonable. Fertilizer use should theoretically increase as farmer practices improve and fertilizer benefits are shown, but as the P and K statuses of the soils improve, the use of these nutrients should logically decrease to levels that maintain the soil fertility in respect to these nutrients. Nitrogen use, on the other hand, should logically increase with improved practices, but then the use should remain at rates that provide the economically optimum yields. Many analysts have incorrectly ascribed the reduced use of N in developed countries to environmental reasons. There should be wider realization that new deficiencies develop with time as the soil reserves of various nutrients are depleted.

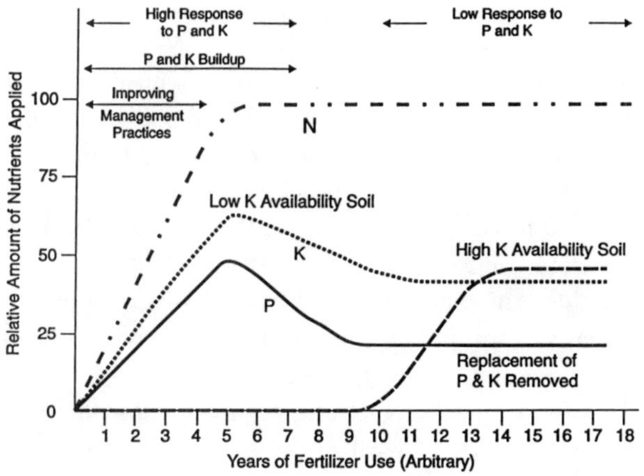

Figure 5.3. Theoretical changes in fertilizer use with time under "ideal" conditions.

There also is confusion in the analysis of fertilizer use trends about imbalance of fertilizer use, which people generally mean to be that too much N is used relative to P and K. Many expect a specific ideal ratio of nutrients is needed for particular cropping systems. Of course, this universal ratio does not exist since the nutrient additions should be related to the nutrient-supplying capacity of the soils. Since soil fertility differs with past cropping practices, offtake, and nutrient inputs from all sources, the ratio of nutrients needed changes with time (Figure 5.3). In the hypothetical Figure 5.3, fertilizers are initially introduced into the system and use increases with better management, weed control, varieties, etc. Use of P and K then should decline as the soils become fertile. In a totally mature cropping system, in which the soil fertility in respect to P and K has been increased (or declined in the case of K in some soils), the amounts needed are only to replace the amounts removed in harvest, erosion, and leaching losses. As N is not stored in mineral forms effectively in soils, the N needs are continued for nonleguminous crops in accord with crop prices and environmental and efficiency concerns.

There are confused issues regarding the stagnating yields in rice production. Unless the yield potential of varieties is increased, yields will logically approach a plateau if nutrient conditions become optimum. Also, average yields may actually decline as more marginal lands come under cultivation. There is also confusion about the decline in fertilizer response as countries have adopted more intensive production, as measured in yield addition from increments of fertilizer use (Hobbs and Morris, 1996). This is a totally expected and desirable development as soils become more fertile and less responsive to applied nutrients. Response to applied nutrients other than N will decrease as the soil fertility status of soils improves. In fact, the goal of soil fertility management is that in a single year of cropping there will be essentially very limited response to nutrients except for N. The desirable buildup of soil fertility and reduced response alarms those that view the efficiency (unit of yield per unit of fertilizer) as evidence that the systems' productivity is declining. If fertilizers are used correctly, there is very little response to either P or K in a particular year except after several cropping seasons with no additional fertilizer applications. Exceptions are with sandy soils, which have little ability to retain and supply K.

CONCLUSIONS

With the current funding constraints on nutrient research, progress toward new knowledge concerning fertilizers and fertilizer management is likely to be slow. The demand for more food production, particularly in the most disadvantaged areas, mandates that we apply existing knowledge more effectively. Fertilizer use adoption is progressing in South America and Asia. In some cases, a mature market situation is approaching, but there must be more attention to the major nutrients other than N. Soil testing services are very much needed to improve the efficiency of fertilizer use in many countries. The private sector fertilizer dealers will become an increasingly important component in effecting use and to furnish the best possible services to farmers. They should be considered as an extension arm to help farmers to understand soil fertility. Fertilizer use in much of Africa is a major concern as the soil fertility of the very nutrient-poor soils continues to decline, and farmers in many areas lack the economic incentives and the opportunities to use fertilizers profitably. Policy initiatives that try to address root causes for the soil fertility decline and soil degradation are urgently needed in Africa (Pinstrup-Andersen and Pandya-Lorch, 1994).

More attention to soil ameliorants to reduce the degradation of irrigated land from soil salinity and waterlogging is needed. In many cases, policy interventions will be required to avoid long-term decline in food production capabilities in irrigated areas.

There will be an increasing need to become more innovative in trying to close the nutrient cycles by utilizing various agricultural and human wastes more efficiently to maintain soil productivity in tropical regions. This will help to maintain soil fertility in the long term, efficiently utilize resources, and avoid environmental problems caused by disposal of the wastes.

As fertilizer use increases and cropping intensification continues in an attempt to meet increasing food needs, new nutrient problems will develop, and trained individuals with the analytical tools to make good judgments will be needed. In many cases the international research community will be called upon to help countries meet their urgent needs by better application of existing knowledge.

The international scientific community should try to educate donor groups about where nutrients originate, as some seem to believe that nutrients simply appear from somewhere, even unto soils that are extremely nutrient deficient in their native state. In addition, they often have difficulty distinguishing between the environmental effects of applying excessive rates of fertilizers with the use of fertilizers in general (Easterbrook, 1997). Correct perspectives of environmental problems and how they vary in different environments are sometimes difficult for many policy makers and administrators to comprehend when they lack basic science and agricultural backgrounds. In these times of poor financial support, it is very important that the available money be spent to address real rather than imagined problems.

REFERENCES

Beets, W.C. *Raising and Sustaining Productivity of Smallholder Farming Systems in the Tropics.* AgBe' Publishing, Alkmaar, The Netherlands, 1990.

Blackie, M.J. Maize Productivity for the 21st Century: The African Challenge, in *Proc. Fourth Eastern and Southern Africa Regional Maize Conference, Harare, Zimbabwe, 28 March–1 April, 1994*, Jewell, D.C., Waddington, S.R., Ranson, J.K., and Pixley, K.V., Eds., Mexico City, 1995, pp. 11–22.

Breman, H. *Opportunities and Constraints for Sustainable Development in Semi-Arid Africa. The United Nations University, Institute for New Technologies (UNU/INTECH) Working Paper No. 18*, Maastricht, The Netherlands, 1995, p. 23.

Bronson, K.F. and A.R. Mosier. Effect of Encapsulated Calcium Carbide on Dinitrogen, Nitrous Oxide, Methane, and Carbon Dioxide Emissions in Flooded Rice. *Biol. and Fert. of Soils*, 11:116–120, 1991.

Bumb, B.L. and C.A. Baanante. *The Role of Fertilizer in Sustaining Food Security and Protecting the Environment to 2020, Food, Agriculture and the Environment Discussion Paper 17*, IFPRI, Washington, D.C., 1996a, p. 54.

Bumb, B.L. and C.A. Baanante. *World Trends in Fertilizer Use and Projections to 2020, 2020 Brief 38*, IFPRI, Washington, D.C., 1996b, p. 2.

Byrnes, B.H. and J.R. Freney. Recent Developments on the Use of Urease Inhibitors in the Tropics. *Fertilizer Research.* 42, pp. 251–259, 1996.

Daftardar, S.Y., S.M. Wagle, and N.K. Savant. Management of Urea Briquettes Containing Diammonium Phosphate Increases Rice Yields of Small Farmers, *International Rice Research Newsletter* (in press).

Easterbrook, G. Forgotten Benefactors of Humanity. *The Atlantic Monthly.* 279(1), pp. 74–82, 1997.

Engelstad, O.P. and D.T. Hellums. *Water Solubility of Phosphate Fertilizers: Agronomic Aspects—A Literature Review. Paper 17*, International Fertilizer Development Center, Muscle Shoals, AL, 1993.

Hobbs, P. and M. Morris. *Meeting South Asia's Future Food Requirements from Rice-Wheat Cropping Systems: Priority Issues Facing Researchers in the Post-Green Revolution Era. NRG Paper 96-01,* CIMMYT, Mexico City, 1996.

Nambiar, K.K.M. Long-Term Field Trials Reveal Importance of Plant Nutrients in Sustaining Productivity. *Better Crops* 79(1), pp. 26-27, 1995.

Ofori, C.S. Need for Long Term Research in Africa, in *Soil Management: Experimental Basis for Sustainability and Environmental Quality*, Lal, R. and Stewart, B.A. Eds., CRC Press. Inc., Boca Raton , FL, 1995.

Olsen, R.K. The Future Context of Sustainable Agriculture: Planning for Uncertainty, in *Integrating Sustainable Agriculture, Ecology, and Environmental Policy*, Olsen, R.K., Ed., Haworth Press, Inc., Binghamton, NY, 1992.

Pinstrup-Andersen, P. and R. Pandya-Lorch. *Alleviating Poverty, Intensifying Agriculture, and Effectively Managing Natural Resources. Food, Agriculture, and Environment Discussion Paper 1*, International Food Policy Research Institute, Washington, D.C., 1994.

Pouzet, D. Profitability of Fertilizer Use Options in Ghana, in *Fertilizer Use at the Village Level: Constraints and Impacts, Summary Proceedings of Workshop*. IFDC, Lomé, Togo, 1991, pp. 45–47.

Repetto, R. Economic Incentives for Sustainable Production, in *Environmental Management and Economic Development*, Schramm, G. and Warford, J.J., Eds., The World Bank, Washington, D.C., 1989, pp. 69–86.

Savant, N.K. and P.J. Stangel. Recent Developments in Urea Briquette Use for Transplanted Rice. *Fertiliser News.* 40(8), pp. 27–33, 1995.

Savant, N.K., G.H. Snyder, and L.E. Datnoff. *Depletion of Plant-Available Silicon in Soils: Possible Cause of Declining Yields of Irrigated Rice in the Tropics. Paper presented at the 26th Annual Rice Technology Working Group Meeting, Feb. 25–28, 1996, San Antonio, TX*, 1996, p. 7.

Savant, N.K. Increased Rice Production with Less Fertilizer. *Low External-Input and Sustainable Agriculture Newsletter*, 11(1), pp. 24–25, 1995.

Sawant, A.S., V.H. Patil, and N.K. Savant. Effect of Rice Hull Ash Applied to Seedbed on Deadhearts in Transplanted Rice. *International Rice Research Notes.* 19(4), pp. 21–22, 1994.

Scherr, S.J., and S. Yadav. *Land Degradation in the Developing World: Implications for Food, Agriculture, and the Environment. Discussion Paper 14.* International Food Policy Research Institute, Washington, D.C., 1996, p. 36.

Schultz, J.J., D.I. Gregory, and O.P. Engelstad. *Phosphate Fertilizers and the Environment—A Discussion Paper. Paper 16.* International Fertilizer Development Center, Muscle Shoals, AL, 1993, p. 103.

Shah, S.M., W. Mohammad, H. Nawaz, and M.M. Iqbal. Genetic Potential and Nutritional Requirements of Newly Developed Wheat Lines. *Rachis* 13(1/2), pp. 51–53, 1994.

Smaling, E.M.A. The Balance May Look Fine When There is Nothing to Mine: Nutrient Stocks and Flows in West African Soils, in *Use of Phosphate Rock for Sustainable Agriculture in West Africa*, Gerner, H. and Mokwunye, A.U., Eds., International Fertilizer Development Center, Muscle Shoals, AL, 1995, pp. 10–20.

Umali, D.L. *Irrigation-Induced Salinity. World Bank Technical Paper No. 215*. The World Bank, Washington, D.C., 1993, p. 78.

Van Kauwenbergh, S.J. Overview of the Global Phosphate Rock Production Situation, in *Direct Application of Phosphate Rock and Appropriate Technology Fertilizers in Asia—What Hinders Acceptance and Growth. Proc. Workshop IFS and IFDC, Kandy, Sri Lanka, Feb. 20-25, 1995*, Van Kauwenbergh, S.J., and Hellums, D.T., Eds., 1995, pp. 15–28.

Van Reuler, H. (1996). Nutrient Management Over Extended Cropping Periods in the Shifting Cultivation System of South-West Côte d' Ivoire. Ph. D. thesis, Agricultural University, Wageningen, The Netherlands.

Vlek, P.L.G. The Role of Fertilizers in Sustaining Agriculture in Sub-Saharan Africa. *Fert. Res.* 26, pp. 327–339, 1990.

Vlek, P.L.G. Strategies for Sustaining Agriculture in Sub-Saharan Africa: The Fertilizer Technology Issue, in *Technologies for Sustainable Agriculture in the Tropics. ASA Spec. Publ. 56*, Ragland, J., and Lal, R., Eds., ASA, CSSA, and SSSA, Madison, WI, 1993, pp. 265–277.

SIX

Sustainable Management of Land and Water Resources in Irrigated Semiarid and Tropical Regions

Jacob W. Kijne and Marcel Kuper

INTRODUCTION

The world's irrigated area per capita has decreased from a peak of 48 ha/1000 people in the late 1970s to 45 ha/1000 people in 1989, with further decreases because of population growth exceeding the rate of growth of irrigated area. Without irrigation, the present area under cultivation in the world would only meet the minimum food needs of less than half of the world's present population as traditional agriculture is estimated to need 0.6 ha of arable land to meet the per capita minimum dietary requirements (Ghassemi et al., 1995). These two facts combined mean that there has been no alternative but to increase the output of existing arable land through technological innovations in order to feed the world's growing population. This is still true today. The technological innovations that made it possible to provide enough food included the introduction of high yielding varieties of seeds, and high inputs of water, fertilizers and pesticides. However, there is concern that for a variety of reasons, yields from irrigated lands in the tropics have stagnated or declined. The case of stagnating yields from the rice-wheat production system in the Indian subcontinent is well documented (see for example, Abrol and Gill, 1993).

This is particularly alarming in view of the increased scarcity of water. The availability of water for agriculture is threatened by competition from domestic and industrial requirements. Many countries which were thought to have secure water supplies are now found to be short of drinking water in their mega-cities during at least part of the year. The standard indicator of water scarcity defines 1,700 cubic meters per capita per year as the level below which water shortages are likely to occur (Gleick, 1993). This index ignores variations in the importance of irrigated agriculture among countries. About 70% of the total withdrawals of water in the world are for irrigated agriculture and the figure, although not precisely known, is expected to be even higher, on the order of 90%, for many developing countries. Seckler and de Silva (1996) have developed an indicator of relative water scarcity based on a country's water supply and demand for irrigated agriculture. From their analysis it appears that accounting for seasonal differences in supply is essential when assessing whether a country is water-short; e.g., in India over 70% of the total

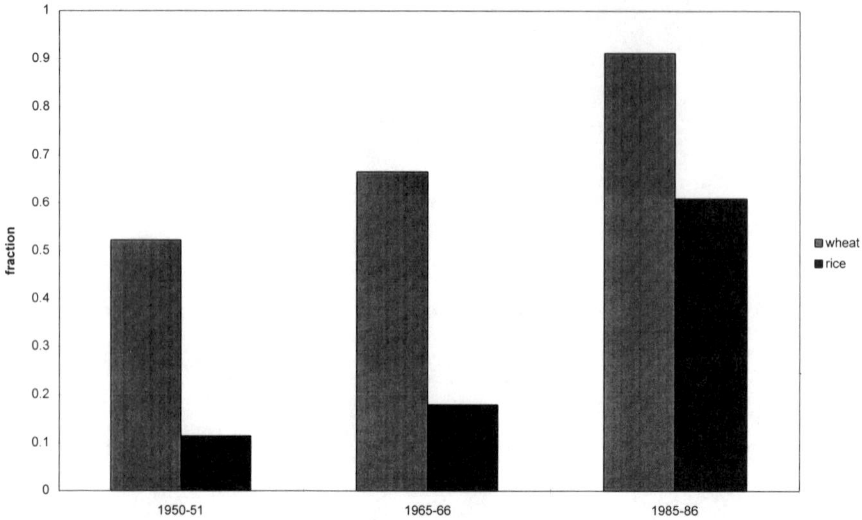

Figure 6.1. Fraction of irrigated land in India under wheat and rice.

supply occurs in the three monsoon months of June, July, and August, when most of it floods out to the sea. Moreover, country-wide data ignore regional differences in water supply and withdrawal within a country. The tentative conclusion is that many countries do not have a surplus of irrigation water, i.e., they do not have sufficient annual water withdrawal to irrigate their potential gross irrigated area even at a high basin irrigation efficiency[1] of 80%. Whatever indicator is used, more than half of the world's population lives in countries with varying degrees of water scarcity.

With the advent of high-yielding wheat and rice varieties, the production of these crops in the Indian Punjab has more than tripled from 1975 to 1987. The irrigated area with wheat and rice has also grown both in absolute and relative terms. Figure 6.1 shows the increase in the fraction of the irrigated land occupied with wheat during the winter season (*rabi*) and with rice during the monsoon summer season (*kharif*) (from data reported by Singh and Paroda, 1993). A similar increase took place in Pakistan with the result that pressure on available water resources in the Indian subcontinent has increased greatly.

Water requirements for paddy cultivation vary and depend largely on soil type. The average requirement under the conditions of Pakistan's Punjab is considered to be around 1600 mm, which includes water needed for land preparation, intentional drainage and deep percolation (Bhatti and Kijne, 1992). However, the mean irrigation application was found to be around 1300 mm from measurements on a fairly large number of farms, which means that on average, yield is only 0.14 kg rice per m^3 water diverted from the source (i.e., canal or groundwater). It has been shown that considerable amounts of water can be 'saved'[2] by reducing the period of continuous flooding without adverse effect on the yield (Narang and Singh, 1988). Continuous flooding is done also for weed control.

[1]Basin irrigation efficiency includes all reuse of drainage water and is considerably higher than system irrigation efficiency, if drainage flow from one system is used for irrigation again downstream in another system. Basin-wide irrigation efficiency for the Nile basin in Egypt has been calculated to be about 90% (Keller and Keller, 1995).

[2]Saved between quote marks, because a reduction in water diverted from the source does not mean that less water is consumed by the crop through transpiration. A shorter duration of flooding may indeed save water that would otherwise have been evaporated from the ponded surface.

Farmers tend to take as much water as available because of general unreliability of canal supplies. Dry seeding of rice, other than expected perhaps, tends to increase consumptive water use compared with growing the paddy seedlings in a nursery because of the much larger area (about eight times) that needs to be irrigated from the time of sowing the crop. Not all percolation or flooding water is wasted, however, as rice is grown in some areas as reclamation crop. Deep percolation during rice cultivation helps to leach the salts from the root zone and restore a more favorable salt level for the subsequent wheat crop. The high tolerance of rice to exchangeable sodium arises chiefly from its ability to withstand a layer of water on the field throughout the growing season. Also, the high pH of sodic soils is reduced under continuous flooding.

The objective of this chapter is to examine whether irrigated agriculture in water-scarce environments has sufficient resilience to increase productivity while sustaining the natural resources of soil and water. Sustainable management of irrigated land implies that the irrigation system generates outputs that are valued and used by people, and that the potential for production of available natural resources is not reduced by short-term deficiencies in operation and maintenance of the systems. It has been suggested that the performance of an irrigation system may temporarily decline, but that as long as the negative trends in performance are reversible within normal economic criteria, the productive potential of the system has been maintained for future generations (Kijne, 1994). It is realized that this concept is a bit nebulous, hence the occasional use of the term resilience (see also Lal, 1994). Resilience expresses perhaps better than sustainability that irrigated agriculture is a dynamic system, but it may also be argued that resilience is subsumed in sustainability. A system is not sustainable unless it is resilient. The irrigated agriculture system is now threatened by competition for one of its resources—water—while another resource—soil—seems to be in decline. Is it sustainable under these conditions? Or, in other words, is it resilient enough to cope with the competition and apparent decline *and* increase output per unit of water diverted for irrigation?

The main focus in this chapter is on Pakistan, with shorter descriptions of Egypt and Thailand for comparison. We are not in a position to answer the question about sustainability of irrigated agriculture for all semiarid tropical or subtropical countries. However, we believe that the issues discussed in this chapter are of relevance for many other water-scarce (developing) nations.

METHODOLOGY

For the analysis of sustainability of irrigated agriculture, data are required on changes over time in yield and total factor input, trends in soil salinity and other aspects of soil degradation [e.g., change in infiltration rates or saturated permeability due to sodicity, reduced (micro-) nutrient availability and organic matter content], changes in available water (e.g., of canal supplies, and change in water table depth where part of the crop water requirement is met through capillary rise from a shallow water table into the root zone, or through low lift pumping by tube wells from groundwater), and also changes in water quality. Some of this information is available in journal articles and review publications (e.g., Ghassemi et al., 1995; Gill, 1993 with respect to the rice-wheat farming system), and some only in the grey literature of conference proceedings, etc.

With respect to Pakistan, we draw heavily on data collected by the International Irrigation Management Institute (IIMI) country program there. As we intend to show in this chapter that water quality is of paramount importance in the analysis of sustainability of irrigated agriculture (which in Pakistan is increasingly dependent on pumped groundwater), frequent reference is made to several IIMI publications which report in depth on the incidence and significance of salinity, especially in Pakistan's Punjab (e.g., Kijne and Vander Velde, 1992; Kijne and Kuper, 1995; Kijne, 1996). Site selection and research methodology are described in detail in those papers.

Salinity and Waterlogging

Two issues threaten the sustainability of irrigated agriculture: water shortage and salinity. The two are linked in several ways. Irrigated agriculture concentrates salts because water is taken up by the crop and salts are left behind in the root zone. Proper management must ensure that the salt is concentrated outside the root zone, and away from future water supplies for irrigation, domestic, or industrial use. Hence water is required to transport salt from the root zone, in addition to consumptive requirements of the crop, and also land is required to store the salt. Water shortage prevents remedial action on salt-affected lands because leaching of salts from the root zone requires additional water over and above the amount required to satisfy the evaporative demand of the crop.

Salinization often also occurs naturally and not only as a consequence of poor irrigation practice.For example, part of the Indian subcontinent including the Indus Plain was formed from sediments transported by rivers into a shallow sea. The receding sea left salts behind in the soil profile. Whether the salinity of the aquifer is also a remnant of this ancient sea is disputed. Isotopic evidence as presented by Sajjad et al. (1993) indicates that salinity of the groundwater is mainly due to transport of salts from the soil profile by infiltrating fresh water. Application of fertilizer and irrigation continue to add salts to the system. Each year, irrigation with good quality river water adds 1 to 2 tons of salt per ha (e.g., an annual application of 800 mm of water with a salinity of 250 mg/L adds 1.6 T/ha). The obvious solution to prevent this accumulation of salt in the soil profile is to apply more water than can be taken up by the crops and leach the salts out of the root zone. Unless there is an adequate drainage system, this practice leads to rising water tables and waterlogged conditions. High water tables actually increase the salt transport into the root zone because of capillary rise from the water table.

Managing the water and salt balances of irrigated agriculture well has proved to be difficult. It is not only a matter of minimizing the amount of water required to transport salt from the root zone, but also minimizing the land area required to store the salt temporarily or permanently. Although human-induced salinity problems can develop swiftly, solutions can be very time-consuming and expensive. Installation of subsurface drainage, the most common physical solution in many countries, costs between $1500 and $3000 per ha.

The global extent of salt-affected lands is considerable. However, there is wide divergence in the figures reported by different institutions. For example, estimates for India range from 7 to 26 Mha, or between 17 and 60% of the irrigated land. For Pakistan the most likely figure is some 40%, Israel 13%, Australia 20%, China 15%, Iraq 50%, and for Egypt 30% of the irrigated land (Gleick, 1993 and Ghassemi et al., 1995). Human-induced (or secondary) salinization occurs in large and small irrigation schemes alike: in recent years many farmers have been abandoning their rice fields in Sahelian irrigation schemes due to the incidence of salinity. The countries affected by secondary salinization are predominantly but not exclusively located in arid and semiarid regions. Other activities, such as land clearing and replacement of native trees with shallow-rooted crops contributed to the development of so-called dryland salinity. This has been well documented for areas in Australia, the USA, Canada and Thailand, among others. Salinization also occurs when crops are irrigated with pumped groundwater of marginal or poor quality, and as a result of sea-water intrusion when water tables have been lowered by mining of groundwater in coastal areas, as has occurred for example in Bangladesh and in the State of Gujarat in India. Various improvements in irrigation and agronomic practices can be introduced, dependent on the type of salinity; i.e., the presence of excessive amounts of sodium (sodicity) or carbonates and bicarbonates (alkalinity) in the water, and the cause of the accumulation of these salts to harmful levels in the root zone.

We will return to the topic of water and salt balances in the discussion, after describing three case studies in the next paragraphs.

Case Studies

Pakistan

Pakistan is ranked fifth in the world on the basis of irrigated area with 16.2 million ha (Postel, 1993). The average annual canal withdrawal of the Indus basin in Pakistan is estimated to be about 129 billion m^3. The annual usable groundwater, defined as water of a salinity level of less than 1000 mg/L, is only about 29 billion m^3, which is about half the average annual recharge to the aquifers which underlie the Indus plains. However, conjunctive use of canal water and groundwater has become essential for irrigated agriculture throughout much of the Indus Basin. The first experiments with tube wells for water table control took place in the 1950s and was followed by the implementation of large-scale Salinity Control and Reclamation Projects (SCARPs) in the 1960s. Over the past 30 years, more than 12,500 public tube wells have been installed in various SCARP project areas. The primary objective of these projects was to combat waterlogging and associated salinity. A secondary objective, however, became supplementing canal water with pumped groundwater for irrigation, usually discharging directly into the existing watercourse network.

Groundwater quality in the Indus Basin varies in quality in a rather complicated way, depending as it does on the extent of the recharge. The occurrence of recharge is largely governed by topography, old and existing river beds, rainfall, and irrigation practices. The general pattern is for the groundwater quality to decrease going downstream in the Indus valley. The best quality is found in the northeastern part of the country where the rainfall is highest. Fresh groundwater is also found close to the river beds of the Indus and its tributaries, and poor quality groundwater in the central parts of the interfluvial regions. Locally, the spatial variability in quality of pumped groundwater is greater than this general picture would make one think. Even tube wells in close proximity are known to yield different quality waters, partly depending on the depth from which they are pumping. Since the early 1960s, groundwater development has been exponential, especially in Punjab, through private tube wells, of which according to the latest estimates there are about 400,000 in all of Pakistan. A national survey undertaken in 1991 shows that at present about 46 billion m^3 of groundwater is used in the Indus Basin for irrigation, of which 85% comes from private tube wells (NESPAK/SGI, 1991). This exceeds the annual usable groundwater recharge by more than 50%. Clearly, a nonsustainable situation.

By reducing waterlogging, as was done in large parts of Punjab where groundwater was pumped to lower water tables, it was hoped that also the problem of salinity had largely been solved. This is not the case. Firstly, large quantities of salts were present in the soils before the inception of the Indus Basin Irrigation System. An extensive survey of soil profiles in the Indus basin carried out before the introduction of private tube wells, classified nearly one-quarter of the soils as saline-alkaline, 3.5% as alkaline, and 10.7% as saline (Ghassemi et al., 1995). Secondly, pumped groundwater contains far more salts than canal water. When irrigation depends to a large extent on water of marginal quality, each year vast quantities of salts are left behind in the root zone after the water has evaporated or been transpired by the crops. Leaching fractions (the amount of water applied to the fields in excess of the crop water requirement) are usually not sufficient to leach these salts out of the root zone.

The significance of conjunctive use of canal supplies and pumped groundwater for irrigation in Pakistan can be judged from the fact that cropping intensities in the Punjab now commonly exceed 125%, with high values in some watercourse command areas of as much as 190% (IIMI, unpublished data). These canal commands were designed in the late nineteenth century with the expectation of annual cropping intensities ranging from 50% to 75%, two-thirds of which were to occur in rabi and one-third in kharif. The actual amount of irrigation water available at tertiary (watercourse) level has now increased from 2.5 mm/day to about 6 mm/day, especially in head reaches of canals where supplies are at design level or higher.

All groundwater in the Punjab contains sodium (Na), but in varying amounts. Whether Na from the percolating water will be retained in the top layers of the soil profile or leached to greater depths depends on many factors, including evapotranspiration, the dissolution and precipitation of salts, mineral weathering, and organic matter decomposition. There may be a loss of calcium (Ca) and magnesium (Mg) from solution owing to the precipitation of Ca and Mg carbonates, or a gain of these ions in solution from the dissolution of the earth carbonates, processes which in turn are governed by relative concentrations of these salts, CO_2 pressure and pH in the root zone (Rhoades and Loveday, 1990). It is now generally recognized that the presence of Na causes sealing of the soil surface, lowers the basic infiltration rate, and reduces the saturated permeability of the soils, even of light textured soils, and at lower values of the Sodium Adsorption ratio (SAR) than was previously assumed (Sumner, 1993).

Primary measurement data have been collected by IIMI and its research collaborators in Pakistan for a large number of sample watercourses in the Punjab. The data include actual canal discharges and their spatial and temporal variabilities in relation with their design values, public and private tube well operations, the extent of groundwater use at the watercourse and farm levels, salinity of pumped groundwater, soil salinity, crops grown and cropping intensities during the two main seasons. Elsewhere (Kijne and Kuper, 1995) data have been reported on a number of these aspects as collected in two sample areas of IIMI in Pakistan, i.e., the command areas of Upper and Lower Gugera Branch canal in the area (*doab*) between the Chenab and Ravi rivers in central Punjab, and of the Fordwah and Azim distributaries in southeast Punjab (also referred to in the text as Chishtian subdivision). Here we will review more recent data, particularly of the Chishtian subdivision, where now a longer record is available.

The command areas of Azim and Fordwah distributaries is located in an area confined by the Sutlej river in the northwest, the Indian border in the east, and by the Cholistan desert in the southeast. Average rainfall in the Chishtian subdivision is about 260 mm. Annual potential evaporation, in excess of 2000 mm, far exceeds annual rainfall in the area. The soils in the area are light to medium textured: in the Azim area mainly sandy and silt loams, and in the Fordwah command area mainly loamy sands. The dominant crops are cotton in kharif and wheat during rabi. Water tables vary a great deal in the area, from 2 to 10 m depth, and a major drainage project is under construction. Basic data of eight sample watercourses in the Chishtian subdivision are presented in Table 6.1. (RSC is the Residual Carbonate Content, see below.)

In the command area of the Azim distributary, the canal supply and the tube well water quality decrease from head to tail (the number of the watercourse indicates the distance from the head of the distributary). From the low availability of canal water, it is obvious that farmers depend to a large extent on pumped groundwater. The profile electri-

Table 6.1. Basic Data of Sample Watercourses in the Chishtian Subdivision

Watercourse	Equivalent Canal Supply (mm/day)	Tube Well Water Quality EC (ds/m)	SAR	RSC
Azim 20L	1.7	0.77	2.17	neg.
Azim 43L	1.3	0.79	1.19	0.04
Azim 63L	0.9	0.76	1.45	0.33
Azim 111L	0.2	1.02	4.99	0.78
Fordwah 14R	1.9	1.93	8.21	neg.
Fordwah 46R	2.2	0.82	2.62	0.94
Fordwah 62R	2.4	1.04	5.65	2.04
Fordwah 130R	1.9	1.21	8.17	2.56

RSC = residual carbonate content; SAR = sodium absorption ratio.

cal conductivity, measured in the saturation extract, ECe, of soil samples collected in the sample watercourses generally increases from head to tail watercourse, as shown in Figure 6.2. The exception is the ECe of FD14. From Table 6.1 it can be seen that tube well water quality in this command area is worse than in the other Fordwah watercourses. However, tubewell water use is higher: an average of 185 mm during kharif in FD14R versus 64 mm for FD46R, and 138 mm for FD 62R. Figure 6.3 shows a similar trend of deteriorating quality, this time expressed as SAR, going from head to tail in the command area.

Negative RSC values indicate a relative excess of calcium and magnesium ions over carbonate and bicarbonate ions. Hence it could point toward the presence of calcite and gypsum in the water. It is interesting to note that the number of tube wells with negative RSC values decreases toward the tail end of the distributary: the Azim watercourse command areas have, respectively, 5 out of 5, 11 out of 12, 7 out of 9, and 4 out of 13 tube wells with negative RSC values. In the Fordwah watercourse command areas, going from head to tail watercourse command areas, the number of tube wells with negative RSC values are, respectively, 16 out of 17, 3 out of 10, 5 out of 16, and 3 out of 27. These effects combined show a negative trend in water quality going toward the tail end of the distributary command. This trend has been observed before (e.g., Kijne and Vander Velde, 1992) and is found to be associated, among other things, with greater distance

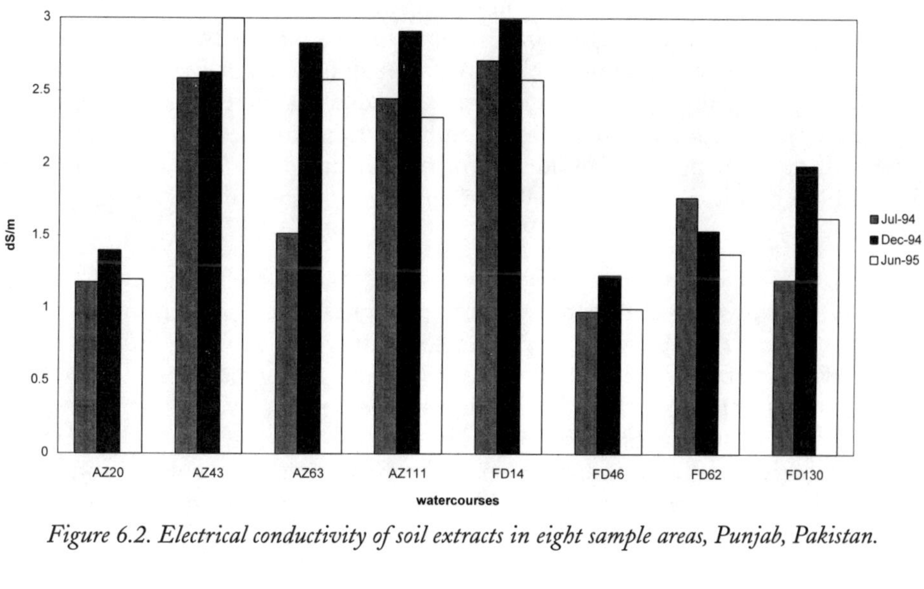

Figure 6.2. Electrical conductivity of soil extracts in eight sample areas, Punjab, Pakistan.

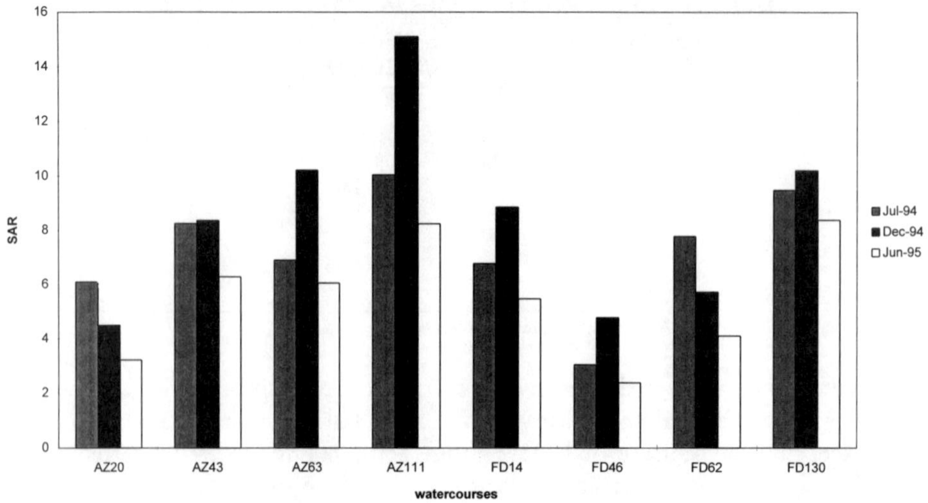

Figure 6.3. Sodium adsorption ratios of soil extracts in eight sample areas, Punjab, Pakistan.

from recharge areas, such as link canals, river arms, etc., and with relatively greater proportions of irrigation water derived from pumped groundwater. The chemical composition of the original groundwater also plays a major role in the spatial variability of pumped groundwater. Where shallow hand pumps and deeper tube wells were close together, it has been found that in about half the cases the salt gradient, expected to be toward greater salt concentrations at greater depths, had been reversed. The aquifer is known to have clay lenses, and vertical mixing is apparently restricted. Hence two processes seem to be operating simultaneously: upconing of more saline water from greater depth in the aquifer due to localized extraction of groundwater by pumping, and recycling of increasingly saline water that percolated from the root zone back to the top of the aquifer. One would expect general deterioration of groundwater to occur over time under these conditions. It has indeed been reported that tube well water pumped from the shallow fresh water zone overlying saline 'native' groundwater, i.e., undisturbed groundwater from before the introduction of irrigation (which is saline because the geologic formation is of marine origin) has become more saline with time (Hafeez et al., 1986). Also many of the deep tube wells have been shut down at the request of farmers because of the poor quality of the pumped water. In sum, the increasing dependence of irrigated agriculture on pumped groundwater and the attendant increase in cropping intensity carry with it the danger of degradation of the soil and water resources and therefore the risk that the production system itself can no longer be sustained at the present level.

Egypt

About 96% of the country is desert and without the flow of the Nile, agriculture would be rain-fed and only possible during the winter period near the Mediterranean coast. The annual flow of the Nile is 84 billion m^3/year, compared with 207 billion m^3/year for the Indus (Van der Molen, no date). Traditional seasonal irrigation has been transformed into a perennial system through the construction of storage reservoirs, notably the High Aswan dam. The winter crops are mainly Egyptian clover (berseem) and wheat, and in summer maize, rice, and cotton are grown. Lake Nasser, the reservoir behind the High Aswan dam, has a live storage of some 120 billion m^3, equivalent to two years supply. It extends deeply into the Sudan and could not have been built without an agreement between the two countries on the use of the Nile waters. Sudan has not taken up its full share yet, and Egypt has been able to use the extra water. The Nile river basin encompasses the two sub-basins of the White and Blue Niles. The White Nile flows through eight of the nine riparian states (Rwanda, Burundi, Tanzania, Zaire, Kenya, Uganda, Ethiopia and Sudan), before it joins the Blue Nile at Khartoum and proceeds northward into Egypt and the Mediterranean Sea. The Blue Nile rises almost exclusively from the Ethiopian highlands, and only three riparian states are concerned with its management: Ethiopia, Sudan, and Egypt (Whittington and McClelland, 1992). The period of relative water abundance in the Nile Basin is coming to an end as population and development pressures in the riparian states intensify.

In a thoughtful paper, Whittington and McClelland (1992) describe the changing economics of using Nile water in Egypt to reclaim desert land for agricultural cultivation. Until recently, it was not economic to do so because of the high costs of pumping water from the Nile Valley up to the higher elevations of the desert. Higher agricultural productivity, improved irrigation technologies, better crop varieties for desert conditions and lower energy prices, however, have changed the equation. As a result, Egypt is now rapidly pressing ahead with the desert reclamation option. It is unlikely that Egyptian current water supplies, including extensive drainage water reuse, could cover the increased water use arising from the planned reclamation of 580,000 ha. Whittington and McClelland (1992) quote an estimated increase of agricultural water use in Egypt by over 10 billion cubic meters by the year 2000. The political and economic implications of the increased demand for water in Egypt are immense but outside the scope of this chapter. Whittington and McClelland (1992) argue convincingly that all countries in the basin stand to gain from cooperation in the development and management of the Nile water resources.

In the past, serious salt problems have not been associated with the large irrigation area of Egypt (number 17 in the list of countries with large irrigated areas), in spite of its continuous use for thousands of years. It was only after the widespread introduction of perennial irrigation that measures to counteract salinization were needed. Factors that contributed to worsening the problem include the horizontal expansion of irrigated agriculture into sandy or light textured soils and the fact that much of this land lies on the Nile Valley fringes of higher elevation; greater seepage from perennial irrigation; and an increase in rice and sugarcane growing areas requiring high water application. The estimates of the irrigated area affected by salinity vary widely. A reasonable estimate seems to be about one-third of the area (Kishk, 1986).

Upstream of Cairo, nearly all drainage water is considered to be of good quality and is returned to the Nile. Near Cairo, much of the drainage water is still good enough for reuse and is pumped back into the irrigation canals and mixed with Nile water. Estimates of the drainage water from the Nile delta range from 11 to 16 billion cubic meters per year. A number of drains discharge nearly 3 billion m^3/year in either of the two branches of the Nile or in main canals for reuse. Expected drainage water reuse in 2000 is 5.4 billion m^3/year.[3] The remainder goes to the Mediterranean Sea or to lakes used as evaporation ponds, close to the sea. One of the problems associated with the operation of evaporation ponds is leakage to the groundwater and to adjacent fields. Another problem is the

[3]All excess water applied to the fields in upper and middle Egypt is recovered and recycled downstream. Drainage reuse is thus already widespread and not easily identified. The simple arithmetic of 'project' efficiencies of around 40% and a basin efficiency of 90% suggests that water is applied at least twice on average. Drainage reuse is thus already in the order of 40 billion m^3/year. The figure quoted here refers simply to water leaving the system.

long-term environmental hazard once an evaporation pond has been decommissioned. From analysis of the salt balance for the Nile basin within Egypt, it appears that a large amount of salt is entering the drains in the northern part of the Delta, where the groundwater is salty due to the marine origin of the deposits, and where water movement is upward through the clay layer covering the aquifer. Here drainage water becomes unfit for any further use in agriculture (Van der Molen, no date).

In short, like Pakistan, Egypt is a water-short country. There are important differences in soil type, as the soils in Egypt are generally of heavier texture than those in Pakistan's Punjab. This means that the effects of degradation take longer to become apparent, because the soil acts as a buffer. Clay types differ as well, with more montmorillonite in Egyptian soils, and more illites in Pakistan. The former lead to swelling and shrinking, the latter are prone to surface sealing. Lighter soils are more easily leached, so reclamation of salt-affected land should in principle be easier in Pakistan than in Egypt. Provided, of course, that water for leaching is available. Also, the extent to which the irrigated land is drained differs considerably between the two countries. Egypt has probably the largest drainage coverage of all major irrigated countries and most of it is through subsurface tile-drainage discharging into open drains. The unstable structure of the lighter and often alkaline soils of Pakistan's Punjab causes problems with open drains (Van der Molen, no date). Side slopes of the open drains often erode quickly, even under the prevailing low rainfall conditions of Pakistan. Unstable subsoil necessitates the installation of filter envelopes around subsurface drains. The choice of the appropriate envelop material is difficult, and makes the installation of subsurface drains expensive. Pakistan has large areas with vertical drainage through deep tube wells in the SCARPs, but the area covered is a much smaller part of the irrigated land than in Egypt.

Thailand

The average annual rainfall in Thailand is about 1485 mm, so it is not a semiarid country. Nevertheless, it has problems with periodic water shortage and salinity. Some 85% of the rain occurs during the northeast monsoon from May to October. Rainfall is not evenly distributed throughout the country, but increases from north to south. The aquifer underlying Bangkok has been depleted by excessive pumping and in 1982, the water table was more than 50 m below the ground surface. The land has been subsiding for years, and the rate at which Bangkok subsides is about 5 to 10 cm/year. Before 2000, another 2 m fall is expected unless pumping is greatly reduced (Ghassemi et al., 1995).

Thailand is the ninth ranking country with more than 4 million ha irrigated (Postel, 1993). It has three types of salt-affected lands. Three-quarters of these lands are to be found in the northeast region. Here the soils are generally sandy and have low fertility. Rainfall is erratic but much of the land is under rainfed cultivation with rice during the monsoon and crops such as cassava, kenaf, and tobacco during the dry season (Ghassemi et al., 1995). With irrigation, rice, sweet corn, soybeans and peanuts can be grown. According to Ghassemi et al. (1995), the salinity in the area is partly of natural origin, but aggravated by deforestation, construction of reservoirs, salt-making, and irrigation. Deforestation has decreased the evapotranspiration and increased recharge to the groundwater. As a result, water tables have risen which salinized the land through capillary action as the groundwater is saline (although highly variable with 200 - 60,000 mg/L). Coastal saline soils are heavy clays affected by sea water intrusion, resulting from depletion of the groundwater resource near the sea. Acid sulfate soils, with their characteristic low pH values, also occur in coastal regions. Discussion of the management and reclamation of these soils is beyond the scope of this chapter.

In summary, problems associated with (periodic) water deficiency occur also in countries with high rainfall. The scale of the water deficiency is less than in the semiarid countries but nevertheless of considerable regional importance. Population pressure and the introduction of irrigation have aggravated the problems, particularly when the change in farming practices has led to increased salinity of the land. Examples of increased salinity due to seepage from nonirrigated, deforested saline lands or saline groundwater seeps can be found in many other tropical countries (e.g., Sri Lanka, Dr. C. Panabokke, personal communication, 1995). In the past it was easier to abandon land that became salt-affected; nowadays, land and water both are more and more seen as scarce resources.

DISCUSSION

Water and Salt Balances

Two issues have frequently come up in the above descriptions of irrigated agriculture in Pakistan, Egypt and Thailand, both of which threaten the sustainability of irrigated agriculture: water shortage and salinity. The processes of salinization of irrigated and nonirrigated lands are reasonably well understood, and preventive and remedial measures are generally well known (see, e.g., Rhoades and Loveday, 1990; Van Hoorn and van Alphen, 1994). However, salinization, and particularly enrichment with sodium, is an insidious process and early symptoms are often ignored. The cases described in the previous section indicate that general solutions are not readily available. Canal irrigation doesn't necessarily lead to waterlogging, but under some conditions it does. Irrigation with tube well water doesn't necessarily lead to depletion of the groundwater resource, but under some conditions it does. Careful monitoring of salt and water balances in irrigated areas would help to provide a scientific basis for deciding on the need and urgency of remedial actions. Moreover, monitoring should be done at field or farm level and at system level, and also for the entire river basin. The example of Egypt shows that information at farm level, without considering the reuse of drainage water may give the wrong impression that water can be saved for the development of irrigation in hitherto nonirrigated land.

Measuring the components of the water balance is easier than evaluating the salt balance. The water balance approach, e.g., has often been used at field level for the determination of the contribution from a shallow water table to the water supply of an irrigated crop. Ragab and Amer (1986), for example, described two independent methods for determining the water supply through capillary rise to maize at the field research station of the University of Alexandria, Egypt, which is underlain by a shallow but fluctuating water table. They found that for an average depth of the water table of some 30 cm below the root zone, capillary rise contributed more than 4 mm/day—from the fourth day after an irrigation until the next irrigation—for part of the growing season (40 days after planting till harvest). These studies, though important in understanding water uptake by crops, do not by themselves provide insight in what needs to be done to sustain agricultural productivity.

Unfortunately, reliable data relating to the components of the water and salt balances are usually not collected when changes in irrigation practice and management are considered, which would—if implemented—have considerable impact on the water and salinity relationships. The interactions between the various sources of water (rainfall, canal water, and pumped groundwater) and the salt contents of each of these are complex. However, a number of reasonable, simplifying assumptions can be made in order to establish water and salinity balances of entire irrigation systems or subsystems, such as distributary command areas. Analyzing these water and salinity balances is informative with respect to the potential impact on the sustainability of irrigated agriculture of continuing current irrigation and agronomic practices in the system (Kijne, 1996).

Information needs to be available (or reasonable guesses need to be made) with respect to the percentage of canal inflow that is lost to escapes, the percentage lost to seepage in canals, and the percentage of inflows from canals lost to seepage in watercourses. The field application efficiency, i.e., the percentage of field deliveries from watercourses that is used by the crop, need to be known or estimated, as well as the field losses going to drains and the percentage of all losses (except runoff) going to nonbeneficial evapotranspiration. The percentage of rainfall on unirrigated areas that is evaporated and the percentage of the noneffective rain that goes to drains need to be estimated. With respect to groundwater use, data or estimates are required on the percentage of water that goes to the groundwater and is recovered through pumping, the percentage of water going to drains and recovered through pumping, and the efficiency with which pumped groundwater is used by the crop (which may be higher than the field efficiency of surface deliveries, as it is generally a more expensive source of irrigation water and farmers may be expected to apply it more carefully).

In systems or command areas where groundwater and surface water are used conjunctively, the percentage of water going to groundwater from seepage from canals, watercourses, drains, and in the fields is an extremely important figure. It usually cannot be known through an independent assessment and needs to be stipulated for the first round of calculations. If the water balance calculations lead to unrealistic estimates of the irrigation allocation from canal supplies, rainfall, and pumped groundwater, changes need to be made in the values used for the seepage losses and in the estimated groundwater pumping until the figures agree with known irrigation practice. In other words, crop water requirements need to be known or be evaluated through CROPWAT, the FAO approach to the calculation of crop evapotranspiration (see Doorenbos and Pruitt, 1977), and some idea about the ratio of irrigation supplies from canal water and pumped groundwater is necessary.

The salt balance model that has been used by IIMI regards the root zone as one layer with a homogeneous distribution of water and salt (Van Hoorn and van Alphen, 1994). It is assumed that the irrigation water is thoroughly mixed with the soil water in the root zone, and, hence, that the salt concentration of the soil water at field capacity equals the salt concentration of the water that percolates from the root zone. The leaching fraction is defined as the ratio of net deep percolation and the irrigation supply. In addition to the information required for the water balance, the salinity (for example, the electrical conductivity) of the irrigation water also needs to be known. It should be noted that this approach does not take into account the type of salts in the root zone or in the irrigation water. As has often been observed (e.g., So and Aylmore, 1993), the sodium component of the total salinity is of particular importance because of its effect on the soil structural stability under irrigated conditions.

From the analysis of water and salt balances of three irrigated areas in Pakistan, it was concluded that at one site overirrigation took place and that waterlogging would be inevitable, and that at two other sites more salts were added to the root zone than were removed, if current irrigation practices were to continue (Kijne, 1996). It was concluded that at the sites where salts were accumulating less tube well water should be used, and that this would only be feasible if crop intensities were drastically reduced (in the order of 25% to 45% less). The implication for the dominant rice-wheat farming system in one of the sample sites (Upper and Lower Gugera command area), discussed in the previous paper (Kijne, 1996), would be that less rice can be grown during kharif season. However, for the Chishtian subdivision this earlier analysis was based on generalized data over a

Table 6.2. Various Management Options and Interventions

Category	Options
Engineering	construct additional storage facilities (dams and reservoirs) improve maintenance of irrigation infrastructure conserve water in catchment, and rain in irrigated areas construct drainage facilities improve maintenance of existing (incl. natural) drains reuse waste and drain water, and find alternative ways to dispose drainage effluent, industrial and municipal waste water
Agronomic	grow different crops; i.e., less water-demanding, or more drought- and salt-tolerant ones reduce irrigated area (use more water per unit land) on-farm watercourse improvement and precision land-leveling
Policy	introduce water and power pricing to make water more expensive introduce transferable water entitlements provide incentives for land reclamation; e.g., subsidizing gypsum
Management	improve the operation of existing irrigation and drainage infrastructure through introduction of management information systems, etc. enhance farmers' involvement in management and maintenance of irrigation and drainage facilities

relatively short period of time. Now more detailed data on eight watercourses are available and one of the most striking features is the spatial variability of soil and water quality. If the average electrical conductivity of all tube wells in the eight watercourse command areas is used for the analysis it appears that the salt balance is reasonably well maintained, better in kharif than in rabi as more canal water is available during kharif, particularly in the Azim watercourses. However, current irrigation practices using tube well water from just over one-quarter of the tube wells would lead to salt accumulation in the root zone. There is thus a need for a dynamic management approach which recognizes the spatial differences of soil and water resources. In the next section, we want to consider what other management options, if any, are available to farmers and irrigation officials, other than reducing the crop intensity. In the final section we will come back to the need for different management strategies of Pakistan's irrigation systems.

Management Options

Several types of management interventions are presented in Table 6.2. They have been categorized as engineering, agronomic, policy, and management options. However, the distinction between them is not watertight. Often before engineering or agronomic options can be implemented, certain policy decisions need to be made. Nevertheless, for clarity of presentation, we will discuss the options in terms of these various types.

The first category, engineering options, includes interventions aimed at enhancing the supply of good quality water and the removal of poor quality water. Dam construction, with its environmental and social implications, is not popular with donors, nor with environmental action groups in the countries in need of more water. Moreover, the high costs associated with the construction of water storage facilities in countries such as Pakistan, where the easier and less costly sites have all been used, requires priority setting at government level. The options aimed at the removal of poor quality water include construction of drainage facilities, and improved maintenance of existing (natural) drains; alternative disposal of drainage effluent, industrial and urban waste such that it does not contaminate water supplies; and rehabilitation of irrigation canals to enhance the reliability and quantity of supplies (e.g., improve the hydraulic performance of irrigation canals through desilting).[4] The implementation of any of these engineering options (whose benefits mature slowly) depends, just like the construction of storage facilities, on strong governmental support both in terms of incentives and sanctions as well as in economic prioritizing. Often there is no immediate prospect of enhancing the supply of good quality water to the system and/or of reducing the amount of poor quality water that remains in the system.

The agronomic options presuppose that scarcity of water and salinity are here to stay and that one has to learn how to live with them. One of the key agronomic options consists of crop diversification. Different crops have different evapotranspiration patterns and lead to different rates of recharge of the groundwater. In areas with rising water tables, planting sufficient trees can reduce or even reverse this trend. Varieties and species with higher salt tolerance can be planted to replace those that yield less when the salt content of the available irrigation water increases; for example, when more of the irrigation supplies are drawn from pumped groundwater. Of course this option is only feasible if there is a market for these crops and farmers who grow them can make a decent living. Fodder crops that have successfully been introduced on salt-affected land in the Indian subcontinent include kallar grass (*Leptochloa fusca*) and sesbania (*Sesbania aculeata*). Improving soil fertility by addition of organic matter through green manuring can also help to restore a more favorable root zone environment. Other agronomic interventions that can be exercised at farm level include precision land leveling to remove high spots in fields, such that a thinner layer of water would cover the entire field. Experience in the Murray-Darling Basin has shown that laser leveling of the land can greatly improve net salt accumulation in the root zone (private communication by Dr. S.A. Prathapar, September 1996).

Perhaps the last agronomic option a government would be prepared to support is to reduce deliberately the area that is cropped each season to match demands for good quality water with available supplies. This could be done by reducing crop intensities, while maintaining the total command areas, through the introduction of periodic fallows. Rotation of canal supplies and restricting the use of pumped groundwater are the instruments to achieve it. It can also be done by closing off parts of command areas, with the result that those areas would depend completely on tube well water, and would most likely continue to degrade. The socioeconomic and political implications are huge. The government may have to stimulate alternative sources of employment in rural areas to provide acceptable livelihoods for the rural population affected by these changes.

Other policy options available to governments in order to change current agronomic and irrigation practices include water and power pricing, issuing transferable water entitlements, subsidizing gypsum for reclamation of sodic soils,[5] and better coordinate the activities of various ministries and departments dealing with water and power supplies.[6]

---

[4]Computer modeling prior to desilting is recommended to assess the likely effect of the exercise in terms of improved equity of distribution of the irrigation water and evaluate the costs and benefits of the desilting operation (see, e.g., Murray-Rust and Vander Velde, 1993)

[5]The application of gypsum is more widely practiced in India than in Pakistan, partly because of subsidies and better distribution in India. One drawback of widespread use of gypsum would be enhanced deterioration of shallow groundwater as soluble salts are leached.

[6]For example, irrigation and drainage responsibilities in Pakistan's Punjab are not coordinated at field level, only at the level of the office of the Chief Engineer, probably overseeing around 1 million ha. Also co-ordination between power and water supplies in rural areas is lacking, with the result that power cuts occur when farmers need to use their electrically powered tube wells to water their crops. Another example is the well known lack of coordination between Departments of Agriculture and Irrigation.

Increasing charges for water would tend to reduce application per unit area and hence exacerbate the problem of salinity.

Management options that come to mind include improving the management (operation and maintenance) of irrigation canals, e.g., through the introduction of management information systems (decision support systems). These information systems would require that data on water supplies and demands throughout the system, and the breakdown of infrastructure (e.g., breaches) are available and quickly communicated to the decision center. Traditionally much of this information system was in place in a country such as Pakistan but has since been allowed to break down and is not in use any longer. Current communication techniques and computer modeling allow vast improvements over the present decision support systems to be made. Another important management option is to involve the irrigation community to a much greater extent in the management of the natural resources water and soil, including their quality (salinity) aspects. Many governments, often spurred on by the World Bank, are moving toward greater involvement of farmers in the management of (parts of) irrigation systems.

In the following section, we consider how these options can help in finding an answer to the sustainability issue in Pakistan.

Sustainability of Pakistan's Irrigated Agriculture

One of the most persistent problems of the irrigation systems in Pakistan's Punjab is the unreliability of the canal supplies. For example, the number of missed turns (farm level irrigation applications) was recorded for the sample watercourse FD 14R during the kharif season of 1994. It turned out that each week the farmers on this watercourse missed between 15 and 95 turns (up to around 80% of the number of turns), and consequently could not irrigate or had to revert to tube well water. The minimum time allocation in the rotational system practiced in Pakistan's Punjab (called *warabandi*) was less than 10 minute per acre (0.4 hr/ha) in the eight sample watercourses; the average in all Fordwah watercourses was just over 1 hr/ha. Small holdings and thus short water delivery times at the farm gate contribute to poor distribution of water within the farm. In a 1995 survey, 20% of the outlets from distributary canals were found to be broken out of a total of 410 outlets in the sample area, another 20% had been enlarged and 5% made smaller than authorized. Of a total of 14 inlets from the main canal to distributary canals in the Chishtian subdivision, the average actual target discharge was 3.5% lower than the official full discharge; however, the discrepancy ranged from 22% less to more than double the officially sanctioned discharge (unpublished IIMI data, 1996). All of this points to a greater actual inequity in present-day water distribution than ever before.[7]

The unreliability of supplies generally increases toward the tail end of the irrigation systems. It has been shown before (e.g., Murray-Rust and Vander Velde, 1993) that the cropping decisions of farmers are influenced strongly by access to surface water of good quality and relatively high dependability. Tube well water is available in larger amounts but is of poorer quality and needs to be obtained through equipment or power sources of dubious reliability. Murray-Rust and Vander Velde (1993) found in the command area of the Upper Gugera Canal that cropping intensities decreased from 80% in head-end wa-

[7]Equity in distribution was intended as equity in sharing a seasonal shortfall of canal water; it should not be taken to mean that everyone received an equal share of the canal water per unit land. In the past, so-called leaching shoots (additional allocations of canal water to leach salts from affected fields) were authorized by the Irrigation Department staff. This inequitable allocation made good sense as some lands are more affected by salinity than others. Unfortunately, this practice has all but stopped because powerful farmers succeed in claiming unauthorized additional discharges, e.g. to irrigate their orchards.

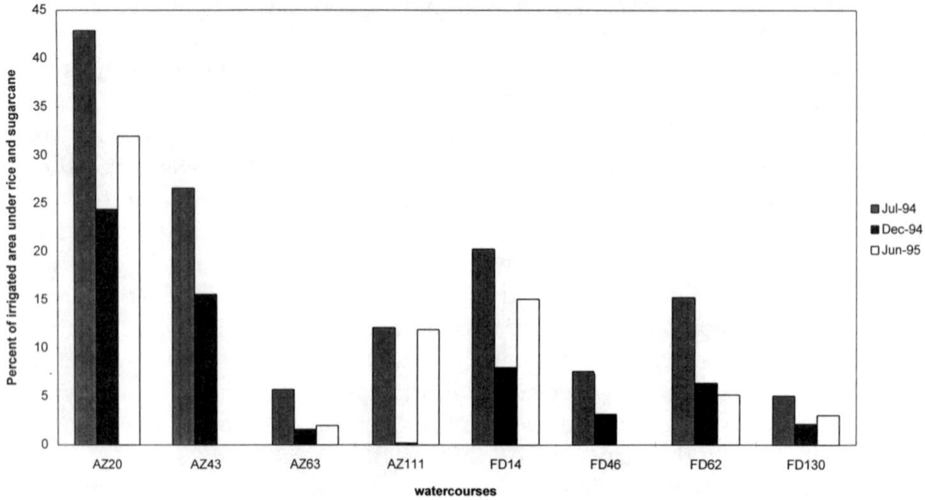

Figure 6.4. Areas under rice and sugarcane in eight sample areas, Punjab, Pakistan.

tercourses to about 60% at the tail, and that the area under rice dropped from nearly 40% of the cultivated land in the head-end watercourses to about 20% at the tail. In the Chishtian area, cropping intensities follow a similar trend in the watercourses of the Fordwah distributary, and—with the exception of Azim 20—also in the Azim distributary. The area under rice and sugarcane, the two most water-demanding crops in the cropping system, decreased sharply toward the tail end (Figure 6.4), and is well correlated with access to canal water. The correlation coefficient of the area planted with rice and sugarcane and the amount of canal water available in the watercourse is 0.71.

It could be argued that if the reliability of canal supplies were to be improved, more farmers in the head reaches would grow rice and sugarcane at the expense of farmers at the tail end. This is indeed likely to happen unless the farmers on a watercourse realize that it is in the long-term interest to all of them to spread the scarce, good quality canal water over the entire watercourse command area. Crop diversification, and the gradual introduction of more salt-tolerant crops should go hand in hand with a reduction in the area under rice and sugarcane. Several other crops are grown in the area, but preference is given to rice, sugarcane, and cotton.

The two objectives of the development of groundwater, first through the SCARPs and later through private tube well installation, were to lower the watertables and to increase cropping intensities by making more water available at the farm gate. These objectives have been achieved to a large extent. It could be argued, though, that increasing the cropping intensity has gone too far, and there is a need for a policy measure aimed at limiting the exploitation of groundwater. The options open to farmers would then be to irrigate more efficiently (produce more crop per unit of water), or reduce the cropping intensity. The management option to improve farmers' practices received much attention during the1960s and 1970s through USAID-supported On-Farm-Water-Management projects. Gradually the aim of these on-farm activities, including the so-called Command Water Management projects, became limited to the lining of watercourses. It should now be realized that where canal water and groundwater are used conjunctively for irrigation, on-farm water 'savings' are not equal to reductions in consumptive use of water; at most they have the potential to reduce the supply of water diverted from the source. The analysis of the water and salt balance clearly suggest that it is necessary to reduce the irrigated area where the available canal water is inadequate to maintain the salt balance, and the quality of the tube well water is such that continuous use would lead to further soil and water degradation (e.g., in the watercourse commands of Azim 111 and Fordwah 130). All of these options, i.e., selectively reducing the area under rice and sugarcane, reducing the total area that can be irrigated, and introducing different crops, are not easily implemented. It should be mentioned again that the salt balance only considers the total amount of salt that is added or removed from the root zone. The change in sodicity and alkalinity of soil and water cannot readily be modeled, but it is of such importance that constant monitoring is required.

At present, advanced plans for institutional reform exist in Pakistan; i.e., to transform the provincial irrigation departments into financially autonomous Provincial Irrigation and Drainage Authorities, and to decentralize irrigation management with the creation of area water boards at the level of the distributary canal command area. Farmers would be represented in the water boards. However, it is questionable whether these institutional changes will suffice to strengthen the sentiment that the common interest with respect to the preservation of the natural resources must prevail over the short-term interests of today's farmers. Moreover, infrastructural improvements must accompany institutional change.

In conclusion, then, we favor a management approach, which recognizes the variability of the land and water resources even over relatively small distances. Management should recognize that there is a distribution problem (equity, timeliness, reliability), which may be addressed if better and more timely information is available on what is happening in the command area. Management information systems and decision support systems for irrigation system managers have been developed with just that purpose in mind. Permanent reallocation of water should not *a priori* be ruled out if, through reallocation, agricultural productivity can be enhanced. For example, it has been shown (IIMI, unpublished data) that a heavy preplanting irrigation with canal water is very important for salt management. If planting were to be staggered for the main kharif crop in the command area, all farmers could be supplied with water for a preplanting irrigation. Several other specific management interventions have also been mentioned in the past, one of which is the restoration of leaching shoots to provide timely additional supplies of water to fields which, because of prolonged application of pumped groundwater, threaten to become saline and where the production level is seriously diminished. However, crop diversification and reduced cropping intensities are inevitable for those areas where leaching shoots are not feasible or where soil and water degradation have proceeded too far already.

How can these changes be initiated, if the institutional changes currently under consideration are unlikely to bring them about? Experience elsewhere indicates that strict adherence to a set of regulations of allowable recharge to the groundwater and maintenance of an acceptable salt balance in irrigated areas is probably necessary to ensure the sustainability of irrigated agriculture. In the Murrumbidgee Valley of Australia, for example, acceptable levels of recharge to the groundwater and of environmental damage have been stipulated for each farm. Farmers have been informed about the consequences of various cropping patterns in terms of recharge and salt accumulation. They can thus make well-informed choices about the crops to be grown and the cropping intensities but *within* a strict set of rules. Excess recharge to the groundwater and/or excess salt accumulation are penalized through a tiered pricing structure of the irrigation water (private communication by Dr. S.A. Prathapar). Local control in regulating water use has also been attempted in Nebraska, where the rate of groundwater withdrawal from the Ogalalla aquifer is much more than the rate of recharge. Also, residues of agrochemicals, i.e., fertilizers and pesticides, have led to increased levels of nitrogen and other chemicals in the local water supplies. These problems reflect the inability of the existing institutional structure to regulate the use of 'common pool' resources. Peterson et al. (1993) argue convincingly that in the case of the Nebraska irrigators, recourse to the judicial system has been critical in overcoming local resistance to change in the *status quo* property rights. Regulations on the amount of water farmers are authorized to use, as well as on the application of nitrogen fertilizer have been put into place. For the most part, these regulations have been developed by local agencies run by the farmers themselves. However, the effectiveness of the local governmental institutions charged with implementing these policies in managing groundwater resources depends critically on institutions at the state and federal levels. These examples show that coordination in regulation and enforcement is necessary, involving farmer organizations as well as state and federal institutions. An adequate monitoring system and the willingness to comply to the rules for the long-term benefit of all are necessary. It will not be easy to introduce this type of regulation in Pakistan's Punjab where social control in rural areas is nowhere as well developed as in Australia and the USA, and where the compliance to rules and regulations in society at large is weak at best.

It is difficult to say, then, whether these irrigation systems are sustainable, considering the demands that are placed on them. The soil and water resources are at risk of degradation, the competition for water between agriculture, industry, and domestic use is likely to increase, and there is an increasing demand for reliable water supplies in order to produce food and fiber for the present and future population. Inherently, we think, the systems are sustainable, as they have proved to be in the past, but it will require an effort by all concerned. Whether the systems and the production system turn out to be sustainable probably depends on the political will of the government and the ability of the managers to address the issues urgently and in an integrated manner.

REFERENCES

Abrol, I.P. and M.S. Gill. Sustainability of Rice-Wheat Production Systems in India, in *Regional Expert Consultation on the Sustainability of the Rice-Wheat Production in Asia.* FAO, RAPA, Bangkok, FAO, Rome, 1993.

Bhatti, M.A. and J.W. Kijne. Irrigation Management Potential of Paddy/Rice Production in Punjab in Pakistan, in *Soil and Water Engineering for Paddy Field Management: Proceedings Intern. Workshop, AIT, Bangkok, 1992,* AIT, Bangkok, Thailand, 1992, pp. 355–366.

Doorenbos, J. and W.O. Pruitt. *Guidelines for Predicting Crop Water Requirements.* FAO Handbook 24, FAO, Rome, 1977.

Ghassemi, F., A.J. Jakeman, and H.A. Nix. *Salinisation of Land and Water Resources: Human Causes, Management and Case Studies.* CABI/ Univ. New South Wales Press Ltd., 1995.

Gill, K.S. Sustainability Issues Related to Rice-Wheat Production System in Asia, in *Regional Expert Consultation on the Sustainability of the Rice-Wheat Production in Asia,* FAO, RAPA, Bangkok, FAO, Rome, 1993.

Gleick, P.H. *Water in Crisis. A Guide to the World's Fresh Water Resources.* Pacific Institute for Studies in Development, Environment, and Security. Stockholm Environmental Institute, Oxford University Press, 1993.

Hafeez, A., Z.A. Piracha, and N. Ahmed. *Multi-Strainer Tubewells for Skimming Top Layer of Fresh Water Underlain by Saline Aquifer,* Mona Reclamation Experiment Project, WAPDA, Bhalwal. Report no. 152, 1986.

Hoorn, J.W. van, and J.G. van Alphen. Salinity Control, in *Drainage Principles and Applications,* 2nd ed., Ritzema, H.P., Ed., ILRI Publication 16, Wageningen, 1994.

Keller, A. A. and J. Keller. *Effective Efficiency: A Water Use Efficiency Concept for Allocating Freshwa-*

ter Resources, Centre for Economic Policy Studies, Winrock International, Arlington, VA, 1995.

Kijne, J.W. Role of Irrigation Management in Sustainable Farming Systems, in *Transactions 15th World Congress of Soil Science,* Commission VI, Symposia, Volume 7a. 1994, pp. 72–86.

Kijne, J.W. *Water and Salinity Balances for Irrigated Agriculture in Pakistan*, Research Report, International Irrigation Management Institute, Colombo, Sri Lanka, 1996.

Kijne, J.W. and M. Kuper. Salinity and Sodicity in Pakistan's Punjab: A Threat to Sustainability of Irrigated Agriculture? *Water Resources Development.* 11(1), pp. 73–86, 1995.

Kijne, J.W. and E.J. Vander Velde. Salinity in Punjab Watercourse Commands and Irrigation System Operations, in *Advancement of IIMI's Research 1989–1991*, International Irrigation Management Institute, Colombo, Sri Lanka, 1992.

Kishk, M.A. Land Degradation in the Nile Valley. *Ambio.* 15(4), pp. 226–230, 1986.

Lal, R. Sustainable Land Use Systems and Soil Resilience, in *Soil Resilience and Sustainable Land Use*, Greenland, D.J. and Szabolcs, I., Eds., CABI, Wallingford, 1994.

Molen, W.H. van der. *Water Management in Egypt and in Pakistan: Similarities and Differences and Their Consequences for Soils and Agriculture.* Lecture Notes, Wageningen Agricultural University, Wageningen (n.d.).

Murray-Rust, D.H. and E.J. Vander Velde. Conjunctive Use of Canal Water and Groundwater in Punjab, Pakistan: Management and Policy Options, in *Advancements in IIMI's Research 1992.* International Irrigation Management Institute, Colombo, Sri Lanka, 1993.

Narangh, R.S. and N. Singh. Water Management Studies in Rice. *Intern. Rice Research Letter.* 13(4), p. 43, 1988.

NESPAK-SGI. *Contribution of Private Tubewells in the Development of Water Potential.* Final Report, Planning and Development Division, Ministry of Planning and Development, Islamabad, Government of Pakistan, 1991.

Peterson, E.W.F., J.D. Aiken, and B.B. Johnson. Property Rights and Groundwater in Nebraska. *Agriculture and Human Values.* 10(4), pp. 41–49, 1993.

Postel, S. Water and Agriculture, in *Water in Crisis. A Guide to the World's Fresh Water Resources,* Gleick, P.H., Ed., Pacific Institute for Studies in Development, Environment, and Security. Stockholm Environmental Institute. Oxford University Press, 1993.

Ragab, R.A. and F. Amer. Estimating Watertable Contribution to the Water Supply of Maize. *Agric. Water Management.* 11, pp. 221–230, 1986.

Rhoades, J.D. and J. Loveday. Salinity in Irrigated Agriculture, in *Irrigation of Agricultural Crops*, Monograph 30, American Society of Agronomy, Madison, WI, 1990.

Sajjad, M.I., S.D. Hussain, M.A. Tasneem, M.A. Ahmad. W. Akram, and I.H. Khan. Isotopic Evidence of the Source of Groundwater Salinity in the Indus Basin, Pakistan. *The Science of the Total Environment.* 129(3), pp. 311–318, 1993.

Seckler, D. and R. de Silva. *The IIMI Indicator of International Water Scarcity*, Research Report, International Irrigation Management Institute, Colombo, Sri Lanka, 1996.

Singh, R.B. and R.S. Paroda. Sustainability and Productivity of Rice-Wheat Systems in the Asia-Pacific Region: Research and Technology Development Needs, in *Regional Expert Consultation on the Sustainability of the Rice-Wheat Production in Asia*, FAO, RAPA, Bangkok, FAO, Rome, 1993.

So, H.B. and A.G. Aylmore. How Do Sodic Soils Behave? The Effect of Sodicity on Soil Physical Behaviour. *Aust. J. Soil Res.* 31, pp. 761–77, 1993.

Sumner, M.E. Sodic soils: New perspectives. *Aust. J. Soil Res.* 31: 683–750, 1993.

Whittington, D. and E. McClelland. Opportunities for Regional and International Co-operation in the Nile Basin, in *Water International.* 17, pp. 144–154, 1992.

SEVEN

Addressing Key Natural Resource Management Challenges in the Humid Tropics Through Agroforestry Research

Dennis P. Garrity

INTRODUCTION

The countries in tropical Asia have begun to lay the foundation for sustained economic growth. Their remarkable progress has been heralded as a source of encouragement for developing countries throughout the world. This optimism is well-deserved, but should not be misinterpreted as victory achieved. Hundreds of millions of people remain in poverty, and the momentum of development has frequently accelerated the degradation of the natural resource base. These pressures are particularly acute in uplands of the region. The rate of forest degradation and conversion has increased in recent years in those countries where significant forests remain. In other countries, such as the Philippines, Thailand, and Vietnam, deforestation rates seem to have declined, but this is due to the fact that these nations have virtually exhausted their exploitable natural forests (Figure 7.1). The expansion of degraded *Imperata cylindrica* grasslands is often the consequence of this rapid conversion process. The human populations inhabiting the uplands continue to increase. They are accelerating pressure on the fragile sloping uplands to provide their food and cash income.

Policy and technological change can have a major role in assisting national governments and farming communities to better cope with the imperative to manage natural resources more sustainably. The International Center for Research in Agroforestry's (ICRAF's) Southeast Asian Regional Research Program is engaged in both policy and technology development with a range of collaborators. Our aim is to exploit the recognized potential of agroforestry to help mitigate deforestation and land depletion, and reduce rural poverty. The two broad issues we focus on are the development of alternatives to unsustainable slash-and-burn, and the rehabilitation of degraded or degrading upland ecosystems.

Agroforestry systems already provide a major contribution to the sustainability of agriculture in the Asian tropics, and in the conservation of protected ecosystems. However, their role has only recently been given wide recognition. Agroforestry solutions must be carefully assessed. The research base for prescriptive action is still quite weak. This has led

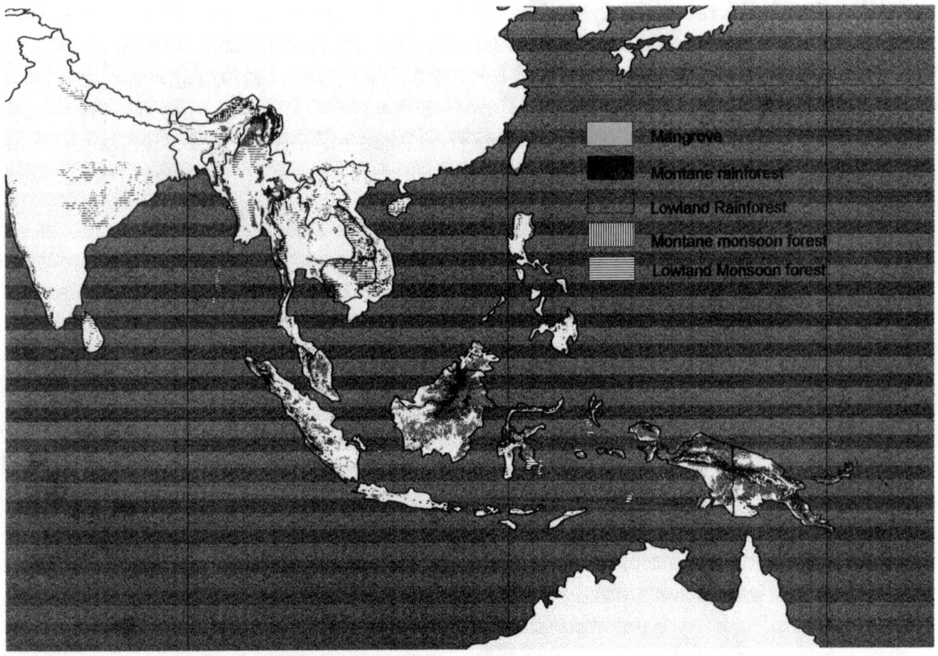

Figure 7.1. Forest cover in Southeast Asia (data from the World Conservation Monitoring Center).

to overoptimism in promoting some technologies resulting in disappointing results, such as with alley cropping (Sanchez, 1995). Agroforestry has only recently been approached as a scientific pursuit. We must rapidly deepen the base of practical knowledge to guide the implementation of agroforestry-based interventions.

The conventional view of agroforestry is that it is "the deliberate cultivation of woody perennials with agricultural crops on the same unit of land in some form of spatial mixture or sequence." This has led many people to see it merely as a set of distinct prescriptions for land use. This limits its ultimate potential. We now see agroforestry as the increasing integration of trees in land-use systems and conceive of it as the evolution of a more mature agroecosystem of increasing ecological integrity. Leakey (1996) has proposed that agroforestry be considered as a "dynamic, ecologically based, natural resource management system that, through the integration of trees in farm and range land, diversifies and sustains smallholder production for increased social, economic, and environmental benefits." This definition is currently being refined by ICRAF as a more holistic concept of agroforestry. It evokes the process of integrating the variety of current agroforestry practices into productive and sustainable land-use systems. Land use becomes progressively more complex, biodiverse, and ecologically and economically resilient. This new vision of agroforestry is transforming ICRAF's approach. This chapter will show how it is imbedded into our research agenda in Southeast Asia.

The chapter reviews the issues being tackled and the progress made to employ agroforestry as a means to increase productivity while protecting the land and biodiversity resources of the humid tropics. Our work in tropical Asia is targeted toward three distinct upland ecosystems: the forest margins, *Imperata* grasslands, and hill-slope farmlands. Our work is enhanced through participation in two global programs: the CGIAR (Consultative Group on International Agricultural Research) systems-wide program on Alternatives to Slash-and-Burn (ASB) (van Noordwijk et al., 1995a), and the Sustainable Agriculture and Natural Resources Management Program (SANREM, 1996).

The first part of the chapter focuses on the hypothesis that agroforests are uniquely suited to substitute for the environmental protection benefits of natural forests when forest conversion occurs. It also examines the role of agroforestry in buffer zone management on the boundaries of protected forests and ecosystems. The second part examines key research issues that may guide rehabilitation of *Imperata* grasslands toward more intensive land use. The last part reviews the role of agroforestry in the sustainable cultivation of sloping lands.

THE FOREST MARGINS: EVOLVING COMPATIBLE LAND USE SYSTEMS FOR FOREST PROTECTION

Successful action to mitigate tropical deforestation depends upon a more comprehensive understanding of the forces driving this process. Research in Brazil, Cameroon, and Indonesia (the core countries involved in the ASB program) has clarified that these driving forces are dramatically different across the continents, and that they vary locally within regions of individual countries as well. Mitigation efforts must address the interacting forces within each nation and locality.

We are examining two basic issues: Where forest conversion is inevitable, or has already occurred, what are the preferred land uses that best protect watersheds and provide the environmental services that natural forests would have provided? And, where forest ecosystems have been designated for full conservation, how can the boundaries best be protected? We examine the agenda for each of these cases in turn.

Retaining Environmental Benefits when Forest Conversion is Inevitable

In the Indonesian context we observe that a substantial amount of large-scale conversion of natural forests to other land uses is inevitable in the future. This conversion is driven dominantly by pressures to convert forest land to large-scale agricultural and forest plantation enterprises, and partially by the unplanned objectives of swelling populations of local smallholders who seek to make a living through the conversion of forests to smallhold farms. We note, however, that the new land uses vary greatly in their ability to substitute for the environmental services provided by natural forests.

Agroforests on Production Forest Land

In Indonesia, much of the land designated as Production Forest has been so degraded by inappropriate logging practices that regeneration of forest has not occurred. Devastating fires have intensified the degradative processes (e.g., 1994, when an estimated 5 million hectares were burned). The Ministry of Forestry has sought to keep land designated as Production Forest from being settled by smallholders. However, hundreds of villages already existed traditionally on lands that were classified as Production Forest only in recent decades. These villagers have often evolved complex agroforest land use systems by cultivating mixed perennials with their food crops after slash-and-burn as a logical part of their livelihood strategy. These agroforests are predominantly based on rubber, dipterocarp resin, or fruit species, with timber species husbanded as a component.

Farmer-evolved agroforests often resemble natural secondary forest systems in structure and ecology. The trees provide food, fuel, and cash income. The agroforest accumulates a carbon stock that in some systems may be maintained indefinitely. There are many examples of agroforests in the humid tropics. An outstanding case is the 'damar' agroforest system in Lampung, Sumatra, Indonesia (Michon et al., 1995). Over the past century, local populations have extended the cultivation of the dipterocarp tree *Shorea javanica*, which is tapped to yield a resin that is sold for industrial products on the national and international markets. This man-made forest now extends over some 10,000 hectares on production forest land in Lampung. It harbors a major proportion of the natural rainforest flora and fauna species (Michon et al., 1995). There are many indirect sources of evidence that soil fertility levels are maintained over long periods, and that soil organic matter levels increase in these mixed systems (see the review of literature by Torquebiau [1992]).

We hypothesize that these mixed tree-crop systems, particularly complex agroforestry systems or 'agroforests,' provide a superior alternative to other land uses in protecting watersheds from soil erosion and flooding risk, conserve a greater amount of biodiversity, and provide a greater sustained source of income generation for local communities than other forms of crop or tree monocultures. Our research on these systems, and that of others, is accumulating empirical evidence that supports this (e.g., de Foresta and Michon, 1997; Salafsky, 1993; Momberg, 1993; Mary and Michon, 1987). Thus, we find that the objectives of smallholder communities practicing such systems may be more compatible with those of national governments in protecting watersheds and biodiversity than has previously been assumed.

The most widespread agroforest system in Indonesia is rubber agroforestry (or 'jungle rubber') which occupies over 2.5 million hectares (Gouyon et al., 1993). In this system rubber trees are the main component, but many other species of fruit and timber trees are combined with rubber, either intentionally or through natural regeneration. The rubber seedlings are established as intercrops planted along with food crops in a slash-and-burn cultivation system. After the 1-2 year cropping phase the plot is left alone and the rubber trees mature along with the secondary forest regrowth. Biodiversity levels often approach those of natural secondary forest (de Foresta and Michon, 1997; Thiollay, 1995).

Tenure reform. Official recognition of local land and tree tenure systems would underpin their security and enhance the development and expansion of agroforests. Public sector assistance would further strengthen the trend toward smallholder tree-based land use systems. Smallholder communities could then contribute substantively to the national production objectives for which the production forests exist. Complex agroforests are one of the most promising solutions to transforming unsustainable slash-and-burn systems.

Research and case study experiences are needed as a foundation for policy reform to enable communities to obtain more secure tenure and to make aggregate contributions to the economy. Detailed protocols for developing management agreements between local populations and the national government are essential. ICRAF and its partners in the ASB Program are examining how viable instruments may be designed at three levels: the household, the community, and the regional level. We are focusing on management mechanisms at the community level that provide an appropriate degree of tenurial security while ensuring that the environmental and production objectives of the national government are met. ICRAF is coordinating a project with the University of Indonesia, LATIN (an NGO), CIFOR (Centre for International Forestry Research), the Ministry of Forestry, and local government to develop new instruments of tenurial security for the *damar* agroforest communities of Krui in Lampung Province. The work is based on a solid scientific understanding of the agroecology of these agroforests developed during past several years (Michon et al., 1995). A successful outcome will provide a basis for implementation of similar tenurial instruments with the communities that are the *de facto* local managers of Production Forest land in enclaves all over the country.

Figure 7.2. Example of a mature agroforest designed for both profitability and sustainability in smallholders' conditions. From de Foresta and Michon (1997).

Environmental impacts. Data on the comparability of the environmental services of agroforests vis-à-vis natural or plantation forests will be required. The ASB Program is estimating the implications of alternative land uses, including various types of agroforests, on carbon sequestration, greenhouse gas emissions, and soil erosion (van Noordwijk et al., 1995a). It has been shown, for example, that agroforests are significant sinks for methane, a greenhouse gas generated by wetland rice and other annual crop systems.

De Foresta and Michon (1997) have emphasized that agroforests are successful only when they meet smallholders' income needs. They note that such a system is usually composed of two sets of commercial tree species suited to local conditions, one set providing regular cash income (e.g., rubber, resin) and the other providing seasonal or irregular cash income. Such composition ensures economic and ecological viability of the forest in the long run, provided that clear tenurial rights on the basic units are recognized. Figure 7.2 is a cross-section of an Indonesian smallholder agroforest that illustrates these objectives.

Segregate or Integrate Nature and Agriculture?

The issue of partially conserving biodiversity outside protected areas through complex agroforests needs to be critically assessed. Conflicts between 'nature' and 'agriculture' can

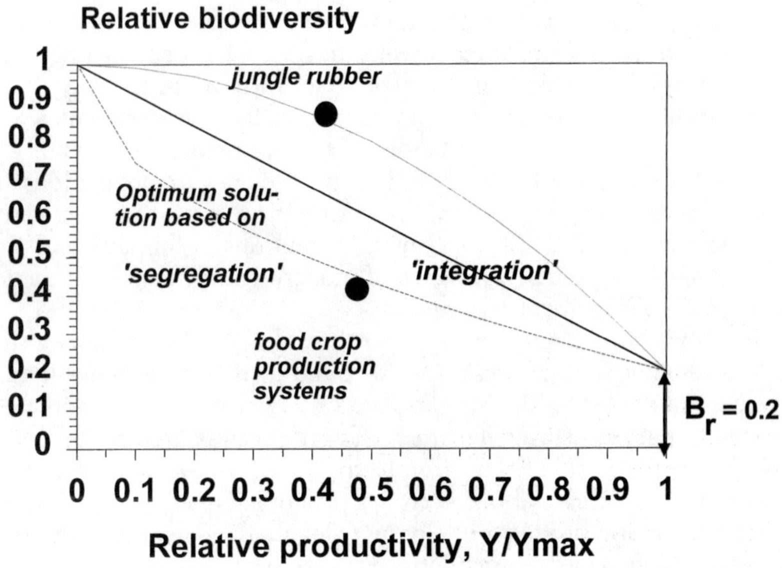

Figure 7.3. Hypothesized relationship between relative productivity and relative biodiversity for different land use systems. From van Noordwijk et al. (1995b).

be resolved by segregating nature from agricultural land by designating areas of full protection, or by integrating nature into the agricultural landscape through production systems that ensure conservation of a major part of the biodiversity of the natural forest (van Noordwijk et al., 1995b). Multifunctional forests and agroforests are examples of the 'integrate' option. Mixed strategies may be envisioned where nature reserves coexist with agriculture. Biodiversity will then vary among the facets of the landscape.

Our research is putting strong emphasis on assessing the viability of the 'integrate' option. Van Noordwijk et al. (1995b) proposed a simple model to examine the decision framework (Figure 7.3). The figure illustrates the hypothesis that as relative productivity of different agricultural systems increases, the rate biodiversity loss will vary among systems. It illustrates the presumption that relative biodiversity tends to decline more drastically with productivity increases in food crop systems than in jungle rubber agroforestry. This is because the rubber-based system tends to recreate a forest structure that nurtures a range of niches that can be filled by other natural species without detrimental effects on productivity. If this is the case, then relative agricultural intensification in such types of agroforest systems may be compatible with a major degree of biodiversity conservation. We are currently examining the implications of this model by surveying the level and nature of the biodiversity maintained in a wide range of agroforest types in Sumatra.

Can smallholder jungle rubber systems be made more productive while retaining substantial biodiversity value? ICRAF and GAPKINDO (the Rubber Association of Indonesia) are investigating this issue in a collaborative project implemented with smallholder rubber farmers in several parts of Sumatra and Kalimantan (Penot, 1996). Eighty-four percent of the rubber area in Indonesia is cultivated by smallholders, but the latex yields are low (generally about 500 kg/ha/annum). The most important innovation to increase productivity is the use of improved rubber germ plasm, particularly modern clones. The public sector has encouraged the adoption of estate-type monoculture systems, but these require levels of investment that are so high that project support has only touched about 10–15% of the smallholder population during the past 25 years.

We hypothesize that smallholders can substantially increase their rubber yields by incorporating new germ plasm into their present jungle rubber or mixed agroforestry systems, and that this would enable a much broader advance in productivity than investing in intensive monoculture systems. This concept for rubber development would also be more conducive to the retention of the biodiversity conservation benefits of these mixed systems. But currently no one knows how improved clones or populations perform in such systems. We've developed a network of trials with scores of farmers across a spectrum of Indonesia's smallholder rubber growing conditions in Sumatra and Kalimantan, in which we are comparing the performance of modern germ plasm with the local material under a range of management systems.

Allied with this work are a number of in-depth studies on the component interactions in rubber agroforestry systems. These seek to build a predictive understanding of the competitive environment for light, nutrients, and water between rubber and the cultivated and natural forest species associated with it. These species interactions determine the total system's productivity of rubber agroforestry. Field research on agroforest systems must cope with cycles that extend over decades. There is typically a gap of several years between establishment and production. Computer modeling is particularly well suited to assist in the design and analysis of such long-term research efforts. We have a major simulation modeling effort embedded in this work to guide and extrapolate the results of research on such systems.

At this point we have examined the issue of land uses that may best provide favorable environmental services when forest conversion inevitably occurs. What can agroforestry research contribute to protecting the boundaries of natural areas designated for full protection? In such situations we must view the research approach from a very different perspective.

Agroforestry for the Buffer Zones of Protected Ecosystems

National parks and nature reserves are the last-ditch bulwarks of protection for the priceless biodiversity resources of the humid tropical forests. They are under enormous threat of encroachment virtually everywhere. The classical method of preserving a natural area has always been to declare it off-limits and to enforce exclusion by local people. Boundaries were delineated and guards patrolled. Unsurprisingly, this isn't working. It has often resulted in serious conflicts of interest and hostility between the enforcement agency and the local communities. Enforcement alone didn't work in most countries, either because population pressure on the land was too great, the gain captured by local elites by encroachment was too lucrative, or the costs of enforcement were too high. What might be the role of agroforestry in preventing these protected forests from degradation or ultimate conversion?

MacKinnon et al. (1986) defined buffer zones as "areas adjacent to protected areas on which land use is partially restricted to give an added layer of protection to the protected area while providing valued benefits to neighbouring rural communities." Wind and Prins (1989) defined them simply as "areas outside of parks that are designed to protect parks." The term has also been used to describe almost any initiative involving people that takes place near a protected area.

In this situation we have a domain of protected forest, a boundary, and land outside the boundary (converted forestland or agricultural land). We can specify the problem as identification of a suite of conditions that reduce or eliminate the economic 'necessity' for smallholders (or large-scale operations) to encroach across the protected-forest boundary. There are now many projects in the tropics called integrated conservation-development projects (ICDPs) that are attempting to save particular natural areas using this approach. Unfortunately, there is a widespread assumption among practitioners of the ICDP approach that people made better off by a development project will refrain from illegal exploitation of a reserve area, even if no enforcement is practiced. Wells and Brandon's (1992) global review of ICDPs found absolutely no evidence to support this. Such expectations are naïve: a social contract between communities and outside stakeholders must include enforcement mechanisms in tandem with the development benefits received.

Compensation to communities in terms of development activities may take many forms. Most projects attempt to encourage improved natural resource management practices in the areas outside the reserve. The objectives are to increase people's incomes, and to intensify their production systems away from the more extensive, environmentally-degrading systems they may currently practice. There is growing interest in the development of more intensive land use systems on the margins of protected forests and to identify policy and technology directions to underpin these efforts.

Let us first examine the simple case which assumes that no net immigration is occurring. In this case, boundary protection may be conceived as a function of reasonable enforcement combined with provision of more favorable conditions for the intensification of land use. The hypothesis is that if local livelihood systems can be improved, then the propensity to encroach into the protected area will be reduced. Since there is some risk to local people associated with encroachment, advances in the alternative livelihoods outside the boundary increases the perception of their gains exceeding their risks. The threshold risk is a 'cost' to the encroacher, and these costs increase relative to the returns in other enterprises.

After examining the base conditions of the local actors and the land use systems currently practiced, a range of alternatives (or combination of alternatives) may be proposed that meet global environmental objectives and local economic objectives. A subset of these may then be selected that would evidently reduce the propensity for encroachment. Then the type and level of outside investment to assist local actors to implement these land uses is determined.

So far we have dealt with protection as a function of enforcement and land use intensity. In many real-world situations (but not all) there are two other important factors: migration and off-farm employment. If immigration is occurring, the accelerated population pressure will destabilize the balance between intensification and enforcement. Migration must be controlled in the communities on the boundary. In some areas this has been successfully achieved in mature communities through local land tenure systems (see Cairns, 1994 for an example in mature Minangkabau communities in Indonesia). But in most pioneer communities, local control of migration is problematic. A further complication is that if success is achieved in increasing land productivity, this may actually attract new migrants and exacerbate inmigration pressure.

Conditions in the wider economy play a major role in affecting migration. Migration trends may at some point be reversed, and net outmigration become dominant as the national economy develops. Also, off-farm employment for village residents in the buffer zone may be increased or decreased. Figure 7.4 illustrates the protection nexus as a function of the four factors of intensity (I) of land use, enforcement (E), migration (M), and off-farm employment (OFE). ICDP programs must consider the implications of all of these factors and their interactions. Unfortunately, very little analytical work has been done on approaches to cope with these complexities. The result is that the solutions attempted tend to be too simplistic. To date there are few successful ICDP examples (Wells and Brandon, 1992). Systems thinking and a long-term involvement of highly committed individuals will be critical to making ICDPs a success.

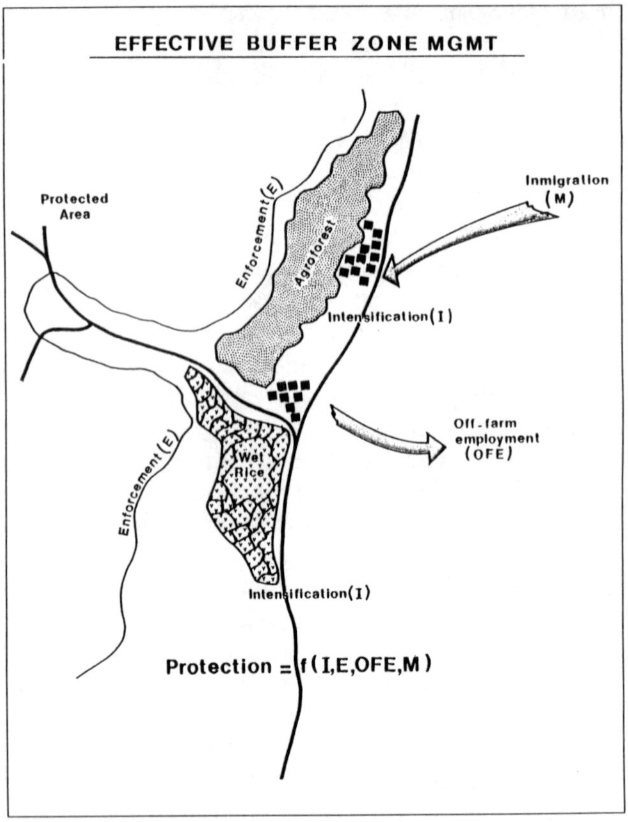

Figure 7.4. The protection of a natural area through effective buffer zone management is a function of land use intensification (I), enforcement (E), off-farm employment (OFE), and migration (M).

With the above caveats as a backdrop, improved agroforestry systems have frequently been cited as a path toward appropriate intensification in the buffer zones of protected areas (Wells and Brandon, 1992; Garrity, 1995; Cairns et al., 1997). Tree seedlings are often a highly desired intervention by recipient communities near protected areas. Provision of seedlings through nursery programs has therefore been one of the most popular ICDP development interventions. Farm families can increase their nutrition and economic welfare through a greater quantity and diversity of fruit and timber trees in the home garden area and on their farms. In many areas there is an encouraging trend toward tree crop farming as an alternative to practicing shifting cultivation (Garrity and Mercado, 1994).

Where there has been a history of tree crop cultivation in the vicinity of a protected area, the environment of the farming zone outside the boundary develops ecologically favorable characteristics for protection, and even extension, of the biological diversity of the park itself. The 'damar' agroforest systems on the boundaries of the Barisan National Park in Lampung, Indonesia, harbor a major proportion of the natural rainforest flora and fauna species (Michon et al., 1995), and effectively act as an extension of the biodiversity of the park itself. Rubber agroforests on the boundary of Kerinci-Seblat National Park in Jambi Province play a similar role (van Noordwijk et al., 1995a). Even in areas where smallholder agroforestry systems do not yield such striking levels of protection or extension for natural biodiversity, the benefits of increased tree cover on the landscape may be important.

Do households on the boundaries of protected areas that cultivate agroforests show less propensity to encroach on protected resources compared to households with simple farm types? We examined this question in a study on the northeastern boundary area of the Kerinci-Seblat National Park in the benchmark area of the Alternatives to Slash-and-Burn Program (ICRAF, 1996; Murniati et al., 1998). The boundary area was comprised largely of village forests that were gradually converted to agriculture, particularly mixed-perennial gardens or agroforests. The upper story of these gardens is dominated by rubber, along with durian (*Durio zibethinus* Murr), jengkol (*Pithecellobium lobatum*), petai (*Kleinhovia hospita* L.), and coconut (*Cocos nucifera*), with fewer individuals of a range of other species. The understory generally contains bananas and coffee. Interviews with 60 respondents indicated that the gathering of forest products inside the park took place on both a routine and an occasional basis. Products routinely gathered were timber (for lumber and poles) and fuelwood. Rattan, incense, palm fibers, game animals (deer and monkeys), and fish were also gathered.

We observed three general farm types outside the park: farms with only wetland rice fields; farms with only mixed-perennial gardens; and farms composed of both of these components. The families interviewed were grouped by farm type and the data set adjusted to remove the effects of differences in total farm size. For farms of similar size (< 1 ha) we found families with both rice fields and mixed gardens extracted 87% less value of products from the park than those with only rice fields. Thus, farmers that integrated rice production with tree crops exhibited only minor extractive pressure on the park, compared to farmers with only one enterprise type. Families having mixed gardens, and no rice fields harvested 24% less value from the national park than those that had only rice fields (Table 7.1). They had more opportunity to supply their wood needs, but were short of cash to buy rice, so they harvested some timber and other products from the park to help supply their cash needs. Harvest of forest products from the park was negatively correlated with nonfarm income across all farm types. Thus, reliance on income from the forest declined if other sources of income were available. Most respondents claimed that an integrated farm, composed of both rice and perennial gardens, was the ideal farm type. This is also the situation most conducive to park protection. The work showed that agroforest farming in the boundary area may contribute significantly to national park protection.

Research will play an increasingly important role in providing options and insights for ICDP development. The Sustainable Agriculture and Natural Resources Management (SANREM) Collaborative Research Support Program is a global program that takes a landscape approach with a strong participatory bias. At the SANREM research location in the Manupali Watershed in Mindanao, Philippines, ICRAF is coordinating a consortium that is developing the elements of a practical social contract for buffer zone management, developing improved agroforestry systems for the buffer zone, and assembling a

Table 7.1. Average Annual Value of Forest Products Gathered by Villagers from Kerinci-Seblat National Park (May 1993–May 1994)[a]

	Annual Value of Harvest ($) per Family	
Farming Types	**Routine**	**Occasional**
Rice field	171 a[b]	8 a
Mixed gardens	131 b	13 a
Rice field + mixed gardens	22 c	8 a

[a]From Murniati et al., 1998.

[b]Amounts with the same letter in a column are not significantly different at the 10% error level.

natural resource management system for the Katanglad National Park. The research team is composed of scientists and practitioners from institutions including ICRAF, NGOs, universities, the tribal community, and local and national government institutions.

We found that the natural resource management strategies of the indigenous Talaandig communities living on the boundary provide a strong foundation for park protection (Cairns, 1995). However, population increase and commercialized vegetable production are causing serious encroachment pressure. The buffer zone area surrounding the park (classified as national production forest land) has high agricultural settlement pressure and is now dominantly grassland and shifting cultivation. The emerging path for household farming systems intensification is small-scale vegetable production combined with timber and fruit tree production. We have surveyed and mapped (Glynn, 1996) the performance of current tree species by elevation in the watershed (200 m to 1800 m), and have initiated trials to evaluate the most promising farm forestry species. We are working with scores of farmers on conservation farming practices and tree nurseries to elucidate more effective methods of diffusing new practices that will sustain crop yields and increase tree cover. Our policy group is tackling the challenge of combining these technical innovations with stronger community-level resource management systems that will support measures to build a 'safety net' of active enforcement of the park's integrity. This entails assisting in the development and implementation of a municipal-level natural resource management plan as well as a management plan for the national park and its buffer zone. The lessons of this approach will be scaled up through partnership with the Integrated Protected Areas Network in the Philippines.

In northern Thailand ICRAF is working with numerous partners in a similar effort to incorporate individual fields as components of a broader landscape management system in a key watershed. The concept is to move beyond individual households to include management functions at a community level (Thomas, 1995). The agroforestry system is a community watershed land-use mosaic that includes forest, tree, and crop components which interact in numerous ways.

IMPERATA GRASSLAND REHABILITATION

When forests are opened for food crop production throughout Southeast Asia the land becomes infested with *Imperata cylindrica* grass within a few years. Farming the grasslands becomes increasingly laborious and crop yields generally decline. Smallholders then leave the grasslands behind and move to the forest margin where their returns to labor are higher. Currently there is about 35 million ha of *Imperata* grasslands distributed among the tropical Asian countries from India through mainland southeast Asia and Indonesia and the Philippines (Garrity et al., 1997). Can the initial degradation into *Imperata* be avoided? Can the grasslands be reclaimed and used more intensively?

Tenure insecurity is a key factor depressing effective conversion of *Imperata* grasslands to more intensive uses (Tomich et al., 1997). The ideal response may be to establish secure property rights, but this may not be practical because of the political sensitivity of national governments over their control of state land. The recent international workshop (Garrity, 1997) on improved agroforestry systems for *Imperata* grassland rehabilitation (sponsored by ICRAF in Indonesia) offered a major policy recommendation on this issue. It recommended that for large blocks of state forest land covered by *Imperata* grassland, farmers who convert that grassland by planting and managing trees in small plots should receive property rights over all their products, including the timber. This policy recommendation would apply only to existing grasslands, not to forests cleared in the future, avoiding perverse incentive to accelerate conversion of forest to grassland. ICRAF is implementing a major Southeast Asian regional policy project that will test this recommendation on a pilot scale, and provide further analysis to assist the Indonesian Ministry of Forestry and other national governments in implementing it on the ground.

Turning to the question of suitable smallholder technologies for intensification in the grasslands, we find that purely annual crop-based production systems have only limited scope to be sustainable under upland conditions prone to infestation by *Imperata* if animal or mechanical tillage is not available (van Noordwijk et al., 1997). However, where farmers have access to animal or tractor draft power, continuous crop production has been sustained in many *Imperata* areas. A wide range of agroforestry systems that evolve from shifting cultivation have shown particular potential (Garrity, 1997). We now review our work on some of the most promising options.

Building on Indigenous Strategies of Intensification

Natural fallows work eminently well in regenerating soil fertility, if the local fallow species diversity and soil quality have not been degraded (e.g., Szott et al., 1991). But where natural fallows in shifting cultivation systems have been degraded to the point where grasses, particularly *Imperata* cylindrica, dominate the abandoned fields, natural regeneration is usually infeasible. Annual nutrient accumulation in *Imperata* fallows levels off after 1–2 years and is far inferior to that of woody vegetation. The result is a fallow incapable of regenerating adequate nutrient accumulation, yet is very laborious to reopen for cultivation.

Farmers prefer to locate their prospective plots in woody vegetation, and avoid grassland whenever possible (Cairns, 1994). *Chromalaena odorata* is an important pioneer fallow species that naturally suppresses *Imperata* in the absence of frequent fires, accumulates many times more biomass, and regenerates crop productivity much more efficiently. Shifting cultivators throughout Southeast Asia find it a highly desirable fallow species (Dove, 1986; de Foresta and Schwartz, 1991), even compared to secondary or primary forest.

Austroeupatorium inulifolium is a similar non-native invasive species common at mid-elevations above 600 m, that has proved very beneficial to farmers practicing shifting cultivation in West Sumatra (Cairns, 1994). It spread widely after its introduction in the late nineteenth century. Farmers found that it halved the necessary fallow period to regenerate soil fertility. This was a major contribution since land pressure was intense (Stoutjesdijk, 1935). Cairns' (1994) work showed that *A. inulifolium* fallows accumulated several times more biomass and nutrients beneficial to crop growth in a two year period (over 150 kg N/ha and 20 kg P/ha) compared to nearby *Imperata* fallows (25 kg N/ha and 6 kg P/ha) of the same age.

We believe that one of the most promising approaches to identifying biophysically workable and socially acceptable technologies is to document, verify, and disseminate cases of indigenous adaptations. If successful indigenous strategies for managing fallow land can be refined and diffused to other upland areas with degrading shifting cultivation, this will enable intensified land use and improved living standards for some of the most marginalized communities in the region. Unfortunately there is little documentation of such innovations to feed into the national and international research agenda or to inform policy makers. Indigenous strategies are either unobserved or misinterpreted.

ICRAF is collaborating with local partner institutions in a regional research initiative on 'Indigenous Strategies for Intensification of Shifting Cultivation.' Teams in several countries are investigating a variety of improved fallow systems that have evolved in different ecozones. These vary from herbacious to shrub-based to tree-based fallows (Figure 7.5). The approach is identifying a range of pragmatic and adoptable solutions for wider extrapolation among communities facing similar swidden degradation problems. Some of the most important indigenous systems combine elements of local practices with new exogenous technology. One fertile example is that of a slash-and-burn community in Indonesia that is harnessing new conservation tillage technology.

Shifting Cultivators Use Herbicides to Manipulate Fallow Vegetation

South Kalimantan has one of the largest areas of sheet *Imperata* in Indonesia (623,000 ha). The grassland lies in a north-south belt on the western side of the Meratus Mountains. Farming in the village of Belangian, Riam Kanan, is based on a manual slash-and burn system. Farmers confine their cultivation to those areas dominated by *Chromalaena odorata* and try to avoid opening *Imperata* due to the enormous amount of work involved. But *Imperata* has spread over most of the village lands, so there is progressively less secondary vegetation available for new fields every year. About three years ago the entire village started using glyphosate herbicide to control *Imperata* when their fallow land was opened for cultivation. The vegetation is first slashed and burned. When the grass grows back to a height of 20-40 cm after 2-4 weeks, the grassy patches are sprayed. The field is burned a second time. Planting is then done by dibbling the seed. All farmers are now using glyphosate as part of their management system. They found that glyphosate enabled them to reopen fields with a substantial amount of *Imperata*. They relate that the use glyphosate not only controlled *Imperata,* enabling zero-tillage annual cropping, but also caused the post-cropping fallow vegetation to shift back to *Chromalaena*. They observe that the area dominated by *Imperata* is now declining.

Successful zero-tillage food crop systems using herbicide are rare in the tropics. Thus, the fact that such a system was developed independently by a village of shifting cultivators is remarkable. The glyphosate system does not replace slash-and-burn; rather, it enhances it through more effective weed management during the cropping cycle, and a favorable vegetation shift during the fallow period. Both aspects improve sustainability and reduce pressure on the local forest boundaries. The extrapolation value of this system is not trivial because *Chromalaena* is such a widely-preferred fallow species for slash-and-burn in Southeast Asia (including western Indonesia, Philippines, Laos, and Thailand). Other species of the *Compositae* family have been shown to play the same role in other environments. We are documenting this in more depth in the context of the Indigenous Strategies thrust of the global program on Alternatives to Slash-and-Burn (ICRAF, 1996).

Food Crop/Tree Crop Systems for Grasslands Without Tillage.

Growing crops or trees in *Imperata* grasslands is difficult. Intensive and repeated tillage is usually essential to maintain annual cropping, or to establish perennial crops or timber plantations. *Imperata cylindrica* regrows rapidly, competes vigorously, and creates a major fire hazard for young plantations. Recent on-farm trials by the Agency for Agricultural Research and Development (A. M. Fagi, pers. comm.) have tested a promising system for food and tree crop establishment based on the judicious use of glyphosate herbicide. The grass is killed by an initial application of the herbicide, and upland rice or other crops are direct-seeded by dibbling into the thick residue mulch. Tree crops such as rubber, fruit trees, or timber trees, are established as intercrops with the food crops. Successive crops of annuals may be grown for two or three years while the canopy cover of the trees develops. More research is needed on a number of aspects. The global Alternatives to Slash-and-Burn Program (ICRAF, 1996) is collaborating with Indonesian researchers to examine the constraints and potential solutions in using herbicides as part of an integrated approach to annual and perennial crop establishment to rehabilitate large parts of Indonesia's nine million hectares of sheet *Imperata*.

Figure 7.5. Spectrum of indigenous approaches to manipulate 'fallow' vegetation in S.E. Asia (Cairns and Garrity, 1997).

In countries such as the Philippines, Vietnam, and Thailand, which are experiencing extreme forest encroachment pressures on the remaining natural areas, there is a concurrent trend toward major increases in the value of farm-grown timber. Smallholders, even shifting cultivators on the frontier, are now engaging in farm forestry for the first time in great numbers, in response to recent strong price incentives (Garrity and Mercado, 1994). This situation dramatically increases the prospect of stimulating smallholder timber production systems as a major vehicle for rapidly increasing overall tree cover in the landscape. In the Philippines, ICRAF is evaluating smallholder timber production systems, and more effective methods to disseminate improved tree germ plasm, through private and village nursery systems.

Pruned-Tree Hedgerow Fallow Rotation Systems

Alley cropping was originally conceived as a sustainable replacement for fallow rotation on degrading lands. However, many Asian farmers that installed pruned-tree hedgerow systems have been observed to fallow them after a period of time. It is now clear that hedgerow systems may extend the cropping cycle beyond that possible with natural fallows, but cannot eliminate fallowing unless fertilizer or manure is used. Our survey of farmers that adopted contour hedgerows in Claveria, Mindanao, Philippines, elucidated two dominant reasons why they eventually fallowed their hedgerow systems: inadequate farm labor to maintain the pruning regime on the hedgerows, or the crop productivity was observed to decline in the hedgerowed fields. The sample of farmers interviewed had tree fallows that ranged from 0.5 to 10 years in duration.

We are conducting studies to characterize the hedgerow fallow system and its prospective sustainability in both the Philippines and Indonesia. Crop productivity and overall profitability from fields with fallowed hedgerows of *Senna spectabilis* (syn. *Cassia spectabilis*) that were reopened for maize cropping increased compared to cropping in adjacent plots with natural fallow vegetation (Suson and Garrity, unpubl. data). We observed a major shift in fallow vegetation in the alleyways as a consequence of hedgerow fallowing. After three years the alleys were intensely shaded and dominated by broad-leaved species. *Imperata cylindrica* was effectively suppressed. Fire is a frequent threat on grassland farms because it is very difficult for the individual cultivator to protect his fields from fires that spread from elsewhere. Experiments in Lampung, Sumatra have shown that *Peltophorum dasyrachis* shows greater promise than *Gliricidia sepium* as the hedgerow component because *Peltophorum* is more resistant to fire (van Noordwijk et al., 1997).

We hypothesize that the extrapolation domain for pruned-tree hedgerow fallowing is most likely to be areas where farmers practice slash-and-burn cultivation in *Imperata* grasslands with manual cultivation and no access to fertilizer inputs. Farmers in these environments may benefit from higher yields and reduced labor in opening the land from *Imperata*, although this is partially offset by the added labor to cut the tree biomass prior to cropping. The system is less likely to be acceptable to farmers with access to animal power. In such situations the control of *Imperata* is manageable with tillage. Farmers with draft power are also less likely to invest their labor in hedgerow pruning than farmers that cultivate manually. Hedgerow fallow systems have received little research attention thus far. Further comprehensive systems analyses of this alternative fallow-rotation concept are needed.

CONSERVATION FARMING SYSTEMS FOR SLOPING LANDS

Conservation tillage is "any tillage system that reduces loss of soil or water relative to conventional tillage" (Lal, 1989). Slash-and-burn farmers were the initial adherents of conservation tillage. As population density increased, however, most farmers obtain an animal for draft power. This enables a household to intensively till a much larger area than is possible by hand hoe, at a fraction of the time and drudgery. They can plow and harrow frequently enough to control *Imperata* cylindrica and can hold their own against the annual grasses that invade frequently-tilled fields. Yet, as they gain the capacity to till their land more frequently, they exacerbate erosion, particularly since most dryland cultivators farm sloping fields. They find that retaining surface residue is impractical with animal power. So their clean tillage accelerates soil loss to typical levels of 50 to 200 tons per hectare per year (Sajjapongse and Syers, 1995), rapidly wasting their soil assets at a rate 10 to 20 times the maximum soil loss tolerance limit.

Clean tillage was the path toward higher yields for smallholders: It was the modern way to farm. But while it spread across the upland landscape, it accelerated soil loss at the farm and catchment level. Today, the sedimentation rates from Southeast Asian river systems are an order of magnitude higher than those of any other part of the world (Milliman and Meade, 1983). One way of conceptualizing the pathway toward conservation tillage is to look at it in relation to both crop productivity and the sustainability of the soil and water resource base (Figure 7.6). Productivity generally increased as farming evolved from shifting cultivation to intensive tillage agriculture, in terms of yield per unit agricultural area. But in the process the health of soil and water resources declined dramatically. The new intensive methods produced more, but jeopardize the resource base. Recent approaches to conservation tillage aim to rebuild the sustainability of the land resources while further improving, or at least maintaining, productivity and profitability. But there is a long way to go.

Smallholders who have farmed sloping lands with clean-tillage for some years are well aware of the threat of soil erosion, and are keen to learn about and apply conservation

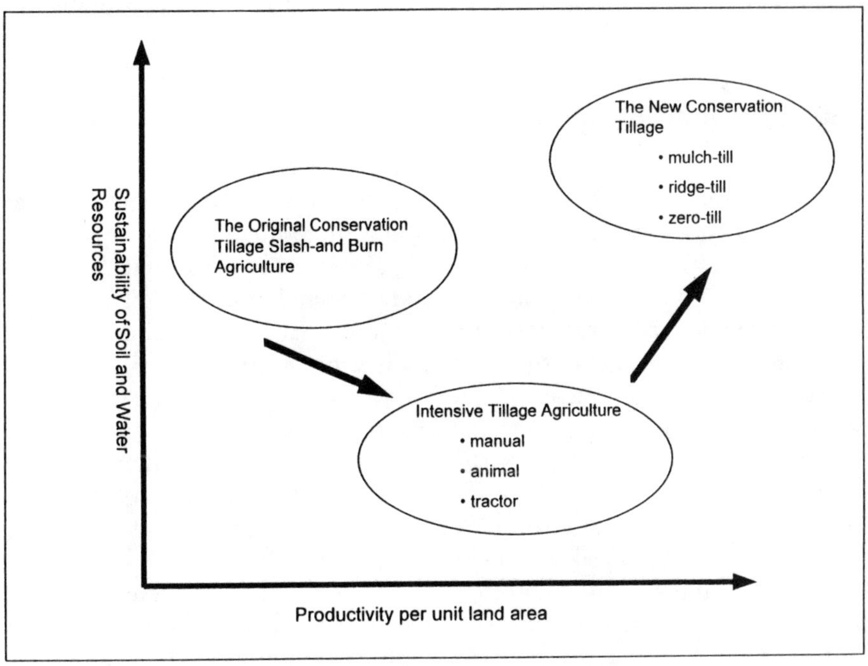

Figure 7.6. Historical evolution of upland tillage systems as related to productivity and the sustainability of soil and water resources.

measures (Fujisaka, 1993), as long as such methods are practical, within their very limited resources and labor. Unfortunately, most proposed methods are not practical, in the farmer's eyes, and are not adopted. But low-labor, low-investment practices that do the job of saving soil are eagerly awaited by small farmers. This section reviews some of the most promising directions toward providing conservation tillage practices that may make sense to Southeast Asian upland farmers.

Contour Hedgerow Systems

The main conservation farming practice prescribed for open-field intensive cultivation systems in Southeast Asia has been contour hedgerow systems (Garrity, 1995). Contour hedgerow farming with leguminous trees has thus become a common feature of extension programs for sustainable agriculture on the sloping uplands in southeast Asia. These systems control soil erosion effectively, even on steep slopes (Kiepe, 1995; Garrity, 1995) Data from the IBSRAM Sloping Lands Network trials in six countries have confirmed that annual soil loss with hedgerow systems is typically reduced 70 to 99% (Sajjapongse and Syers, 1995).

There are also numerous reports of increased yield levels of annual crops when grown between hedgerows of leguminous trees. However, farmer adoption of these systems is very low. Constraints include the tendency for the perennials to compete for growth resources and hence reduce yields of associated annual crops, and the inadequate amounts of phosphorus cycled to the crop in the prunings. But the major problem is the extra labor needed to prune and maintain them. We found that farmers' labor investment to prune their leguminous-tree hedgerows was about 31 days per hectare, or 124 days of annual labor for four prunings (ICRAF, 1996). This increased the total labor for an upland rice an average of 64%. Labor for a maize crop increased 90% due to pruning operations. Such an increase in production costs seldom was rewarded by a commensurate increase in returns. The extra labor didn't pay off.

The Case for Natural Vegetative Strips

In Claveria, Philippines, tree legumes and fodder grasses were tried and adopted by farmers during the first years of a research and farmer-to-farmer training project. Farmers that perceived soil erosion to be a problem were much more interested in vegetative barrier techniques that minimized labor (Fujisaka et al., 1994). Some farmers independently developed the practice of laying out contour strips that were left unplanted, and were revegetated by native grasses and forbs. Researchers found that these natural vegetative strips had many desirable qualities (Garrity, 1993). They needed much less pruning maintenance compared with fodder grasses or tree hedgerows, and offered little competition to the adjacent annual crops compared to the introduced species (Ramiaramanana, 1993). They were very efficient in minimizing soil loss (Agus, 1993). And they did not show a tendency to cause greater weed problems for the associated annual crops (Moody, 1992, pers. comm.). Natural vegetative strips (NVS) were also found to be an indigenous practice on a few farms in other localities, including Batangas and Leyte Provinces.

Installing NVS is quite simple. Once contour lines are laid out there is no further investment in planting materials or labor. The vegetative strips do not need to conform closely to the contour; they act as filter strips rather than bunds. Their biomass production, and economic value as fodder is lower than many other hedgerow options, but labor is minimized. *Vetiver* grass fills a similar niche as a low value-added but effective hedgerow species. But for *vetiver* or any other introduced hedgerow species the planting materials must be obtained and planted out, requiring extra labor.

One limitation of low maintenance hedgerows is that they do not enhance the nutrient supply to the crops. In this respect they do not differ from many other hedgerow enterprises, including fodder grasses or perennial cash crops like coffee. With continuous cropping, NVS or other low management hedgerow options can only be sustainable with fertilization. They have proved to be popular in northern Mindanao and have been adopted by hundreds of farmers in recent years.

Terrace Development and Soil Sustainability

One of the most-cited advantages of contour hedgerow systems is their tendency to create 'natural' terraces through a soil redeposition process. Visual evidence and field level soil loss measurements indicate that the soil moves within the alley, but not off the field. But there is clear evidence that the development of these terraces may cause a serious decline in crop productivity in the upper alleyway.

There have been few continuous observations of crop yields in contour hedgerow systems on slopes. But where these are reported, the yield response tends to become skewed after a few years (Garrity et al., 1995). In 5-year-old hedgerow systems of either leguminous trees (*Leucaena leucocephala* + *Cajanus cajan*) or grass (*Paspalum conjugatum*) on an Ustic Kandihumult of 21-35% slope in Thailand, Turkelboom et al. (1993) reported reductions in rice yields of greater than 50% in the upper part of the alleyways, compared to the middle and lower areas. Solera (1993) excluded both aboveground and belowground hedgerow competition by intensive pruning and installation of 50-cm deep plastic barriers. He found that the spatial variability of yields from the upper to the lower alleyways was still present. Surveys by Peden (1994, pers. comm.) and coworkers on hillsides in Uganda indicate that the effects may be quite long-lasting. They have documented cases of steep front-facing terraces on slopes >50% in which upper-alley crop productivity remains drastically lower than in the lower alleys, after cultivation for at least 50 years (ICRAF, 1994).

Soil Spatial Fertility Gradients

Soil may be very rapidly redistributed across the alleyways. Basri et al. (1990) reported a 60-cm drop in soil level between the alleys after 2 1/2 years on a field with 25 % slope. Fujisaka (1993) observed similar rates of terrace development on several farmers' fields in Claveria, Philippines. A tendency toward rapid terrace riser development was observed in trials on a range of other soil types (Sajjapongse, 1992). Rapid terrace development is associated with upper alley yield depressions in many cases. We now have evidence as to why. In a *Cassia spectabilis* hedgerow experiment (Garrity et al., 1995), soil organic carbon was found to vary from 1.7% near the hedgerow in the upper alley to 2.8% near the lower hedgerow. Available P was twice as high in the lower zone compared to the upper. Soil pH was unchanged but exchangeable aluminium increased. These patterns were observed in the subsoil (15-30 cm) as well as in the topsoil. Agus (1993) sampled 5 points across the alleyways between *Gliricidia* contour hedgerows, and observed a linear soil fertility gradient. Exchangeable calcium, organic C, pH, and Bray-2 extractable P increased from the upper to the lower alleyway, exchangeable Al decreased, while K and Mg were unchanged. Samzussaman's (1994) row-wise study of soil properties elucidated linear increases in organic carbon (from 2% to 3 %) and nitrogen (0.20% to 0.27%) on an Oxic Palehumult. Turkelboom et al. (1993) monitored soil organic carbon on an entire slope transect. They observed a "saw-tooth" pattern, with much higher SOM levels in the lowest part of the alleyway and under the hedgerow, and a tendency for these differences across the alleys to be accentuated on the terraces located furthest downslope.

Upper alley scouring, and rapid soil displacement to the lower alley seem particularly prominent where draft animal plow systems are used. But the scouring process also occurs with hand-hoe farming in Thailand on steep slopes (Turkelboom et al., 1993). However, it is not always observed, particularly when no-tillage is practiced continuously, or when hand-hoeing is done superficially (as in shifting cultivation). The effects of scouring raise serious issues concerning the short-term and long-term agricultural sustainability of natural terraces. We need to know whether scouring effects are a short-term or a long-term phenomenon, and how they can be effectively avoided or alleviated. There are three basic ways to try to cope with scouring: to avoid it, alleviate it, or change the cropping system to live with it. The research agenda for each is outlined below.

1. Adjust Biomass and Nutrient Inputs

To alleviate scouring it is logical to consider changing the spatial pattern of nutrient inputs. The application of crop residues, hedgerow prunings, animal manure, and fertilizers may be biased to the upper alley zones. Since scouring tends to redistribute topsoil nutrients downward naturally, application to the upper alley may ensure a more even distribution in the long run. In recent surveys we found that Claveria farmers who have experienced significant scouring have tried many of these options and some observed impressive success in rehabilitating their soils. Yields in the upper part of alleys once again approached those in lower zones (Stark, pers. comm.). We have initiated experiments to elucidate these fertility management issues, and gain a predictive knowledge about them.

2. Change the Cropping (or Hedgerow) System

Another strategy is to accept the scouring effect and to alter the cropping system to adjust to the changes (particularly if they are drastic). Substitution of perennials (e.g., bananas, timber trees, or MPTs) for annual crops, or more tolerant annuals (e.g., cassava), are practices that farmers are experimenting with. Researchers and farmers must learn how to better exploit the degraded upper alley zone that a conservation technology may create.

3. Reducing the Intensity of Tillage

To avoid scouring, a farmer must avoid frequent primary tillage. In zero-tillage systems there is little tendency for hedgerow risers to develop, or for soil scouring to occur. Unfortunately, they have not proved practical for annual crops in the tropics. But there are many variations of reduced tillage that smallholders can adopt. ICRAF is developing a practical ridge tillage system for smallholders. In ridge-tilled fields, unplowed strips are maintained indefinitely, and the annual crops are planted consistently in these strips. This keeps the soil in place.

A Ridge Tillage System for Animal-Draft Farming

As we searched for practical ways to farm hedgerow systems to avoid the movement of soil downslope by tillage, we examined the principles of permanent-ridge tillage as practiced in the United States. The principle of the permanent-ridge tillage system is to maintain alternate strips of untilled and tilled land in a row-cropped field: the untilled strip (the ridge) is where the crop is planted in the same exact row position in each successive season; the inter-row area is where cultivation is practiced for weed control and hilling-up is done.

As adapted to animal-powered systems, the sequence of operations is to make a shallow furrow through the stubble of the previous crop and plant in the same row; cultivate the crop with the moldboard plow in the standard way that maize farmers use in the Philippines (usually an off-barring operation at about 14 days after emergence, and hilling-up at about 28 days after emergence) to create the ridges; after harvest, replant the next crop through the stubble in the same rows as the previous one, without any plowing or harrowing operations. If there is considerable weed growth on the ridges, band-spray a broad-spectrum systemic herbicide (usually glyphosate is preferred) on the ridges to eliminate early weed competition. Inter-row cultivation controls weeds and rebuilds the ridges. The ridge-till field looks little different than a conventional row-crop field. The ridges are but slightly more pronounced than those in conventional hilled-up maize. They tend to slump down during and between crops. At planting, the ridges may be only slightly elevated above the base level of the inter-row furrows.

What are the hypothesized advantages of such a system? First, many studies in the U.S. have shown that the permanent ridge system dramatically reduces soil loss. The ridges act as a partial barrier to the surface flow of water, but their major distinction is that they act as a zone of greater infiltration. The no-tillage area tends to accumulate organic matter and macropores, due to soil biological activity and root channels. Since primary and secondary tillage operations are not practiced for land preparation between crops, the land is less subject to erosion in the off-season. Labor and expense in land preparation is eliminated. Preplant weed control is accomplished by judicious use of a herbicide.

We recently completed a four-year study of permanent-ridge till systems on two farmers' fields in Claveria, Philippines (Thapa et al., 1996). The soils were very fine kaolinitic clays classified as Lithic and Rhodic Hupludox, with strongly acidic surface soils (pH of 4.1 to 4.4), organic carbon contents of 1.6%. The conventional cropping system in this area is two crops of maize per year. Most cropland has slopes in the range of 10 to 60%. Annual rainfall is about 2000 mm. Farmers plow and harrow twice to prepare the land for each crop using a single draft-animal and moldboard plow. Planting is done by hand in furrows made with the plowpoint. Seeds, usually single-cross hybrids, are dibbled in the furrow and covered by foot. Inter-row cultivation is also done with the same moldboard plow. We used the same local equipment for the permanent-ridge till system.

In conventional maize production in Claveria, farmers generally plant on the contour, more or less. We compared the conventional system with ridge tillage, and with natural vegetative strips (NVS) laid out at an 8 m distance from each other, a difference of 1.5 meters in vertical drop. The fourth treatment was a combination of ridge tillage in the alleyways between natural vegetative strips. Hybrid maize was produced in all treatments in rows spaced 60 cm apart.

The slopes of the on-farm experiments were 14% at Patrocinio village and 20% in Anei. The annual soil loss was 85.5 tons/hectare on bare, uncropped soil. Ridge tillage reduced soil loss by 49% in Patrocinio, and 58% in Anei (Table 7.2). Natural vegetative strips reduced soil loss even more: 97% in Patrocinio and 90% in Anei. When the two conservation tillage systems were combined, annual soil loss was 0.3 t/ha in Anei, 1.1 t/ha in Patrocinio. Clearly, both systems were effective measures to dramatically reduce erosion. When combined, they proved exceedingly effective.

The permanent ridges in the ridge-till treatments had infiltration rates of 49 cm/hr, 30% higher than in the rows of the conventional treatment. Infiltration rates were also high for the zone just above the natural grass strips (59 cm/hr). The high infiltration rates

Table 7.2. Water- and Tillage-Induced Soil Erosion as Affected by Management System in Claveria, Misamis Oriental, Philippines[a]

Soil Mgt. Systems	1993	1994	1995	Mean
		Anei site (soil loss ton/ha)		
Contour plowing	26.7	26.1	17.1	23.4a
Ridge tillage (RT)	9.9	12.5	7.1	9.8 b
Natural grass strip (NGS)	2.2	1.8	2.5	2.2c
RT plus NGS	1.2	1.0	1.2	1.1c
Bare	0.0	75.2	95.7	85.5
CV% = 39.9				
		Patrocinio site (soil loss ton/ha)		
Contour plowing	6.0	10.0	5.7	7.2a
Ridge tillage (RT)	3.3	5.1	2.7	3.7b
Natural grass strip (NGS)	0.3	0.3	0.1	0.2c
RT plus NGS	0.4	0.3	0.1	0.3c
CV% = 75.9				

[a]From Thapa, Cassel, Garrity, and Mercado, 1996.

reduced runoff from row-to-row in the ridge-till system, and reduced runoff through the grass barriers in the NVS. Kiepe (1995) demonstrated that a much higher infiltration rate in the vicinity of contour hedgerows is the major factor explaining the exceptional ability of contour hedgerows systems to reduce runoff and off-field soil losses.

Mean grain yields of six crops over the three-year period were the same for the conventional system and ridge-till (3.9 t/ha and 3.8 t/ha, difference nonsignificant). The grass strips occupied 12.6% of the field area. This resulted in grain yields that were about 10% lower than those for the conventional system or the ridge-till system alone (Table 7.3). Thus, ridge tillage maintained maize yields, although it drastically reduced the amount of labor invested in tillage and weed control. It was not as effective as NVS in erosion control, but did not cost the farmer any penalty in crop yield compared to conventional maize production.

Did permanent-ridge tillage reduce downslope soil scouring from the upper alleyways to the lower alleyways? This was the issue that first prompted our interest in adapting this

Table 7.3. Corn Grain Yield as Affected by Soil Management System in Claveria, Misamis Oriental, Philippines[a]

	1993 Crop Season		1994 Crop Season		1995 Crop Season		
Soil Mgt. Systems	1st	2nd	1st	2nd	1st	2nd	Mean
Patrocinio site (grain yield ton/ha)							
Contour plowing	5.6	3.8	5.0	2.8	3.4	3.7	4.1
Ridge tillage (RT)	5.4	3.7	5.2	3.0	3.3	4.5	4.2
Natural grass strip (NGS)	4.9	3.6	5.6	2.7	3.0	3.4	3.7
RT plus NGS	5.2	3.4	4.4	2.8	3.1	4.1	3.8
Standard deviation = 0.9 - 1.0							
Anei site (grain yield, ton/ha)							
Contour plowing	4.4	3.0	4.4	2.1	2.8	4.0	3.8
Ridge tillage (RT)	5.5	2.8	4.5	1.6	2.8	4.0	3.5
Natural grass strip (NGS)	4.4	2.7	3.8	1.9	2.6	3.6	3.4
RT plus NGS	4.0	2.6	3.8	1.4	2.5	4.1	3.1
Standard deviation = 0.9–1.1							
CV% = 75.9							

[a]From Thapa, Cassel, Garrity, and Mercado, 1996.

unique tillage system to animal-powered farming systems. Preliminary examination of the data suggests that ridge tillage did indeed dramatically slow down the downslope movement of soil across the alley ways. This would slow the rate at which topsoil is removed on the upper alleyways and is deposited above the grass strip downslope. The practice, therefore, appears promising. It could be the most effective way to manage contour hedgerow systems on land with quite shallow soils, such as in areas where the topsoil is often only 20–30 cm deep over limestone rock, as in some parts of Southeast Asia. Under these conditions the scouring effect can be devastating to long-term crop production if hedgerows are installed.

Weed control is the key challenge in converting to a permanent-ridge till system. In superimposed trials we tested various methods of weed management that compared the use of glyphosate with nonherbicide alternatives. In this humid tropical environment there is profound potential for weed regrowth in the period between wet-season crops, and during the dry season. We found that the only practical way to plant on permanent ridges without prior tillage was to use glyphosate herbicide before planting. Fortunately, it is one of the most environmentally safe agricultural chemicals.

Will small farmers adopt herbicides in the future as part of their crop production system? No one knows the full answer, but it is clear that many already are using them. Many farmers in Claveria now use glyphosate and other herbicides for specific applications to complement their animal-powered weed management, especially where it reduces manual labor. Our studies in Indonesia have shown that many upland farmers are using glyphosate in opening land that has been in grass fallow. As discussed above, even villages of shifting cultivators are integrating herbicides into their crop production system to expand the potential to use *Imperata cylindrica*-infested land, and to save labor in cultivation. The cost of labor relative to other production factors is expected to continue to increase. Therefore, it is quite likely that use of herbicides by upland farmers will accelerate in the coming decade. Such a trend would parallel the experience in wetland rice in the Philippines, where herbicide use was unknown in the early 1970s, but was ubiquitous by the late 1980s.

CAPACITY-BUILDING FOR NATURAL RESOURCES MANAGEMENT RESEARCH

Research on natural resource management requires skills in systems thinking, and new research methods and perspectives are essential. ICRAF's training activities in Southeast Asia focus on capacity-building through hands-on research conducted jointly with national program colleagues. We also host graduate students working for degrees at universities throughout the region, and some working at advanced institutes in western countries. We currently have 12–15 students working with our scientists at any one time. This number is expected to increase to 25–30 during the next two years.

Our more formal training courses tend to focus on the specialized research skills and methods associated with our research themes. For the 2-year period from 1994 to 1996, we conducted courses on soil organic matter modeling, soil biodiversity research methods, assessing the effects of global change on complex agroecosystems, and rapid rural appraisal methods for land use intensification, among others. We plan to begin conducting the Agroforestry Research for Development Course in the region on a regular basis in 1998. We are also discussing with IRRI, SEARCA, and other institutions, a joint annual course on natural resource management research and development for upland ecosystems. We collaborate closely with the Asia-Pacific Agroforestry Network (APAN), an FAO-supported activity involving 10 countries, in a wide range of agroforestry training and information activities. Together with major universities in four countries, we are working to establish a new network to support agroforestry education.

CONCLUSION

This chapter reviewed ICRAF's current agroforestry research thrusts in the context of a broader concept of agroforestry as a natural resource management system that involves the increasing integration of trees into the agricultural landscape. The discussion emphasized work on both policy and technology in our three key ecosystems: the forest margins, *Imperata* grasslands, and sloping permanent farmlands. Agroforestry is a highly integrative field on the interface between the agricultural, forestry, social, and environmental sciences. We actively seek to strengthen and expand our collaborative partnerships with research and development institutions to tackle these problems as holistically and comprehensively as possible.

On the forest margins we work on systems based on smallholder agroforests for communities located in production forest environments, where forest conversion is inevitable or has already occurred. We are also developing agroforestry systems to intensify land use outside the boundaries of protected forests, in combination with enforcement and other development measures. Appropriate forms of secure land and tree tenure are a key foundation to better land use. Case studies with affected communities are building feasible options for tenure policy reform. Research on the ecology and improved productivity of complex agroforests supports the expansion of these systems to stabilize shifting cultivation, and to improve the sustainability of permanent farming systems. Comparative environmental benefits of agroforests vis-à-vis natural forests and other land uses are being assessed. In the *Imperata* grasslands we are emphasizing the identification, refinement, and extrapolation of indigenous strategies to intensify shifting cultivation. Smallholders practicing slash-and-burn in natural grass fallows are beginning to integrate herbicides in their farming system to manipulate the fallow vegetation from undesirable species (viz., *Imperata*) toward more desirable ones (viz., *Chromalaena*). Research and development could build upon this innovation to enable many other farmers to enjoy more productive fallows that better sustain their livelihood, and eliminate the need to abandon the land due to weed pressure. Leguminous tree fallow systems have been developed by some farming communities. We are investigating hedgerow fallow rotation systems for low-input grassland farming. Contour hedgerow systems have proven potential to control soil losses in crop production even on fairly steep slopes (20–50%). A promising low-labor farmer-developed solution is natural vegetative strips on the contour. As these gain widespread acceptance in the southern Philippines we are investigating ways to avoid or alleviate testing farmer-to-farmer methods for rapid dissemination. We recently adapted permanent-ridge tillage to animal-power systems in Southeast Asia.

REFERENCES

Agus F. Soil Processes and Crop Production under Contour Hedgerow Systems on Sloping Oxisols. Ph.D. dissertation, North Carolina State University, Raleigh, NC, USA, 1993.

Basri, I.H., Mercado, A.R. Jr., and Garrity, D.P. Upland Rice Cultivation Using Leguminous Tree Hedgerows on Strongly Acid Soils. IRRI Saturday Seminar Paper, March 31. Agronomy, Plant Physiology, and Agroecology Division, International Rice Research Institute, Los Banos, Philippines, 1990.

Cairns M., Murniati, Otsuka, M. and Garrity, D.P. 1997. Characterization of the Air Dingin-Muara Labuh Area of the Kerinci Seblat National Park: Farm and National Park Interactions,

In *Proceedings of the Workshop on Alternatives to Slash and Burn*, Central Research Institute for Food Crops, Bogor, Indonesia, June, 1995, 1997, pp. 135–172 (in press).

Cairns, M. Shifting Cultivation on the Perimeter of a National Park. M.Sc. thesis, York University, Toronto, Canada, 1994.

Cairns, M. Ancestral Domain and National Park Protection: A Mutually Supportive Paradigm? International Center for Research in Agroforestry, Southeast Asian Regional Research Program, Bogor, Indonesia, 1995.

Cairns, M. and Garrity, D.P. Indigenous Fallow Systems in Southeast Asia: New Research Attention on Promising Old Technic. Paper presented at the International Workshop on Improved Fallow Systems: Principles and Practice, Lilongwe, Malawi, March, 1997.

de Foresta, H. and Michon, G. The Agroforest Alternative to *Imperata* Grasslands. In Garrity D.P., Ed. Agroforestry Innovations to Rehabilitate *Imperata* Grasslands. Agroforestry Systems special issue (in press), 1997.

de Foresta, H. and Schwartz, D. *Chromalaena oderata* and Disturbance of Natural Succession after Shifting Cultivation: An Example from Mayonbe, Congo, Central Africa, In R. Nuniappau and P. Ferrar, Eds. Ecology and Management of *Chromalaena odorata*. BIOTROP Spec Publ. No 44:23–41, 1991.

Dove, M.R. The Practical Reason of Weeds in Indonesia: Peasant vs State Views of *Imperata* and *Chromalaena. Human Ecol.* 14:163–190, 1986.

Fujisaka, S., Jayson, E., and Dupasala, A. Trees, Grasses, and Weeds: Species Choice in Farmer-Developed Contour Hedgerows. *Agroforestry Sys.* 25:13–24, 1994.

Fujisaka, S. A Case of Farmer Adaptation and Adoption of Contour Hedgerows for Soil Conservation. *Exp. Agric.* 29:97–105, 1993.

Garrity, D.P., Ed. Agroforestry Innovations for *Imperata* Grassland Rehabilitation. Agroforestry Systems Special Issue (in press), 1997.

Garrity, D.P. and Agustin, P.C. Historical Land Use Evolution in a Tropical Acid Upland Agroecosystem. *Agric., Ecosys. Environ. J.* 53:83–95, 1995.

Garrity, D.P. and Mercado, A. Reforestation through Agroforestry: Market-driven Small-holder timber Production on the Frontier. In Raintree, J.B. and Francisco, H.A., Eds. *Marketing of Multipurpose Tree Products in Asia,* Winrock International, Morrilton, AR, USA, 1994, pp. 265–268.

Garrity, D.P., Soekardi, M., Van Noordwijk, M., de La Cruz, R., Pathak, P.S., Gunasena, H.P.M., So, N. van, Huijin, Guo and Majid, N.M. The *Imperata* Grasslands of Tropical Asia: Area, Distribution, and Typology. In Garrity, D.P., Ed. Agroforestry Innovations to Rehabilitate *Imperata* Grasslands. Proceedings International Workshop on Agroforestry Innovations for *Imperata* Grassland Rehabilitation. Banjarmasin, January 1995. Agroforestry Systems Special Issue (in press), 1997.

Garrity, D.P., Mercado, A. Jr., and Solera, C. The Nature of Species Interference and Soil Changes in Contour Hedgerow Systems on Sloping Acidic Lands. In Kang, B.T., Ed. Proceedings of the International Conference on Alley Farming, International Institute of Tropical Agriculture, Ibadan, Nigeria, 1995.

Garrity, D.P. Conservation Tillage: A Southeast Asian Perspective. International Center for Research in Agroforestry, Southeast Asian Regional Research Program, Bogor, Indonesia, 1997.

Garrity, D.P. Sustainable Land-Use Systems for Sloping Uplands in Southeast Asia. In Technologies for Sustainable Agriculture in the Tropics. American Society of Agronomy Special Publication 56, Madison, WI, USA, 1993, pp. 41–66.

Garrity, D.P. Improved Agroforestry Technologies for Conservation Farming: Pathways Toward Sustainability. In Proc International Workshop on Conservation Farming for Sloping Uplands in Southeast Asia: Challenges, Opportunities and Prospects. IBSRAM, Bangkok, Thailand. Proceedings No. 14, 1995, pp. 145–168.

Garrity, D.P. Buffer Zone Management and Agroforestry: Some Lessons from a Global Perspective. International Center for Research in Agroforestry, Southeast Asian Regional Research Program, Bogor, Indonesia, 1995.

Glynn, C. Overcoming Constraints to Agroforestry Adoption in Tropical Highlands: Part I: An

Investigation of Performance by Elevation Patterns for Some Commonly Grown Timber Species in the Manupaly Watershed, Bukidnon, Philippines. M.Sc. thesis, Tropical and Subtropical Horticulture and Crop Science, Wye College, University of London, 1996.

Gouyon, A., de Foresta, H., and Levang, P. Does "Jungle Rubber" Deserve its Name? An Analysis of Rubber Agroforestry Systems in Southeast Sumatra. *Agroforestry Sys.* 22, pp. 181–206, 1993.

ICRAF. Annual report for 1993. International Center for Research in Agroforestry, Nairobi, Kenya, 1994.

ICRAF. Annual report for 1995. International Center for Research in Agroforestry, Nairobi, Kenya, 1996

Kiepe, P. No Runoff, No Soil Loss: Soil and Water Conservation in Hedgerow Barrier Systems. Wageningen Agricultural University, Wageningen, The Netherlands, 1995.

Lal, R. Conservation Tillage for Sustainable Agriculture: Tropics Versus Temperate Environments. *Adv. Agron.* 42:85–197, 1989.

Lal, R. Ridge-Tillage. *Soil Tillage Res.* 18:107–111, 1990.

Leakey, R. Definition of Agroforestry Revisited. *Agroforestry Today* 8(1):5–6, 1996.

MacKinnon, J., MacKinnon, K., Child, G., and Thorsell, J. Managing Protected Areas in the Tropics. International Union for Conservation of Nature and Natural Resources (IUCN), Gland, Switzerland, 1986.

Mary, F., and Michon, G. When Agroforests Drive Back Natural Forests: A Socio-Economic Analysis of a Rice/Agroforest System in South Sumatra. *Agroforestry Sys.* 5, pp. 27–55, 1987.

Michon G., de Foresta, H., and Aliadi, A. An Agroforestry Strategy for the Re-Appropriation of Forest Resources by Local Communities: The Case Study of Damar Agroforests in West Lampung, Sumatra. International Center for Research in Agroforestry, Southeast Asian Regional Research Program, Bogor, Indonesia, 1995.

Milliman, J.D., and Meade, R.H. World-Wide Delivery of River Sediment to the Oceans. *J. Geol.* 91:1–21, 1983.

Momberg, F. Indigenous Knowledge Systems. Potentials for Social Forestry Development: Resource Management of Land-Dayaks in West Kalimantan. M.Sc. thesis, Technische Universitat Berlin, Germany, 1993.

Murniati, Garrity, D. P., and Gintings, A. N. The Contribution of Mixed Gardens and Community Forest to Farmers' Reduced Reliance on Extraction from a National Park. *Agroforestry Sys.* (submitted), 1997.

Penot, E. Sustainability Through Productivity Improvement of Indonesian Rubber-Based Agroforestry Systems. International Center for Research in Agroforestry, Southeast Asian Regional Research Program, Bogor, Indonesia, 1996.

Ramiaramanana, D.M. Crop-Hedgerow Interactions with Natural Vegetative Filter Strips on Sloping Acidic Land. M.Sc. thesis, University of the Philippines at Los Banos, 1993.

Sajjapongse, A. and Syers, K. Tangible Outcomes and Impacts from the ASIALAND Management of Sloping Lands Network. In *Proc International Workshop on Conservation Farming for Sloping Uplands in Southeast Asia: Challenges, Opportunities and Prospects.* IBSRAM, Bangkok, Thailand. Proceedings No. 14, 1995, pp. 3–14.

Sajjapongse, A. Management of Sloping Lands for Sustainable Agriculture in Asia: An Overview. In *Technical Report on the Management of Sloping Lands for Sustainable Agriculture in Asia, Phase I (1988–1991).* Network Document 2, International Board for Soil Research and Management, Bangkok, Thailand, 1992, pp. 1–15.

Salafsky, N. The Forest Garden Project: An Ecological and Economic Study of a Locally Developed Land-Use System in West Kalimantan, Indonesia. Ph. D. thesis, Duke University, Durham, North Carolina, USA, 1993.

Samzussaman, S . Effectiveness of Alternative Management Practices in Different Hedgerow-Based Alley Cropping Systems. Ph.D. dissertation, University of the Philippines at Los Banos, 1994.

Sanchez, P.A. Science in Agroforestry. *Agroforestry Sys.* 30:5–55, 1995.

SANREM. 1995 Annual Report. Sustainable Agriculture and Natural Resources Management Collaborative Research Support Program, University of Georgia, Griffin, 1996.

Solera, C.R. Determinants of Competition Between Hedgerow and Alley Species in a Contour Hedgerow Intercropping System. Ph.D. dissertation, University of the Philippines at Los Banos, 1993.

Stoutjesdijk, J.A.J.H. "*Eupatorium pallescens* DC op Sumatra's westkust" (*Eupatorium pallescens* DC on the West Coast of Sumatra). *Tectona* 28:919–926, 1935.

Szott, L.T., Palm, C.A. and Sanchez, P.A. Agroforestry in Acid Soils of the Humid Tropics. *Adv. Agron.* 45:275–300, 1991.

Thapa, B., Cassel, D.K., Garrity, D..P, and Mercado, A.J. Sustainable Soil Management Systems on Acid Steepland Soils in the Humid Tropics. Paper presented at the 12th Annual Scientific Conference of the Federation of Crop Science Societies of the Philippines, Davao, 13-18 May. International Centre for Research in Agroforestry, Philippines Program, Los Banos, Laguna, Philippines, 1996.

Thiollay, J.M. The Role of Traditional Agroforests in the Conservation of Rain Forest Bird Diversity in Sumatra.*Conserv. Biol.* 9 (2), pp. 335–353, 1995.

Thomas, D.E. Opportunities and Limitations for Agroforestry Systems in the Highlands of North Thailand. International Center for Research in Agroforestry, Chiangmai, Thailand, 1995.

Tomich, et al. *Imperata* Economics and Policy. In Garrity, D.P., Ed. Agroforestry Innovations for *Imperata* Grassland Rehabilitation. Agroforestry Systems Special Issue (in press), 1997.

Torquebiau, E. Man-Made Dipterocarp Forest in Sumatra. *Agroforestry Sys.* 2(2):103–128, 1984.

Torquebiau, E. Are Tropical Home Gardens Sustainable? *Agric. Ecosys. Envir.* 41:189–207, 1992.

Turkelboom, F., Ongprasert, S., Taejajai, U. Alley Cropping on Steep Slopes: Soil Fertility Gradients and Sustainability. Paper presented at the International Workshop on Sustainable Agricultural Development: Concepts and Measures, 1993.

van Noordwijk, M., Hairiah, K., Partoharjono, S., Labios, R.V. and Garrity, D.P. Food-Crop Based Production Systems as Sustainable Alternatives for *Imperata* Grasslands? In Garrity, D.P., Ed. Agroforestry Innovations for *Imperata* Grassland Rehabilitation. Agroforestry Systems Special Issue (in press), 1997.

van Noordwijk, M., Tomich, T.P., Winahyu, R., Murdiyarso, D., Partoharjono, S. and Fagi, A.M., Eds. Alternatives to Slash-and-Burn in Indonesia. Summary Report of Phase 1. ASB-Indonesia Report Number 4. International Center for Research in Agroforestry, Southeast Asian Regional Research Program, Bogor, Indonesia, 1995a.

van Noordwijk, M., van Schaik, C.P., de Foresta, H. and Tomich, T.P. Segregate or Integrate: Nature and Agriculture for Biodiversity Conservation. International Center for Research in Agroforestry, Bogor, Indonesia, 1995b.

Wells, M. and Brandon, K. People and Parks: Linking Protected Area Management with Local Communities. The World Bank, Washington, D.C., 1992.

Wind, J. and Prins, H.H.T. Buffer Zone and Research Management for Indonesian National Parks: Inception Report. World Bank National Park Development Project. DHV/RIV Consultancies, Bogor, Indonesia, 1989.

EIGHT

The Biological Management of Tropical Soil Fertility: Some Research and Development Priorities for Africa

Paul L. Woomer, Jean Kotto-Same, Mateete A. Bekunda, and J. Robert Okalebo

INTRODUCTION

Deforestation, land degradation and nutrient depletion present a challenge to agricultural planners throughout sub-Saharan Africa, one that if not met over the next generation will threaten the food security and political stability of the region. Complicating these issues are the past difficulties in formulating agricultural research and development projects which successfully impact upon the livelihoods and lands of smallhold agriculturalists. There are several root causes of these difficulties. The "Green Revolution" may be said to have passed Africa by (Okigbo, 1990), especially in the areas of developing infrastructure to produce agricultural supplies and to create the market-based rural societies described by Harwood (1990). International research organizations appear adverse to the testing and promotion of technologies which others have pioneered. Structural readjustment policies intended to liberalize markets and reduce governmental controls upon trade have also resulted in the removal of subsidies and price controls which benefited many smallhold farmers. National governments frequently place priorities upon commodities, both food and export, and target policies toward larger producers. Agricultural extension remains weak, with officers placed in isolation. Many local and international nongovernmental organizations specialize, both geographically and in terms of technologies disseminated, resulting in strong but scattered impacts. The farmers themselves remain trapped by political and economic difficulties, seemingly content when not harassed by police or rival tribes while remaining able to feed their families and fulfill social obligations. Yet within this disopportune setting, real opportunity for progress is possible at the farm and community levels.

One example of such progress is described by Tiffen et al. (1994) for Machakos, in Eastern Kenya. Population growth and colonial policies forced the Akamba people to practice permanent agriculture on defined land parcels rather than a mixture of shifting cultivation and pastoralism, as was done in the past. At the onset of this change, landscapes were denuded by fuel wood collection, overgrazing, and soil erosion. And then, despite expectations to the contrary, an amazing transformation took place. Population growth resulted in stronger local markets, acted as a magnet for infrastructural and educational improvement and provided a stronger pool of agricultural laborers. Smaller herds of better managed livestock were maintained. Erosion control structures were established. Trees were planted. Local industries emerged based upon traditional woodcarving and weaving skills. Living standards and environmental quality correspondingly improved except for one facet, more people resulted in greater rates and coverage of soil nutrient depletion and a resultant decline in crop performance. As a result of this experience, addressing nutrient depletion in the semiarid croplands of Eastern Kenya has become a research priority for two International Agricultural Research Centers, the Kenya Agricultural Research Institute, and several grassroots organizations, and progress is being made in better managing limited nutrient resources. An important point in the choice of Machakos as an example is that the first step to agricultural and environmental problem-solving was made by the farmers themselves and subsequent research and development activities served to backstop a community, instead of presuming to lead it.

But alas, it appears the seemingly insolvable disasters-in-the-making better attract international media and our own fatiguing attention. To name a few, the nationwide insanity resulting from two different tribes desiring the same high potential lands in Rwanda, social and environmental devastation in Mozambique, and mass starvation resulting from poor rains in Ethiopia, and prolonged civil wars in Somalia, Sudan and Angola. Each of these human disasters has become the focus of major international relief efforts of varying, and occasionally disappointing, success. One might conclude that the less the likelihood that an outside agency is able to positively impact upon disintegrating social fabric and ecologies, the greater the funds spent in trying to do so. But less dramatic, more easily reconciled problems abound in Africa, often unsuccessfully competing for attention with their larger, better publicized counterparts. We offer three real opportunities for positive agricultural research and development impacts in Africa, specifically, the development of alternatives to current slash-and-burn agriculture in Central Africa, the amelioration of *matoke* (banana) decline in Uganda, and nutrient replenishment in the densely settled highlands of East Africa.

IMPROVED LAND CONVERSION IN CAMEROON

Tropical deforestation advances at a rate of 14 million ha per year (Myers, 1989) and is the source of approximately 1.6 Gt of carbon per year lost to the atmosphere accounting for 23% of global CO_2 emissions (Post et al., 1990). Degraded lands now occupy approximately 600 million ha of the original 1500 million ha of humid tropical forests (Sanchez et al., 1994). Tropical deforestation poses a large threat to plant and animal biodiversity, water quality, and the stability of indigenous lifestyles (Sanchez, 1990). Slash-and-burn agriculture as practiced by indigenous peoples in areas of low population pressure allowed for forest regrowth intervals sufficient to reaccumulate system carbon and nutrients (Nye and Greenland, 1960), reestablishment of plant and animal populations and no discernible environmental degradation. But as population pressure increases and lands become less available, fallow intervals shorten, crop productivity decreases, and land degradation ensues. Agenda 21, Chapter 11 recognizes the role of slash-and-burn agriculture in combatting deforestation as "limiting and aiming to halt destructive shifting cultivation by addressing the underlying social and ecological causes."

Policy options aimed at reducing the destructive effects of slash-and-burn agriculture include educational programs to slow population growth, deflection of internal migrants to less fragile ecosystems and the creation of additional parks and protected areas. These options are far-reaching in that it reduces the pressures of deforestation and destructive

Table 8.1. Carbon Dynamics in the Primary Forest in Southern Cameroon (1973–1988)[a]

Total land area	113,531 km^2
Total primary forest (1973)	79,255 km^2
Remaining primary forest (1988)	59,235 km^2
Primary forest loss (1973-1988)	25.3%
Annual deforestation rate	1,335 km^2 yr^{-1}
Total carbon loss (1973-1988)	202,404,900 tons
Average carbon loss	102 tons ha^{-1}
Average carbon loss rate	6.8 tons ha^{-1} yr^{-1}

[a]Analysis conducted by ICRAF GIS Laboratory based on CENADEFOR (1988) map of Cameroon and Kotto-Same, Appolinaire and Woomer (1996) estimates of forest carbon stocks.

slash-and-burn agriculture in the future but in fact these policies do not address the problems encountered by shifting cultivators on a daily basis. How can farmers better manage their lands and forests allowing for extended cropping intervals, greater biomass storage on their lands and protection of useful indigenous species? Brown et al (1993) described several field-level mechanisms with the potential to conserve and sequester carbon in managed and converted forests. The authors cite working with shifting cultivators to implement new agricultural practices as a means of decreasing pressure on remaining forests, reclaiming degraded forests and conserving C resources in lands and soils undergoing conversion to agriculture as a challenging and necessary area of research.

Southern Cameroon is a testing ground for reducing primary forest destruction and developing alternatives to slash-and-burn agriculture. Between 1973 and 1988, 20,020 km^2 of primary humid forest were destroyed, an annual rate of 1.67% which if left unchecked will eliminate remaining unprotected forests within 60 years (Table 8.1). Based on the carbon stocks within the predominant land uses in Southern Cameroon (Kotto-Same et al, 1996), and the changes in those stocks that land uses change, we estimate that 202,404,900 tons of carbon were lost by those primary forests. This rate is equivalent to 102 t ha^{-1} over 15 years or 6.8 t ha^{-1} yr^{-1}.

The implications of complete deforestation and accelerated atmospheric charge are ominous, but not inevitable. Based upon the total system C contained in the croplands, fallows and forests of known ages (Figure 8.1), a mathematical model of C-reaccumulation in Southern Cameroon was identified. On average, 88 t ha^{-1} of C remained in the system during the cropping interval and increased at a rate of 7.7 t C ha^{-1} yr^{-1} following land abandonment. Thus, the rate of carbon reaccumulation is potentially greater than its overall estimated loss.

However, when forests are converted into continuous agriculture, the potential to sequester carbon during future fallow intervals is lost. We assert that an important, alternative mechanism of carbon sequestration is to protect selected C pools at the time of land conversion.

A mechanism is now in place to conduct "best bet" alternative land management research in the humid forest zone of Cameroon. Several international research groups, including the Tropical Soil Biology and Fertility Programme (TSBF), have linked their efforts with national research systems as part of the Alternatives to Slash and Burn Program (ICRAF, 1994). During the past two years, scientists from IRAD, The University of Yaounde, and TSBF have identified representative sites in farmers' fields, characterized the total system C in original forests and adjacent croplands, and developed protocols for limiting nutrient, system carbon and biodiversity assessment. We now stand

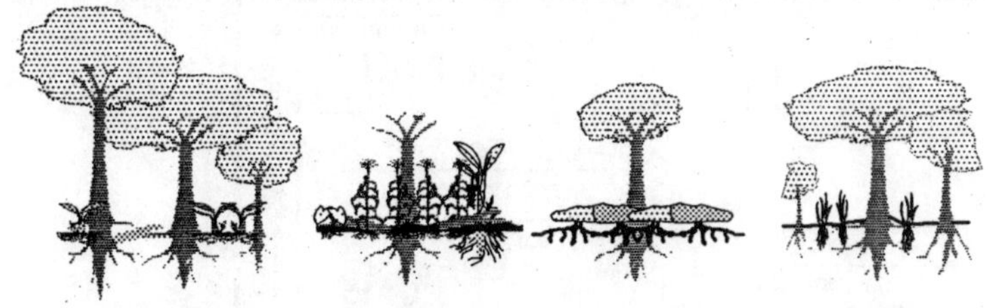

Land Use	Original Forest	Slash & Burn Agriculture	Young Fallow *(Chromolaena sp.)*	Secondary Forest
Age	>160 yrs (year 0)	year 1 to 3	year 4 to 6	year 22
Carbon stock	307 t C/ha	88 t C/ha	87 t C/ha	227 t C/ha
Soil Carbon	76 t C/ha	69 t C/ha	67 t C/ha	77 t C/ha
Plant Biomass	502 t/ha	39 t/ha	42 t/ha	322 t/ha
Plant Biodiversity Indices	1.0 (species) 1.0 (families)	0.97 1.16	0.58 0.80	0.95 1.0

Figure 8.1. Carbon stocks, biomass and biodiversity within the slash-and-burn land use chronosequences of Southern Cameroon.

ready to establish and evaluate "best bet" land management strategies that have evolved from our studies by working closely with slash-and-burn farmers. Briefly, our approach will conserve vegetation carbon by identifying the largest and most valuable trees in the original forest and protecting these during land conversion, fell other trees along the contour to reduce soil erosion, to identify and supplement the nutrient most limiting crop production (Woomer et al, 1993) by addition of farmer-available organic resources (Palm, 1995), and to establish field trials and demonstration plots of mound-based cultivation of root crops (Figure 8.2).

BANANA LAND REHABILITATION IN UGANDA

Banana (*Musa* sp.) is the most important crop in the economy and food security of Uganda. The Ugandan Ministry of Agriculture reports that over 75% of the farming population and 70% of urbanites rely upon cooking banana (*matoke*) as their staple food source (MAAIF, 1992). Bananas and plantains occupy about 1.2 million hectares with an average fruit yield of 5.58 T ha^{-1} yr^{-1} (Rubaihayo, 1992). Uganda produces approximately 9.2 million metric tons of bananas and plantains (*Musa sp.*) annually (FAO, 1995a) and is the world's largest producer and consumer nation (Stover and Simmonds, 1987). Per capita consumption is estimated at 220-460 kg yr^{-1} (Hartman, 1989). In addition to accounting for the largest cultivated land area in the country, the crop provides income and employment to a considerable number of Ugandans. Given the importance of banana production to the economy and people of Uganda, and declining trends in the use of chemical fertilizers within the country over the past 15 years (Gerner and Harris, 1993), improved understanding of soil fertility management through the use of farmer-available organic resources occupies a high priority in the agricultural research agenda of Uganda (Bekunda and Woomer, 1996).

Banana is a giant, perennial herb with large demands for water and nutrients, particu-

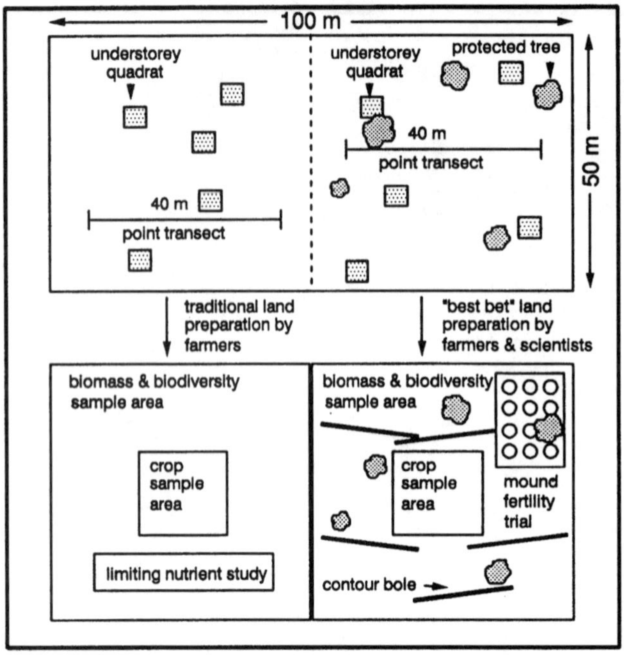

Figure 8.2. Generalized experimental procedure combining original forest, traditional slash-and-burn, and "best bet" alternatives.

larly potassium (Purseglove, 1972). The nutrient requirements of banana are sufficiently great that productive bananas may be regarded as an indicator of fertile soils. Conversely, the opposite is also the case in many smallholdings in Uganda; reduced banana productivity has become symptomatic of severe nutrient depletion. Bwamiki et al (1996) report that banana yields have steadily declined from 8.4 t ha^{-1} in 1971 to approximately 0.5 t ha^{-1} in Mukono District, east of Kampala. Average banana bunch yields of less than 10 kg were reported in Masaka, west of Kampala (Bekunda and Woomer, 1996), while more distant and younger plantations to the north report average bunch weight of >30 kg.

Researchers from Makerere University, the National Agricultural Research Organization, TSBF, IITA, and CIAT, with financial assistance from the Rockefeller Foundation, have joined forces to investigate *matoke* decline in Uganda . To date, extensive household surveys were conducted and resource user groups monitored (Bekunda and Woomer, 1996), banana varieties inventoried, pests identified and integrated pest management experiments initiated, nutrient addition experiments conducted, and farmer-available organic resources characterized. Focus has shifted to better conserving currently productive banana lands as an immediate objective. But we contend that a sufficient knowledge base exists to design "best bet" on-farm experiments targeted at regeneration of severely degraded lands.

We view *matoke* decline as a syndrome resulting from differential degrees of soil fertility decline and intensification of pests and diseases. While it is unclear whether or not plant nutrition and infestations are physiologically interactive, it is apparent that both constraints are operative. One interaction between the two constraints appears when sanitation measures are performed in infested fields. Banana weevils and stem nematodes present in pseudostems at the time of bunch harvest may be physically removed from the banana plantation by using stems and leaves for other purposes, such as mulch in adjacent, nonbanana fields, and as livestock feed. This action stands in contrast to the "farmer traditional practice" of cutting the pseudostem and leaves into large pieces and mulching

Table 8.2. Nutrient Content of Napier Grass, Cattle Manure, Maize Stover, Bean Trash, and Banana Residues

Resource	Nitrogen	Phosphorus	Potassium	Calcium
		%		
Napier grass leaves	1.97	0.14	3.85	0.31
Cattle manure	2.72	0.50	2.92	—
Maize stover	0.67	0.09	0.58	0.25
Bean trash	0.77	0.06	1.31	0.73
Banana stem	0.91	0.06	7.36	—
Banana leaves	2.75	0.12	4.85	—

the mat with them. Our research demonstrates that the farmer traditional practice is no longer as widespread, in large part due to a developing appreciation of pest control measures. However, removal of banana residues has profound impacts upon nutrient cycling as stems and leaves of banana contain large amounts of nitrogen and potassium (Table 8.2). Given the paucity of fertilizer use in Uganda (FAO, 1995b; Tukahirwa, 1992), soil nutrient depletion may be accelerated by pest control measures. Regardless of the set of circumstances driving declining banana yields, we see two basic routes to combatting "matoke decline." Firstly, areas that remain productive may be better conserved through the establishment of soil erosion control, external nutrient inputs, and adoption of integrated pest management (pesticides, lures, and field sanitation). But another, more pressing issue is the rehabilitation of fields that have undergone yield decline (Figure 8.3).

We operate under the assumption that a component of banana yield decline is related to soil fertility depletion. A corollary to Leibig's Law of the Minimum states that the unavailability of one condition necessary for growth is limiting growth and the rate of growth will rise and fall dependent upon that condition's supply until sufficient amounts are available and another condition becomes limiting (Russell, 1976). Simply stated, we have not yet identified that condition for the *matoke* decline areas in Uganda and because of our ignorance we are unable to formulate resource management strategies of use to

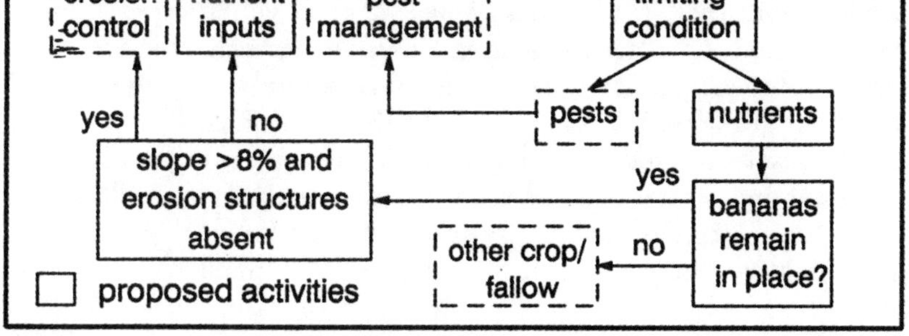

Figure 8.3. Two complementary approaches to combating matoke *decline in Uganda. Greater progress is now being made in better conserving productive banana lands.*

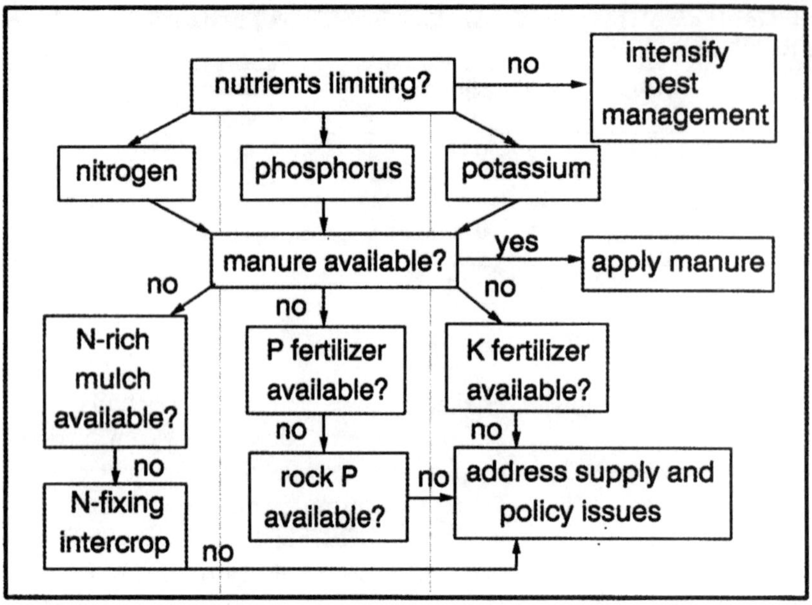

Figure 8.4. A fertility management decision tree based on knowledge of plant limiting nutrients.

farmers. But, let us assume that one of the three macronutrients, nitrogen, phosphorus, or potassium, are limiting (Figure 8.4). We view cattle manure as a key resource that is able to overcome all three soil macronutrient constraints (Table 8.2). Our findings that farmers who apply manure to banana have the greatest reported bunch weights supports this assumption (Bekunda and Woomer, 1996). Also, in the banana production areas of Uganda, manure is not used as a source of fuel (although it is in other areas of Uganda) and has no alternative uses other than as a soil amendment. Nitrogen is biogeochemically cycled and available in N-rich plant residues or from the cultivation of N-fixing plants. Phosphorus and potassium, on the other hand, are undergoing sedimentary processes over geological time, tend to be low in highly weathered soil and are more conveniently replenished as fertilizers than organic inputs. Rock phosphate deposits exist in Uganda and have been mined in the past.

We propose to work with farm households in Masaka and Mukono Districts to test our current ideas concerning "best bet" interventions in degraded banana systems. These farms will be selected alongside sites where limiting nutrient studies are conducted. First a team of researchers and extensionists will meet with each farmer to discuss their interests and impressions of different management options. It is likely these farmers will have swayed toward fertilizer use because of the excessive rates under study in the nearby limiting nutrient trial. We will resist applying large, unaffordable and under-available rates of fertilizer, but rather explore with the farmers such avenues as field sanitation, pest lures, organic input placement, judicious fertilizer application and improved intercrop selection. An example of one such set of management practices for a field severely nutrient depleted and pest infested is presented in Figure 8.5. We will limit the "better banana" area to 0.25 ha farm^{-1} and compare the results to production levels of the current farmer practices. By locating the limiting nutrient and "better banana" trials in close proximity, we will be in a better position to interpret the success or failure of various interventions. Whenever possible, the "better banana" field will be prominently located and identified. We will conduct farmer field days to obtain a cross section of farmer impressions of our attempts at

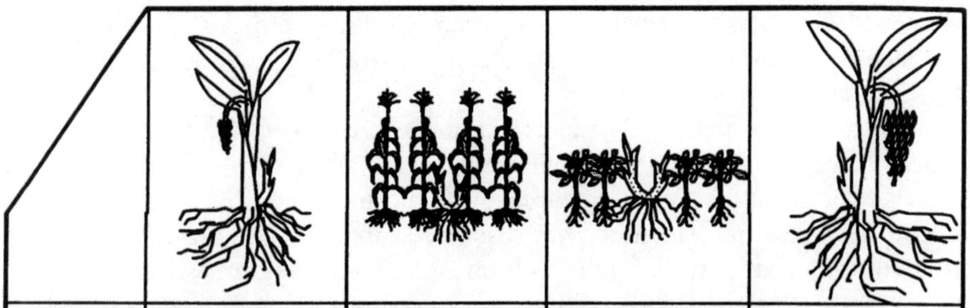

month	0	1	6	18
crop	banana	maize	banana/beans	banana
fertilize	no	30 kg N,P,K	30 kg P	100 kg N,P,K (month 12)
pesticide	all	no	bananas only	as required
activity	select farms, shoot component of yield	limiting nutrient trial, thin banana to 3 shoots/mat	limiting nutrient trial, rhizobial inoculation trial	limiting nutrient trial, harvest begins by month 18.

Figure 8.5. A suggested management strategy for the rehabilitation of severely nutrient depleted and pest infested banana lands.

land regeneration. We anticipate the results from this activity to result in the preparation of guidelines in the rehabilitation of degraded banana smallholds in the Lake Victoria Basin of Uganda.

PHOSPHORUS REPLENISHMENT IN EAST AFRICA

Per capita food production is declining in Africa, particularly in countries with largest population growth rates (Cohen and Maibaum, 1993). In the highlands of Kenya, land fragmentation results in holdings as small as 0.1 ha per household (Ikombo et al., 1994), causing farmers to grow crops for food rather than income generation, which in turn disallows purchases of fertilizers necessary to improve production. Continuous cropping in absence of nutrient replacement results in nutrient depletion characterized by the low levels of organic carbon, nitrogen (N), and phosphorus (P) (Okalebo et al., 1990; 1993). Stoorvogel et al. (1993) have reported that in the major agricultural lands of Eastern African Highlands (Ethiopia, Kenya, Malawi, Rwanda) the annual net losses of N and P were 42 and 3 Kg/ha/yr respectively, from 1982 to 1984 period. In Kisii District, Kenya, the nutrient losses are 112, 3 and 70 kg/ha/yr of N, P, and K, respectively. These nutrient losses occur mainly from harvest and crop residue removal, leaching, denitrification, and water erosion (Smaling et al., 1993). Other estimates of P loss from agricultural soils in Africa by Smaling et al. (1993) range between 8 and 13 kg/ha/yr.

Obviously, East African countries have many good reasons for developing their rock phosphate deposits. First, phosphorus is often a limiting nutrient in highly weathered and volcanic soils (Deckers, 1993; Okalebo et al. 1990; 1993) and it is essential for farmers to apply P to many soils in order to increase yields (FURP, 1994). Second, national interests seek to develop domestic or regional resources as a means of stimulating local economies and assist in long-range forecasts for agricultural development. Domestic production of fertilizers reduces demands for foreign exchange investment in agriculture and buffers local economies from fluctuation in currency exchange rates and world fertilizer shortages

(McClellan, 1991). Finally, domestic production of agrochemicals has the potential to grow and become export commodities, further strengthening national economies and regional trade linkages.

Deposits of rock phosphate are widespread in East Africa. Those that occur in Kenya's eastern shore of Lake Victoria are low yielding (Van Kauwenberg, 1991), but two higher quality deposits occur in northern Tanzania and eastern Uganda (Kisitu, 1991). The Minjingu deposit near Arusha, Tanzania, was discovered in 1956 and ground rock phosphate production commenced in 1983. Total rock phosphate reserves are estimated to be 4 million tons and annual production in 1986 was 21,000 tons (Van Kauwenberg, 1991). Processing the rock phosphate is a dry process involving crushing, screening and drying. The Sukulu deposit in eastern Uganda was mined between 1962–1978 and then abandoned during the prolonged civil unrest. This carbonatic deposit rich in apatite is approximately 4 km in diameter and is estimated to contain 2.3×10^7 tons of phosphate rock. In 1968, production levels were 23,000 tons of apatite concentrate and single superphosphate. Processing the phosphate rock requires a wet-magnetic separator before a high quality concentrate can be obtained. Chemical analyses of Kenyan (Rangwa), Minjingu, and Sukulu raw rock phosphates are presented in Table 8.3. From the above evidence, we conclude that strong potential exists to further develop the rock phosphate mining and processing operations in nearby Tanzania and Uganda and that likely beneficiaries of increased fertilizer P supply are western Kenyan farmers who cultivate highly-weathered, P-deficient soils.

Currently in Kenya, fertilizer recommendations are generated for individual crops at a District level (FURP, 1994). Representative soils and their appropriate fertilization are described in booklets distributed by the Kenya Agricultural Research Institute (KARI) to extension officers and development organizations. These recommendations resulted from an intensive, multisite investigation by KARI and represent a considerable improvement over the previous, country-wide recommendations. Addition of P fertilizers is recommended for Kakamega (Volume 2), Kisii (Volume 8), Busia (Volume 9), and Bungoma (Volume 11) districts. Based on the results from various trial sites, fertilizer recommendations, economic risks, and average yield increases are provided with interpretations and guidelines to conduct economic analyses for different priced fertilizers and commodity costs.

But is this level of resolution sufficient? The Tropical Soil Biology and Fertility Programme (TSBF) conducted limiting nutrient trials in 20 farmers' fields in Kakamega, Western Kenya. Under the assumption that nutrient limitations may correspond to soil taxonomy, 20 farms were selected in Kabras, Kakamega with five farms each located on Acrisols, Cambisols, Ferralsols and Nitisols. An unreplicated trial was installed at each

Table 8.3. Chemical Analyses of Rangwa, Minjingu and Sukulu Phosphate Rock[a]

	Rangwa	Minjingu %	Sukulu
P_2O_5	4.8	25.0	12.5
CaO	50.3	38.3	14.5
CO_3	34.9	6.9	0.5
K_2O	0.04	1.2	0.2
MgO	0.4	2.6	0.2
SiO_2	1.5	12.5	16.6
Fe_2O_3	5.8	1.3	28.6

[a]From Van Kauwenberg, 1991.

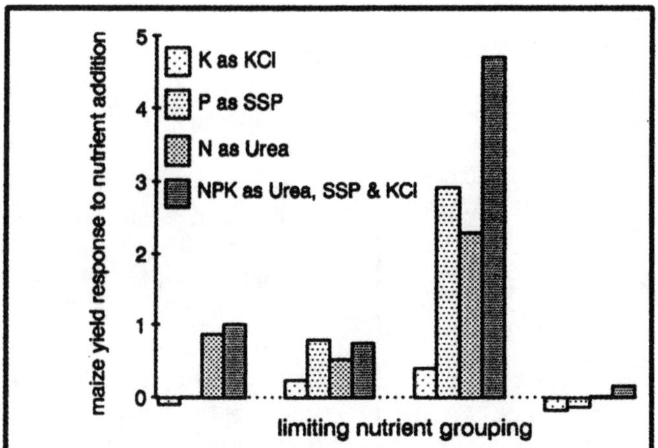

	nitrogen	phosphorus	nitrogen + phosphorus	nutrients not limiting
frequency	0.38	0.22	0.16	0.22
DAP return ratio	0.9	12.7	7.3	-0.9
Ca (cmol/kg)	2.0	2.6	0.9	1.7
soil organic C (%)	1.2	1.2	1.2	1.4

Figure 8.6. Limiting nutrient categories identified on 20 farms in Kakamega, Western Kenya.

farm which tested singular and combined effects of applied N, P, and K on maize yields. Soil samples were collected immediately before the trials were installed and later analyzed for plant nutrients, soil organic, carbon and soil pH. The nutrient most limiting to maize was identified by comparing the relative growth increase for the individual and combined nutrient applications.

Maize grain data was obtained from 18 of the 20 farms (above) with one experiment lost to weeds and another to hungry cattle. The observed limiting nutrients were not clearly defined by soil order although Cambisols, Ferralsols and Nitisols tended to be N-limited while Acrisols were either P- or N and P-limited (Figure 8.6). This observation agrees with large maize N and P responses obtained from on-farm studies in an Acrisol at Kabras (Okalebo and Woomer, 1994). When yield data from TSBF scientists (above) are grouped into nutrient-limitation categories, 38% were N-limited, 22% P-limited, 16% N and P-limited, with 22% of the farms not responding to the single or combined fertilizers (Figure 8.7). Extractable calcium (Ca) acted as a strong indicator of those soils severely limited by both N and P. Those farms which did not respond to nutrient additions had yields of 5555 and 6504 Kg maize grain/ha in the unamended and the NPK treatments, respectively. Maize grain increased from 580 to 4419 Kg/ha in response to nutrient additions to those soils limited by both N and P. The application of 50 Kg DAP/ha offered favorable economic returns to farmers with P and combined N and P limitations, but not to those with N-limitations. These results suggest a degree of site-specificity in nutrient response that extends beyond soil type (Smaling and van de Weg, 1990).

We assert that an important research and development priority be targeted toward the better exploitation of East African rock P by smallhold farmers. In doing so, we recognize that the solubility and availability of rock P for plant uptake are influenced by soil properties, particularly the pH, P, and Ca status of soils, clay, and Fe_2O_3 contents and the P-sorption characteristics (Smyth and Sanchez, 1982; Floor and Kimambo, 1988). There is strong evidence of effectiveness of rock P in increasing cereal, legume, and cotton yields

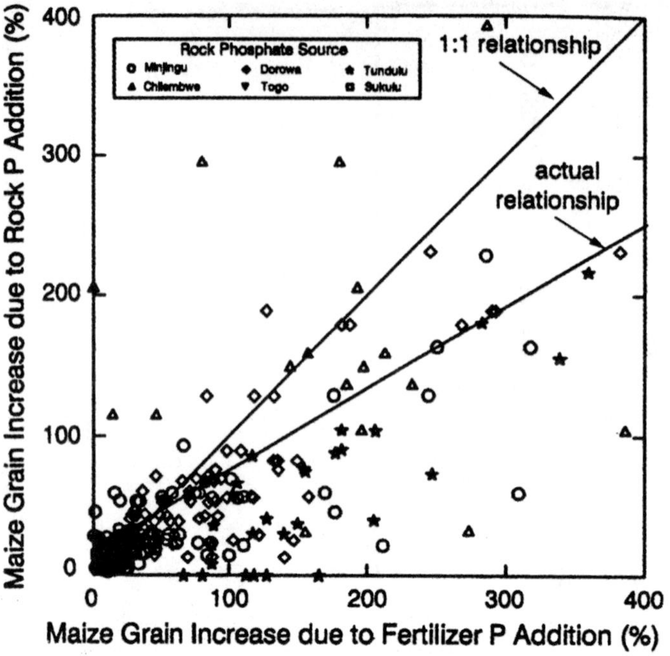

Figure 8.7. Comparison on maize yield response to rock and fertilizer P applied to equal rates.

in East and Southern Africa (Ssali, 1988-92). When yield improvement of maize with equal amounts of P as rock and fertilizer are composed, rock P is 69% as effective as fertilizer P (Figure 8.7, Okalebo and Woomer, 1994).

An emerging awareness among soil managers in Africa is the need for recapitalization of nutrient stocks in smallholder fields as a means of assuring future crop productivity. At issue is the replacement of nutrients lost over decades and the inability of smallholders to periodically invest in fertilizers to replace those nutrients (Mokwunye, 1995). We believe that farmers must be assisted in this effort through direct subsidy. A realistic target is to replace P losses incurred over 20 years, approximately 250 kg P/ha, in farmers' fields where P is the most limiting nutrient. Furthermore, the use of rock P is desirable owing to its subregional availability and price advantage. Yet much remains to be done before wide-scale nutrient recapitalization projects may commence.

Uncertainties exist as to how recapitalization may proceed most effectively. How do we best select areas and farms for recapitalization? What rates of materials need be applied, and should farmers be required to meet additional criteria for participation beyond the cultivation of a P-deficient soil? How may nutrient recapitalization be better balanced with the need for other farm improvements, and what additional considerations are required to assure that the poorest farmers on the smallest farms are full beneficiaries? Clearly, many of these issues extend beyond the scope of soil fertility and will be difficult to address within anything less than a full interdisciplinary setting. These issues include:

- **How do we identify geographical areas for recapitalization and select individual farms within them?** A decision tree intended to assist in the selection of areas for P recapitalization using rock phosphate is presented in Figure 8.8.
- **How do we select which nutrient resources to apply and their rates?** All plans to date assume that P limitations are widespread and may be ameliorated through the direct application of phosphate rock. Is this assumption valid?
- **How do individual fields qualify for fertilizer subsidy and to what extent should fertilizer use be supervised?** A candidate field design for on-farm pilot studies has been proposed requiring slightly more than 6000 m^2 (0.6 ha) of which 0.5 ha will receive a one-time application of 250 kg P/ha. Should the plots be scaled up or down depending upon farm size? What adjustments are possible for split or irregularly shaped holdings?
- **Should farmers be required to meet additional criteria for participation such as the plant-**

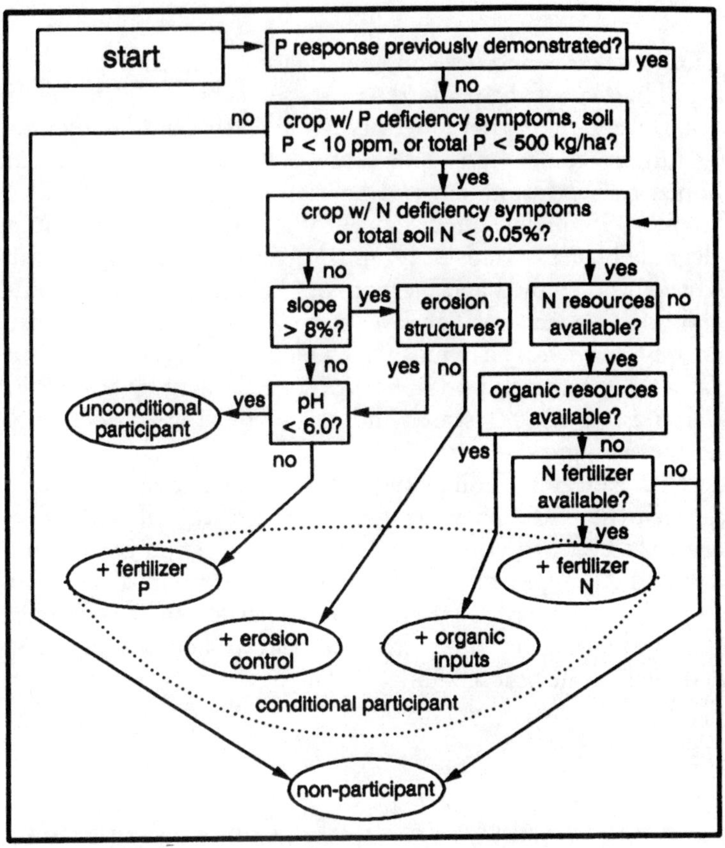

Figure 8.8. A prototype decision tree for GIS applications requires refinement and testing as a means of selecting areas and farms for nutrient replenishment.

ing or construction of erosion control structures, tree planting, organic additions to soils or intercropping with N-fixing crops? Whether or not farmers are required to meet more rigorous criteria other than P-limitation, and the nature of those criteria will be essential to the voluntary participation by farmers and the overall success of a recapitalization effort.

- **Should the project assist farmers in additional and subsequent fertilizer purchases?** Nutrient recapitalization, even as a single, large application, may be viewed as an opportunity for greater access of farmers to fertilizers. Should a recapitalization project provide for subsequent group purchases of fertilizer or other external inputs?
- **How should the relatively high cost of soil analyses be best balanced with fertilizer purchase, for example, if one P analysis costs approximately as much as 100 kg rock P, how many P analyses should we conduct in the farmers' interests?** We estimate the cost of duplicate samples to be measured for extractable P, with accompanying reference samples, controls and blanks, to be approximately $11.40, a cost which is equivalent to providing a farm with 100 kg rock P.
- **How do we balance the current needs of densely populated rural settings against the probability that these will become more urbanized in the near future?** Should densely populated, highly fragmented landscapes be considered for recapitalization considering the likelihood that the importance of agriculture will diminish over the following two decades? On the other hand, recapitalized field size might be scaled to available field area so that operators of smaller farms are not excluded from full participation.
- **How do we assess project impacts and farmer perceptions within large recapitalization project areas?** As this project is intended to provide an interface between soil fertility research and rural development, project impacts must be monitored over time.

CONCLUSIONS

We have presented three priority research and development areas in agriculture that are in the advanced planning stage. The first of these addresses land preparation in the humid forest zone of Central Africa. By lending greater longevity to cultivated lands along forest margins and by facilitating carbon storage and biodiversity protection during the land conversion process, it is hoped that deforestation and habitat loss within the Guineo-Congolian humid forest (White, 1983) may be reduced. Another priority for agricultural improvement rests in the degraded banana lands of Uganda. Over the past few years, the root causes of this decline have been studied and methods are being developed to better manage pests and conserve productive lands. What is yet to be accomplished is the identification of a means to rehabilitate the severely nutrient-depleted and pest-infested banana lands of Uganda. Lastly, we recommend phosphorus replenishment of the densely settled East African Highlands as a priority research and development area. Rock phosphate deposits occur within the subregion, yet are not being effectively mobilized as nutrient inputs to P-deficient soils. Research findings suggest that a one-time, massive application to replace losses occurring over 25 years may offer substantial returns, yet mechanisms of site and farm selection and criteria for farmer participation are yet to be refined. These three research and development priorities have several common elements. All are built on past research efforts, all have a strong on-farm participatory component and all offer substantial returns to farmer livelihood, their respective societies, and to enhancement of environmental quality once successfully concluded.

ACKNOWLEDGMENTS

Many of the ideas presented in this chapter have evolved through discussions with agricultural and social scientists in Africa. Moukam Appolinaire and R. Njomgang, IRAD, Cameroon, C.A. Palm, E. Crowley and S. Carter, TSBF Programme, Nairobi and J.K. Lekasi and D. Bwamiki, Makerere University are gratefully acknowledged for their exchange of ideas with the authors. Geographic analyses for Cameroon were conducted by the GIS Laboratory at ICRAF; we thank John Corbet, Gary Couto and Eric Muchugu for their generous assistance. The research results presented for slash-and-burn systems in Cameroon were obtained through a grant from the Global Environment Facility. Research in Uganda was made possible through financial assistance from the Rockefeller Foundation. Both of these organizations are gratefully acknowledged.

REFERENCES

Bekunda, M.A. and Woomer, P.L. Organic Resource Management in Banana-based Cropping Systems of the Lake Victoria Basin, Uganda. *Agric. Ecosys. Environ.* (in press), 1996.

Brown, S., Gillespie, A.J.R. and Lugo, A.E. Biomass Estimation for Tropical Forests with Applications to Forest Inventory Data. *Forest Sci.* 35:881–902, 1989.

Brown, S., Hall, C.A.S., Knabe, W., Raich, J., Trexler, M.C. and Woomer, P. Tropical Forests: Their Past, Present and Potential Future Roles in the World's Carbon Budget. *Water Air Soil Pollut.* 70:71–94, 1993.

Bwamiki, D.P., Zake, J.Y.K., Bekunda, M.A. and Woomer, P.L. The Use of Coffee Husks as an Organic Amendment to Improve Soil Fertility in Ugandan Banana Production. Presented to the International Symposium "Carbon and Nutrient Dynamics in Natural and Agricultural Tropical Ecosystems," Harare, Zimbabwe, 29 April-4 May 1996 (in press), 1996.

Cohen, A. and Maibaum, W. *Commentary on Food Security and the Role of Fertilizers in Sub-Saharan Africa.* IFC, Vol. 34, 393, 1993.

Deckers, J. Soil Fertility and Environmental Problems in Different Ecological Zones of the Developing Countries in Sub-Saharan African. In H. van Rueler and W.H. Prins, Eds. *The Role of Plant Nutrients and Sustainable Food Crop Production in Sub-Saharan Africa.* Pensen and Looijen, Wageningen, The Netherlands, 1993, pp. 37–52.

Fertilizer Use Recommendation Project (FURP). *Fertilizer Use Recommendations; Volumes 1-22.* Kenya Agricultural Research Institute, Nairobi, Kenya, 1994.

Floor, J. and Kimambo, E.T. Considerations on the Users of Minjingu Rock Phosphate as a Direct Phosphate Fertilizer. *Proc. of Phosphate Meeting, Tanga, Tanzania, 1–3 August, 1988*, 1988, pp. 23–33.

Food and Agriculture Organization of the United Nations (FAO). *FAO Yearbook: Production.* Volume 48. FAO, Rome, 1995a.

Food and Agriculture Organization of the United Nations (FAO). *FAO Yearbook: Fertilizer.* Volume 44. p. 119. FAO, Rome, 1995b.

Gerner, H. and G. Harris. The Use and Supply of Fertilizers in Sub-Saharan Africa. In H. van Reuler and W.H. Prins, Eds. *The Role of Plant Nutrients for Sustainable Crop Production in Sub-Saharan Africa.* Dutch Association of Fertilizer Producers, Leidschendam, The Netherlands, 1993, pp. 107–125.

Hartman, E. *Five Year Food Plan, 1989-1994, and Recommendations for Strengthening Research and Extension Linkages. Vol 1.* USAID/MFAD project, Makerere University Kampala, 1989.

Harwood, R.R. A History of Sustainable Agriculture. In C.A. Edwards et al., Eds. *Sustainable Agricultural Systems*, 1990, pp. 3–19.

Ikombo, B.M., Esilaba, A.O. and Kilewe, A.M. A Diagnostic Survey of the Farming Systems in Kiambu District. *A Report, NARC Muguga, KARI*, 1994, p. 78.

International Centre for Research in Agroforestry (ICRAF). *Alternatives to Slash-and-Burn: A Global Strategy.* ICRAF, Nairobi, 1994.

Kisitu, V.B. Fertilizer Material Occurrences in Uganda. *Fert. Res.* 30, 187–189, 1991.

Kotto-Same. J., Appolinaire, M. and Woomer, P.L. Carbon Dynamics in Slash-and-Burn Forests of Southern Cameroon (in preparation), 1996 .

McClellan, G.H. Use of East and Southeast African Indigenous Agrominerals for Fertilizer Production. *Fert. Res.* 30, 125–126, 1991.

Ministry of Agriculture, Animal Industries and Fisheries (MAAIF). *Agriculture Annual Report 1992.* MAAIF, Kampala, Uganda, 1992.

Mokwunye, A.U. Phosphate Rock as Capital Investment. In H. Gerner and A.U. Mokwunye, Eds. *Use of Phosphate Rock for Sustainable Agriculture in West Africa.* Miscellaneous Fertilizer Studies No. 11. International Fertilizer Development Center. Muscle Shoals, AL, USA, 1995, pp. 100–106.

Myers, N. *Deforestation Rates in Tropical Forests and Their Climatic Implications.* London, Friends of the Earth, 1989.

Nye P.H. and Greenland, D.J. *The Soil Under Shifting Cultivation.* Commonwealth Bureau of Soils Technical Communication 51, Harpenden, UK, 1960.

Okalebo, J.R. and Woomer, P.L. Use of Rock and Fertilizer Phosphorus in Eastern and Southern Africa: A Data Summary. Presented at the *14th Conference of the Soil Science Society of East Africa*, Mbarara, Uganda, November, 1994.

Okalebo, J.R. Schnier, F. and Lekasi, J.K. Maize Response to Fertilizer and Rock P in a High P-fixing Soil of Malava, Kakamega. In *NARC Muguga (KARI) Record of Research,* Ann. Rpt., 1993.

Okalebo, J.R., Simpson, J.R. and Probert, M.E. Phosphorus Status of Cropland Soils in the Semi-Arid Areas of Machakos and Kitui Districts, Kenya. In *A Search for Strategies for Sustainable Dryland Cropping in Eastern Kenya,* M.E. Probert, Ed. ACIAR Proceedings 41, 1990, pp. 50–54.

Okigbo, B.N. Sustainable Agricultural Systems in Tropical Agriculture. In *Sustainable Agricultural Systems.* C.A. Edwards et al., Eds., 1990, pp. 323–352.

Palm, C.A. Contribution of Agroforestry Trees to Nutrient Requirements of Intercropped Plants. *Agroforestry Sys.* 30:105–124, 1995.

Post, W.M, Peng, T., Emmanuel, W.R., King, A.W., Dale, V.H. and DeAngelis, D.L. The Global Carbon Cycle. *Am. Scientist* 78:310–326, 1990.

Purseglove, J.W. *Tropical Crops: Monocotyledons.* J. Wiley and Sons, New York, 1972.

Rubaihayo P.R, *Banana Based Cropping Systems Research.* Bulletin No. 3. Makerere University, Kampala, Uganda, 1992.

Russell, E.W. *Soil Conditions and Plant Growth.* Longman Group Limited, New York, 1976.

Sanchez, P.A. Deforestation Reduction Initiative: An Imperative for World Sustainability in the Twenty-First Century. In *Soils and the Greenhouse Effect,* A.F. Bouwman, Ed. J. Wiley & Sons, Chichester, UK, 1990, pp. 375–382.

Sanchez, P.A., Woomer, P. and Palm, C.A. Agroforestry Approaches for Rehabilitating Degraded Lands After Tropical Deforestation. *JIRCAS International Symposium Series* 1:108–119, 1994.

Smaling, E.M.A. Stoorvogel, J.J. and Windmuijer, P.N. Calculating Soil Nutrient Balances in Africa at Different Scales II: District Scale. *Fert. Res.* 35, 237–250, 1993.

Smyth, T.D. and Sanchez, P.A. Phosphate Rock Dissolution and Availability in Cerrado Soils as Affected by Phosphorus Sorption Capacity. *Soil Sci. Soc. Am. J.* 46, 339–345, 1982.

Stoorvogel, J.J. Smaling, E.M.A. and Janssen, B.H. Calculating Soil Nutrient Balances in Africa at Different Scales. In *An Agro-Ecological Framework for Integrated Nutrient Management with Special Reference to Kenya.* Ph.D. thesis of E.M.A. Smaling. Agricultural University, Wageningen. The Netherlands, 1993, p. 250.

Stover, R.H. and Simmonds, N.W. *Bananas.* 3rd edition. Longman. London, 1987.

Tiffen, M., M. Mortimore, and F. Gichuki. *More People, Less Erosion: Environmental Recovery in Kenya.* J. Wiley and Sons, Chichester, UK, 1994.

Turkahirwa, E.M. *Uganda-Environmental and Natural Resource Management Policy and Law: Issues and Options. II. Documentation.* Institute of Environment and Natural Resources, Makerere University. Kampala, Uganda, 1992.

Van Kauwenberg, S.J. Overview of Phosphate Deposits in East and Southeast Africa. *Fert. Res.* 30, 127–150, 1991.

White, F. *The Vegetation of Africa.* The United Nations Educational Scientific and Cultural Organization, Paris, 1983.

Woomer, P.L., Lekasi, J., Okalebo, R., and C. Palm. Principles and Methods of Soil Fertility Research in Maize-Bean Cropping Systems of the Eastern African Highlands. In *Soil Fertility Research for Maize and Bean Production Systems in Eastern African Highlands.* C. Wortmann and J.K. Ransom, Eds. Network on Bean Research in Africa, Workshop Series No. 21, CIAT, Dar es Salaam, Tanzania, 1993, pp. 21–34.

THREE

Soil Quality Management in Asia

NINE

The Role of Soil Management in Improving Yields in the Rice-Wheat Systems of South Asia

Kevin F. Bronson and Peter R. Hobbs

INTRODUCTION

Rice and wheat grown in the same field in the same year is a rotation that has been practice in South Asia for thousands of years (Woodhead et al. 1994). About 12 million ha, mostly in India, practice rice-wheat (RW) systems, as shown in Figure 9.1 (Huke and Huke, 1992). These alluvial soils of the Indo-Gangetic flood plains are predominantly cambisols or inceptisols (Fig. 9.2, FAO, 1992) and were at one time highly fertile soils.[1] Their micaceous mineralogy, for instance, means that potassium supply is high in these soils. The predominant silty texture is ideal for most crops. However, thousands of years of intensive cropping, with less and less organic matter being recycled, have led to degradation of these soils, especially in terms of soil organic matter content and thus nitrogen supply capacity.

As a result of "Green Revolution" technologies entailing the introduction of modern semidwarf rice and wheat cultivars, together with a large increase in inputs of fertilizers and water, rice and wheat production in South Asian countries like India have recently reached all-time highs. However, the continuance of these upward trends is now in doubt. Hobbs and Morris (1996) reviewed yield trends and area trends of rice and wheat for India, Bangladesh, Nepal, and Pakistan. Rice areas have plateaued after increasing the last few decades in South Asia. Wheat areas likewise have stopped increasing, with the exception of Pakistan. Rice yields in Bangladesh have only begun to increase appreciably the last few years with the success of boro or winter rice. In India and in Pakistan, rice yields have leveled off, and Nepal is still waiting for its "take-off"in rice yields. Only in India have wheat yields continued to increase the last few years.

Distinctions need to be made between "favorable" and "unfavaorable" rice and wheat growing environments. Northwest India, e.g., Punjab, is considered a "favorable" environment for rice and wheat, due to its high solar radiation and well-developed irrigation in-

[1]Soil fertility refers to the amounts of nutrients available for plant growth.

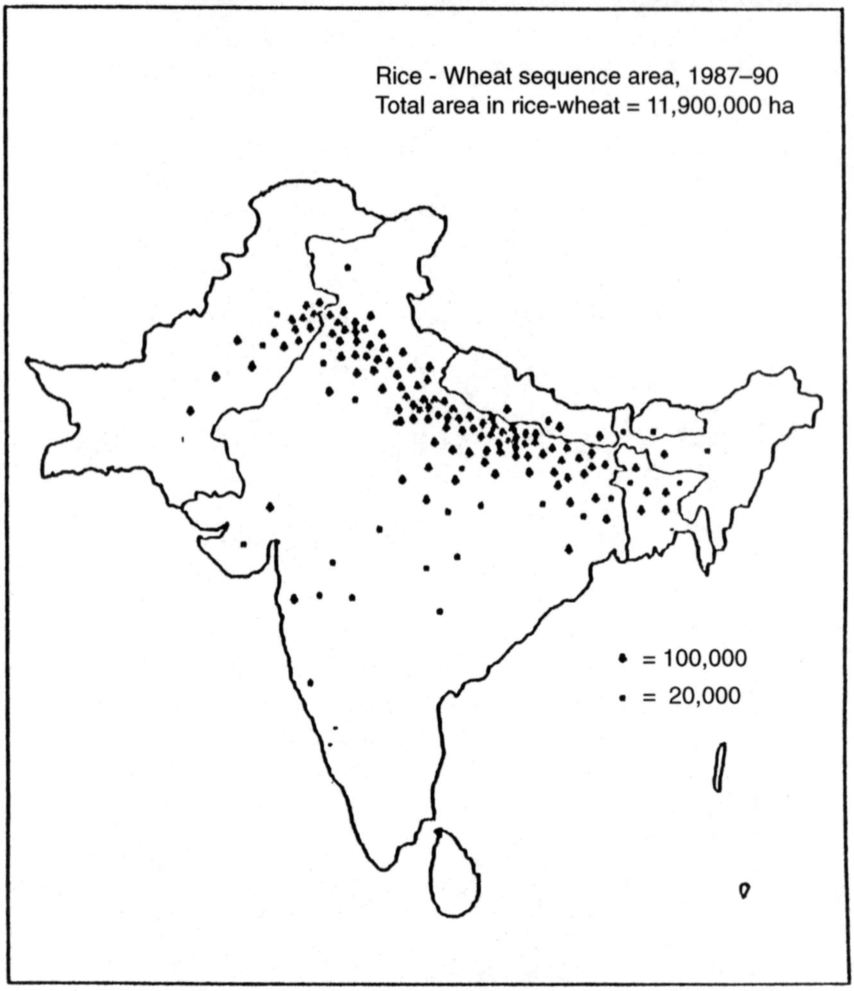

Figure 9.1. Extent of rice–wheat cropping in South Asia (Huke and Huke, 1992).

frastructure. The rainfed and flood-prone areas of East India and Bangladesh are considered unfavorable environments for monsoon rice.

Yields per ha have stagnated in many of the favorable, productive areas, and reports of declining factor productivity with respect to nutrient inputs have been reported in the Indian Punjab for rice (Cassman and Pinglali, 1995) and for wheat (Sidhu and Byerlee, 1992). In other words, farmers are having to apply higher rates of fertilizer to achieve the same yields. These disturbing yield trends are from on-farm and experiment station data throughout the region, and strongly indicate a decline in soil fertility. The problem of land degradation due to soil erosion, declining water tables, waterlogging, and salinity are important, but these problems are restricted to certain regions of South Asia (FAO, 1994). Loss of soil fertility could be associated with declining soil organic matter (SOM), soil erosion, or a general loss of "soil health" that comes with intensive, continuous cereal cropping. Of all the constraints to productivity in RW systems, low SOM levels is probably the most critical.

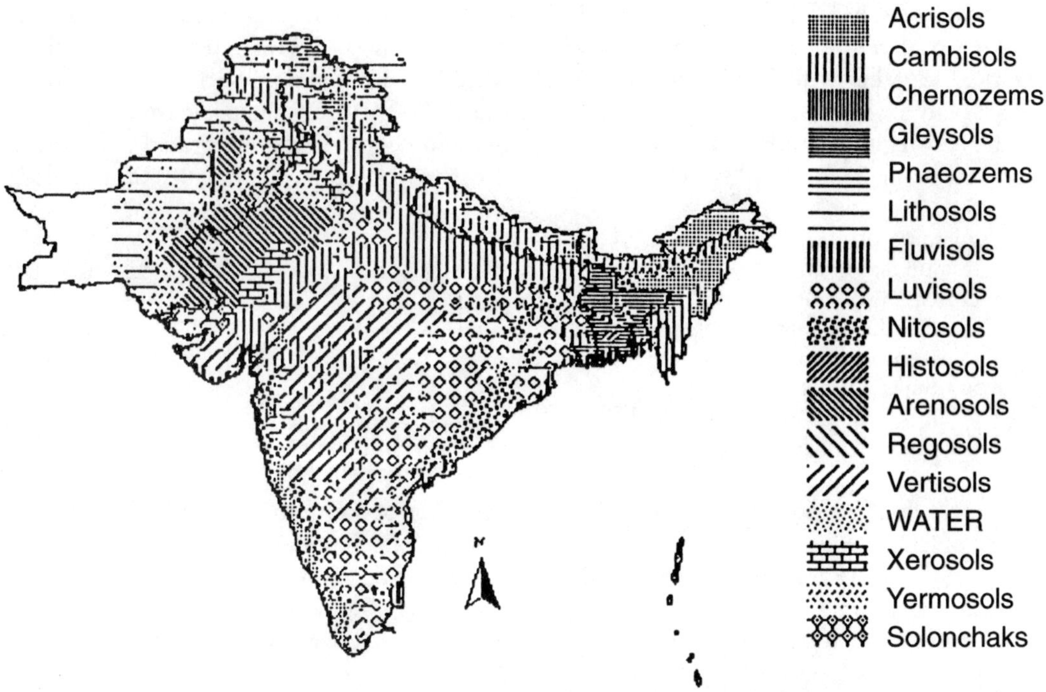

Figure 9.2. FAO major soil groups of South Asia (FAO, 1992).

NUTRIENT MANAGEMENT RESEARCH IN THE RW CONSORTIUM FOR THE INDO-GANGETIC PLAINS

In 1994, a "Rice-Wheat Consortium for the Indo-Gangetic Plains" (RWCIGP) was formed. In this new model for a consortium, NARS (National Agricultural Research System) from the participating countries (India, Nepal, Bangladesh, and Pakistan) have the leadership, in partnership with the IARCs (International Agricultural Research Centers) (IRRI, CIMMYT [International Center for Maize and Wheat], ICRISAT [International Crop Research Institute for Semi-Arid Tropics], and IIMI [International Irrigation Management Institute]). The RWCIGP is managed through a Regional Steering Committee (RSC) comprised of the Directors General from the national research programs of each of the participating countries, one representative from all participating IARCs, a representative of the donor community, and a regionally based coordinator. A regional technical coordinating committee composed of NARS and IARC scientists meets annually to provide the RSC with progress reports of the consortium and to submit work plans for RSC approval. Dr. I.P. Abrol is the Facilitator of the consortium, and the Facilitation Unit is based in New Delhi. He is responsible for fostering linkages between sites and scientists to ensure efficient, high quality research. The Facilitation Unit also plays an active role in encouraging information exchange, and provides facilities for publishing and human development.

Within the consortium, IRRI has been asked to take a lead in developing Nutrient Management Research and Technologies. Likewise, CIMMYT, ICRISAT, and IIMI are the lead institutions in tillage and crop establishment, grain legumes, and water management. A workshop addressing Nutrient Management Research issues within the consortium was held at IRRI on September 21–23, 1995. Local problems were identified by the NARS representatives as a starting point for the workshop. NARS representatives then worked together with RW scientists from IRRI, CIMMYT, and ICRISAT to develop a work plan for Nutrient Management Research in RW. The proposed program is 3-tiered: local, eco-regional, and South Asia, participatory approaches being integrated at all these levels. The main soil fertility-related issues that the workshop participants cited included:

- organic matter (OM) decline,
- residue management,
- decreased soil nutrient supply capacity,
- declining soil quality/health,
- micronutrient deficiencies,
- nutrient imbalances,
- declining factor (especially N and P) productivity,
- declining yields (e.g., rice yields in long-term RW trial in Pantnagar).

IRRI's contributions to an improved understanding of soil nutrient issues in RW will focus on the theme of SOM, which is a long-term, strategic issue vital to the entire RW region. The NARS are interested in the approaches being undertaken in the "Reversing trends of declining productivity in intensive irrigated rice systems " Project on soil nutrient supply capacity as well as soil organic matter characterization, but this will require outside donor support.

RECENT RESEARCH ON FERTILIZER AND ORGANIC MATTER MANAGEMENT IN RICE-WHEAT SYSTEMS

In several on-station long-term RW nutrient management experiments in South Asia, disturbing trends of declining yields are present. These yield declines are similar to those reported by Cassman et al. (1995, 1996) for the double and triple-cropped rice systems in the Philippines. Some of the most prominent examples include the rice phase of rice-wheat experiments in Pantnagar, India (Nambiar, 1995) and the rice phases of a rice-rice-wheat experiment in Bhairahawa, Nepal (Regmi, 1994). These declines occur even with the recommended fertilizer applications (usually 120-60-40). Wheat yields are low and stagnant at these sites.

Declining partial factor productivity with respect to fertilizer has been documented in the Indian Punjab for rice (Cassman and Pingali, 1995) and for wheat (Sidu and Byerlee, 1992). There are similar reports in the rice-rice systems of Bangladesh (Pagiola, 1995) and in the Philippines (Cassman et al., 1995) that higher N fertilizer rates are needed to achieve previous yield levels. This strongly suggests a degradation of the soil base.

Organic matter is important for the mineralization of nutrients (N, P, and S), favorable water relations, and aggregate stability. Although soil physical properties and associated water relations are generally more important in wheat, they are critical for rice as well in the lighter-textured soils of N.W. India and Pakistan. Decline of SOM is prevalent throughout the RW belt (Table 9.1; Nambiar, 1994), while in rice-rice and rice-rice-rice systems, it is conserved or built up (Cassman et al., 1995). In the RW belt, straw is almost always removed and used for animal feed or bedding (Bhardwaj, 1995). Applications of farmyard manure in long-term experiments (LTE) in RW in India and Nepal result in lower rates of OM declines (Table 9.1, Nambiar, 1995) and higher (though declining) grain yields compared to NPKZn treatments (Nambiar, 1995). This is in contrast to the case in the Broadbalk LTE in wheat in the UK, where building up SOM with manuring

Table 9.1. Effect of Long-Term Fertilization and Manuring on Soil Organic C (g kg^{-1}) for Different Cropping Systems in India[a]

		Treatment		
Cropping System, Location	**Initial Value**	**Control**	**NPK**	**NPK + FYM**
Rice-rice, Hyderabad	5.1	6.6	8.2	12.5
Rice-rice, Bhubaneswar	2.7	4.1	5.9	7.6
Rice-wheat-jute, Barrackpore	7.1	4.2	4.5	5.2
Rice-wheat-cowpea, Pantnagar	14.8	6.0	9.0	14.4
Maize-wheat-cowpea, Ludhiana	2.1	2.5	2.7	3.7
Maize-wheat, Palampur	7.9	6.2	8.3	12.0
Millet-cowpea-maize, Bangalore	5.5	3.4	4.5	4.8

[a]From Nambiar, 1994

does not result in different wheat yields from NPK treatments. The LTEs in rice-rice at IRRI indicate that maintaining or building up SOM does not necessarily result in sustained or higher N supply. Will the same hold true for RW? Or if this is true for the rice is it true for the upland crop wheat? The whole role of SOM in RW systems needs to be evaluated. Do we need to begin finding ways to build SOM back up in RW soils? The options for this are extremely limited in South Asia where most residue is removed and utilized. The total size of the SOM stocks in soils is probably not as important as the size and activity levels of the active fraction that is involved in nutrient cycling. Quality of SOM is probably related to the active fraction size. Cassman et al. (1995) and Olk et al. (1996) have hypothesized that changes in the quality of soil organic matter as a result of continuous flooding lead to decreased soil nitrogen supply capacity (SNSC). Consistent with this is his observation that at the IRRI farm, yields of wet season rice are affected by the length and water status (dry or flooded) of the dry season fallow (Cassman et al., 1996). Long, dry season fallows boosted wet season rice yields, while long flooded or short dry fallows depressed yields. Is something similar occurring in the RW system, so that including nonrice crops like jute and sugarcane in the RW rotation would positively affect rice and wheat yields? Obviously, many questions remain on the importance and role of SOM in RW systems.

Although the yield declines in the experiments at Pantangar and Bhairahawa are alarming, do they represent the situation in farmers' fields? In Uttar Pradesh, where Pantangar is located, sugarcane and sugarcane-ratoon is usually rotated into the RW every few years (Hobbs et al., 1992). In Bangladesh, likewise, jute is often grown following wheat. On-station experiments, therefore, were started 4–5 years ago in Pantangar and in Nashipur, Bangladesh to examine yield trend and nutrient management of RW systems with common "break" or cash crops included. Table 9.2 shows 0.4 t/ha benefit in rice yield to a RW sequence that includes rapeseed, sugarcane, and sugarcane ratoon, compared to a continuous RW sequence (Singh, 1995). This could be due to factors other than nitrogen, however, such as a break in disease and pest pressure.

In Table 9.3 we see results of a rice-rice-wheat experiment in Nashipur, Bangladesh (Badaruddin and Razu, 1995). Yields of early rice are much lower than the main monsoon rice, as is observed in Nepal (Regmi, 1994). Temperatures in South Asia are very hot in the premonsoon summer, leading to rapid water loss and heat damage. Additionally, disease pressure on the relatively small areas of early rice are high. Green manures, grown for short periods on early monsoon rains, benefit the following main rice crop, tremendously. Residual benefits of green manures to wheat are usually small (Table 9.3). Farmyard manure applications, applied before either crop, are effective in building up SOM (Table 9.1; Nambiar, 1994) and boosting yields of both rice and wheat.

Table 9.2. Rice Yields as Affected by Rotation, IPM, and Nutrient Management in a Rice-Wheat System, Pantangar, 1995[a]

Treatment	Yield (kg ha^{-1})
Rotation	
Rice-lahi-sugar-sugar-ratoon-wheat	3437
Rice-wheat	2996
IPM	
Control	3029
Chem. weed control	3288
Disease and insect control	3215
IPM	3314
Soil Fertility	
120-60-40	2984
220-6-40	3215
GM + FYM + 55-0-0	3450

[a]Adapted from Singh et al., 1995.

Table 9.3. Grain Yield of Aus (Early Rice), Aman (Main Rice Crop) and Wheat as Affected by Rotation and Nutrient Management, Nashipur, 1992–1995[a]

Treatment	Aus	Aman	Wheat
		Grain yield (t ha^{-1})	
R-R-W 120-60-60	1.50	3.52	3.32
R-R-W 180-60-90	1.73	3.67	3.55
R-R-W 120-60-60 + FYM	1.54	3.87	3.64
R-R-W 180-60-60 + FYM	1.64	3.79	3.69
R-W-J 120-60-60 + FYM	—	3.58	3.60
R-W-GM 120-60-60	—	3.81	3.51

[a]Adapted from Badaruddin et al., 1995.

NITROGEN EFFICIENCY OF WHEAT IN RW SYSTEMS

Hundreds of N use-efficiency studies using Nitrogen-15 have been done in Asia, primarily by IRRI, on flooded rice. However, very few N efficiency studies have been done on wheat in the Asian subtropics or in RW systems. Nitrogen efficiency in wheat in these systems has the potential to be even lower than in flooded rice. Nitrogen applied to flooded soils remains in the reduced, NH_4 form. Leaching and denitrification losses are therefore minimal. Wheat, on the other hand is grown in South Asia in the cool, dry season, with the benefit of 2–4 flood irrigations. Ammonium-based fertilizers, such as urea, have ample chance to oxidize to the NO_3 form, which can be leached or denitrify upon flood irrigations. IRRI, in collaboration with the NARS in South Asia are measuring N use-efficiency using Nitrogen-15 in both the rice and wheat phases of the RW system.

FUTURE DIRECTION OF NUTRIENT MANAGEMENT IN RICE-WHEAT SYSTEMS

Blanket fertilizer recommendations for rice and wheat production in South Asia have been the standard for decades. In most cases these recommendations have not been reviewed in this long period. For example, most of India recommends 120-60-40 (N - P_2O_5 - K_2O) for monsoonal rice or wheat, while Bangladesh and Nepal recommend 80-60-40 for the same. This is the case although experiment station research shows responses of N up to 160 kg N/ha in Pantnagar, India, for instance (Y. Singh, personal communication).

Although standard fertilizer recommendations need revising in most locations, at the same time agronomists, soil scientists, and farmers need to move beyond "blanket recommendations" to more "site-specific" nutrient management. Site-specific management uses knowledge of the tremendous variability of the soil nutrient supply capacity of different fields (Dobermann,1996). This philosophy has been fully developed in North America and in Europe, and has been refined to where variability within the field is acknowledged. Fertilizer application rates are thereby adjusted accordingly, based on the knowledge of, for example, a soil test. Thus for some fields the standard fertilizer recommendation may be sufficient, while for others it may be too much or too little. Thus nutrient management becomes more "knowledge intensive," as described by Byerlee (1996).

New tools are required, however, to provide the information to the farmers on site-specific nutrient supply capacity of the soil. Traditional soil tests have the distinct disadvantage of having a slow "turnaround time" before the results are back to the farmer. This is even more of a problem in developing countries. Additionally, the traditional soil test such as "Olson-extactable P" or "ammonium acetate-extractable K" are static tests that only indicate plant available P or K at the time of sampling. Dobermann et al. (1995) have adopted the use of ion exchange resins that Skogley et al. (1990) used for "dynamic soil tests" or "phytoavailability soil tests" (PST). The dynamic soil test gives information on the magnitude, pattern, and kinetics of nutrient release, from a 1 and 7 day incubation. The PST also has the big advantage of being "multi-nutrient." Plans are underway too for collaborative IRRI-NARS trials on use of the PST on wheat.

Another tool for rapid, simple nutrient assessment that IRRI has worked on for rice is the chlorophyll meter. Used primarily in the United States on maize (Piekielek and Fox, 1992), the chlorophyll meter measures the "greenness" of a leaf blade, which is in turn directly related to nitrogen content. A "critical chlorophyll meter reading," therefore, can be determined and used for precisely timing N applications to when the plant is "hungry." Work by Peng et al. (1996) in the dry season at IRRI have shown that 80 kg N/ha applied according to chlorophyll meter readings gave equal yields to 120 kg N/ha applied at set timings (Peng et al., 1996). Presently, NARS in South Asia are trying the chlorophyll meter as an aide in N management on monsoonal rice, with plans to extend this to wheat as well.

CROP ESTABLISHMENT

Timely planting of wheat is critical for good yields of wheat in the subtropics of South Asia. Research on dates of planting across the region usually indicate that late November is the optimal planting time (Table 9.4; Nur-E-Elahi et al., 1995). Late planting, however, is common in the region as harvest of rice and late monsoon rains often push planting dates into December and January. Yields are drastically reduced as the growth duration of wheat is shortened.

Table 9.4. Grain Yield of Wheat as Affected by Preceding Rice Varieties, Chaudanga, Bangladesh, 1992–94[a]

		1992-93			1993-94		
Rice Variety	Wheat Variety	Wheat Sowing Date	Rice Grain Yield ($t\ ha^{-1}$)	Wheat Grain Yield ($t\ ha^{-1}$)	Wheat Sowing Date	Rice Grain Yield ($t\ ha^{-1}$)	Wheat Grain Yield ($t\ ha^{-1}$)
BR1867-20-1-4	Kanchan	Nov 27	2.77	3.30	Nov 29	4.60	2.97
BR1870-89-1-1	Kanchan	Dec 3	2.67	2.80	Dec 5	4.30	2.83
BR30	Kanchan	Dec 7	2.53	2.70	Dec 10	3.88	2.50
BR31	Kanchan	Nov 27	2.67	3.20	Nov 29	3.90	2.93
BR11	Kanchan	Dec 7	2.50	2.60	Dec 10	3.56	2.33

[a]Adapted from Nur-E-Elahi et al., 1994.

Table 9.5. Effect of Tillage System on Wheat Following Rice, Pantnagar, India[a]

Tillage Treatment	Spikes m^2	Grains $Spike^{-1}$	1000 Grain Wt (g)	Grain Yield ($kg\ ha^{-1}$)
Zero-till	136	67	34.6	5620
Rotavator twice	110	46	35.3	4960
Harrow twice, cultivator + cold-buster thrice	128	52	34.3	4890

[a]Adapted from Modgal et al., 1995.

Generally, farmers have to wait for the land to dry out enough for conventional harrowing and seeding of wheat. Alternative tillage and crop establishment practices are being introduced by CIMMYT, the lead institution in this field within the RWCIGP. Prototype zero-till drills in Pantangar have shown that large improvements in wheat yields can be obtained with the zero-till planting immediately after rice harvest (Table 9.5; Modgal et al., 1995). CIMMYT has also introduced the practice of "surface seeding" or broadcasting wheat seed by hand into wet soil just after rice harvest. Additionally, CIMMYT is encouraging the practice of growing wheat on raised beds, which has advantages in water, weed, and fertilizer management.

Most of the rice in the R-W tract is transplanted. As labor becomes more expensive and water becomes less available, farmers will have to switch to other methods of rice establishment such as direct seeding, both wet or dry. Direct dry-seeding of rice benefits the following wheat crop, which is deeper rooted than rice (Singh, 1995). Not puddling the soil allows soil aggregation to begin, which further develops during the wheat season.

WATER MANAGEMENT

Water use in rice is currently very high. About 8,000 $m^3\ ha^{-1}$ of water are consumed to produce a rice crop, compared to 4,000 $m^3\ ha^{-1}$ for a wheat crop (IRRI, 1995). Research at IRRI has shown that losses of water seepage and percolation in the levees can be more than that through the puddled area (IRRI, 1994). Dry-seeding of rice is a management option that can save substantial amounts of water, since about one-third of the season's water use in puddled, transplanted rice is in the land preparation and puddling process. Direct wet seeding was found to use significantly less water than transplanted rice, because of less time spent in the land preparation activities of the former (IRRI, 1993).

Supplemental irrigation in rice, especially at the crucial reproductive stage can boost production substantially (IRRI, 1992). However, in South Asia, irrigating monsoonal rice is not a common practice (Hobbs et al., 1996; Ahmed et al., 1993; Hobbs et al., 1992). Most of the wheat in South Asia is irrigated, but in many cases farmers are only irrigating twice even if irrigation water is available (Hobbs et al., 1996). Sattar (1984) showed that 4 or 5 irrigations gave highest yields in N.W. Bangladesh.

Waterlogging is a serious problem in the region, with about 6 million ha of irrigated land in India known to be affected (IRRI, 1995). Salinity is often a direct consequence of waterlogging in semiarid and arid regions, and crop production is therefore affected.

Water scarcity is going to become more and more of an issue in Asia as both water quality and availability declines over time.

CONCLUSION

A large amount of good, agronomic and soils research has been carried out by the NARS in RW systems the last several years. However, several questions about the sustainability of the RW stand out. Are yields declining in farmers' fields? Are farmers applying more fertilizer to obtain the same yields? Do the low levels of soil organic matter in RW soils need to be built back up? Can nitrogen fertilizer use-efficiency be improved in RW systems? Nutrient management in South Asia will likely have to become more "site-specific" than in the past. This means that farmers will need "knowledge-intensive" tools that will allow them to decide what rates of fertilizer to apply, how to handle their residue, etc. These and other questions related to the applied problems and long-term sustainability issues of the RW system are currently being studied by the NARS and the IARCS together in the Rice-Wheat Consortium for the Indo-Gangetic Plains.

REFERENCES

Ahmed, N.U., M.M. Alam, A.M. Bhuiyan, R. Islam, and M.U. Ghani. Rice-Wheat Diagnostic Survey in Greater Kushtia District of Bangladesh. Bangladesh Rice Research Institute, Gazipur, Bangladesh, 1993.

Badaruddin, M. and R.A. Razu. 1995. Crop Management Research. Wheat Research Center. Bangladesh Agricultural Research Institute. Nashipur, Bangladesh, 1994–95.

Bhardwaj, K.KR. Recycling of Crop Residues, Oil Cakes, and Other Plant Products in Agriculture. In *Recycling of Crop, Animal, Human, and Industrial Wastes in Agriculture*. H.L.S. Tandon, Ed. Fertiliser Development and Consultation Organization, New Delhi, 1995, pp. 9–30.

Byerlee, D. Knowledge-Intensive Crop Management Technologies: Concepts, Impacts and Prospects in Asian Agriculture. Paper presented at the International Rice Research Conference, Bangkok, Thailand, 3–5 June, 1996.

Cassman, K. G. and P. L. Pingali. Extrapolating Trends from Long -Term Experiments to Farmers' Fields: The Case of Irrigated Rice Systems in Asia. In *Agricultural Sustainability, Economic, Environmental and Statistical Considerations*. V. Barnett, R. Payne, and R. Steiner, Eds. John Wiley & Sons, Chichester, 1995.

Cassman, K.G., S.K. De Datta, D.C. Olk, J. Alcantara, M. Samson, J. Descalsota, and M. Dizon. Yield Decline and Nitrogen Economy of Long-Term Experiments on Continuous, Irrigated Rice Systems in the Tropics. *Adv. Soil Sci.* 181–222, 1995.

Cassman, K.G., A. Dobermann, P.C. Sta. Cruz, G.C. Gines, M.I. Samson, J.P. Descalsota, J.M. Alcantara, M.A. Dizon, and D.C. Olk. Soil Organic Matter and the Indigenous Soil Nitrogen Supply of Intensive Irrigated Rice Systems in the Tropics. *Plant and Soil* (in press), 1996.

Dobermann, A. Perspectives for 1997–2010 in the Sustainability and Management of Intensive Rice Systems. Paper presented at the India-IRRI Dialogue, 27–29 September, 1996.

Dobermann, A., K.G. Cassman, P.C. Sta. Cruz, H.U. Neue, E.O. Skogley, M.F. Pampolino, and M.A.A. Adviento. Dynamic Soil Tests for Rice. In *Fragile Lives in Fragile Ecosystems*. Proceedings of the International Rice Research Conference, 13–17 February, 1995. International Rice Research Institute, Manila, Philippines, 1995, pp. 343–365.

FAO. World Soil Resources Report 67 (2–7). The Digitized Soil Map of the World. Food and Agriculture Organization of the United Nations, 1992.

FAO. Land Degradation in South Asia: Its Severity, Causes and Effects Upon the People. World Soil Resources Reports 78. FAO, Rome, 1994.

Hobbs, P.R., G.P. Hettel, R.K. Singh, R.P. Singh, L.W. Harrington, V.P. Singh, and K.G. Pillai, Eds. Rice-Wheat Cropping Systems in Faizabad District of Uttar Pradesh, India: Exploratory Surveys of Farmers' Practices and Problems and Needs for Further Research. Mexico, D.F.:CIMMYT, 1992.

Hobbs, P.R., L.W. Harrington, C. Adhikary, G.S. Giri, S.R. Upadhyay, and B. Adhikary. Wheat and Rice in the Nepal Terai: Farm Resources and Production Practices in Rupaandehi District. Mexico, D.F.;NARC and CIMMYT, 1996.

Hobbs, P.R. and M. Morris. Meeting South Asia's Future Food Requirements from Rice-Wheat Cropping Systems, Priority Issues Facing Researchers in the Post-Green Revolution Era. National Resource Group Paper 96-01. Mexico, D.F.: CIMMYT, 1996.

Huke, R. and E. Huke. Rice-Wheat Atlas of South Asia. International Rice Research Institute, 1992.

IRRI. Program Report for 1991. International Rice Research Institute, Los Banos, Philippines, 1992.

IRRI. Program Report for 1992. International Rice Research Institute, Los Banos, Philippines, 1993.

IRRI. Program Report for 1993. International Rice Research Institute, Los Banos, Philippines, 1994.

IRRI. Water A Looming Crisis. IRRI 1994–1995. International Rice Research Institute, Los Banos, Philippines, 1995.

Modgal, S.C., P.P. Singh, Y. Singh, B. Singh, R.P. Tripathi, P.C. Pandey, and A.K. Bhardwaj. Tillage and Sustainability of Rice-Wheat Cropping System. G.B. Pant University of Agriculture and Technology, Pantnagar, U.P., India, 1995.

Nambiar, K.K.M. Major Cropping Systems in India. In *Agricultural Sustainability, Economic, Environmental and Statistical Considerations*. V. Barnett, R. Payne, and R. Steiner, Eds. John Wiley & Sons, Chichester, 1995.

Nur-E-Elahi, N.U. Ahmed, M.M. Alam, A.M. Bhuiyan, P.C. Bhattacharya, M.R. Islam, M.U. Ghani, T. Islam, and M. Nuruzzaman. Achievements in Rice-Wheat Research in Bangladesh Rice Research Institute. In *Sustainability of Rice-Wheat Systems in Bangladesh*. M.A. Razzaque, M. Badaruddin, and C.A. Meisner, Eds. BARI, BRRI, CIMMYT, 1995.

Olk, D.C., Cassman, K.G., Randall, E.W., Kinchesh, P., Sanger, L.J., and Anderson, J.M. Changes in Chemical Properties of Organic Matter with Intensified Rice Cropping in Tropical Lowland Soil. *Eur. J. Soil Sci.* (in press), 1996.

Pagiola, S. Environmental and Natural Resource Degradation in Intensive Agriculture in Bangladesh. Environmental Economics Series Paper No. 15. The World Bank. Washington, D.C., 1995.

Peng, S., F.V. Garcia, R.C. Laza, A.L. Sanico, R.M. Visperas and K.G. Cassman. Increased Nitrogen Use Efficiency at High Yields Levels Using a Chlorophyll Meter on Irrigated Rice. *Field Crops Res.* (In press), 1996.

Piekielek, W.P. and R.H. Fox. Use of a Chlorophyll Meter to Predict Sidedress Nitrogen Requirements for Maize. *Agron. J.* 84:59–65,1992.

Regmi, A.P. Long-Term Effects of Organic Amendment and Mineral Fertilizers on Soil Fertility in a Rice-Rice-Wheat Cropping System in Nepal. M.S. thesis, University of the Philippines at Los Banos, 1994.

Sattar, M.A. A Water Management Profile of the Thakurgaon Tubewell Project in Bangladesh. M.S. thesis, University of the Philippines, Los Banos, Languna, Philippines, 1984.

Sidhu, D.S. and D. Byerlee. Technical Change and Wheat Productivity in the Indian Punjab in the Post-Green Revolution Period. CIMMYT Economics Working Paper 92-02. Mexico, D.F. CIMMYT, 1992.

Singh, Y. Rice-Wheat Cropping System NARP Basic Research Sub-Project. Progress Report 1994–95. G.B. Pant University of Agriculture and Technology, Pantnagar, U.P., India, 1995.

Skogley, E.O., S.J. Georgitis, S.J. Yang, J.E. and B.E. Schaff. The Phytoavailability Soil Test-PST. *Commun. Soil Sci. Plant. Anal.* 19:1477–1493, 1990.

Woodhead, T., R. Huke, and E. Huke. Areas, Locations, and Ongoing Collaborative Research for the Rice-Wheat Systems of Asia. In *Sustainability of Rice-Wheat Production Systems in Asia.* R.S. Paroda, T. Woodhead, and R.B. Singh, Eds. Food and Agriculture Organization of the United Nations, Regional Office for Asia and the Pacific, Bangkok, Thailand, 1994, pp. 126–130.

TEN

Sustainable Management of Soil and Water Resources in the Upper and Trans-Gangetic Plains of India

B.S. Ghuman and R.S. Narang

Rivers, in addition to other agents, play a significant role in the formation and development of soils of a country. The river Ganges is one of several prominent rivers of North India. It originates in the Central Himalayas and flows through Uttar Pradesh (U.P.), Bihar and West Bengal. Therefore, western U.P., Punjab, and Haryana form the Upper and Trans-Gangetic Plains (Figure 10.1).

Punjab State is located between 29°30′ to 32°32′N latitudes and 73°55′ to 76°55′E longitudes; Haryana between 27°39′ to 30°55′N latitudes and 74°27′ to 77°36′E longitudes; and Uttar Pradesh between 23°50′ to 31°28′N latitudes and 77°04′ to 84°38′E longitudes (Singh, 1991). The summers are extremely hot and winters quite cold. During May and June, the temperature rises to 40 to 45°C. The coldest months in the region are January and February. During these months, mean minimum temperature in general falls below 5°C in Punjab, Haryana, and Uttar Pradesh. In Haryana, temperature falls to an extent of −2 to −3°C for a few days. Occasionally, frost also occurs in the region. The mean annual rainfall ranges from 30 to 140 cm in Punjab, 30 to 120 cm in Haryana, and 60 to 220 cm in Uttar Pradesh. The open-pan evaporation varies from about 1 mm day^{-1} in winter to about 15 mm day^{-1} in summer. There are two main growing seasons in the region, summer (locally called *kharif*) from June to September, and winter (*rabi*) from October to April. This area produces almost every crop—wheat (*Triticum aestivum*), rice (*Oryza sativa* L.), maize (*Zea mays* L.), gram (*Phaseolus mungo* Roxb.), pulses, oilseeds, cotton (*Gossypium arboreum* L.), sugarcane (*Saccharum officinarum* L.), potato (*Solanum tuberosum*), and several vegetables and provides food security to the country. With increasing human and animal population of the country, demands for food and forage are also increasing, which put tremendous strains on land and water resources. To maintain the quality of these resources for a long time to come, sustainable management of them is a priority requirement.

In this region, water resources are unevenly distributed and the soils range from very productive to unproductive. Punjab and Haryana have also sizeable semiarid area (Table 10.1). Punjab has the largest (94%) and Uttar Pradesh the smallest (59%) irrigated area in the region.

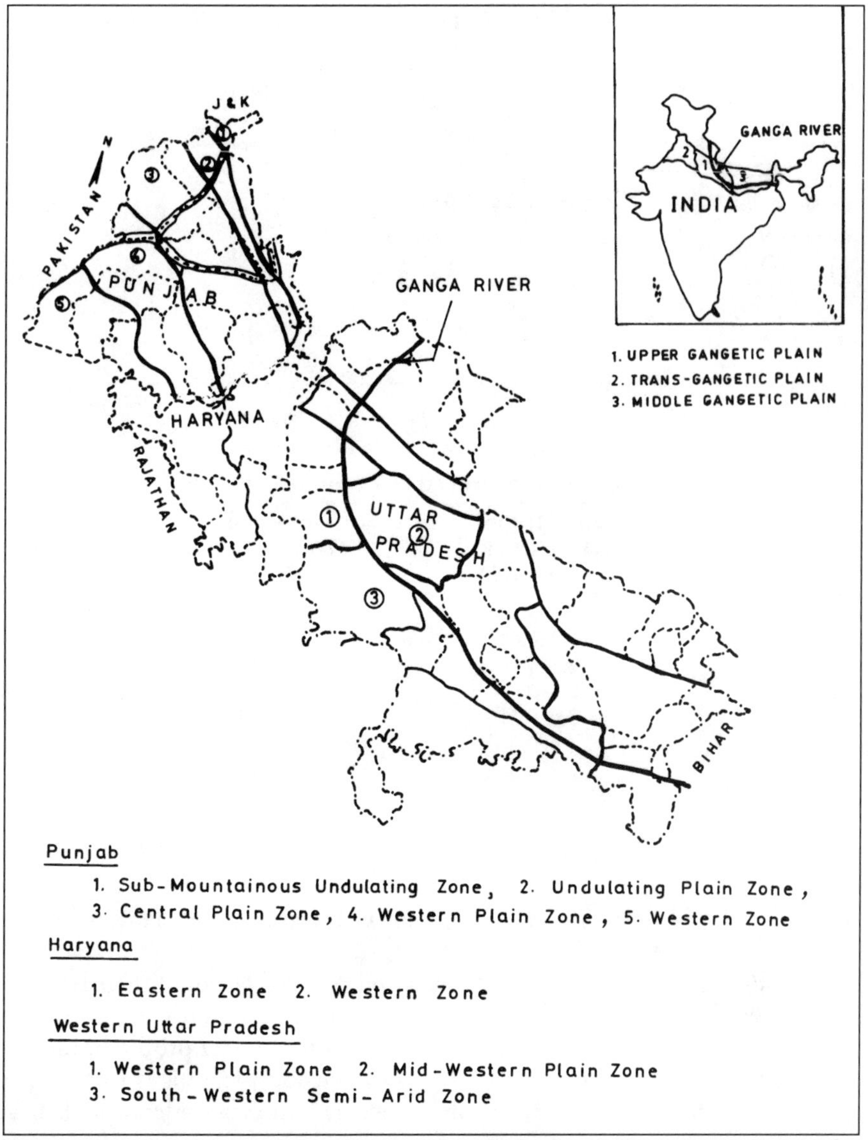

Figure 10.1. Agro-climatic zones of Punjab, Haryana, and Western Uttar Pradesh.

LAND RESOURCES AND USES

Arid Zone

There is no hot and arid regime in Uttar Pradesh, but Punjab and Haryana account for 9% of the hot desert (32 Mha) of the country (329 Mha). Soils are light-colored and have been developed under arid and semiarid hot climate from parent material provided by the River Indus system, and wind-eroded and deposited sediments. Sand dunes are present in southern parts of Punjab and Haryana. Aridisols are calcareous throughout this region and contain calcium carbonate nodules or gypsum in lower layers. These soils are characterized by the ochric epipedon underlain by cambic, calcic, or gypsic horizons. The pre-

Table 10.1. Some Statistics of Punjab, Haryana and Uttar Pradesh (U.P.) regions of Northwestern India[a]

Particulars	Punjab	Haryana	U.P.
Geographical area (Mha)	5.0	4.4	29.4
Arid area (%)	26.8	29.3	Nil
Semiarid area (%)	60.2	59.7	7.0
Forest area (%)	4.4	3.8	12.2
Net sown area (Mha)	4.2	3.5	17.2
Cropping intensity (%)	176	169	147
Net irrigated area (%)	94	73	59
Nutrients use (kg/ha)	168	106	89
Population density (per km^2)	401	369	491

[a]From Singh (1994).

Table 10.2. Properties (Range) of Some Arid Zone Soils[a]

Soil Property	Sand Dune (Ustic Torri-psamments)	Interdune (Ustrochreptic Camborthids)	Terrace (Ustrochreptic Camborthids)
Sand (%)	87.5–96.7	13.2–81.9	14.2–60.1
Silt (%)	0.2–6.6	6.7–57.7	24.2–68.7
Clay (%)	2.5–5.9	10.9–29.1	15.0–32.0
$CaCO_3$ (%)	0.1–0.5	0.3–20.0	0.8–19.3
pH (1:2)	7.8–8.5	7.8–8.5	7.6–8.4
EC (dS/m)	0.1–0.8	0.1–1.3	0.2–0.5
CEC (cmol[C^+]/kg)	2.8–4.0	6.0–16.3	6.2–17.8

[a]From Sharma et al. (1996).

dominant soil order is Entisol and in some areas Inceptisol, i.e., Psamments, Fluvents, Ustochrepts etc.

In Punjab, sand dunes are seen in Bathinda and parts of Faridkot, Ferozepur, and Sangrur districts. Sand dune soils contain the highest and terrace soils the lowest amount of sand (Table 10.2). Clay content increases with depth in the soil profile of interdunal and terrace soils but not in dune soils. Soils are nonsaline and pH ranges from 7.6 to 8.5. In Haryana, they are found in Sirsa, Hisar, Bhiwani, and Mahindergarh districts. Sand dunes cover 29% of the arid zone. In this area, pearl millet (*Pennisetum typhoideum* Pers.), castor (*Ricinus communis* L.), clusterbean (*Cyamopsis tetragonoloba*), grain legumes, and oilseeds are the predominant crops. In canal-irrigated areas, small dunes have been leveled to irrigate the fields, and cotton and wheat are predominant crops. Groundwater in the arid areas is highly brackish and, therefore, precludes raising crops (which have high water requirements) without canal irrigation. In canal-irrigated areas, rice now has also spread to considerable extent, which further aggravates the situation.

The arid zone soils are low in available N, medium in available P, and have a wide range of K content (Table 10.3). The fertility status of these soils is closely related to soil texture, and these require liberal fertilization for sustained production.

Semiarid Zone

A large part of Punjab, Haryana, and Western Uttar Pradesh lies in the semiarid zone. A number of rivers flowing in the region have deposited alluvium, the source of which de-

Table 10.3. Available Nutrients in Typical Arid Zone Soils[a]

Soil Series	Organic Carbon (%)	Mean Available P (kg/ha)	Mean Available K (kg/ha)	Mean Available Fe (mg/kg)	Mean Available Mn (mg/kg)	Mean Available Zn (mg/kg)
Dune and sandy plain soils (Typic Torripsamment)						
Dune	0.05	14	158	8.3	5.8	0.7
Molasar	0.12	17	189	10.7	14.9	0.1
Brown loam soils (Typic Camborthids)						
Parbatsar	0.15	22	176	11.1	22.1	1.2
Gajsing pura	0.25	11	239	8.8	18.4	0.8

[a]From Joshi (1985).

Table 10.4. Physicochemical Properties and Available Nutrients in Some Soils of Ghaggar Basin[a]

Profile No.	Depth (cm)	pH	ECe (dS/m)	$CaCO_3$ (%)	N (kg/ha)	P (kg/ha)	K (kg/ha)	Clay (%)
1	0–178	8.0–8.5	0.32–0.84	–	33–36	15–26	124–216	5.9–12.6
2	0–190	7.8–8.2	0.35–0.99	0.1–1.0	50–78	13–22	25–312	12.5–31.4
3	0–110	8.1–8.8	0.62–0.99	0.5–1.0	32–356	4–26	32–252	3.0–27.7
4	0–156	9.2–10.4	0.88–25.8	0.5–14.4	22–56	23–36	160–388	10.1–12.6
5	0–220	8.5–8.8	0.52–1.12	2.7–3.7	28–64	8–16	216–294	4.0– 9.0

[a]From Singh and Ahuja (1990).

termines the nature of soils developed. Topography, climate, and time have further differentiated the soils developed in the semiarid zone. These soils belong to orders Entisol, Inceptisol, Aridisol and Alfisol with a large number of great groups. Properties of five soil profiles from the Ghaggar basin in Punjab and Haryana are shown in Table 10.4. These soils have wide variation in plant nutrients, clay and calcium carbonate distribution within and between the profiles. This region also has salt-affected soils as shown by profile number 4. The available P content of salt-affected soils is higher than that of normal soils. Irrigated soils of the region are highly productive. Major crops grown are rice, wheat, sugarcane, cotton, oilseeds, fodders, fruits, and vegetables. Wherever supplemental irrigation is not feasible, sugarcane and rice are omitted and the low-water-requiring crops are grown.

WATER RESOURCES

The water resources of the region are rainfall and groundwater. Mean annual rainfall is 56, 61, and 84 cm in Haryana, Punjab, and Western Uttar Pradesh, respectively, with a coefficient of variation often exceeding 20% (Parthasarthy et al., 1987). The uncertain rainfall, high evaporative demand (annual potential evaporation > 180 cm) and very low water-holding capacity of sandy soils expose the vegetation to droughts, often causing complete crop failure. The groundwater resources of the region are highly deficient in quality as well as quantity. Aquifers are of sweet water and at some places of saline water. The occurrence of the saline groundwater aquifer in the region is closely related to aridity of the area, proximity to source of salts and the hydrological conditions. The salts in soils and groundwaters are products of either weathering or fossil origin.

Table 10.5. Distribution and Use of Water Resources in the Region[a]

State	Irrigation Benefit from Major and Medium Schemes by the End of 1984–85 (Mha)		Groundwater Resources (Mha-m/annum)			
	Potential	Utilizable	Usable	Net Drawdown	Potential Available	Use of Low Quality Waters
Punjab	1.568	1.508	1.31	0.93	0.38	0.38
Haryana	1.929	1.745	0.88	0.61	0.27	0.38
U.P.	6.813	5.513	9.27	2.68	6.59	1.28

[a]From Anonymous (1992).

The semiarid zone is relatively better placed than the arid zone in respect to quantity and quality of groundwater aquifers. Tubewell irrigation is well developed in the good quality water zone in Punjab, Haryana, and Western Uttar Pradesh, which makes it the major food grains, sugarcane, and cotton surplus area of the country. The aquifers are generally of good quality water except for the incidence of high residual sodium carbonate (RSC). High RSC waters are common in central and southwest Punjab, and Karnal, Kaithal, Panipat, and Gurgaon districts of Haryana. It is imperative to use gypsum with such waters.

Both arid and semiarid regions of north India experience water scarcity, necessitating dependence on canal water for sustained and stable crop production. The potential of canal water use is quite impressive in the Upper and Trans-Gangetic Plains. In Uttar Pradesh, groundwater potential remains underutilized (Table 10.5). Substantial quantities of low quality groundwaters are also used in this region.

SOIL AND WATER RELATED PROBLEMS

Arid and semiarid zones have a variety of problems associated with soil and water management (Table 10.6). The causative factors are natural as well as man-made. Even under the best soil and water management practices, one or the other soil degradation process operates at a perceptible pace. Major problems of soil and water management are:

Wind Erosion

Wind erosion in the arid region is prevalent due to loose dry soil, lack of or scarce vegetation cover, and turbulent winds. There are two forms of wind erosion: (a) dust storms carrying fine soil particles in suspended form and affecting large areas, and (b) sand drifts from unstable sand dunes and loose sandy soils into neighboring areas. Almost 18% of the area of Haryana and a similar area of Punjab are affected by wind erosion.

Table 10.6. Some Soil Problems in the Region (Expressed as % of Total Area)

Nature of Problem	Punjab	Haryana	U.P.
Waterlogging	4.0	6.1	6.7
Salt-affected soils	10.4	10.3	4.4
Coarse-textured soils	13.2	18.7	—
Erosion	0.9	36.0	N.A.[a]

[a]N.A. = Data not available.

Water Erosion

Rainfall per se is not an important factor in water erosion in low rainfall areas of the region. Most of the soil erosion occurs through floodwaters emanating from hilly areas or when rivers and rivulets break through banks during flash floods. This process is quite predominant in the Shiwalik hills and submountainous plain region, called *Kandi* area (0.46 Mha) in the Punjab and Haryana, and Kumaon and Garhwal Hills of Uttar Pradesh.

Waterlogging and Soil Salinization

Irrigation is required to successfully raise crops in the arid and semiarid zones. As a result, an impressive network of canals; e.g., Bhakra-Nangal canal system, have come into existence in the region. But with water percolation from canals, distributaries, and watercourses and fields, the water table is rising. Overirrigation is another factor of water table rise. In Haryana, rate of water table rise has been 0.14 to 1.0 m yr^{-1} and so far more than 0.4 Mha of land has the water table within 3 m of soil surface. Likewise in southwestern Punjab, a steady rise in the water table is occurring, particularly in the low-lying and poorly drained area of the Faridkot district. More and more areas are being salinized through excessive use of canal water and irrigation with saline water (Table 10.7), which was necessary to meet the enhanced water demands of paddy cultivation.

Groundwater Quality

The high amount of salts in groundwater in the arid region is a major limitation to its agricultural and domestic use. In almost 60% of the area in Haryana, southwest Punjab, and western Uttar Pradesh, the groundwater is brackish. Groundwater in the arid and semiarid regions also contains high residual sodium carbonate (RSC) (Table 10.8). Irrigation with such water has rendered a large area of land unproductive.

Receding Groundwater in the Good Quality Zone

With increase in area under rice and sugarcane, a large number of shallow tubewells have been installed to meet high water demands of these crops. This causes an excessive overdraft of groundwater from good quality aquifers. As a result, water tables are going down. At present, Punjab, where tubewells irrigate 68% of the area, has 0.8 M tubewells. Annual draft by these tubewells is 1.91 Mha m while the groundwater recharge is 1.44 Mha m (Prihar et al., 1992). If the 0.19 Mha m saline water is also added, the annual water deficit amounts to 0.66 Mha m. This deficit causes an average decline in the water table

Table 10.7. Waterlogged and Salt-Affected Area in Canal Commands of the Region[a]

	Waterlogged Area (Mha)			Salt-Affected Area (Mha)		
State	Canal Command	Outside Command	Total	Canal Command	Outside Command	Total
Punjab	NA[b]	0.199	0.199	0.393	0.127	0.520
Haryana	0.230	0.045	0.275	0.455	NA	0.455
U.P.	0.455	1.525	1.980	1.061	0.234	1.295

[a]From Singh (1994).
[b]NA = Data not available.

Table 10.8. Proportion of Different Quality Groundwater in the Region[a]

	Groundwater Quality			Incidence of High RSC Water		
State	**Good**	**Marginal**	**Poor**	**Saline**	**Sodic**	**Saline-Sodic**
Punjab	59	22	19	22	54	24
Haryana	37	8	55	24	30	46
U.P.	37	20	43	NA	NA	NA

[a]From Manchanda et al. (1989).

of 23 cm per annum in good quality aquifers. In Haryana, where number of tubewells has increased 15-fold in the last 20 years, the average annual decline in the water table has been 14 to 33 cm in different areas. An indirect effect of decline in the water table is the ingress of water from the adjoining less-exploited high-water table aquifers with marginal to poor quality waters.

Soil Fertility Depletion

Soils of arid and semiarid regions are low in N, medium in available P, and medium to high in available K (Tables 10.3 and 10.4). Fertilizers with single nutrient or at best two nutrients have so far sustained high production in the rice-wheat cropping system. For example, Ludhiana district in Punjab is using 233 kg ha^{-1} of nutrients, largely N and P, to produce 4.05 and 6.22 Mg ha^{-1} of wheat and paddy, respectively, besides removing 600–700 kg NPK ha^{-1} yr^{-1} (Narang et al., 1990; Narang and Gill, 1996). This level of productivity also depletes the soil of the other essential macro- and micro-nutrients. In Haryana, all soils are deficient in N, while 62% are medium in P. About 74% of the soils of the state show deficiency of Zn, 9.5% of Fe, and 1.0% of Cu. Zinc deficiency is widespread in Punjab soils as well as in Haryana. Deficiency of S has been reported in wheat and mustard grown on coarse-textured soils. Lately, deficiency of Mn in wheat and berseem has emerged as a serious problem in highly permeable soils brought under rice cultivation.

Soil Structural Status

With the introduction of high yielding varieties of wheat and rice during 1960s and 1970s, respectively, the cropping system has changed from low water requiring crops like pearl millet, pulses, and oilseeds in favor of these crops. This change is causing considerable losses of water by evaporation and nutrients by leaching in sandy soils of the region. In fine-textured soils, water stagnates as infiltration capacity is low and crop yields decline. There are also soils with a compact subsoil horizon. The bulk density of this zone can be 1.65 to 1.8 Mg m^{-3} (Sur and Singh, 1972) which prevents root penetration and causes excessive moisture and anaerobic conditions in the soil profile. Soil crusting due to low organic matter and silica cementation upon drying of dispersed soil particles after a rainfall is a serious problem because it affects seedling emergence in structurally unstable soils of the region.

Drought

Droughts are common in the region due to low amount and uneven distribution of rainfall. It lowers crop and fodder production and also causes shortages of drinking water. According to an appraisal by the World Bank, 60 Mha in 72 districts of the country are drought-prone, and the majority of them fall in northwestern India. Drought causes migration of cattle to better rainfall areas for fodder, and a number of socioeconomic problems.

SUSTAINABLE MANAGEMENT OF LAND AND WATER RESOURCES

With the urgency of producing more food for feeding increasing population and increasing concern for environmental safety, the issue of land and water management assumes great significance. Some thoughts based on research findings are presented here to manage the resources of the region on a sustainable basis.

Reducing Wind Erosion

Since high wind velocities (up to even 40 km hr^{-1}) from May to July are common features of the arid zone, huge amounts of sand can move from bare sandy plains and unstable sand dunes. Gupta (1990) reported that 1449 Mg of sand was removed per hectare from bare sandy plains in 70 days. In the same period, 5560 Mg ha^{-1} of sand was removed from bare unstable sand dunes. Reducing wind velocity, protecting the soil with vegetation cover, or maintaining the surface soil in a least erodible state can effectively reduce wind erosion. For instance, Gupta and Gupta (1981) showed that a medium-textured plowed soil having 42% clods of more than 5 mm size lost only 0.5 Mg ha^{-1} of soil as against 40 Mg ha^{-1} soil loss from the field which was planked after plowing that reduced clod size. Abrol and Venkateswarlu (1995) reported that shelter belt plantations of species *Prosopis juliflora* or *Acacia tortillis* significantly reduced wind velocity and soil loss. In areas where severe wind erosion does not allow arable land use, the land should be put under grasses, trees, or a mixture of both. When sandy plains were planted to grass, 13.5 Mg ha^{-1} of sand was deposited on the field in 70 days during April to June (Gupta, 1990). Furthermore, agriculturally suitable areas should be amended with the addition of organic matter, pond sediments, etc., to improve soil structure. Therefore, stubble-mulch farming and leaving crop residues on the surface and keeping the soil surface cloddy or uncultivated during windy weather help reduce wind erosion.

Some techniques for stabilizing sand dunes have been standardized by the Central Arid Zone Research Institute (CAZRI), Jodhpur (Rajasthan), India. These consist of fencing the area for protection against biotic interference, establishing windbreaks with suitable plant species (Table 10.9), and later on avoiding overgrazing. One limitation in

Table 10.9. Planting Methods and Suitable Plant Species for Arid Habitats[a]

Land Form	Planting/Afforestation Method	Suitable Tree/Shrub/Grass Species
Sand dunes	Six to eight month saplings deep planted in 5mx5m between strips of microwind breaks	*Acacia tortillis, Prosopis juliflora, Cordia Rothi, Prosopis cineraria*
Sandy plains	Planting in 45 cm^3 pits in 5m × 5m spacings	*Acacia tortillis, Zizyphus mauritiana Albizzia lebbek.*
Shallow soils (with hard pan	Planting to be done in 60 cm^3 pits after breaking the hard layer	*Eucalyptus tesselaris, E. melanophloa Acacia nilotica, Zizyphus mauritiana*
Saline patches	Planting on raised beds	*Prosopis juliflora*

[a]From Singh (1994).

planting and later sustaining young saplings during hot weather in arid zone is the inadequate availability of good quality water. Some tree species, e.g., *Prosopis juliflora*, *Acacia tortillis*, *Eucalyptus camaldulensis*, *E. hybrid*, *Leocena leucocephala*, *Azadiracta indica*, are hardy enough to tolerate the locally available saline well water (salinity 3–9 dS m^{-1}). Afforestation with these tree species may facilitate stabilization of sandy belts in arid areas.

Checking Water Erosion and Floods

Rainwater accumulates in the foothills and piedmont areas which often become a source of excessive runoff and flash floods. Therefore, one efficient way of checking water erosion in the plains is to conserve and retain rainwater in situ in these areas. Furthermore, soil conservation techniques, i.e., provision of check dams, and waste weirs, gully plugs, drop spillways, vegetative barriers, waterways, and staggered trenches may also be used to store the rainwater uphill and enhance recharge of aquifers. In uneven terrains of the plains, contour bunding and graded bunding, combined with land leveling, effectively check runoff and water erosion. Maintaining a groundcover with crop residue reduces impact of raindrops, improves water infiltration into soil, and reduces runoff and soil losses. Gullied areas need protection with small check dams and afforestation with suitable hardy tree species. Watershed development programs and making water storage tanks are also dependable techniques when applied on a community basis (Grewal and Mishra, 1982).

Controlling Waterlogging and Soil Salinization

Reclamation of waterlogged and salinized soils involves surface and subsurface drainage. Wherever groundwater quality permits, subsoil drainage should be preferred over surface drainage.

For reclaiming alkali soils, reclamation technology has been well standardized, and considerable area has been reclaimed in the region (Singh et al., 1981). Now this area is contributing significantly to rice and wheat production in Punjab and Haryana. Post-reclamation management of such soils is necessary, without which these can again get degraded quickly. Such soils should never be left bare for long periods because the salts which were leached down during the reclamation process can again come to the surface.

In canal command areas, a rise in the water table precedes secondary salinization, which could be controlled by minimizing water losses from the conveyance system by lining canals, distributaries, and field watercourses. To check seepage losses, 15,200 km of watercourses in Punjab and 17,700 km in Haryana are being lined. Also, canal lining in-

Table 10.10. Cropping Intensity and Irrigation Time as Influenced by Lining of Watercourses[a]

	Before Lining		After Lining	
Location of Farm at the Watercourse	Cropping Intensity (%)	Irrigation Time (min[b]/ha)	Cropping Intensity (%)	Irrigation Time (min[b]/ha)
Head	135	150	150	140
Middle	116	170	146	148
Tail	85	210	142	175

[a]From Khepar and Kaushal (1983).
[b]Minutes.

creases availability of irrigation water and thus the cropping intensity (Table 10.10). Maximum advantage of lining accrues to the farmers at the tail end of the canal in the form of increased water supplies, making high yield realization and improvement in cropping intensity feasible.

Field water losses can be minimized in a number of ways. Rice and sugarcane have high water requirements. In areas where groundwater is poor, cultivation of these crops should be discouraged, otherwise waterlogging and secondary salinization may jeopardize the productivity of these soils.

Following a well-developed irrigation schedule for different crops can reduce irrigation water requirement and increase water use efficiency. Research shows that there is no need to keep water standing for 3 weeks after the transplanting of rice (as recommended earlier), and that the duration of ponding could be reduced to about 2 weeks with no adverse effect on the yield of rice (Narang and Singh, 1988). Likewise, there is also no need for continuous ponding of water during the post-seedling establishment phase in the rice fields on these coarse to medium textured soils. Draining soil for 1–2 days before the next irrigation can reduce water requirement by 45 to 65 cm without affecting rice yield (Table 10.11). Further saving in water use in rice fields is made by ceasing irrigation about 2 to 3 weeks prior to harvest. Thus a rice crop can be raised with 160–180 cm of seasonal water use.

Substantial amounts of irrigation water (25–30%) can be saved by an alternate furrow irrigation method in sugarcane and other broad-row crops grown on ridges. Reducing plot size in wheat cultivation can also save water without losing crop yield (Table 10.12) and prevent overirrigation and eventually, waterlogging. On undulating and sandy soils, sprinkler irrigation is a desirable irrigation practice. Drip irrigation and microsprinklers are highly useful in orchard irrigation. Besides seepage, overirrigation is a serious problem

Table 10.11. Saving in Irrigation Water by Temporarily Withholding Irrigation to Rice in Punjab[a]

Irrigation Treatment	Irrigation Water Applied (cm)	Mean Paddy Yield(Mg/ha) (1974–77)
Continous submergence	190	5.51
1-Day drainage[b]	145	5.44
2-Day drainage	125	5.53
3-Day drainage	113	5.15
4-Day drainage	96	5.20

[a]From Sandhu et al. (1980).

[b]Drainage period is the time lapse between complete infiltration of ponded water and next irrigation.

Table 10.12. Effect of Plot Size on Irrigation Water Use and Yield of Wheat[a]

Plot Size (m × m)	Irrigation Water Applied (cm)	Wheat Yield (Mg/ha)
60 × 14	21.5	3.46
30 × 14	19.3	3.49
60 × 7	18.9	3.40
30 × 7	15.7	3.40

[a]From Singh et al. (1981).

Table 10.13. Effect of Sodic Water, With and Without Gypsum, on Mean pH, SAR and ESP in 180 cm Soil Profile and Water Infiltration Rate (I) after 10 Years of Irrigation[a]

Treatment	pH(1:2)	ESP_c	SAR_c	I (cm/hr)
	Rice-wheat rotation			
Good water (canal)	8.4	6.0	0.8	1.40
Sodic water	9.2	38.3	27.8	0.35
Sodic water + gypsum	8.7	28.1	15.9	1.20
	Millet-wheat rotation			
Good water (canal)	8.4	6.2	1.2	4.40
Sodic water	9.0	26.7	21.7	0.8
Sodic water + gypsum	8.5	22.1	11.7	1.5

[a]From Bajwa and Josan (1989). ESP = exchangeable sodium percentage; SAR = sodium absorption ratio.

in canal-command areas. Farmers tend to overirrigate as a security against delay in their next turn of water, since canal water is supplied on 'supply-driven basis' on rotational schedule of 7, 10, or 14 days by turn. The uncertainty in canal water supply should be removed through proper planning of canal water supplies on demand-driven or need-based schedules.

Proper Use of Poor Quality Groundwater

In areas where good quality water is scarce, crops are irrigated with saline water but on a limited scale. Farmers have been using saline water from wells for a few irrigations applied to a winter crop. The field is then left fallow for one or two rainy seasons for effective leaching of salts. This practice is not successful with high RSC waters, in which case application of gypsum is required to offset adverse effects of sodium on the soil. In Punjab, Haryana, and Uttar Pradesh, large areas have become unproductive due to misuse of sodic waters. Bajwa and Josan (1989) showed that soil degradation is caused more by adopting crop rotations requiring more water and where gypsum was not applied to the sodic water used for irrigation (Table 10.13). Saline water irrigation should not be considered as an alternative to fresh water irrigation unless it is used occasionally with due precautions and used in conjunction with recommended amendments.

Conjunctive use of canal and groundwater for irrigation is an effective method of making use of brackish water. The limitation in adopting this strategy is the quality of groundwater. Mixing surface water does improve the quality of groundwater temporarily, but over a period of time it leads to buildup of salts in the rooting depth, hence necessitating the use of gypsum for reclamation. There also occurs dilution of brackish water and making its use through shallow skimming wells, as is evident in the transition zone of good quality and poor quality aquifers in Punjab, Haryana, and Uttar Pradesh. In these areas, salinity of rising groundwaters has been diluted by rain and canal water.

Managing Receding Water Tables

Farmers in the good quality groundwater zone irrigate their crops excessively with the view to produce high yields. But this practice has resulted in lowering of the water table (WT). This problem can be tackled in two ways: (a) using less groundwater, and (b) recharging groundwater aquifers. Groundwater use can be reduced by changing the cropping pattern. Predominant crop rotation in these areas is rice-wheat. Serious efforts should be made to decrease the area under rice and replace it with an equally remunerative alternate crop such as soybeans (*Glycine max*), hybrid maize, or dairy feed and fodder crops in Punjab and Haryana. As indicated earlier, rice fields could be drained for 1–2 days before ponding them again. In this way also, considerable irrigation water can be saved. It has been observed that some farmers in Punjab and Haryana have started transplanting rice earlier, i.e., in late April and May, than its optimum period from June 15 to 30, due to easily available manual labor and electricity. But potential evaporation during this hot and dry period (April and May) is very high, and need for water by early-transplanted crop increases over timely-planted crop by 25 to 30% (Narang and Gill, 1993), which puts unnecessary extra stress on groundwater reserves. By transplanting rice during the optimum period, rate of fall in the WT can be decreased.

Groundwater recharge can be increased through cavity wells, percolation tanks, and spreading water of surface drains during the rainy season onto waste lands. Detention of water in foothills through small reservoirs can help recharge ground aquifers in the plains. Recharging of soil profiles with runoff water from rains, however, is not perused in this region as is done in inherently water-deficient states of Gujarat and Maharashtra in western India.

Improving and Sustaining Soil Fertility Status

Integrated management of plant nutrients using fertilizers, organic manures and other farm by-products, biofertilizers, and introducing legumes in rotation with cereals can be practiced for preventing decline in the fertility of intensively cultivated soils. Introduction of a break crop on a regulated basis (15–20% per annum) in a rice-wheat system to give respite to badly "fatigued" rice lands is being advocated to rejuvenate the soil health and its physical, chemical, and biological equilibrium and reduce buildup of soil/plant-borne diseases (on alternate host plants) of the rice crop in particular. However, this hypothesis needs to be validated experimentally. Farmers should be advised to give rest to their land at least by replacing one of the cereals with a legume or even a cereal crop during the main crop season. Some scientists advocate the use of green manures in rice-wheat rotation (Meelu, 1996). But the period during which a green manure crop can be raised coincides with the hot months of April, May, and June when evapotranspiration is very high and growing a green-manure crop entails 4–5 additional irrigations. Such a recommendation is debatable if the objective is to save scarce water.

Improving Soil Physical Properties

Predominant limitations of coarse-textured soils are low water and nutrient retention capacities. By addition of organic manures, and recycling of crop residues (estimated availability nearly 30 million tonnes and now being burned), both moisture and nutrient retentive capacities of these soils can be enhanced. In fine-textured soils, organic manuring helps in soil aggregation, thereby creating more congenial air and water regimes for plant growth. Seed-bed preparation at an optimum moisture content and preventing buildup of hard plow-sole layers by occasional deep tillage are very helpful in creating a good root environment. Turning under of crop residues and green manuring also helps in improving soil structure. In rice-wheat rotation, where a compact sublayer is likely to be formed during repeated puddling, deep plowing or chiseling could be used to break the hard layer and improve crop production (Gajri et al., 1992).

Drought Management

Proper soil and water management has a great role in mitigating the drought. These measures include development and management of irrigation sources; in situ water conservation; water storage in low-lying places or in specially constructed tanks; breeding crop varieties tolerant to insects, pests and diseases, and extremes of moisture stress; afforestation with multipurpose tree species; introduction of alternate crops and cropping systems; efficient crop management practices like early sowing, optimum plant population and moderate use of fertilizers tailored to the profile-stored water and irrigation water availability.

Managing Resources in Dryland Farming

For dry lands, increased efficiency of rainwater use is essential and it can be achieved by adopting the following strategies:

1. retaining rainfall in situ and minimizing the runoff,
2. reducing soil evaporation in relation to transpiration,
3. using drought-tolerant crops that fit in the rainfall pattern,
4. recycling the runoff water after water harvesting and drainage, and
5. using watershed management concept for maximization of rainwater absorption and retention in situ.

In fact, it is the use of a combination of all these practices that ensures best results with minimum degradation of the environment. The most critical factors of the rainfed zone are the moisture stress and nutrient (deficiency) stress. An integrated technology is needed for alleviating these stresses for achieving stable agricultural production.

The majority of soils in dryland areas are coarse-textured. Rather than fallowing during the monsoon period for recharging the profile with rainwater, these soils can be cultivated to grow a green-manure crop like sunnhemp (*Crotolaria juncea*), dhaincha (*Sesbania aculeata*), vetches or beans etc. Results in Table 10.14 show that three years of green-manuring with sunnhemp significantly increased soil water retention in the top 0.10 m layer. These results have great significance for seeding of dryland winter crops, such as wheat, where success or failure of a crop depends upon achieving successful germination and an early crop stand which, in turn, is affected by seedbed moisture content. Beneficial effects of green-manuring through increased moisture retention and that, too, for a longer period, significantly increased wheat grain yield (by 300–500 kg ha^{-1}) under dryland conditions (Table 10.15).

Mulching soil with crop residue is another practice for managing and conserving soil moisture in situ. For the *kandi* area of Punjab, the following in situ moisture conservation practices have been deveolped (Singh et al., 1983)

Table 10.14. Effect of Green Manuring (after Three Seasons) with Sunnhemp on Soil Water Retention (m^3/m^3) at Different Suctions[a]

Suction	Control		Treated Plots		LSD (0.05)	
(kPa)	0–0.1 m	0.1–0.2 m	0–0.1 m	0.1–0.2 m	0–0.1 m	0.1–0.2 m
0	0.34	0.34	0.35	0.35	0.005	0.006
10	0.17	0.17	0.19	0.18	0.011	NS
33	0.09	0.09	0.10	0.09	0.006	NS
1500	0.02	0.04	0.03	0.04	0.005	NS

[a]From Ghuman et al. (1997).

Table 10.15. Effect of Green Manuring with Sunnhemp and N Rates on the Yield of Dryland Wheat (Mg/ha)[a]

	N Rate (kg/ha)					
	0		40		80	
Year	C[b]	GM	C	GM	C	GM
1990-91	3.00	3.12	3.19	3.45	3.89	3.83
1991-92	3.12	3.71	3.65	4.43	3.84	4.46
1992-93	0.24	0.58	0.35	0.77	0.71	1.35
1993-94	0.92	1.04	1.04	1.12	0.95	1.19
Mean	1.82	2.11	2.06	2.44	2.35	2.71
LSD (0.05): GM = 0.11; N rate = 0.07; GM x N = 0.10						

[a]From Ghuman et al. (1997).
[b]C stands for control, GM for green manuring.

- Level the land, bund the field, and provide outlets for overflow.
- Plow before the onset of monsoons to enhance infiltration.
- Plow the fallow land occasionally during the summer or monsoon season or grow a green manure crop of sunnhemp so as to keep the weed population in check.
- Use locally available mulching material in the standing crop of maize during the last week of August to conserve the stored soil moisture for sowing wheat in November.
- Prepare soil soon after the harvest of summer crops for sowing winter crops.
- Plant the winter crop as early as possible to provide an early ground cover, and use the conserved moisture effectively.

REFERENCES

Abrol, I.P. and J. Venkateswarlu. Sustainable Development of Arid Areas in India with Particular Reference to Western Rajasthan. In *Land Degradation and Desertificationin Asia and the Pacific Regions.* A.K. Sen and A. Kar, Eds. Scientific Publishers, Jodhpur, India, 1995, pp. 135–153.

Anonymous. History of Irrigation in Indus Basin. Publi. 230. Central Board of Irrigation and Power, New Delhi, 1992, pp. 65.

Bajwa, M.S. and A.S. Josan. Prediction of Sustained Sodic Irrigation Effects on Soil Sodium Saturation and Crop Yields. *Agric. Water Manage.* 16:217–228, 1989.

Gajri, P.R., V.K. Arora and S.S. Prihar. Tillage Management for Efficient Water and Nitrogen Use in Wheat Following Rice. *Soil Tillage Res.* 24:167–182, 1992.

Ghuman, B.S., H.S. Sur, and Kulwant Lal. Effect of Green Manuring on Soil Properties and Yield of Wheat Under Dryland Conditions. *J. Indian Soc. Soil Sci.* 45:99–103, 1997.

Grewal, S.S. and P.R. Mishra. Sukhomajri Model of Development—A Boon for Foothill Farmers. *Indian Farming* 32:31–33, 1982.

Gupta, J.P. Sand Dunes and Their Stabilization. In *Techniques for Waste Lands Development,* I.P. Abrol & V.V. Dhruvanarayana, Eds. Indian Council of Agricultural Research, New Delhi, India, 1990, pp. 59–68.

Gupta, J.P. and G.N. Gupta. A Note on Wind Erosion from a Cultivated Field in Western Rajasthan. *J. Indian Soc. Soil Sci.* 29:278–279, 1981.

Joshi, D.C. Proc. Development and Management of Training Course on Irrigated Agriculture in Arid Areas. Water and Power Consultancy Services (India) Pvt. Ltd., Kailash Towers, New Delhi, 1985.

Khepar, S.D. and M.P. Kaushal. Impact of Lining Canal Water Courses on Farm Economy and Optimum Decision Making. Seminar 'Focus on Lining Water Courses in Canal Command Areas' Organized by Haryana State Minor Irrigation and Tubewell Corporation, Karnal at New Delhi from 7–11 December, 1983.

Manchanda, I I.R., I.C. Gupta and B.L. Jain.. Use of Poor Quality Waters. In *Review of Soil Research on Sandy Soils*. Int. Symp. Managing Sandy Soils. Central Arid Zone Research Institute, Jodhpur. February 6–10, 1989, pp. 362–383.

Meelu, O.P. Integrated Nutrient Management for Ecologically Sustainable Agriculture. *J. Indian Soc. Soil Sci.* 44:582–592, 1996.

Narang, R.S. and M.S. Gill. Water Management Constraints in Rice-Wheat Rotation in India. *Proc. Int. Conf. Vol. II. Wheatin Warm Area Rice-Wheat Farming System*, Dinajpur, Bangladesh, Organized by UNDP, ARC, BARI, CIMMYT, 1993, pp.328–338.

Narang, R.S. and M.S. Gill. Yield and Soil Fertility as Influenced by Rice-Wheat Cropping System in Long-Term Maximum Yield Research Studies. Paper presented in rice-wheat consortium for Indo-Gangetic Plains. Organized by ICRISAT at Surajkund, Haryana, India, 14–18 Oct., 1996.

Narang, R.S and N. Singh. Water Requirement Studies in Rice. *IRRI Newsl. Los Banos* 13:43, 1988.

Narang, R.S., S.S. Cheema, D.S. Grewal, H.S. Grewal and B.D. Sharma. Yield, Nutrient Removals and Changes in Soil Fertility Under Intensive Rice-Wheat Cropping System. *Indian J. Agron.* 35:113–119, 1990.

Parthasarathy, B., N.A. Sontokke, A.A. Munot, and D.R. Kothwale. Droughts/floods in the Summer Monsoon Season over Different Meteorological Sub-Divisions of India for the Period of 1871–1984. *J. Climatology* 7:57–70, 1987.

Prihar, S.S., B.S. Sandhu, M.P. Kaushal and M.S. Bajwa. Bull. No. PAU/1992/R/524 on Efficient On-Farm Management of Irrigation Water, Punjab Agril. Univ., Ludhiana, India, 1992. p. 160.

Sandhu, B.S., K.L. Khera, S.S. Prihar, and B. Singh. Irrigation Needs and Yield of Rice on a Sandy Loam Soil as Affected by Continuous and Intermittent Submergence. *Indian J. Agric. Sci.* 50:492–496, 1980.

Sharma, B.D., P.S. Sidhu, Gurmel Singh, and S.S. Mukhopadhyay. Elemental Distribution and Mineralogy of Arid Zone Soils of Punjab. *J. Indian Soc. Soil Sci.* 44:746–752, 1996.

Singh, Atar. Agricultural Characteristics in the Agro-Climatic Zones of Punjab and Haryana. In *Agro-Climatic Zone Specific Research*, S.P. Gosh, Ed. Indian Council of Agricultural Research, N. Delhi, India, 1991, p. 539.

Singh, K. and R.L. Ahuja. Distribution of Primary Nutrients in Relation to Soil Characteristics in the Ghaggar River Basin. *J. Indian Soc. Soil Sci.* 38:733–735, 1990.

Singh, N.T. Land and Water Management in North and Northwestern India. In *Management of Land and Water Resources for Sustainable Agriculture and Environment*. Indian Soc. Soil Sci. Public. New Delhi, India, 1994, pp. 93–120.

Singh, N.T., A.C. Vig, and R. Singh. Soil Water Plant Relationships. Res. Bull. Punjab Agril. Univ., Ludhiana, India, 1981, p. 114.

Singh, N.T., G.S. Hira and M.S. Bajwa. Use of Amendments in Reclamation of Alkali Soils in India. *Agrochemica es Talajtan Tom (Suppl.)*. 30:153–177, 1981.

Singh, M. Singh, N. Singh, and S. Sagar. Dryland Cropping in the Submontane Punjab (1971–81). Res. Bull. of Punjab Agril. Univ., Ludhiana, India, 1983, p. 47.

Sur, H.S. and N.T. Singh. Morphological and Physical Characteristics of Soils of the Pilot Project Patiala. *J. Res. (PAU)* 9:586–597, 1972.

ELEVEN

Soil Fertility Management for Sustained Production in the West Asia-North Africa Region: Need for Long-Term Research

John Ryan and Mohamed Abdel Monem

INTRODUCTION

Since the dawn of mankind's existence, the quest to provide enough food has been a major preoccupation. While many people in today's crowded world are not directly concerned with agriculture, the challenge of balancing the food production-consumption equation is still relevant to the world as a whole. For brief periods over the past century, the dire predictions of Malthus were downplayed by global food reserves, largely generated by the developed west. With current food stocks at precariously low levels and with dim prospects for improvement in Africa and parts of Asia, the "prophet of doom" is more credible again. Predicted population growth rates leave little grounds for optimism. But agriculture and the scientists that serve it must respond to the challenge.

It is ironic that the Middle East area, the reputed center of origin of many of the world food crops, cereals, and legumes, and once a major producer of these crops (Harris, 1995), should now find itself unable to feed its burgeoning populations. Despite some countries of the region being resource-rich, the prospects for the region as a whole are gloomy (Nordblom and Shomo, 1994). Nevertheless, a ray of hope is represented by the establishment in 1977 of the International Center for Agricultural Research in the Dry Areas (ICARDA) in Aleppo, Syria. This center, affiliated to the global network of International Agricultural Research Centers seeks, through its collaboration with the region's National Agricultural Research Systems (Ryan et al., 1995), to improve output of crops and livestock through applied research and institution-building, at the same time protecting the fragile environment in this drought-prone region. The area of ICARDA's mandate is the West Asia-North Africa (WANA) region (Fig. 11.1), a vast and harsh region extending from Morocco in the east, and from Turkey as far south as Ethiopia. With the exception of the Nile Valley in Egypt and Sudan and the Tigris-Euphrates in Iraq and Syria, most of the region is semiarid drylands with limited potential for irrigation.

Agriculture in WANA is largely subsistence farming and rainfed production is low; it is labor-intensive with minimum inputs of new technology. The farm holdings are small—often only a few hectares and frequently in fragmented parcels. Effective change in land

Figure 11.1. Outline of the West Asia-North Africa region, ICARDA's primary mandate area.

management is not always promoted by traditional inheritance laws, tribal and common lands, and nomadism. Most farmers have little formal education. Support services are less than satisfactory for most rural communities, i.e., limited credit, poor roads and distribution systems, weak marketing and research and structures, and, in most cases, ineffective extension services. The private commercial sector is poorly developed in most countries. Such socioeconomic constraints are often as insurmountable as the biophysical ones, but an understanding of the social context in which farmers operate is essential to exploiting the region's natural resources and improving people's lives. As a background to considering the issue of long-term trials, some brief comments on the natural resources—weather, soils, water resources and the socioeconomic and institutional aspects—are in order.

THE WEST ASIA-NORTH AFRICA ENVIRONMENT

Climate

The WANA region is mainly characterized by a Mediterranean climate, with cool to cold wet winters and warm to hot arid summers (Kassam, 1981). However, local conditions are modified by topography and nearness to the sea. Thus, the countries of North Africa are milder due to maritime influences and low elevation, and are typically Mediterranean, while those of West Asia and higher elevations and in larger land masses are characterized by a Mediterranean to a continental climate. Rainfall, though generally low, is highly variable in space and time.

Variability over space is illustrated (Fig. 11.2) across a short transect from 150–600 mm in northwestern Syria. Fortunately, this facilitates the study of cropping systems in different rainfall zones in the immediate environs of ICARDA's main station at Tel Hadya. Similarly, a sharp variation in rainfall occurs in Morocco with favorable conditions in the north (<600 mm) to more stressed conditions (<300 mm) in the barley-growing south.

The range of rainfall in most countries is about 100–600 mm, the extremes being in desert areas and high mountains. Snow is common in high plateaus and mountainous areas, often lasting for several months of winter and early spring. The winter rains normally commence in mid-September to early October, reach a peak in January-February,

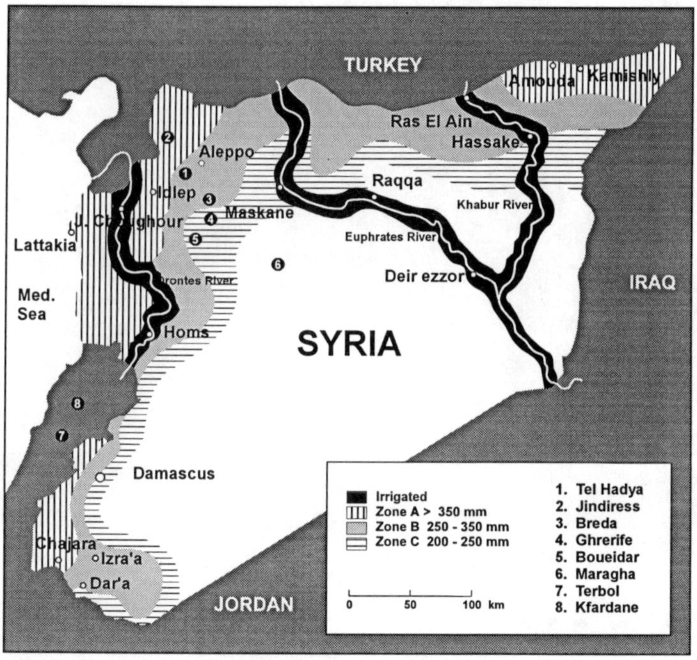

Figure 11.2. Rainfall zones in Syria in relation to some of ICARDA's agricultural experimental stations.

and fall off rapidly in April (Harris, 1995). The pattern changes a bit toward the east, i.e., Turkey and Iran, where the peak of rainfall is later (May), and rain can extend in June and July. Not only does the total seasonal rainfall vary from year to year, but so also does its distribution. Rain is often delayed by a few months, it may finish much earlier than usual, or it may be dry for long spells during the season. Severe droughts which result in partial or complete crop future are common. For instance, in the early 1980s Morocco experienced the worst drought in history, and again another one of comparable magnitude in 1995. Similarly, Syria was badly hit by drought in 1988–1989.

Topography and maritime conditions similarly influence temperatures, especially in winter (Harris, 1995). Thus, winters are milder in the lowland areas of West Asia and in North Africa and severe in highland areas of West Asia. The typical climatic parameters for the Mediterranean region is illustrated for Tel Hadya near Aleppo in a moderate rainfall belt, i.e., 330 mm yr^{-1} (Fig. 11.3). This shows the inverse pattern between rainfall and maximum and minimum rainfall and the relationship with evapotranspiration which, in turn, influences crop water use. Both climatic variables dictate the extent to which rainfed cropping is possible for the region. Crops depend on stored soil water to complete their life cycle, and invariable suffer some degree of moisture stress during the grain-filling stage. What limited rainfall that does occur is not always used by the crop, depending on the pattern of rainfall. Harris (1995) reported that rain that comes in small showers of 5 mm or less is lost through evaporation and has little or no influence on soil water at Tel Hadya; about 10% of the normal rain falls as less than 2 mm showers, with a further 25% between 2 and 5 mm. Where Class A pan evaporation is about 2 mm, such rains would have little effect.

Given the importance of ICARDA's research stations to support its applied research for the region as a whole, and as some of our long-term trials are conducted there (Matar et al., 1992), some climatic details from those sites are pertinent. For instance, in the northern area of Syria, three subzones are recognized. Data from Brown et al. (1987) show long-term rainfall averages of 476 mm at Jindiress and 286 mm at Breda. At both

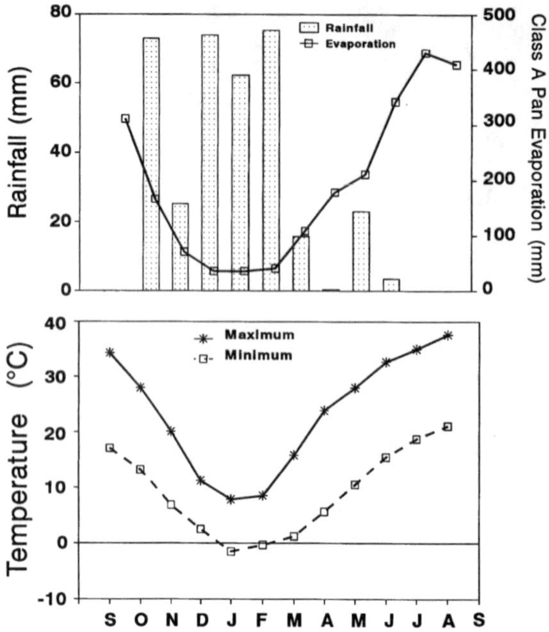

Figure 11.3. Weather data, characteristic of a Mediterranean climate, from Tel Hadya, Aleppo, Syria.

sites, monthly rainfall was highest in January, i.e., 102 mm at Jindiress and 47.2 mm at Breda. The number of frost days at these sites was 33 and 63, respectively. At Jindiress, average temperatures ranged from a low of 6.6°C in January to 27.6°C in July; data for the Breda site were comparable. As cropping and water loss are dependent on the extent of evaporation, which is an outcome of temperature and rainfall, evaporation data are generally recorded. Monthly Class A pan evaporation was on the order of 34 to 37 mm (December, January, February) to 355 to 492 mm in June and July, with little difference between sites. Tel Hadya, with 330 mm/year, is intermediate between the wetter Jindiress and drier Breda.

Soil Resources

While the agriculture of the Middle East region is dominated by climate, specifically limited water, the quality of the region's soils is also of vital concern. Historically, civilizations have flourished in areas of the Mediterranean basin where both soil conditions were favorable and where water was adequate. The soil resources of the region are as variable as in other parts of the world, reflecting variation in climate and topography. Although soils of arid regions have unique features, about half of the world's soil orders are found in such areas: Entisols, Mollisols, Vertisols, Inceptisols, and Aridisols in extremely dry areas. Kassam (1981) indicated that semiarid cropland in North Africa is dominated by Inceptisols (20%), Lithosols or shallow soils (15.5%), Entisols (11%), and Aridisols (16.2%). In West Asia, the landscape is dominated by Lithosols and Aridisols. Mollisols, with relatively high levels of organic matter in the soil profile, are not widespread, but are important locally in countries such as Morocco. No single soil map covers the WANA region; the only common one is the generalized map of FAO- UNESCO (1974) which depicts the region in several sheets for Africa and Asia. While the FAO classification used by Kassam (1981) does not correspond exactly to the U.S. Soil Taxonomy, it is apparent that wide diversity exists at lower levels of classification.

While broad soil variation can be expected as a result of wide climatic variation in Syria (Ilaiwi, 1985), a diversity of soils can occur over a small range. For instance, experimental sites in northern Syria varied at the order level, ranging from Calciorthids in the drier end of the 200 mm rainfall spectrum to Calcixerollic Xerochrepts at Tel Hadya and Breda, and Chromoxererts in the wetter Jindiress station (Ryan et al., 1997); a close relationship exists between rainfall and OM and soil nutrients, but it is modified by the extent of cultivation (Ryan et al; 1996). Even more extreme variation was noted in Morocco where, around Settat (380 mm), one can find Petrocalcic Palexerolls, Typic Chromoxererts, and Vertic Calcixerolls (Shroyer et al., 1990).

While soils of rainfed regions in general are not as intensively studied as temperate areas, reviews of such areas (Dregne 1976) permit some broad generalization about dryland areas such as the Mediterranean. According to Matar et al. (1992) such soils range in texture from sands to clays, most are shallow and have serious inherent or external drawbacks. Limitation on depth, in turn, limits the water-holding capacity of the soil—a major factor, since with infrequent rains most rainfed crops survive on residual soil moisture. Also, shallow soils are particularly vulnerable to soil erosion. While clay soils, i.e., Vertisols are deep and inherently productive with high water-holding capacity of the soil—a major factor, since with infrequent rains most rainfed crops survive on residual soil moisture. Also, shallow soils are particularly vulnerable to soil erosion, while clay soils, i.e., Vertisols, are deep and inherently productive, with good water-holding capacity, the range at which they can be tilled is limited. Frequently, they are either too wet or too dry to cultivate. Extensive areas of Vertisols are found from Morocco to Syria. Crop yields from these soils are usually more than double those of shallower soils in the same rainfall zone (Shroyer et al., 1990). In North Africa, 1.2 million ha are under cultivation, while 4.1 million ha are cultivated in West Asia. Vertisols are difficult to hand-cultivate (Kassam, 1981), and are usually worked mechanically.

Dryland soils are usually low in organic matter (OM), which, in turn, limits soil structure and chemical fertility. Arid soils usually contain from 0.1% to 1% OM, while semiarid soils range from 1% to 3% (Hagin and Tucker, 1982). Organic matter serves as a nutrient reserve, particularly for nitrogen (N) and, to a lesser extent, P. With cultivation and intensification of agriculture, declines in OM invariably occur. As a consequence, P behavior in dryland soils is dominated by inorganic soil compounds. As most dryland soils are calcareous, solubility relationships dictated by high pH and $CaCO_3$ combine to reduce available P in soils. As a result, most dryland soils which have not been fertilized are P-deficient.

Thus, inherent soil properties dictate nutrient behavior and fertilizer use; as a consequence, N is invariably deficient (Ryan and Matar, 1992). Prior to the advent of commercial fertilization, P deficiency was also widespread (Matar et al., 1992). These deficiencies reflected many centuries of exhaustive cropping, with little or no return of nutrients, since crop residues were usually grazed to the ground. While potassium is rarely deficient in the soils of Mediterranean region—a result of the parent materials and the low weathering intensity—increasingly, there is evidence of other nutrient stresses being locally important, e.g., zinc deficiency (Materon and Ryan, 1995), and boron toxicity (Yau et al., 1995). The Soil Test Calibration Network, involving ICARDA and the national research systems, did much to stimulate awareness of potential value of soil testing as a basis for fertilizer recommendations (Ryan and Matar, 1990; 1992) and to improve the performance standards in laboratories that conduct those tests (Ryan and Garabet, 1994). However, despite such developments, the impact at farm level is still far from satisfactory in terms of yield increases. Thus, it is pertinent to comment on how farmers cope with their biophysical constraints and socioeconomic constraints.

FARMING SYSTEMS: TRADITIONAL AND CURRENT

Agriculture as we know it evolved in the Middle East; the soil of the "cradle of civilization" has been tilled for millenia, and its landscape and vegetation have been ravished in the process. The overview of the region's agriculture (Clawson et al., 1971) indicates that it is still very much traditional, having evolved over centuries. Detailed descriptions of rainfed farming in the Middle East are provided by Gibbon (1981) and Cooper et al., (1987), but immense changes have occurred in the last 10–15 years. Today, there is evidence of change, but much of the traditional character remains. The more obvious change is the inexorable shift from animal traction to mechanical cultivation and from hand broadcasting of seed to drilling. Similarly, the change from hand-harvesting to combine harvesting is nearly complete in some countries—the harvesting of lentils (*Lens culinaris* L.) is still the exception. The last few decades have witnessed increased use of chemical inputs, i.e., fertilizers and, to a lesser extent, pesticides. Although dryland farming dominates the WANA region, and will continue to do so, supplementary irrigation is being increasingly introduced in order to stabilize rainfed yields in areas where groundwater or surface water sources can be tapped.

Dryland cropping in most cultivated areas with winter rainfall is dominated by cereals, i.e., wheat (*Triticum* spp) and barley (*Hordeum vulgare* L.); cereal production and livestock are combined enterprises (Keatinge et al., 1986). The relative importance of breadwheat (T. *aestivum* L.) and durum wheat (T. *turgidum* var durum L.) depends on rainfall and the particular country; breadwheat tends to occupy the more favorable rainfall areas. In general, barley tends to dominate the lower end of the rainfall range (Cooper et al., 1987); some limited areas of triticale are grown mainly in North Africa (Mergoum et al., 1992). A close relationship exists between crop yields in general and rainfall (Keatinge et al., 1986), the effectiveness of which is modified by the soil's water storage capacity and the evapotranspiration rate (Smith and Harris, 1981). Data from Morocco (Ryan et al; 1991) illustrate this relationship; barley grain yield was 0.90 t ha^{-1} at Skour Rehamna (270 mm), 1.75 t ha^{-1} at Settat (370 mm) and 2.31 t ha^{-1} at Khouribga (396 mm).

Associated with cereal production are food legumes, i.e., chickpea (*Cicer arietinum* L.) lentil faba bean (*Vicia faba*), and peas (*Pisum sativus* L.). Forage legumes are also common i.e., vetch (*V. sativa*) for hay and Medicago for grazing. Oilseed crops such as sunflower (*Helianthus annus*), safflower (*Carthamus tinctorius*), and rape (*Brassica rapa*) are of minor importance. As all rainfed crops in the WANA region are invariably limited by drought to some degree in most years, the cropping "strategy" that has evolved to mitigate this constraint is the use of rotations, i.e., growing of crops in a particular sequence.

Rotations

From the brief description by Harris (1995) of the historical origins of agriculture in the Middle East, it was evident that the farmers of several millennia ago were aware of the value of legumes on fallow for improving cereal yields. This perception has provided the basis for the cropping systems common in the region today. In the wetter part of the rainfall spectrum (>350 mm), wheat, both bread and durum are alternated with fababean, lentil, and chickpea as winter-sown species, or melon (*Citrullus vulgaris*), sunflower, cotton (*Gossypium hirsutum*), maize (*Zea mays*), sorghum (*Sorghum bicolor*), sesame (*Sesamum indicum*), and vegetables sown in spring or summer. Barley replaces wheat in drier areas with a more limited range of alternative crops. While fallowing was common, especially for barley, land-use pressure is now such that continuous cereal cropping is more common. Notwithstanding such cropping changes, the rationale for, and the benefits of, rotational cropping systems are as valid as ever.

It is ironic that as land use pressure should be induced to shift away from traditional rotations in the Middle East, the concept of rotations should be reasserting itself again in the West, after having fallen into disuse for decades, largely because of the obscuring effects of chemical fertilizers on crop productivity. This resurgence of interest in rotations is fueled by conservation ethics (Blevins and Frye, 1993) and environmental concerns. Thus, in the context of soil quality, it is important to bear in mind the proven benefits of rotational systems on soil physical, chemical, and biological properties (Karlen et al., 1994): improvements in soil N supply through biological N fixation; enhanced soil structure, i.e., aggregation and bulk density; increased soil organic matter, stimulated microbial activity; and increased soil water availability, in addition to weed and disease control. While the latter benefits may be devalued by pesticides, Karlen *et al.* (1994) concluded that "no amount of chemical fertilizer or pesticide can fully compensate for crop rotation effects." While the relative importance of any soil property in crop rotation systems will vary with the region of the world, OM assumes a pivotal role in semiarid rainfed areas because of its disproportionate influence on water and nutrient availability and on yield stability (Rasmussen and Collins, 1991). Rotations involve a many-faceted system and have a long-term dimension, thus requiring a change in perceptions of soil fertility research.

Soil Fertility Research: Need for Change

The traditional concept of soil fertility research in the Middle East region, as in other areas of the world, centers around a relationship between soil nutrient levels at any point in time in relation to constraints to crop growth. Furthermore, the discipline of soil chemistry-fertility was seen as separate from physical properties or physics and agronomy. *A soil with adequate nutrients but poor structure or limited biological life is not a productive soil.* A more comprehensive view of soil as a medium for crop production must consider factors such as texture, a *fixed* soil component, and aggregation which is more *dynamic* and which impacts on infiltration, permeability, and aeration; these in turn influence soil crusting, rooting and seed bed preparation, which can be modified by management, i.e., tillage and residues. Similarly, in any soil that supports legumes, the presence of rhizobia and other soil organisms, such as mycorrhizae, is essential.

In moisture-stressed conditions as in dryland cropping, the expanded view of fertility must accommodate the role of water as a mediator of all biological processes in the soil. While much has been achieved by applied research in the WANA region (Ryan and Matar, 1990; 1992), the dimension of *spatial variability* and *change with time* is, as yet, poorly emphasized. With more stress on applied research by ICARDA in collaboration with the WANA region's national programs, and a focus on assessing impacts of technology, the trend toward problem-oriented research is gaining momentum. Similarly, the embracive concept of *soil quality* is only beginning to take root in soil fertility research. The concepts of *holistic* and *positive soil fertility management* and soil "quality" in the context of effective technology transfer can only be fully expressed through long-term trials, whether at experiment stations or in farmers' fields.

LONG-TERM TRIALS

Despite the antiquity of rotations in the Middle East, the literature on rotations emanates primarily from the West, where the presumed benefits of rotations may not apply to the Middle East, largely because of differing soils, environment, and cropping conditions. Indeed in a global directory of long-term trials (Steiner and Herdt, 1993) only two countries in WANA were represented, e.g., the 1912 cotton experiment in Egypt, and more recently the vetch/barley and N cycling trials in Cyprus, which will be discussed later. How sustainable are current cropping systems? The answers can only come from realistic long-term studies within the region. As ICARDA began to address the agricultural constraints and potential of its mandate area, it became evident that what was needed was a comprehensive or systems approach to agriculture and an awareness of both the short-term and long-term effects of any crop or farming practice on sustainable production (Jones, 1993). Thus, the issue of sustainability underpinned all of the rotation trials, regardless of whether the focus was crops, fertilizer response, grazing intensity, or tillage systems.

The main issue in most long-term trials is to provide alternatives to fallow (Papastylianou and Jones, 1988), i.e., either a food legume or one for forage or grazing, since sheep raising is a unique feature of dryland farming in the Middle East. The impact of the grazing animal can have a significant impact on soil properties, especially the nutrient status (Haynes and Williams, 1993). Given the dominance of environment on biological processes in soil and the unique features of the Mediterranean climate, long-term studies which assess soil quality and nutrient dynamics are, of necessity, site-specific. Valid generalizations can only be made based on a network of such trials across the range of environments in the region Thus, a brief overview is presented on multi-year or long-term studies in selected countries of the WANA region, followed by a focus on ICARDA's dryland trials as well as its new initiative with primarily irrigated long-term trials in Egypt.

Crop Sequence Studies in the Mediterranean Region

Given the significance of crop rotations in the region, it was hardly surprising that many national agricultural systems had begun to undertake long-term field trials, either independent of or in collaboration with, ICARDA. Few were more than 10 to 15 years duration (Table 11.1). A brief reference to some of them is pertinent.

Prior to the Mid-America International Agricultural Consortium project in collaboration with the Institute National de la Recherche in Morocco's rainfed zone, virtually all field experimentation was confined to experimental stations, most of which were unrepresentative of the region's soil types (Ryan et al., 1990). Subsequently, a series of field trials over three years on a shallow (Petrocalcic Palexeroll), moderately deep (Calcixeroll), and a deep soil (Chromoxerert) demonstrated the dominant effect of soil depth on yield, as well as showing that, on these soils, where the previous crop was a legume, the response to N was reduced. Thus, the legume has contributed to soil N for the wheat crop and was reflected in increased NO_3-N levels in the profile (Abdel Monem et al., 1990).

These on-farm studies highlighted the importance of spatial variability of nutrient levels in farmers' fields, a factor to be considered in soil sampling for fertilizer recommendation. While these initial on-farm studies were not long-term trials per se, they did involve representative rotations on representative soils for a few years but not necessarily in the same field each year. Despite the potential importance of these findings, the enormous costs and the logistic problems of conducting long-term on-farm trials only allowed for such trials to be conducted in a relatively small area and with small plots at experiment stations (Sidi El-Aydi, Jemat Shaim, Jemat Riah). A more ambitious fenced grazing rotation trial was established at Ain N'Zagh, but soil variability, both in terms of slope and soil depth, is a major flaw in the design.

Over the past two decades, long-term trials in Cyprus, at low and medium rainfall

Table 11.1. General Details of Some Long-Term Trials in the WANA Region, Excluding ICARDA's Trials

Country	Location	Year Established	Rotation	Purpose
Cyprus	Dromalaxia	1982	Barley/barley	Crop yield, fertilizer response
	Laxia	1981	Fallow/vetch	
Turkey	Haymana	1982	Wheat/legumes	Yield sustainability, water use efficiency
Algeria	Al Kroub	1992	Wheat/legumes	Yield trends, grazing
Morocco	Settat, Abda	1989	Medicago/cereals	Pasture productivity
		1986	Legumes/cereals	Tillage methods
Egypt	Delta Coast Reclaimed desert	1996	Wheat, cotton, rice, berseem, beans	Fertility buildup and maintenance, Water use, salinity control
Jordan	Mushaggar, Maro Ramtha	1991	Lentil, chickpea, vetch/cereal	Tillage methods
Lebanon	Terbol Kfardane	1990	Wheat/legumes	Assess yield and biological N fixation

sites, compared the productivity of barley when grown after vetch or fallow or in continuous barley (Papastylianou, 1993). The vetch-barley, and barley-fallow rotations were more productive than continuous barley. The required N fertilizer rates for these systems were zero (vetch), 30 kg ha^{-1} (fallow), and 45 to 60 kg ha^{-1} for continuous barley. Both soil N increases and improved water storage were attributed to the vetch and fallow rotations.

In neighboring Turkey, an ongoing trial at the Haymana station near Ankara in the Anatolian Plateau is examining the impact of a wide range of alternative crops in relation to subsequent wheat yields, soil N, and moisture availability (AvÇin and AvÇi, 1990). Early results show that there was little difference between fallow, cumin, chickpea, winter and spring lentil; the lowest yields were with wheat after wheat and after safflower. Thus, some legumes are a good alternative to fallow, which yields a crop only once in two years. Currently, a long-term grazing rotation trial has been established at Al-Kroub in Algeria in collaboration between ICARDA and ITGC (Institute Techniques des Grandes Cultures), Algeria's national research agency.

Though much research has been conducted in Lebanon through the auspices of the American University of Beirut, especially at its field station in the Beka'a Valley, the research program did not involve long-term studies. Neither were such trials conducted by the Ministry of Agriculture. Nevertheless, many of the short-term studies did contribute to knowledge of the behavior of soils of the Middle East, e.g., reactions of P fertilizer (Ryan et al., 1985a, 1985b), potential for volatile loss of ammonia from soil-applied urea (Ryan et al., 1981), and micronutrient availability (Curtin et al., 1980).

ICARDA's Long-Term Dryland Research

Though Tel Hadya (330 mm), the Center's headquarters and main station, is representative of semiarid areas in general, the variation in climate, notably rainfall, within the normal range for rainfall cropping, is large. The importance of rainfall zones was readily es-

Table 11.2. Long-Term Trials Conducted by ICARDA in Syria

Location/Trial Name	Year Established	Crop Rotation	Purpose
Tel Hadya (330 min)			
Biological N fixation	1990	Food, forage legumes	Assess yield and biological N fixation
Grazing management	1985	Wheat, barley/medicago	Yield, N response, water use, grazing intensity
Tillage timing	1978	Wheat/lentil	Impact of early, middle, late tillage (conventional), zero tillage.
Tillage systems	1985	Wheat/lentil, chickpea, melon	Compare deep (chisel, disc) with minimum and no till.
"Cropping systems"	1983	Wheat/chickpea, lentil vetch, medic, fallow, wheat	Sustainable productivity, water and fertilizer-use efficiency.
"New rotation"	1982	Barley/vetch, fallow	Sustainable productivity water and fertilizer-use efficiency.
Breda (280 min)			
"New rotation"	1996	Barley/vetch, fallow	Sustainable productivity
Jindiress (460 min)			
Residual/Yearly P Application[a,b]	1986	Wheat/lentil	Assess yields, available P, rotations

[a]Others include the atriplex/barley trial at Gherife (1990); stubble burning/seedbed preparation at Tel Hadya (1987); continuous barley at Tel Hadya and Breda (1986); hill grazing-phosphate trial at Tel Hadya (1984); rangeland trial at Maragha (1989); and straw/compost management trial at Tel Hadya (1996).
[b]Also established at Tel Hadya and Breda.

tablished (Dennett, 1987) and was the basis for the establishment of substations to represent the higher and lower rainfall zones (Keatinge *et al.*, 1985; 1986). The early research at ICARDA provided the rationale for a series of long-term trials at Tel Hadya and Jindiress in the wet area (>400 mm yr^{-1}) and Breda in the drier area (>300 mm yr^{-1}). While there is some overlapping in some of these trials (a cross section of the main trials is shown in Table 11.2), the main focus of each is different.

Now after 10 years, these trials have begun to yield valuable data in terms of crop production, soil water use, and fertilizer parameters. Most of these trials are scheduled to continue, but others may be phased out due to funding constraints or the absence of significant findings or trends. Results of some of these trials have been reported, e.g., the barley/legume trial at Breda and Tel Hadya (Jones and Singh, 1995) and the main two-course "Cropping Systems Productivity" (up to 1991) and tillage trials (Harris, 1995). This brief overview focuses on fertility aspects of selected trials with addition of the latest (1991–95) data from the main "Cropping Systems" trial.

Residual/Yearly Phosphorus Application

As P is a key element in crop production in the region, and its efficiency low (Matar *et al.*, 1992), a trial was established to examine the value of residual and currently applied P under a cereal/legume rotational system (Ryan *et al.*, 1994) and to establish what is the minimum amount of P fertilizer necessary to maintain adequate soil P levels. The study showed that despite the relatively low range of available P (0.5 *M* $NaHCO_3$) levels initially (Tel Hadya, 6.0 ppm; Jindiress 4.0 ppm; Breda, 2.4 ppm), there was no crop response to P at any location for the first 3 years, but the initial heavy P dressing (residual) and the yearly application significantly raised soil test levels especially in the top 20 cm, with little evidence of P movement down the profile. Subsequently, native P levels were drawn down sufficiently to evoke a response to yearly applied P. Therefore, residual P levels (up to 90 kg P ha^{-1}) had no impact on crop yield and did not persist in the soil beyond the fourth or fifth years. None of the rotations at any site had a differential effect on P response or availability. However, yearly additions of more than that gradually caused a buildup in the soil. As little as 15 kg P ha^{-1} was sufficient to maintain available P values at their initial levels. With little justification for continuing, the experiment was terminated.

Biological Nitrogen Fixation

This long-term trial was established in 1990/1991 at Tel Hadya and at cooler, higher elevation and rainfall sites in Lebanon, e.g., Terbol (>550 mm) and Kfardane (350 mm). The rotation involved wheat with a range of feed and forage legumes, with and without inoculation with *Rhizobia*, with the broad objective of assessing the impact of specific crops and temperatures and rainfall on N fixation and its subsequent value for the cereal crop. So far, the results indicate that forage legumes are, in general, more efficient than food legumes; *Vicia villosa* spp. *dasycarpa* produced highest values of 100 kg N ha^{-1} at Terbol and 60 kg at Tel Hadya (Afandi et al., 1997). Wheat following the legumes were on a par with yield of wheat following fallow and significantly higher than yield of wheat after wheat.

Grazing Management

While several of the Center's trials involve an element of grazing, this trial established in 1985/1986 focused on grazing intensity of medic (*Medicago spp*) grown in rotation with wheat with and without N fertilizer. Some of early findings were reported by White et al., (1994), who examined changes in total soil N and organic matter. Both medic and vetch had, within the 6 years of the trial, increased N and organic matter, but lentil had no effect. However, yields after medic were lower due to higher water use. For the fallow treatment, total N remained constant but OM tended to decrease with time. The positive effect of N on soil OM was small but consistent. Similarly, as grazing intensity increased, OM decreased from 1.28% to 1.11%. The trial has subsequently been modified with a replacement of wheat by barley. While the off-station trial at Kamishly is on a similar deep clay soil, it is in a higher rainfall zone, is simple in design, and is operated in collaboration with the Syrian national research program.

New Rotation

The focus of this 1982 trial at Tel Hadya and Breda (Jones and Singh, 1995) was barley alternated with feed legumes (vetch, lathyrus), in pure or mixed stand, and with fallow or another barley crop in combination with N and P. In yield terms, barley after fallow had the highest yield, followed by vetch after barley, and least with continuous barley. Such yield data and N off-take point to the value of fertilization with N and P, as well as using legumes in the rotation, especially in drier areas. Other trends emerging show that fertilization tended to increase soil OM, which also tended to be higher with continuous barley than with fallow.

Other Trials

Although some of the remaining long-term trials deal with tillage systems and timing of tillage and residue management, no clear fertility trends are yet apparent except in the case of minimum or zero tillage. In such cases, the content of soil OM and total N was higher by comparison with conventional tillage. No apparent effect of stubble burning, in comparison to stubble incorporation, is yet apparent with respect to soil chemical properties.

Cropping Systems Productivity

Details of the treatments, design, and experimental conditions of this, the most prominent of ICARDA's trials, were reported by Harris (1995), with preliminary data up to 1991. The trial measurements not only involved biomass and grain yield in terms of cropping sustainability and economics, but also water balance and use efficiency, disease and pest control, N dynamics, and aggregate stability and bulk density. In essence, it evaluated wheat in rotation with seven alternatives (fallow, wheat, food and forage legumes) and both N fertilization and grazing or stubble management.

Yield

Data for the last 4 years exhibited a similar pattern with respect to treatment as reported by Harris (1995). The overall yield trends for the full 10 years are shown in Table 11.3. Seasonal rainfall had a major effect on overall yields of cereals and the alternative crops. The three crop sequences which rely on current season's rainfall yielded lowest, i.e., wheat after medic, wheat/chickpea, and continuous wheat the least. Wheat after lentil and vetch, both of which have a sparse growth habit yielded about 80% of that of the fallow or summercrop. Significantly, the fallow yields were over double those of the continuous wheat.

Response to N in the wheat phase varied with the year and its rainfall (Table 11.4). In a dry year, responses occurred only at the lowest N rate (30 kg ha^{-1}) but increased to 90 kg ha^{-1} in the wettest year. Responses were also highest under fallow and least with continuous wheat, and with legumes, especially medic; again, *this is attributed to moisture availability as the limiting factor.*

Fertility Trends

Mineral N values showed a seasonal trend, being highest after the first rains in the autumn and decreasing as the cropping season progressed; values from the medic plots were always higher than from continuous cereal. When measured after the alternative crop phase, i.e., in the autumn, differences in mineral N were not consistent between rotations, but after the cereal phase, which receives N fertilizer, differences were obvious, with medic being highest and wheat/wheat and wheat/fallow, the lowest. However, as these values are averaged across N treatments, they may reflect residual N unused by the previous drought-restricted crop.

The total N data, which followed the same pattern as mineral N, provide stronger evi-

Table 11.3. Total Biomass Yield Grain and Straw of Durum Wheat in the "Cropping Systems" Rotation for 10 Years at Tel Hadya

	W/F	W/W	W/L	W/C	W/S	W/V	W/M	L.S.D.
Year				t ha^{-1}				
85-86	6.52	3.66	5.01	5.16	6.51	5.36	3.77	1.06
86-87	5.25	2.87	4.93	3.88	5.36	5.00	3.64	0.82
87-88	10.6		10.1	7.67	9.36	10.19	8.45	2.06
88-89	5.37	1.26	3.02	1.57	3.26	3.28	1.67	0.78
89-90	3.69	2.26	2.45	2.28	4.29	2.67	2.35	0.42
90-91	5.52	3.91	4.86	4.46	6.08	4.77	4.89	1.18
91-92	8.39	5.22	6.65	7.05	8.07	6.75	6.93	1.07
92-93	6.24	2.77	5.91	5.05	5.23	5.56	5.40	1.00
93-94	7.35	5.32	7.19	5.73	6.13	7.43	7.77	1.11
94-95	7.90	4.05	6.36	4.60	8.41	6.39	7.43	0.99
Mean	6.68	3.48	5.56	4.75	6.27	5.74	5.23	

Notation: F= Fallow; W= Wheat; L= Lentil; C= Chickpea; S= Summercrop (melon); V= Vetch; M= Medic.

Table 11.4. Overall Effect of Nitrogen Application Rates on Wheat Grain Yield from Each Rotation

	N, kg ha^{-1}				
	0	30	60	90	Mean
Rotation	t ha^{-1}				
Wheat/wheat	0.86	1.25	1.45	1.46	1.23
Wheat/fallow	1.93	2.37	2.71	2.92	2.48
Wheat/melon	1.95	2.29	2.57	2.83	2.41
Wheat/lentil	1.73	1.96	2.20	2.31	2.05
Wheat/chickpea	1.31	1.65	1.71	1.92	1.65
Wheat/vetch	1.78	2.11	2.20	2.22	2.08
Wheat/medic	1.62	1.72	1.78	1.85	1.73
Mean	1.60	1.91	2.09	2.26	1.96

dence of an actual contribution from the legume to soil N at about 950–1000 mg kg^{-1}. Nitrogen values from medic were significantly higher than the wheat/wheat or wheat/fallow rotations (Fig. 11.4). Indeed, N-mineralization data of Hasbany (1997) clearly showed that medic plots released considerable more N than from either cropped or fallow plots from the same trial. Similarly, total soil N values increased as the level of fertilizer N increased.

This trial was the site of a more detailed N cycling study involving Reading University and ICARDA. When ^{15}N-labeled urea or ammonium sulfate were used in macroplots, N recovery in wheat after wheat, fallow, chickpea, and lentil was low, i.e., from 8.7 to 26.4%; this was either attributed to volatile loss or immobilization, since only about 25% of the applied N was recovered from the soil. Thus, volatile loss as ammonia or denitrification presumably accounted for the sizable loss from the system. However, as Garabet et al. (1998) found that about 90% of the applied ^{15}N could be recovered in an adjacent field, these losses may be due to the method of application, i.e., mist spray vs. solution.

A significant development, reflected in soil quality, is the impact of the rotations on

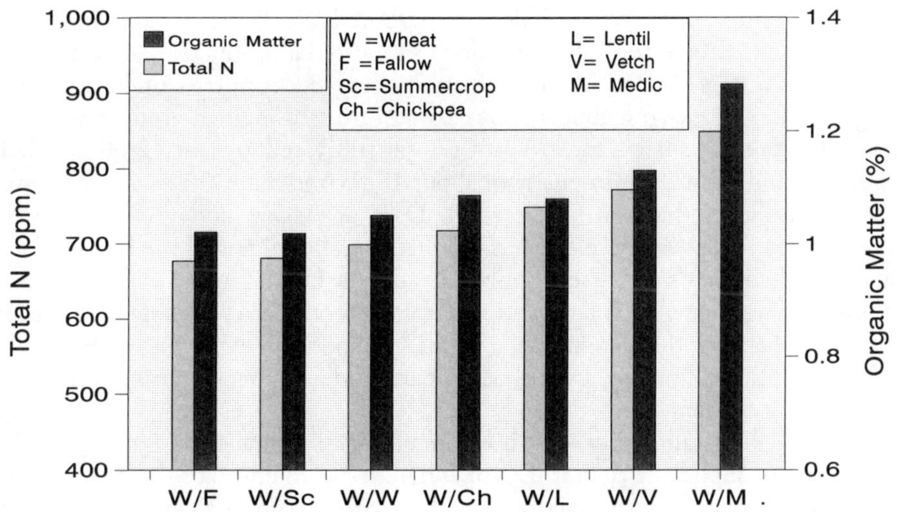

Figure 11.4. Mean effect of rotations over 10 years on total soil nitrogen and organic matter.

OM (Fig. 11.4). While there were variations from year to year, the effects were consistent. Values after medic were consistently about 20 to 35% higher than after wheat or fallow; the other legumes were intermediate. Organic matter also increased as the fertilizer N level increased, probably due to greater root biomass; for the 0, 30, 60, and 90 kg ha^{-1} N rates, the respective OM values, averaged over all years, were 1.1, 1.3, 1.8, and 1.9%. The enrichment of OM in the *Medicago* plots extended to about 50–60 cm depth. Such differences in organic matter were translated into increased aggregation and improved hydraulic conductivity (Masri et al., 1996). Compared with fallow, the medic plots had a lower degree of dispersion. Dispersion decreased with added N, but surprisingly, values from the zero grazing plots were somewhat more dispersed than from the heavily grazed plots. Only in the last 2 years had grazing intensity any significant effect on N and organic matter.

Soil Moisture

As detailed measurements have been made on soil moisture throughout each season in all plots, a number of conclusions made by Harris (1995) are still pertinent.

1. The dominant and consistent influence on yield is rainfall, and therefore, soil moisture.
2. Response to applied N is governed by soil moisture.
3. The soil profile exhibits a typical pattern of recharge from October onward, reaching a maximum in January–February, followed by a discharge or drying-out over the subsequent months.
4. Only in years of exceptional rainfall, i.e., 1987–1988 with 480 mm, was the entire profile wetted. Thus, rarely is the soil storage capacity a limiting factor, but rather the water supply. As a consequence of limited water movement down the profile, leaching is rarely a loss mechanism for N.
5. In years of average rainfall, the amount of soil moisture accumulated under fallow for the subsequent wheat crop is small. Only in exceptionally high rainfall years does fallow contribute substantially to residual soil moisture. In such years, lentil also accumulates some residual soil moisture.
6. When yield responses occurred for applied N, water-use efficiency also increases.

Resource Management in Egypt

Historically, the lands bordering the Nile have harbored one of the world's oldest civilizations; the rich sediments of the valley and delta supported a productive and thriving agriculture, with the waters of the Nile for irrigation being the country's lifeline. Today's intensive agricultural sector contributes about 20% of gross domestic product and exports, and occupies 40% of the labor force. With a total cultivated area of 12 million feddans (about 5 M ha) and a cropping intensity of about 180%, the main crops are cotton, wheat, rice, maize, and clover (berseem). The High Dam at Aswan and the loss of seasonal flooding, with the consequent use of chemical fertilizer, has changed the face of Egyptian agriculture. The country's massive population and its rapid growth has put inevitable pressure on existing agricultural land and water resources. Agricultural land is inexorably giving way to industry and urbanization; a similar competition exists for water. While loss of land from traditional farming areas is being partly offset by reclamation of the desert, a new range of concerns emerge.

The concept of "sustainability" has been in vogue for some time (Jones, 1993), but nowhere is it more applicable than in Egypt. There it has to embrace the notion of maintaining a trend of increasing production, i.e., *food security with a future dimension*, and at the same time, *protecting the resource base* that generates such production. Why should a system of agriculture, which has stood the test of time for millennia, now not be considered so? The threats to sustainability, real or apparent, are many. While many of these stem from changes induced by the High Dam and population growth, about which agricultural researchers can do little, the following issues can undermine future land use:

- Salinity is a major factor in about one-third of Egypt's soils; this arises from increased use of drainage water and seepage from the sea and saline lakes.
- Excessive N fertilizer use has serious implications for water quality; some wells already have higher nitrate levels than permissible by international standards. Deregulation of the market with respect to inputs and crop prices has, partly at least, contributed to this situation.
- Wastes are the by-products of industry and urbanization. As elsewhere in the developed world, wastewater and sewage sludge, with their load of biological, chemical, and heavy-metal contaminants, are being used in Egypt.
- Because of inefficient water delivery systems, crop water use rarely exceeds 50% of that applied.
- Lack of nutrients, poor structure, together with heat and drought stress are specific constraints to the new lands.

These constraints all hinge around soil fertility and are interrelated. None of these concerns can adequately be addressed by conventional field trials. Only long-term trials can provide some of the answers to the questions raised. Thus, given its lead role in conducting such studies in the context of sustainability, ICARDA's Farm Resource Management Program, in collaboration with Egypt's National Agricultural Research Center and funded by the European Union, initiated a long-term program for sustainable productivity based on protection of the resource base (land, water) through good management (Abdel Monem et al., 1996).

Trial Sites: Establishment and Monitoring

The program's broad objective would be achieved through basic intensive long-term on-station research, and on-farm extensive monitoring of farmers with respect to resource utilization and their decision-making processes. Preparatory studies sought to (a) define the farming systems within the main agroecological environments, (b) identify and prioritize the constraints to continued resource utilization, and (c) provide a basis for long-term research activities. The initial phase of information gathering involved inventory studies, rapid rural appraisals, and multidisciplinary surveys.

To be successful, long-term trials need careful planning. Thus, based on information gleaned from the preparatory phase, five sites were chosen for long-term resource management activities (Table 5). Four were irrigated and one rainfed. While most had many common features, i.e., water quantity and quality, inputs (fertilizers, manure), and rotations, there were concerns specific to each one. *Fertility and its maintenance were dominant themes in each.* The broad goal is to improve the cropping-forage system as a whole within each situation.

The old irrigated site at Sids, with its highly productive crop-animal production system, focuses on water-use efficiency with intensive high-input rotations, i.e., onion/cotton, wheat/soybean, and fababean/tomato in place of prevailing rotations involving berseem and maize. Three sites were chosen on newly reclaimed land. At the Nubaria, with its highly calcareous soil, issues such as crust formation, fertility imbalance, and ineffective biological N fixation are among the concerns, as well as introducing an appropriate rotation. At the sandy Bustan site, *fertility and organic matter buildup* are issues, together with water-table and salinity control. The El Serw site on a reclaimed heavy clay

Table 11.5. Details of Long-Term Trials Established in Egypt

Area	Site	Research Issues	Rotation
1. Nile Valley (irrigated clay)	Sids	— Fresh vs drainage water use — Use of fertilizers, manure	Berseem/cotton, wheat/maize, fababean/ tomato vs. Onion/cotton, wheat/soybean-sunflower, fababean/tomato
2. Delta (clay, saline)	El Serw	Salinity, drainage: water quality	Berseem/rice, wheat/rice, berseem/maize vs. Berseem/rice, berseem/rice, berseem/rice
3. Reclaimed Desert (calcareous)	Nubaria	Water requirements, fertilizer and manure - fertility buildup	Berseem/maize, wheat/maize, wheat/tomato vs. Berseem/sunflower, tomato; sugarbeet/maize wheat/soybean, maize
4. Reclaimed Desert (sandy)	Bustan	Fertility buildup: fertilizers, manure	Wheat/groundnut, wheat/groundnut, berseem/groudnut vs. Wheat/groundnut, lentil/sesame, berseem/ millet; potato
5. North Sinai	El Barth	Fertility buildup, water harvesting, efficient fertilizer use	Barley/wheat monoculture vs. Barley/lentil, barley/lathyrus, barley/fallow

saline soil, focuses on reducing the high water-table, controlling water quality, i.e., salinity, and *improving N use efficiency*, and promoting intensive, but suitable rotations (berseem/rice). The focus of the rainfed trial at El-Barth in North Sinai in the precarious 100–200 mm yr^{-1} rainfall zone is *fertility buildup* through P fertilization and legume establishment with *rhizobial inoculation*. This site is problematic since in some years no crop can be harvested due to drought.

The monitoring activities are planned to be complementary to the long-term trial themselves to establish a broad view of actual farmer's conditions in the environs of these sites. The biophysical and socioeconomic data obtained by a multidisciplinary team will establish a farm management database profile against which future changes can be measured.

CONCLUSIONS: LESSONS LEARNED

It is axiomatic that we live in a world of change. The dramatic changes in global agriculture in the latter half of this century call for changes in how we approach research issues. As farming systems change, so must our concept of how we feed crops change. In the past, much of the focus of soil fertility research was on diagnosis of nutrient deficiencies and behavior of these elements in soils as well as responses to fertilizers in the field. Problems of the real world do not occur in isolation. Fertility must, of necessity, be responsible to soil and crop management changes; in other words, it has to be *dynamic* and not *static*. But dynamism can only be assessed within a long-time framework. The time-honored method to monitor what happens with time is the long-term trial.

While such trials are common in developed countries, e.g., to the famous Rothamstead trials established over a century ago, the long-term dimension is fairly new in semiarid regions where such trials are more problematic; they require large land areas, which often is not available, and considerable financial and labor resources, and an institutional commitment—often a political one—for a minimum specified period. Adapting a long-term trial to accommodate the various cropping options practiced by farmers poses difficulties. Thus, there is a limit on the extent to which long-term rotation can simulate real-world situation. Given these constraints, the conducting of long-term trials is an area where international centers, such as ICARDA, have a comparative advantage.

With the benefit of 10–15 years' experience with long-term trials, new concerns have emerged, while some of the old ones need to be reemphasized:

- Given the costs involved in establishing and maintaining such trials, the utmost effort should be put into their planning. The likely time frame should be established in advance. Changes over time can make a trial obsolete.
- The trials should be relevant over their projected life-span. When ICARDA's trials were being established in the early 1980s, the surrounding areas were totally rainfed, whereas now much of this dryland area is irrigated.
- The trial site should be accurately described in terms of soil type and weather conditions so that the resulting data may be relevant to similar environmental conditions in the region. Greater use could be made of soil maps to characterize sites. Only recently were the sites described and classified.
- The trial design should be flexible enough to accommodate changes in cropping and management with time without invalidating the results; e.g., the switch from wheat to barley as in the "Grazing Management" trial.
- While trials such as the "Cropping Systems Productivity" one contain the range of crops which are commonly related with cereals, in reality, farmers rarely continue with a fixed sequence of crops.
- Although medics were originally perceived as a substitute for fallow, the crops are now seen as suitable only for marginal lands.
- In our trials, grazing intensity is simulated by varying the number of sheep or varying the length of the grazing period. Again, what farmers do is different; they usually put a large number of sheep on the stubble and graze it clear.
- Prior to establishing long-term trials, detailed baseline data should be recorded, i.e., chemical, biological, and physical properties. In our trials, the need for assessment of aggregate stability as well as indices of nutrient availability was only appreciated when the trial was well advanced.
- Yearly monitoring of the trial should be made systematically, i.e., at the same time (before rains) and with the same sampling procedure. This is now being done routinely.
- Some monitoring is advisable throughout the year. While ICARDA regularly monitors the moisture status of the profile in several trials, only in the past 2 years has mineral N been measured during the season.
- As drought is an ever-present threat to cropping, the possibility of crop failure in any one year should be considered in the statistical design.
- To justify the costs of "benchmark" long-term trials, the site should be exploited for as much scientific information as possible. This is the case with the "Cropping Systems" trial. In addition, data from this trial are used in training courses.
- When adequate financial and human resources are available, large complex trials may be justified, depending on the range of objectives. However, where such conditions do not exist, relative simple trials are advisable.
- Long-term trials, often the brainchild of individual researchers, may wither with neglect with a change in personnel. Efforts should be aimed to ensure a smooth transition between scientists, and thus guarantee an "institutional memory," and to periodically evaluate and document progress in the trials.
- While there is a clear need to establish long-term trials in countries of the WANA region,

this should be done through a network of concerned scientists with common objectives and protocols.

- Treatments should be minimized and contrasts between treatments maximized. For instance, in the "Cropping Systems" trial the summer-crop treatment is essentially a mixture of melon and clean fallow since there is usually not sufficient residual moisture every year to grow the crop. Therefore, this rotation could be deleted.
- In developing countries, economic assessment of long-term rotation trials is difficult because of artificial control of costs and by inflation and currency fluctuations.
- Where there is a likelihood of crop failure in semiarid areas, i.e., 3–4 years in 10, rotation trials are futile. Thus, environmental variability is a major issue.

Despite these concerns, the crop yield data and other indicators of soil fertility and quality which are emerging are encouraging. These trials should continue for the foreseeable future until stable equilibrium conditions prevail for these parameters. While prediction is a risky business under most circumstances, it is particularly so with long-term trials. However, the effort of Singh and Jones (1995) to model future yield trends from rotation trials is an attempt to circumvent the time constraint associated with these trials.

ACKNOWLEDGMENT

I thank Drs. Mike Jones and Tom Nordblom for their input on an earlier version of this paper; Dr. Scott Christiansen for his suggestions; Mr. Samir Masri for assistance with figures and tables; and Mrs. Zuka Istanbouli for typing the manuscript.

REFERENCES

Abdel Monem, M., A. Azzaoui, M. El Gharous, J. Ryan, and P.N. Soltanpour. Response of Dryland Wheat to Nitrogen and Phosphorus in Some Moroccan Soils. In *Proc. Third Regional Soil Test Calibration Workshop,* J. Ryan and A. Matar, Eds. Sept. 3–9, 1988. Amman, Jordan. ICARDA, Aleppo, Syria, 1990, pp. 52–65.

Abdel Monem, M., H. Khalifa, and R.A. El Enein. Building and Sustaining a High Production Capacity of Egypt's Irrigation Lands: A Long-Term Research Program. *Abstracts: International Conference on Land Degradation.* June 10–14, Adana, Turkey, 1996, p. 16.

Afandi, F., N. Trabulsi, and M.C. Saxena. Biological Nitrogen Fixation by Cool-Season Food and Feed Legumes: Impact on Wheat Productivity in Syria and Lebanon. In *Accomplishments and Future Challenges in Dryland Soil Fertility Research in the Mediterranean Area.* J. Ryan, Ed., ICARDA, Aleppo, Syria, 1997.

AvÇin, A., and M. AvÇi. Soil Water and Inorganic Nitrogen Accumulation at Sowing Time of Wheat in a Two-Year Rotation as Influenced by Previous Crops Under Central Anatolian Conditions. In *Proc. Fourth Regional Soil Test Calibration Workshop.* J. Ryan and A. Matar, Eds. May 5–10, 1988. Agadir, Morocco. ICARDA, Aleppo, Syria, 1990, pp. 64–70.

Blevins, R.L. and W.W. Frye. Conservation Tillage: An Ecological Approach to Soil Management. *Adv. Agron.* **51**: 33–78, 1993.

Brown, S.C., J.D.H. Keatinge, P.J. Gregory, and P.J.M. Cooper. Effects of Fertilizer, Variety and Location on Barley Production Under Rainfed Conditions in Northern Syria. I. Root and Shoot Growth. *Fields Crops Res.* 16: 53–66, 1987.

Clawson, M., H.H. Landsberg, and L.S. Alexander. *The Agricultural Potential of the Middle East.* Elsevier Sci., New York, 1971.

Cooper, P.J., P.J. Gregory, D. Tully, and H.C., Harris. Improving Water-Use Efficiency of Annual Crops in the Rainfed Farming Systems of West Asia and North Africa. *Expl. Agric.* **23**: 113–158, 1987.

Curtin, D., J. Ryan, and R.A. Chaudhry. Manganese Adsorption and Desorption in Calcareous Lebanese Soils. *Soil Sci. Soc. Am. J.* 44: 947–950, 1980.

Dennett, M.D. Variation in Rainfall—The Background to Soil Water Management in Dryland Regions. *Soil Use and Management* 3: 47–51, 1987.

Dregne, H.E. *Soils of Arid Regions*. Elsevier Science, New York, 1976.

FAO-UNESCO. *Soil Map of the World*. Food and Agriculture Organization, Rome, Italy/United Nations Educational, Scientific and Cultural Organization, Paris, France, 1974.

Garabet, S., M. Wood, and J. Ryan. Nitrogen and Water Effects on Wheat Yield in a Mediterranean-Type Climate. *Field Crop Res.* (in press), 1998.

Gibbon, D. Rainfed Systems in the Mediterranean Region. *Plant Soil* 58: 59–80, 1981.

Hagin, J. and B. Tucker. *Fertilization of Dryland and Irrigated Soils*. Springer-Verlag, New York, 1982.

Harris, H., J. Ryan, T.T. Treacher, and A. Matar. Nitrogen in Dryland Farming Systems Common in Northwestern Syria. In *Livestock and Sustainable Nutrient Cycling in Mixed Farming Systems of Sub-Saharan Africa*. Vol II: Technical papers. J.M. Powell, S. Fernandez-Rivera, T.O. Williams, and C. Renard, Eds. Proc. Int. Conf. Addis Ababa, Ethiopia. Nov. 22–26, 1993. ILCA (Int. Livestock Center for Africa), Addis Ababa, Ethiopia. 1995, pp. 323–335.

Harris, H. Long-Term Trials on Soil and Crop Management at ICARDA. *Adv. Soil Sci.* 19: 447–469, 1995.

Hasbany, R., T. Atallah, S. Garabet, and J. Ryan. Laboratory Assessment of Nitrogen Mineralization in a Wheat-Based Rotation Trial. In *Accomplishments and Future Challenges in Dryland Soil Fertility Research in the Mediterranean Area*. J. Ryan, Ed. ICARDA, Aleppo, Syria, 1997.

Haynes, R.J. and P.H. Williams. Nutrient Cycling and Soil Fertility in the Grazed Pasture System. *Adv. Agron.* 49: 119–199, 1993.

Ilaiwi, M. *Soil Map of Syria and Lebanon*. The Arab Center for the Study of Arid Zones and Dry Lands (ACSAD), Damascus, Syria, 1985.

Jenkinson, D.S. The Rothamstead Long-Term Experiments: Are They Still of Use? *Agron. J.* 83: 2–10, 1991.

Jones, M.J. and M. Singh. Yields of Crop Dry Matter and Nitrogen in Long-Term Barley Rotation Trials at Two Sites in Northern Syria. *J. Agric. Sci. (Camb.)* 124: 389–402, 1995.

Jones, M. Sustainable Agriculture: An Explanation of a Concept. *Crop Protection and Sustainable Agriculture*. John Wiley, Chichester, 1993. pp. 30–47.

Karlen, D.L., G.E., Varvel, D.G., Bullock, and R.M., Cruse. Crop Rotations for the 21st Century. *Adv. Agron.* 53: 1–45, 1994.

Kassam, A.H. Climate, Soil and Land Resources in West Asia and North Africa, *Plant Soil* 58: 1–28, 1981.

Keatinge, J.D.H., M.D. Dennett, and J. Rogers. The Influence of Precipitation Regime on the Management of Three Course Crop Rotations in Northern Syria. *J. Agric. Sci.* (Camb.) 104: 281–287,1985.

Keatinge, J.D.H., M.D. Dennett, and J. Rogers. 1986. The influence of precipitation regime on the crop management in dry areas in Northern Syria. *Field Crops Res.* 12: 239-249.

Masri, Z., J. Ryan, and M. Pala. Specific Effects of a Cereal-Legume Rotation on Soil Physical Properties under Mediterranean-Type Conditions. *Abstracts: International Conference on Land Degradation.* June 10–14. Adana, Turkey, 1996, p. 51.

Matar, A., J. Torrent, and J. Ryan. Soil and fertilizer phosphorus and crop responses in the dryland Mediterranean zone. *Adv. Soil Sci.* 18: 82–146, 1992.

Materon, L.A. and J. Ryan. Rhizobial Inoculation and Phosphorus and Zinc Nutrition for Annual Medics Adapted to Mediterranean Environments. *Agron J.* 87: 692–698, 1995.

Mergoum, M., J. Ryan, J.P. Shroyer, and M. Abdel Monem. Potential for Adopting Triticale in Morocco. *J. Nat. Resour. Life Sci. Educ.* 21: 137–141, 1992.

Nordblom, T. and F. Shomo. Food and Feed Prospect to 2020 in West Asia/North Africa Region. Workshop on "Sustainable Agricultural Growth in the Major Ecoregions of the Developing World Prospects for 2020," Washington, D.C., 7–8 Nov. 1994. Int. Food Policy Res. Inst., Washington, D.C., 1994.

Papastylianou, I. Productivity Requirements of Barley in Rotation Systems in Rainfed Mediterranean Conditions. *Eur. J. Agron.* 2(2): 119–129, 1993.

Papastylianou, I. and M. Jones. Replacement of Fallow in the Rainfed Areas of the Mediterranean Region. *Proc. Intl. Conf. on Dryland Farming "Challenges in Dryland Agriculture: a Global Perspective"* USDA. Bushland, Texas, 1988, pp. 822–825.

Rasmussen, P.E. and H.P. Collins. Long-Term Impacts of Tillage, Fertilizer and Crop Residue on Soil Organic Matter in Temperate Semiarid Regions. *Adv. Agron.* 45: 93–134, 1991.

Ryan, J., D. Curtin, and I. Safi. Ammonia Volatilization as Influenced by Calcium Carbonate Particle Size and Iron Oxides. *Soil Sci. Soc. Am. J.* 45: 338–341, 1981.

Ryan, J., D. Curtin, and M.A. Cheema. Significance of Iron Oxides and Calcium Carbonate Particle Size in Phosphorus Sorption and Desorption in Calcareous Soils. *Soil Sci. Soc. Am. J.* 49: 74–76, 1985a.

Ryan, J., H. Hassan, M. Baasiri, and H.S. Tabbara. Availability and Transformation of Applied Phosphorus with Time in Calcareous Lebanese Soils. *Soil Sci. Soc. Am. J.* 49: 1215–1220, 1985b.

Ryan, J., M. Abdel Monem, and M. El Gharous. Soil Fertility Assessment at Agricultural Experiment Stations in Chaouia, Abda, and Doukkala. *Al-Awamia* 72: 1–47, 1990.

Ryan, J. and A. Matar. *Soil Test Calibration in West Asia and North Africa:* Proc. Third Regional Workshop, Amman, Jordan Sept. 3–9, 1989. ICARDA, Aleppo, Syria, 1990.

Ryan, J. and A., Matar. *Fertilizer Use Efficiency under Rainfed Agriculture in West Asia and North Africa. Proc. Fourth Regional Workshop*. May 5–10, 1991. Agadir, Morocco. ICARDA, Aleppo, Syria, 1992.

Ryan, J., M. Abdel Monem, H. Mergoum, and M. El Gharous. Comparative Triticale and Barley Responses to Nitrogen at Locations with Varying Rainfall in Morocco's Dryland Zone. *Rachis*: 10 (2): 3–7, 1991.

Ryan, J. and S., Garabet. Soil Test Standardization in West Asia-North Africa. *Commun. Soil Sci. Plant Anal.* 25: 1641–1655, 1994.

Ryan, J., S. Masri, L. Habib, and M. Pala. Long-Term Phosphorus Fertilization over a Rainfall Gradient in Dryland Farming Systems in Northern Syria. *Proc. Int. Soil Sci. Cong.*, Acapulco, Mexico, 1994, pp. 369–370.

Ryan, J., L.A. Materon, and S. Christiansen. Networks for Research Collaboration in the Dryland West Asia-North Africa Region. *J. Nat. Resour. Life Sci. Educ.* 24: 155–160, 1995.

Ryan, J., S. Masri, S. Garabet, H. Habib, and J. Diekmann. *Soils of ICARDA's Agricultural Experiment Stations and Sites: Climate, Classification, Physical-Chemical Properties, and Land Use.* ICARDA Tech. Bull., Aleppo, Syria, 1997.

Ryan, J., S. Masri, and S. Garabet. Geographical Distribution of Soil Test Values in Syria and Their Relationship with Crop Response. *Commun. Soil Sci. Plant Anal.* 27(5–8): 1579–1593, 1996b.

Shroyer, J.P., J. Ryan, M. Abdel Monem, and M. E. Mourid. Production of Fall-Planted Wheat in Morocco and Technology for its Improvement. *J. Agron. Educ.* 19:32–40, 1990.

Singh, M., and M. Jones. Estimating Time to Detect Trends in Continuous Cropping. *J. Appli. Stat.* 24:651–661, 1997.

Steiner, R.A. and R.W. Herdt. *A Global Directory of Long-Term Agronomic Experiments.* The Rockefeller Foundation, New York, 1993.

Smith, R.C.G. and H. Harris. Environmental Resources Constraints to Agricultural Production in a Mediterranean Environment. *Plant Soil* 58: 31–57, 1981.

White, P.F., N.K. Nersoyan, and S. Christiansen. Nitrogen Cycling in a Semiarid Mediterranean Region: Changes in Soil N and Organic Matter under Several Crop/Livestock Production Systems. *Aust. J. Agric. Res.* 45: 1293–1307, 1994.

Yau, S.K., M.M., Nachit, J. Ryan, and J. Hamblin. Phenotypic Variation of Boron Tolerance Toxicity in Durum Wheat at Seedling Stage. *Euphytica* 83: 185–191, 1995.

FOUR

Soil Quality Management in Africa

TWELVE

Soil Conservation and Sustainability of Agricultural Systems in Sub-Saharan Africa

B. Kayombo and G. C. Mrema

INTRODUCTION

In its major review of the economic developments in sub-Saharan Africa (SSA), the World Bank notes that deforestation is clearly a major problem of agriculture in Africa which calls for urgent attention (IBRD, 1989). Of the 703 million hectares of forests in SSA in 1989, about 3.7 million hectares (or 0.6%) were being cleared each year. The bank notes that 80–90% of Africa's rangelands and 80% of cropped land in drier areas may be affected by soil degradation.

Soil conservation has been an issue of major environmental concern in SSA since the beginning of this century. This concern has of late (in particular, since the end of the cold war) become a major preoccupation of the bilateral and multilateral donor agencies assisting SSA countries in their development efforts. Indeed, the priority of the major donor agencies involved in SSA has significantly changed over the past eight or so years. Priority areas for the donor funding now include, among others, protecting the environment, building democratic governments, stabilizing population growth, and protecting human health. The development assistance policy of most western nations has of late increasingly become "green," and environmental concerns are now of top priority. We certainly agree that the environment, particularly the fragile one found in SSA, requires to be protected and a sustainable system of agricultural production needs to evolve. We are, however, not sure if the development strategies and priorities emerging out of the think tanks and review panels of the major donor agencies are going to solve SSA environmental problems. Furthermore, the sustainability of the natural environment cannot be assured if the people are poverty-stricken and starving, as they will pass their suffering to the soil (Maher, 1950; Rowland, 1957).

The objective of this chapter is to review the status of physical soil conservation in SSA, presented in the following sequence:

1. Overview of soil conservation in the world;
2. Overview of soil conservation strategies adopted by the colonial authorities in SSA in the first 50 years of this century and those adopted by independent African governments over the last 40 years;
3. Prospects for soil conservation in SSA into the 21st century.

OVERVIEW OF SOIL CONSERVATION IN THE WORLD

That soil erosion is a major environmental issue in sub-Saharan Africa is unquestionable. Concern on the rate of soil erosion in SSA has accelerated over the past 60 years or so, and perhaps the major catalyst to the realization that it was a major environmental hazard in SSA was not events on this continent, but those which occurred in North America in the early 1930s. The Dust Bowl in the United States which led to the creation of the Soil Conservation Service (SCS) of the Department of Agriculture in 1934 (Troeh et al., 1980) played a major role in creating awareness of the soil erosion problem on this continent to agricultural officers, farmers, and colonial authorities. The SCS has played a critical and leading role in the world in advancing our understanding of the science and practice of soil conservation. It remains a major source of knowledge on soil conservation in the world up to today. Indeed, many of the early soil conservation scientists/practitioners from the British colonies could not claim to be experts in the science and practice of soil conservation if they had not visited the United States (Maher, 1950; Rowland, 1957; Hudson, 1957).

Chemical or Physical Soil Erosion

However, we need to distinguish between chemical and physical soil erosion. Chemical soil erosion refers to the depletion of soil fertility either through leaching of the important nutrients or through uptake of the same by plants, without their replenishment. Chemical soil conservation has always been part of agriculture ever since permanent settled agriculture started over 10,000 years ago.

Developments in plant breeding since the beginning of this century led to the introduction of high yielding varieties (HYV) of the major crops (cereals), and these being high yielding, they extracted even more nutrients from the soils. Fortunately for world agriculture, these developments in plant breeding occurred almost concurrently with the developments in fertilizer technology.

Chemical soil erosion is the invariable precursor of physical soil erosion. As it has been argued, a fertile soil, wherever it is formed has many of the properties of a sponge—it can absorb quantities of water and possesses considerable cohesion. An infertile soil loses this water absorbing capacity and cohesiveness and breaks down to a mass of separate particles, in which condition it is very readily washed away by water or blown by wind. Physical soil erosion is therefore the detachment of the soil particles and their transportation by either runoff water or wind away from the land, thus leading to the disappearance of the top soil. Most of the soil conservation efforts in the world have concentrated on controlling physical soil erosion.

In the developed world, soil exhaustion or decline in soil fertility has been prevented by massive application of chemical fertilizers. With fertilizer use being so low in SSA (Table 12.1) soil fertility has been maintained through the so-called shifting cultivation. With the adoption of a more sedentary type of life due to economic and social development by a larger proportion of the rural dwellers, the so-called bush fallow system was adopted. Under this system however, with increasing population pressures, the fallow periods, during which the soil fertility was rejuvenated, have become increasingly shorter leading to a faster decline in soil fertility. The alternative, of course, for the farmers is for them either to locate their fields far away from their homesteads, or use inorganic fertilizers. The traditional system is sustainable in the medium term only as long as there are no population pressures and the traditional crop varieties are grown. The cultivation of such traditional varieties and using crop husbandry techniques such as intercropping and use of livestock manures may generate a site-specific nutrient balance and thus create a

Table 12.1. Comparison and Extent of Utilization of Physical Technologies in Agriculture in SSA and Asia (1985)

	SSA	India	China
a) Total cultivated land (permanent crops and arable land) (millions)[a]	146	169	101
b) Economic active population in agriculture (millions)[a]	129.1	199.8	438.5
c) Draft oxen in use (millions)[b]	11.3	80.5	53.0
d) Tractors in use (thousands)[b]	133.6	607.8	873.0
e) Irrigated land (as % total cultivated land)[c]	2.3	26	49
f) Hectares of cultivated land per tractor (a/d)	1094	278	116
g) Hectares of cultivated arable land per agric. worker (a/b)	1.13	0.86	0.23
h) Fertilizer use (kg/ha)(1989/90)[d]	8.9	68.7	261.9

[a]From FAO Production Year Book (1986) Vol. 40.
[b]From Starkey (1988) and Misra (1991).
[c]Samad et al. (1992).
[d]Fertilizer use (kilogram of plant nutrients per ha of arable land) (IBRD, 1993).

low input agriculture, as advocated by many NGOs from the affluent industrialized countries (Reijutjes, 1994; Netting, 1994).

However, in the medium to long-term, it is unlikely that such a low input agriculture can produce the agricultural surpluses required to feed the ever-increasing population of SSA. A major problem of SSA agriculture is the low productivity both per unit of labor and land input. Given the population pressures and rural-urban migration, agricultural production is, in the next several decades, going to rely on fewer and fewer farmers who have to significantly increase their productivity through adoption of HYV coupled with increased utilization of chemical fertilizers in order to feed the ever-increasing urban population (Table 12.2). The problem of physical soil erosion cannot therefore be solved without finding a solution to the chemical soil erosion one. Unfortunately, most of the soil conservation efforts undertaken in SSA have aimed at solving the physical soil erosion problem.

The Science and Practice of Soil Erosion Control

As pointed out in the preceding section of this chapter, the USA is really the birthplace of the science and practice of soil erosion control. The soil erosion control measures which have been used in SSA are essentially imports of what was developed in the USA (Maher, 1950; Hudson, 1957; Stocking, 1988). However, the technologies developed in the USA were and have remained highly capital-intensive, requiring huge expenditures in machinery and equipment. Initially, mechanical conservation works were used which aimed essentially at managing runoff. These capital-intensive methods of mechanical soil conservation

Table 12.2. Urban and Rural Population in SSA, 1960–2025

Year	1960	1985	1990	2000	2010	2025
Total population (millions)	206.8	417.7	498.0	678	920	1294
Urban population	26.8	104.4	144.4	244.1	404.8	698.8
(as % total population)	(13%)	(25%)	(29%)	(36%)	(44%)	(54%)
Rural population	179.0	313.2	353.6	433.9	515.2	595.2
(as % total population)	(87%)	(75%)	(71%)	(64%)	(56%)	(46%)

Source: IBRD (1989); WRI (1986); Bulatao et al. (1990).

were adopted in SSA in the 1940s only in areas where there were large settlements of European farmers—mostly in Kenya and Southern Rhodesia (now Zimbabwe).

Subsequently, biological soil conservation techniques were adopted and these are based largely on conservation tillage methods (Pieters and McKee, 1938; Hudson, 1957; Pereira, 1973). In SSA, however, quite a number of tribes, due to the necessity of having to cultivate on steep sloping land, evolved technologies of soil and water conservation which were quite effective in reducing soil erosion to a tolerable level, and hence created sustainable agricultural systems. Notable among these were the Matengo pit system of southern Tanzania; ridging and tie-ridging on the islands in Lake Victoria and in northern Nigeria; and the agroforestry systems of the Chagga and Kikuyu on the slopes of Mount Kilimanjaro and Mount Kenya, respectively. These indigenous technologies were developed long before the SCS was established in the 1930s. However, the techniques developed by indigenous African tribes are highly labor-intensive and were developed in areas with quite fertile soils, cultivating either high-yielding food crops and/or high-valued cash crops (Stenhouse, 1944; Allan, 1965).

The American methods were initially based on trial and error until the 1960s when Wischmeier and Smith (1965) developed the so-called Universal Soil Loss Equation (USLE) [i.e., A = RKLSCP, where A is the average annual rate of erosion from sheet and rill erosion; R is the factor for climatic erosivity; K is a factor for soil erodibility; LS is a factor for topography; and CP is a factor for land use]. To apply the equation, detailed guidelines have been developed (Wischmeier and Smith, 1978). However, the utility of the USLE in situations such as those found in SSA is rather limited.

Consequently, therefore, in SSA there is a division between those who support the wholesale adoption of the USLE and those who believe that the USLE parameters are not appropriate to African conditions. The former group has largely concentrated on determination of the representative quantitative data for the various terms in the USLE. These include work on rainfall erosivity by, among others, Kowal and Kassam (1976) and Lal (1976a) in Nigeria; Stocking and Elwell (1973a, b) in Zimbabwe; and Roose (1973) in Niger, Ivory Coast, and Senegal. Research work has also been undertaken on erodibility of African soils by Lal (1976b) and Roose (1977) on West African soils. Work on cropping and management effects has been undertaken by Hudson (1957) and Elwell and Stocking (1974) in Zimbabwe; Fournier (1967), Lal (1974, 1976b) and Roose (1977) in West Africa. Fournier (1967), Kowal (1970), Kowal and Stockinger (1973), among others, have researched on different erosion control practices used in West Africa. The latter group has been largely influenced by work done in Zimbabwe starting from the 1950s by, among others, Hudson (1957, 1963), Elwell and Stocking (1975, 1976). The efforts of this group led to the development of the so-called Soil Loss Estimation Model for Southern Africa (SLEMSA). The SLEMSA model is Z = KXC, where Z is the mean annual soil loss; K is the mean annual soil loss from a standard field plot for a soil of known erodibility under bare fallow; C is the ratio of soil loss from cropped plots to that lost from bare fallow; and X is the topographical factor (Elwell, 1977).

In Europe, soil erosion research has been more limited than either in SSA or USA largely because erosion rates have been at a much lower level. In the recent past, however, some regions of the European Union have been experiencing soil losses of up to 10 tonnes/ha per year, as opposed to what is regarded as a tolerable rate of 1 tonne/ha per year (Morgan et al., 1994). The concern in Europe is more with the pollution effect of soil erosion rather than the loss of soil on agricultural fields. The European scientists have thus concentrated on developing models which are event-based and which are capable of providing more accurate information on the timing and source of sediment production

(Stehlik, 1975; Morgan et al., 1984). All these models are computer-based and require considerable amounts of field data before they can be effectively utilized.

The main objective of soil conservation would seem to us to be "to maintain a sustainable level of production in the long term from a given piece of land while keeping the soil loss at a level which allows the natural rate of soil formation coupled with the utilization of external inputs to be in equilibrium with the rate of soil erosion." It is unlikely that an agricultural system which can reduce soil erosion to zero will ever be developed. Consequently, the key issue is to reduce soil erosion to a level where a long-term sustainable agricultural production system can be maintained. Unfortunately, even where soil loss tolerance (SLT) values have been estimated (Schertz, 1983; Morgan, 1986), there can be no generalization, and the best guide is likely to be the actual measurement in the field.

SOIL EROSION RESEARCH IN SSA

Estimation of Soil Loss Through Erosion

Alarming rates of soil erosion have been reported throughout tropical Africa. In the semiarid and Sahel regions of West Africa, lack of adequate vegetation cover in the beginning of the rainy season is reported to be the main cause of severe erosion on arable lands (Lal, 1989). Accidental fires also destroy the scanty vegetation cover of the West African savanna woodlands. In humid West Africa, deforestation is reported to be the main cause of severe soil erosion (Lal, 1986).

Soil erosion research has also been undertaken in East Africa by among others, Rapp et al. (1972) and Rapp (1975). Rapp (1975) reported that annual sediment yields from the semiarid plains of the Dodoma and Arusha districts in Tanzania are as high as 200-730 m^3/km^2. Christiansson (1978) attributed this high sediment load to excessive grazing and lack of a protective vegetative cover. In Kenya, Dunne (1977, 1979) and Edwards and Blackie (1981) monitored sediment load from catchments of various sizes and observed that the long-term geological rate of erosion in these tropical environments for undisturbed catchments is 20–200 tonnes/km^2/year. The load is excessive, however, for heavily grazed and cultivated catchments.

In Lesotho, Rapp (1975) reported an annual sediment yield of as much as 1800 tonnes/km^2. It is perhaps in Southern Africa that there has been considerable and innovative research on estimation of soil loss through erosion. This research coordinated by the Southern African Regional Commission for Conservation and Utilization of the Soil (SARCCUS) has, as one of its products, SLEMSA. The development of SLEMSA occurred in 1976 (Rowland, 1994). SLEMSA, however, deals only with sheet erosion. Much work still needs to be done on it by multidisciplinary teams to refine it for ubiquitous utilization. Furthermore, a significant amount of the data which was used to develop SLEMSA was obtained from the European farm areas in Zimbabwe/South Africa and use of this model in the low input, small holder and peasant agriculture common in most of SSA may thus be limited.

Effect of Deforestation and Change of Land Use

One hundred and four million hectares of forest land in SSA have been deforested between 1965 and 1993 (Table 12.3). Of these, 45 million hectares have been converted to cropland while 59 million hectares have been converted to pastures (which is largely rough grazing). The total land area under cropland in SSA was only 6 and 8% in 1965 and 1993, respectively, with the remaining 92-94% being under forests (32 and 27%),

Table 12.3. Land Use in SSA[a]

	Area × 10^6 ha. (% of Total Land Area)		Increase (+ve) Decrease (-ve)
	1965	1993	Area × 10^6 (% of 1965)
Cropland	142 (6%)	187 (8%)	+45 (+32%)
Pastures	575 (25%)	785 (34%)	+210 (+37%)
Forests	740 (32%)	636 (27%)	−104 (−14%)
Other lands	859 (37%)	708 (31%)	−151 (−18%)
Populations (Total)	234 × 10^6	562 × 10^6	+328 x 10^6 (+140%)
TOTAL	2316 (100%)	2316 (100%)	

[a]Includes 46 countries in SSA (excludes Arab Africa). Source: FAO Production Yearbook Vol. 19, 1965; FAO Production Yearbook Vol. 48, 1994.

pastures (25 and 34%), and other lands (37 and 31%). Given that the increase in cropland between 1965 and 1993 has been only 45 million hectares (or 32% of cropland in 1965) despite the doubling of population during this period, it is unlikely that we can attribute the environmental degradation in SSA to conversion of forests to cropland. If anything, our concern should be on what has happened to the 59 million hectares deforested during this period, as well as the 151 million hectares of other lands (which appear to have been converted to pastures).

If it is assumed that a significant proportion of the additional 45 million hectares put under crop production between 1965-93 was through clearing of forests and this clearing was done using the traditional methods (e.g., slash and burn, etc.) or using hand tool technology, then it is unlikely that this has contributed to increase of soil erosion in SSA. Experiments conducted, by among others, Lal (1989) in Ibadan, Nigeria do show that although deforestation does lead to increased runoff and erosion rates, the method of clearing and postclearing soil management techniques do significantly affect the amount of soil loss (see Table 12.4). The traditional farming methods and manual clearing both registered soil losses of less than 5 tonnes/ha, which would seem to be within the tolerable limits.

Even when shear blade is used for clearing and this is followed by no-tillage type of postclearing soil management, the amount of soil loss per hectare is still considerably low at less than 5 tonnes/ha (see Table 12.4). However, the amount of runoff increases considerably with forest clearing. Sediment loads in mechanically cleared plots were much higher than in normally cleared plots. Soil degradation caused by mechanical clearing can be drastically reduced, however, through subsequent management—through growing cover crops, and adopting appropriate tillage methods which enhance infiltration as well as construction of appropriate structures to direct any runoff from the fields without causing rill erosion and gulleys.

It would seem to us that policy makers should be less concerned about deforestation which opens up forest lands to sustainable and balanced crop production. In any case, at only 45 million hectares, the new forest land opened up for crop production in SSA over the past 28 years is so small (about 1.9% of total land area) as to warrant the attention being paid to it by the donor community, especially when one compares this to the rate of population increase. Such forest lands constitute a huge area of SSA—at 636 million ha (in 1993) they are more than three times the total cultivated land (Table 12.3). Controlling the use of such public lands is quite difficult, whatever form of legislation governments enact. Given the Economical Structural Adjustment Programmes (ESAP) being

Table 12.4. Effects of Methods of Deforestation and Postclearing Soil Management on Runoff and Erosion from an Alfisol[a,b]

Treatments	Basin Area (ha)	Runoff (mm)		Soil Erosion tonnes/ha	
		1979	1979–1981	1979	1979—1981
Forest	15.0	T[c]	T	T	T
Traditional farming	2.6	3.0	6.6	0.01	0.02
Manual clearing/ no tillage	3.1	16.0	16.1	0.4	0.4
Manual clearing/ conventional tillage	3.2	54.0	79.7	5.0	9.8
Shear blade clearing/ no tillage	2.7	86.0	104.8	4.0	4.8
Tree pusher-root rake/no tillage	3.2	153.0	170.0	15.0	15.7
Tree pusher-root rake/ conventional tillage	4.0	250.0	330.6	20.0	24.3

[a]From Lal (1989).
[b]Land was cleared in 1979. Crop rotation schedule from 1979 to 1981 was maize-cassava-maize-cowpea.
[c]T = Unmeasurable trace.

implemented by most SSA countries, it is unlikely that governments will be able to find resources to pay and equip the large numbers of forest rangers who can control the exploitation of these natural forests. In fact, deforestation is a vital issue for economic and social development. Some 80% of the population of SSA rely on wood for fuel. Unless the root cause of underdevelopment—crippling poverty—is addressed, environmental degradation will deepen. It is only when agricultural productivity and overall production increase significantly that living standards will rise, and this in turn will lead to both the rural and urban dwellers being able to afford alternative fuels as well as alternative means of land clearing and management.

Soil Erosion and Crop Productivity

The effects of soil erosion on crop yield depend on, among other factors, soil profile characteristics, prevailing micro and meso climate. In most African soils the nutrient reserves are concentrated in the two surface horizons, and this is particularly the case for the soils most prone to erosion. On the other hand, crop yield is an integrated response to many interacting factors, and hence a one to one cause-effect relationship between rates of erosion and crop yield is difficult to establish. Whereas chemical soil erosion may result, in the short to medium term, in reduction in yields, physical soil erosion may not have that readily noticeable effect.

In SSA, little research has been undertaken on the effects of soil erosion on reduction in soil productivity. Stocking (1986) estimated, on the basis of data collected at the Henderson Research Station in Zimbabwe, that soil erosion nationally carries away an annual average of 1.6 million tonnes of nitrogen, 15.6 million tonnes of organic matter, and 0.24 tonnes of phosphorus. He estimated that the financial cost of soil erosion ranged between US $20–50 per ha on arable lands and US $10–80 on grazing lands. The implication of such figures, if they can be validated, is that soil erosion is a massive hidden cost on the economy of Zimbabwe. The costs of erosion damage have similarly been estimated in Ethiopia (Dregne, 1990). In West Africa, Lal (1985) suggested, as an inference from results of decade-long experiments at Ibadan, Nigeria, that SLT level of such soils was as low as 0.5 tonne/ha/year. In the absence of any defined SLT limits in SSA, however, these figures are, at best, broad approximations.

In addition, one has also to bear in mind the cost of controlling soil erosion against the losses in productivity which may occur if control structures are not constructed. Construction of mechanical control structures is quite expensive and where machinery is not available, requires quite a lot of labor. Likewise, adopting the biological control methods in most cases increases the labor demand to the farmer if there are no mechanization inputs. Tiffen et al. (1994) estimate the cost of terracing in Machakos District of Kenya as between US $1–2 per meter length of terrace. Given that the total length of terraces may be anything between 200–2000 meters long per hectare, this translates to US $300–3000 per hectare if an average cost of US $1.5 per meter length is taken. It is unlikely that peasant subsistence farmers in areas where there is still plenty of land will be bothered to construct such structures, however hard the political or extension system exhorts them.

A REVIEW OF STRATEGIES FOR CONTROL OF SOIL EROSION IN SSA

Soil conservation work started in most of SSA in the 1930s as a result of events in the United States. The concern about soil erosion was most pronounced in those countries where there were large numbers of European settler farmers, notably Kenya and Zimbabwe. These enclaves, which were opened up at the beginning of the century, were beginning to experience serious erosion problems due to intensive and extensive agriculture practiced.

The areas farmed by 'native' or African farmers, although they had been under cultivation for a much longer period, had not as of the 1930s shown such serious erosion problems, due largely to the shifting type of agriculture practiced there. We shall therefore focus on these two countries (Kenya and Zimbabwe), as comparatively successful soil erosion control efforts have been implemented over the past 60 or so years. They are also the two countries which have had high population increases and land shortages compared to other countries in SSA. Furthermore agricultural production in these two countries increased significantly over the period. The Soil Conservation Service (SCS) in both countries realized early that the key to success in soil conservation work was tied to correct land use, bearing in mind the climate, soils, topography, and markets.

Conservation Work on Large-Scale Farms

The soil conservation work undertaken in the large farm sector of Kenya and Zimbabwe in the 1940s and 1950s was largely based on the American methods (Maher, 1950; Murray, 1949). Focus in the 1940s and 1950s was therefore shifted to assisting commercial farmers in undertaking complete farm planning for soil conservation. Thus the SCS was strengthened in the 1940s by provision of more staff and training to enable them to provide such services.

Services provided by the SCS included land survey, realignment of roads, fields, fence lines and waterways, construction of terraces, and advice on crop rotational systems as well as fertilizer requirements; all this was done to ensure that there was minimum erosion. Furthermore, by the 1950s it became apparent that in order to increase the effectiveness of the soil and water conservation efforts, it was necessary to involve farmers even more than had hitherto been the case. The two countries, therefore, copied the United States' approach of having Intensive Conservation Areas (ICA) Committees, chaired and coordinated by farmers themselves (Rowland, 1957; Robinson, 1953). By the end of the

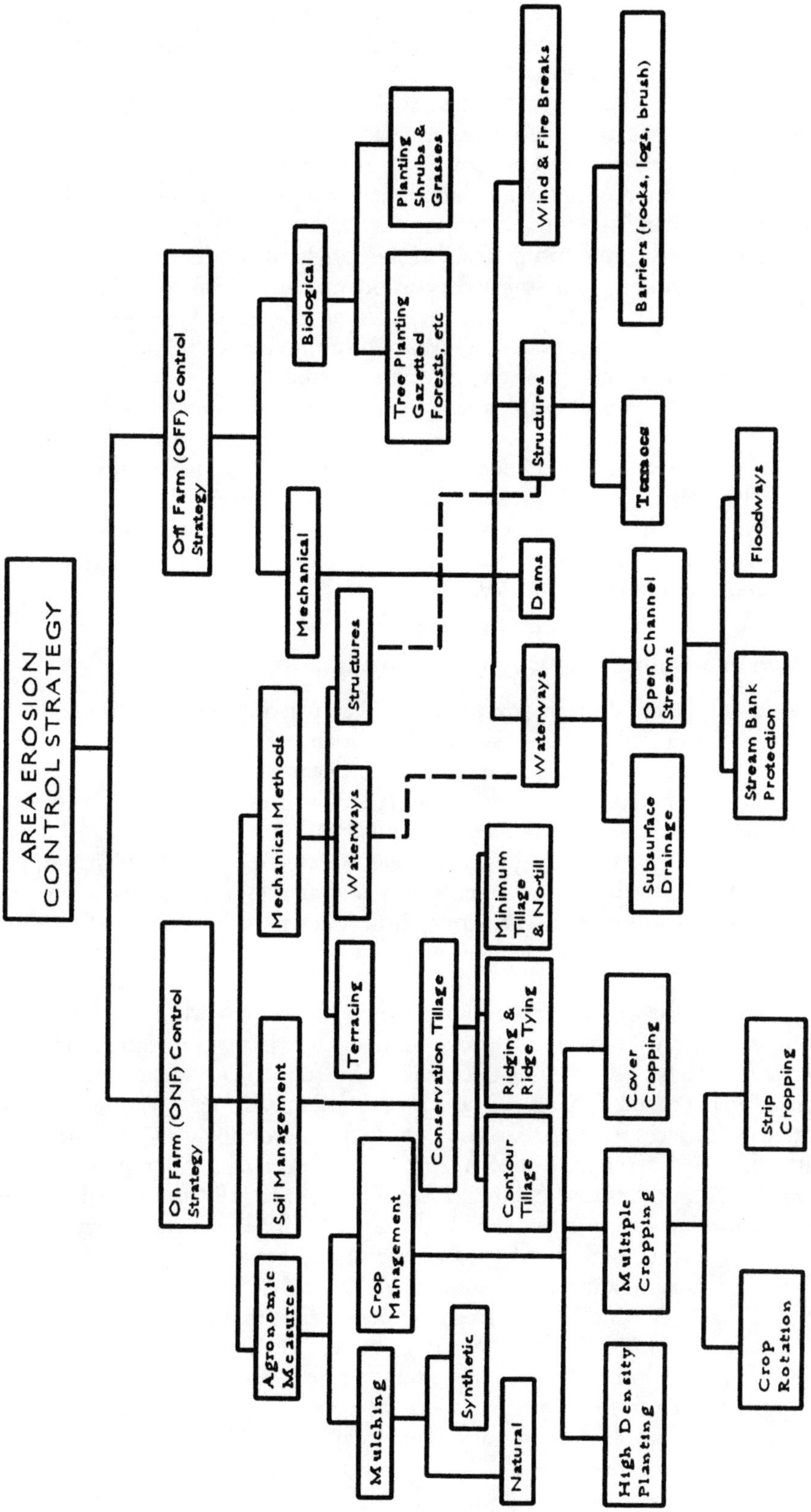

Figure 12.1. Integrated area erosion control strategy (including ONF and OFF areas).

1950s there were hundreds of such ICA committees, each representing a watershed in the European farming lands. There is no doubt these ICA committees played a crucial role in soil conservation work in these areas.

It is quite obvious that the success of these ICA committees and of the overall conservation work in the European farmed areas was largely due to, among other factors, the massive assistance from the colonial governments in the form of extension advice by conservation workers, provision of heavy machinery and equipment for mechanical conservation works, outright financial grants to the farmers, and price support subsidies for their produce through assured minimum prices and monopoly of production in certain products. The land use planning and soil conservation measures combined both On Farm (ONF) and Off Farm (OFF) soil conservation structures (Fig. 12.1). Since most of these farms were in specific areas designated as White farming areas, the farms were in most cases adjoining each other, and the strategy adopted enabled the soil and water conservation planning to be based on a complete watershed thus avoiding the situation found in the native areas where the land between fields is a no-man's-land, which leads to runoff from the fields being diverted to such land and causes severe erosion. Finally the question of land tenure also helped the soil conservation work.

In both Kenya and Zimbabwe, investments in soil and water conservation were more easily undertaken than in Malawi, Tanzania, and Zambia because farmers had freehold leases and felt more secure (Rowland, 1974).

Conservation Work on Peasant (Native) Areas

Concern about soil degradation in peasant or native areas increased significantly in the 1930s and has been described by a number of workers (Rowland, 1957; Allan, 1965; Tiffen et al., 1994). Unlike in the areas where Europeans settled (which were in most cases small enclaves), the African peasant subsistence farmers were scattered over a large area in most parts of Africa, and the intensive conservation approach used in the former areas could not be used. Furthermore, as these peasants were essentially subsistence farmers, they had not yet entered the modern economy, and any resources which could be provided for soil conservation work had to come from elsewhere. The authorities therefore relied to a great extent on compulsion as well as regulatory measures to effect soil erosion control.

These regulatory measures included, among others, declaring certain areas as protected zones; enforcing measures to reduce livestock numbers to what was regarded as the carrying capacity of the rangelands; moving people out of certain areas which were regarded as having been severely eroded; forcing the peasant subsistence farmers to adopt certain cultivation practices (terraces, contour farming, etc.). All these measures were tried by the colonial authorities in several parts of SSA. A few areas received a lot of attention, among them, Machakos district in Kenya (Maher, 1950; Tiffen, et al., 1994); Uluguru mountains in Morogoro, Tanzania (Rapp et al., 1972); Kondoa Irangi in central Tanzania (Allan, 1965); some higher attitude areas of Sierra Leone (Millington, 1987); and the native reserves of Zambia and Zimbabwe (Rowland, 1957, 1974).

Many of the practices which the African farmers were required to follow in order to reduce soil erosion were unpopular as they either increased the demand for labor at critical times (e.g., terracing) or were against cultural norms (e.g., reducing livestock numbers), or stopped people from cultivating areas they considered theirs traditionally (e.g., Uluguru mountains). The nationalist movements exploited this apparent opposition to enforced soil conservation by the African peasant farmers with a promise that it was a colonial move to deny them their freedom, and promised that at independence such measures would be abolished. At independence in the 1960s, therefore, a number of unpopular practices were either banned or were no longer enforced although remaining in the statute books.

The soil conservation strategy adopted by the colonial and subsequent independent government departments of agriculture was essentially regulatory. In most areas there was no attempt made to introduce comprehensive land use planning as was done on the large commercial farms. The peasant farms were scattered all over, thus forming a diaspora of fields with no-man's-land in between them. Any onfarm (ONF) conservation structures constructed (Fig. 12.1) did lead to diversion of runoff to the uncultivated areas adjacent to the farms in those areas where the farmers had no incentive to construct conservation structures. The construction of off farm (OFF) conservation structures in these uncultivated lands was usually not undertaken until huge gullies had been formed and were encroaching on the cultivated areas. In most cases this would be too late to stop serious degradation, and as there were no heavy machinery and implements provided, all the work had to be done using the hand tool technologies, and normally on a voluntary basis. There were no incentives, therefore, for the farmers to construct such OFF erosion control structures. The ONF conservation structures also demanded quite a lot of labor and this being in short supply, made the adoption of such technologies difficult.

As mentioned above, soil conservation work in most of the native areas of SSA was not undertaken as a component of comprehensive land use planning. The only places where there were comprehensive land use plans was in the Kenya Highlands in the 1950s in a few of the new settlement schemes which were established in the 1950s and 1960s, or in areas of high population densities where a system of land use planning has evolved over the years and into which, due to the necessity of soil conservation, this became an integral part of the evolving land use plans (e.g., Chagga farmers on slopes of Mt. Kilimanjaro, Mgeta, Morogoro Tanzania; Kikuyu farmers in Kenya Highlands, Rwanda, Burundi, and some parts of Buganda province in Uganda). The recent study by Tiffen et al. (1994) on the Machakos district shows the influence of population pressures in imposing a land use system into which conservation of the soil is integrated, as opposed to the general view that increase in population is a precursor to environmental degradation.

Conservation Tillage and Soil Erosion

The above review has concentrated on the soil conservation practices which assume that the cultivation methods on the farm will be the conventional ones (i.e., plowing, harrowing, planting, weeding, and harvesting). However, over the past 30 or so years, there has been increasing interest in new cultivation methods which have been adopted to some extent by both small- and large-scale farmers. These include minimum tillage and no tillage. Minimum tillage refers to a number of tillage practices; among others, wheel track planting, rip and disc, rough plowing, ripping, disc plowing with a combination of residue management techniques. All these methods, developed in the USA and Australia, have been tried by the research establishments in the countries in SSA with a large-scale commercial farming sector. Although they gave better soil conservation results in terms of infiltration, and higher organic matter in the soil, as well as reduced erosion, they give lower yields, however, largely due to other problems such as increased pest and weed attack on the crop.

No-tillage refers to the system of cultivation which requires no modification of the soil, with the seed and fertilizer being applied directly—such as spike harrowing, no-till strip-cropping, direct drilling, and broadcasting of seeds into residues, etc. These systems, however, require higher levels of mechanization, and can in most cases be adopted only by large-scale commercial farmers. Elwell (1994) reports that even in Zimbabwe where there has been considerable research and extension work on these conservation tillage practices, only between 5–10% of commercial and 1% of small holder farmers have adopted them. He gives the reasons for the nonadoption as being, among others, conservatism and socioeconomic factors. Although Elwell (1994) seems to be optimistic that economic factors are changing in favor of conservation tillage in Zimbabwe, we are, however, doubtful if these methods will spread into other countries in SSA other than on experimental stations. There are just too many constraints to their widescale adoption in SSA.

SOIL CONSERVATION IN SSA INTO THE 21ST CENTURY

An attempt has been made in the preceding sections of this chapter to objectively review the problem of soil erosion and policies undertaken to solve it in SSA over the past 40 or so years. It is unfortunate that the major donor agencies assisting SSA countries are now narrowing the focus of their assistance to environmental issues. We believe that the environmental problems of SSA are just a small part of the region's overall economic development problems, and the former can only be solved by taking a holistic approach and solving the latter. In our view there are six interrelated factors which have to be considered if the soil erosion problem in SSA is to be adequately tackled.

Macro-Economic Setting

Since the mid 1980s, most of the countries in SSA have been implementing Economic Structural Adjustment Programmes (ESAP) on the advice of the major multilateral financial institutions (World Bank, IMF, ADB) and bilateral donors. The major feature of these ESAP includes, among other measures, reduction in balance of payment deficits, devaluation of currencies (where these were overvalued), reduction in government expenditures, and privatization of public enterprises. The accepted philosophy now includes reduction of the role of government in the economy and enhancement of the role of the private sector through creating an environment where the market can function efficiently in the production and distribution of goods and services. Government's role should be confined to the provision of those goods and services which the private sector cannot provide efficiently and/or is unable to provide (e.g., social welfare, education, health, physical infrastructure).

From an environment point of view, the most significant effect of ESAP is the reduction in government expenditure. Even if SSA governments wanted to spend more on soil conservation programs, there are just no funds to be allocated for such programs. The level of consciousness on environmental concerns is such that it is unlikely that the urban and other sectors of the economy are going to accept higher taxes for soil conservation programs. In any case, with the exception of the few mineral-rich countries in SSA, these other sectors are so small, and agriculture remains the major sector of the economy, producing between 30–50% of the GDP. Therefore, it is the agriculture sector which has to produce the extra money required for soil conservation programs. This can only occur if the sector is more productive. With Peasant Subsistence Farmers (PSF) being the main players in the agricultural sector, relying entirely on the hand tool technology, it is unlikely that they will be in a position to increase their contribution to the government for such conservation programs.

It is for this reason that we agree with Mrema and Odigboh (1993) that the major way through which agricultural productivity and production can increase in SSA is by encouraging the emergence of a Medium Scale Farmers (MSF) group which can produce the surplus required and assist in such endeavors. Donor agencies can of course argue that they will be directing more of their assistance programs to combatting soil erosion and other environmental problems. Our view is that such assistance should be directed at those areas where it is likely to have maximum impact if a sustainable system is to be created. It is such MSF who are likely to take up and utilize such assistance most efficiently. These measures may lead to inequities, but at least there will be a surplus from the agricultural sector, and hopefully the resource base will not be degraded.

Demographic Factors

The second set of major factors insofar as soil conservation (and indeed environmental conservation) is concerned is the demographic issue. The accepted conventional wisdom is that the rapidly increasing population of SSA is a major contributor to the environmental degradation occurring on this continent. The increase in population leads to overexploitation of natural resources and this in turn leads to environmental degradation. Whereas we do agree entirely that reduction in population growth rate in SSA is a critical factor if economic development is to occur, we are, however, not very sure if the increase in population is the main cause of environmental degradation occurring on this continent (especially soil erosion in rural areas). Indeed it is the areas with high population densities where soil conservation programs have been most successful, e.g., the Central Province of Kenya, Kilimanjaro, and Mbinga regions in Tanzania. The recent study by Tiffen et al. (1994) demonstrates that increased population even in the more marginal areas like the Machakos district of Kenya does not necessarily lead to more erosion; if anything, the reverse occurs.

The population of SSA has increased by about two and one-half times between 1960 and 1990 (Table 12.2), with the urban population having increased by five times and the rural population just doubling in the same period from 179 to 354 million. Although the rural population has doubled between 1960 and 1990, the increase in cultivated cropland in SSA between 1965-1993 was only 32% (i.e. from 142×10^6 ha to 187×10^6 ha) (Table 12.3), thus demonstrating that the land cultivated per head of rural population has actually declined from 1965 to 1993 (from 0.72 ha to 0.48 ha). With the exception of a few countries (where population densities are high), the population problem in SSA would seem to be more of rural-urban migration than one of total population increase insofar as soil erosion is concerned. The decline in per capita cropland cultivated from 0.72 ha in 1965 to 0.48 ha in 1993 despite there being large areas still unopened (Table 12.3) demonstrates the dominance of the hand tool technology and low input of the African agricultural systems.

Two demographic trends are likely to have a significant influence on soil conservation in SSA. First is the aging rural population as the young and educated continue to migrate to urban areas. For example, 53% and 79% of heads of rural households in Botswana and Tanzania, respectively, were over 45 years old in 1990 (Mrema and Odigboh, 1993). It is unlikely that such old peasant farmers are going to be bothered with constructing soil conservation structures. This issue of aging rural population is likely to have an even more serious effect on soil conservation in those areas, where the migrants cut all their ties with the rural areas (e.g., some parts of Zambia). In those areas where culturally the urban dwellers maintain strong ties with their rural folks (e.g., Kikuyu and Kamba of Kenya, Batswana of Botswana, and Chagga of Tanzania), it is likely that they will invest some of their savings in improving the rural areas. Indeed the investments made by the Kamba of Machakos district on their land (part of it in soil conservation structures) comes partly from remittances of relatives in urban areas (Alcober et al., 1983; Tiffen et al., 1994). This flow of money and other resources from the urban to the rural areas (as off farm income) appears to be playing a major role in the economies of the rural sectors in such areas as Central Province (Kenya), Kilimanjaro (Tanzania), and Botswana, and may thus serve as an important source of investments for soil and water conservation. The second demographic trend is the impact of the Acquired Immune-Deficiency Syndrome (AIDS) epidemic in SSA. Exactly how this is going to affect the rural population is difficult to predict at this stage. This is likely to have a serious impact in some parts of SSA where the old farmers will be forced to spend a large part of their time caring for orphans of AIDS victims (Mrema and Odigboh, 1993).

The increase in population and migration to urban areas also affects the rate at which the natural forests are cleared for firewood and charcoal. For the rural areas, there is little that can be done to stop the deforestation for firewood and charcoal. For the urban areas it is likely that the economics of using such fuels with low calorific values as firewood and charcoal will force a change to other forms of fuel with higher calorific value. Our survey shows that for Dar-es-Salaam in Tanzania, the price of a 60-kg bag of charcoal has remained the same at the source, at about US $1.00 in real exchange rate between 1974 and 1994—the only difference being that whereas in 1974 charcoal could be obtained at Kibaha (30 km from Dar-es-Salaam city center), in 1994 it has to be obtained at Mdaula (120 km from city center). Because of this distance, the retail price of this one bag of charcoal in the city center has increased from US $2.50 in 1974 to US $7.00 in 1994. This has forced more and more people to use kerosene for cooking. Of course such things as improved charcoal kilns and stoves may help in the short term but in the long term, alternative fuels will have to be found for the ever-increasing urban population. Such fuels can only be afforded by a vast majority of the population if there is economic growth.

Farming Systems

The farming systems common in a larger part of SSA is that which is dominated by peasant subsistence farmers (PSF), who produce almost entirely for subsistence with no surplus for the market. Such low input agriculture cannot, in our opinion, be sustained in the long term even for the sustenance of these peasants. Furthermore with declining government revenues and hence declining allocations for the agricultural extension system (including natural resources conservation) and an aging and poor rural peasantry, it is likely that this will lead to even more soil erosion. Given the population pressures, in particular the need to produce enough food to feed the ever-increasing urban population (Table 12.2), there is need to have a more productive farming system which not only produces enough for those in the rural areas but also a surplus for the urban dwellers.

In our view the success of any On Farm (ONF) conservation measure is very much dependent on there being adequate Off Farm (OFF) conservation structures which will reduce the erosive effect of any diverted runoff on the uncultivated areas adjacent to the farm. Given that the pattern and intensity of rainfall in most of SSA is such that there will always be runoff from the cultivated areas, construction of OFF conservation structures is critical to the success of overall soil conservation measures being undertaken on farms (Fig. 12.1). With labor being in short supply in the peasant subsistence sector, even for constructing the ONF conservation structures, it is unlikely that such farmers will be bothered with OFF conservation structures. Furthermore, construction of such OFF conservation structures requires machinery and implements which the PSF are unlikely to possess.

We therefore agree with Mrema and Odigboh (1993) that there is a need to transform SSA agriculture by an infusion of MSF cadre in between the PSF who are likely to possess the machinery and implements which can be used during the off season not only for maintenance of rural infrastructure but also for construction and maintenance of soil conservation structures. Such a cadre of farmers will have to emerge from the PSF or from urban/dwellers who, if agriculture is made a profitable enterprise, are going to invest their savings in establishing such medium-scale farming enterprises. Governments should therefore assist in the creation of such a cadre of farmers by provision of the necessary incentives and by creating an economic environment whereby they can develop. They are also likely to be the ones who will set up the rural enterprises which will supply agricultural inputs (e.g., fertilizers) and assist in the marketing of any surplus to the urban areas/export markets through formation of cooperatives and/or agricultural marketing companies.

Microeconomic Issues

Establishment of an efficient input supply and output recovery system is crucial to the success of any integrated land use planning which incorporates significant investments in soil and water conservation measures. The success of the soil conservation programs in the Machakos district of Kenya, we would argue, has been caused by among other things, the proximity of Machakos to Nairobi, coupled with investments in trunk and rural roads made by the governments of Kenya over the past 30 or so years, and an efficient transportation system in the form of pickup trucks for both goods and people. Such investments in infrastructure reduce the transaction costs of both input delivery as well as output recovery, hence making inputs such as fertilizers cheaper to the farmers as well as reducing produce marketing costs.

Likewise, farmers in the Kenya highlands and such other areas as Kilimanjaro and Mbinga in Tanzania were able to increase their production and overall productivity in 1930–1960 through the establishment of efficient input and output recovery cooperatives. This in turn led to investments being made on soil conservation structures by both small and medium-scale farmers. Provision of such other amenities as schools and dispensaries plays an important role in convincing those who have migrated to urban areas to invest part of their savings in rural areas. With central government incomes becoming scarce due to ESAP, it will be up to local organizations/governments to shoulder a larger proportion of expenditures for such public investments. This can only occur if there are people who are producing a surplus through which the local governments can raise revenue.

Institutional Issues

Tiffen et al. (1994) have noted the importance of local institutions in the environmental recovery of the Machakos district in Kenya. These include the local county councils, local self-help groups, church/missionary groups, and women's organizations, all of which have played a leading role in soil conservation work. In other areas such as Kilimanjaro and Mbinga, in Tanzania, the local cooperative movements played a leading role in not only increasing agricultural production and productivity but also in soil conservation work (Illife, 1971). At the national level, in most of SSA countries, soil conservation responsibilities have normally fallen under different ministries. These have included ministries of Agriculture, Natural Resources, Lands and Settlements, Rural Development, and Local Government and Lands. It is important in our view, if soil conservation efforts are to succeed, for an integrated approach to be adopted. This will require the soil conservation service to fall under the same ministry as crop and livestock production. In countries such as Zimbabwe and Kenya where the soil conservation service departments have been in existence since the 1950s these have played a leading role in controlling soil erosion (Maher, 1950; Rowland, 1957).

Technology Development and Dissemination

Soil conservation research in SSA and elsewhere has concentrated on those technologies which increase the infiltration of precipitation without causing erosion and/or where there is runoff; on those technologies which reduce the velocity of the runoff and hence reduce erosion. The research has concentrated on what occurs on the farm and there has been little done to see what happens off farm especially in areas where the cultivated fields are scattered all over with uncultivated land in between them. Research is needed on what types of waterways need to be constructed in such uncultivated lands if the ONF soil conservation measures are to have an impact in the long run. Furthermore, the National Agricultural Research Systems (NARS) are quite weak in this area of soil conservation.

A significant proportion of the little research undertaken on soil erosion in SSA has been undertaken by geographers from university departments. The operational strategies adopted by these geographers has tended to focus more on preservation of the biodiversity, and exclusion of people from areas declared as protected zones. To improve the social acceptability of such protected areas, these environmentalists have attempted to establish environmental conservation related programs [often paid for by international nongovernmental agencies such as the International Union for Conservation of Nature (IUCN)] for the people living in the proximity of such protected areas. From our point of view, proponents of this type of approach are more inclined toward preservation rather than conservation of the environment, and we do not think that such an approach is sustainable in the long term in SSA.

The opposite appears to have been the case in the United States, where agricultural engineers and soil physicists were the first to recognize soil erosion as a problem, and hence research on soil and water conservation engineering was integrated into mainline agricultural research. Most of the NARS and extension departments in Africa do not have agricultural engineering divisions/sections, and this may well be the reason why this area has been neglected in our agricultural development programs. Indeed, in the two countries in SSA with strong soil conservation service, i.e. Kenya and Zimbabwe, the agricultural engineering departments were established in the 1950s and have played a major role in controlling soil erosion (Elwell, 1994).

Efforts of these departments have been directed toward developing technologies and disseminating them to farmers. We do not agree with Crosson and Anderson (1993) that what the NARS in Africa require for research on natural resource and environmental issues are more social scientists! What we need are agricultural engineers and soil physicists, as well as agriculturists who are sensitive to socioeconomic issues. A lot of time was wasted in the 1970–80s in the debate between social scientists and engineers on the effects of agricultural mechanization on employment in the Third World (Giles, 1967; ILO 1993). The recent studies by, among others, Misra (1991), Fan and Pardey (1992) and Singh et al. (1994) do conclusively disprove the arguments advanced by social scientists in the 1970s on the unemployment effects of mechanization. We should therefore not waste time on fruitless and academic debates which are unlikely to solve the soil conservation problems of SSA. Mrema and Odigboh (1993) have argued the need to have research, extension, and training systems which balance all the elements of what they call a trinity of technologies in agriculture—biochemical, physical, and socioeconomic. The same arguments apply to soil conservation technology development and dissemination.

CONCLUSION

For SSA a critical environmental protection issue is essentially one of proper land use planning for agricultural production. In this regard, controlling soil erosion is of paramount importance if a sustainable agricultural production system is to be developed in the region. We believe that the adoption of a holistic approach, whereby a broad spectrum of human and environmental issues are considered in an integrated manner, will sustain soil conservation efforts in SSA. In this respect, integrated land use planning—which takes into account the chemical and physical condition of the soil, the terrain, the crops to be produced and the markets—is crucial to the success of any soil conservation program in SSA. Successful programs have occurred where such an integrated approach has been adopted (e.g., the large farming sector enclaves in Zimbabwe and Kenya). Such an integrated land use program, when implemented in the small holder sector, has to consider both on farm (where the crops are actually grown) and off farm (adjacent areas to the farm, which are not cultivated) areas. With declining government revenues (and hence less money for extension programs), and an aging and declining rural population due to a multiplicity of demographic factors (e.g., rural-urban migration; AIDS epidemic) it is important that policy makers address the issue of conservation in such a scenario. Their ability to do so can only be facilitated if the agricultural systems in SSA are transformed to become more productive and are able to finance the capital expenditure required for soil conservation and also for other inputs such as machinery and equipment. Infusion of a more progressive and science-based cadre of farmers who will be producing for the market is likely to play a leading role in such a transformation. Such farmers, who will be medium scale, will need to have their lands properly planned to take into account not only the state of the soils but also the crops and other agricultural enterprises as well as the markets and inputs available. Government policies in SSA should, therefore, be directed at helping such farmers to emerge, and these should not only contribute to increasing output, but also provide such other services to themselves and to the peasant subsistence farmers (such as maintenance of soil conservation structures, help in managing the input and output recovery systems as well as the rural infrastructure). Such farmers cannot be expected to rely on the hand tool technology, and higher levels of mechanization technologies will be essential if they are to succeed.

REFERENCES

Alcober, D.L., J. Cornelius, J. Medland, G.C. Mrema, S. Prayag, and G.O. Sharrock. Agricultural Development Research Priorities for a Semi Arid Area in Kenya: ICRA Bulletin No. 10. ICRA Lawickse Allee 14, Wageningen, The Netherlands, 1983.

Allan, W.C. *The African Husbandman*. The Greenwood Press, Westport, CT, 1965.

Bulatao, R.A., E. Bos, P.W. Stephens, and M.T. Vu. *World Populations Projection; Short and Long Term Estimates.* 1989/90 Edition; World Bank, Washington, D.C., 1990.

Crosson, P. and J.R. Anderson. Concerns for Sustainability: Integration of Natural Resources and Environmental Issues in the Research Agendas of NARS. ISNAR Res. Report No. 4 1993, The Hague, The Netherlands.

Christiansson, C. Relations Between Heavy Grazing, Cultivation, Soil Erosion and Sedimentation in the Semi-Arid Parts of Central Tanzania. Proc. 1st Inter. Rangeland Congr., Denver, CO, 1978, pp. 274–278.

Dregne, H. Erosion and Soil Productivity in Africa. *J. Soil Water Conserv.*, July/August 1990.

Dunne, T. *Studying Patterns of Soil Erosion in Kenya*. FAO Soils Bull. No.33, 1977, pp. 109–122.

Dunne, T. Sediment Yield and Land Use in Tropical Catchments. *J. Hydrol.*, 42: 284–300, 1979.

Edwards, K.A. and J.R. Blackie. *Results of the East African Catchment Experiments 1958–1974.* In

Tropical Agricultural Hydrology, R. Lal and E.W. Russell, Eds. John Wiley and Sons, Chichester, 1981, pp. 163–188.

Elwell, H.A. Soil Loss Estimation System for Southern Africa. Department of Conservation and Extension, Research Bulletin, No.22, Salisbury, Rhodesia, 1977.

Elwell, H.A. Development and Adoption of Conservation Tillage Practices in Zimbabwe. In *Soil Tillage in Africa: Needs and Challenges.* FAO Soils Bulletin No. 69, FAO, Rome, 1994, pp. 129–165.

Elwell, H.A. and M.A. Stocking. Rainfall Parameters and a Cover Model to Predict Runoff and Soil Loss from Grazing Trials in the Rhodesian Sandveld. *Proceedings, Grassland Society of Southern Africa,* 9: 157–164, 1974.

Elwell, H.A. and M.A. Stocking. Parameters for Estimating Annual Runoff and Soil Loss from Agricultural Lands in Rhodesia. *Water Resour. Res.* 11: 601–605, 1975.

Elwell, H.A. and M.A. Stocking. Vegetal Cover to Estimate Soil Erosion Hazard in Rhodesia. *Geoderma,* 15: 61–70, 1976.

Fan, S. and P.G. Pardey. Agricultural Research in China: Its Institutional Development and Impact. ISNAR, The Hague, 1992, p. 96.

Fournier, F. Research on Soil Erosion and Soil Conservation in Africa. *Sols Africains,* 12: 53–96, 1967.

Giles, G.W. Agricultural Power and Equipment: In: "World Food Problems," Vol. VII. Report of the President's Advisory Committee, Washington DC, 1967, pp. 175–216.

Hudson, N.W. Progress Report on Experiments at Henderson Research Station 1953-1956. *The Rhodesian Agric. J.,* 54: 297–323, 1957.

Hudson, N.W. Raindrop Size Distribution in High Intensity Storms. *Rhodesian J. Agric. Res.,* 1: 6–11. 1963.

IBRD. Sub-Saharan Africa: From Crisis to Sustainable Development: A Long Term Perspective. International Bank of Reconstruction and Development, World Bank, Washington D.C., 1989.

IBRD. A Strategy to Develop Agriculture in Sub-Saharan Africa and a Focus for the World Bank. International Bank of Reconstruction and Development, World Bank Technical Paper No. 203, Washington D.C., 1993.

Illife, J. Agricultural Change in Modern Tanganyika. Historical Association of Tanzania: Paper No. 10; E. Africa Publishing House, Nairobi, 1971.

ILO. Mechanization and Employment: Case Studies from Four Continents. International Labour Organization; Geneva, 1993.

Kowal, J. The Hydrology of a Small Catchment Basin at Samaru, Nigeria. Part IV: Assessment of Soil Erosion under Varied Land Management and Vegetal Cover. *Nigerian Agric. J.,* 7: 134–147, 1970.

Kowal, J.M. and A.H. Kassam. Energy Load and Instantaneous Intensity of Rainstorms at Samaru, Northern Nigeria. *Tropical Agric. (Trinidad),* 53: 185–197, 1976.

Kowal, J. and K. Stockinger. Usefulness of Ridge Cultivation in Nigeria Agriculture. *J. Soil Water Conserv.,* 28: 136–137, 1973.

Lal, R. Soil Erosion and Shifting Agriculture. FAO Soils Bulletin No. 24, 1974, pp. 48–71.

Lal, R. Soil Erosion on Alfisols in Western Nigeria, III. Effects of Rainfall Characteristics. *Geoderma,* 16: 389–401, 1976a.

Lal, R. Soil Erosion on Alfisols in Western Nigeria, I. Effects of Slope, Crop Rotation and Residue Management. *Geoderma,* 16: 363–375, 1976b.

Lal, R. A Soil Suitability for Different Tillage Systems in the Tropics. *Soil Tillage Res.,* 5: 179–196, 1985.

Lal, R. Conservation of Tropical Rainforest: Agronomic Potential and Ecological Consequences. *Adv. Agron.,* 39: 173–264, 1986.

Lal, R. Conservation Tillage for Sustainable Agriculture: Tropics Versus Temperate Environments. *Adv. Agron.,* 42: 85–197, 1989.

Maher, C. Soil Conservation in Kenya Colony: Part 1. Factors Affecting Erosion, Soil Charateristics and Methods of Conservation; Part II: Soil Conservation Practice: Organization and

Legislation: Present Position and Outlook. *Empire J. Expt. Agric.*, Vol. XVIII, No.71: 137–149; No. 72: 233–248, 1950.

Millington, A. Environmental Degradation, Soil Conservation and Agricultural Policies in Sierra Leone. In *Conservation in Africa: People, Policies and Practices.* D.M. Anderson and R. Grove, Eds. Cambridge Univ. Press, London, 1987, pp. 229–248.

Misra, S.M. Formulation and Implementation of Agricultural Mechanization Policies in India. In *Agricultural Mechanization Policies and Strategies in Africa: Case Studies from Commonwealth African Countries.* G.C. Mrema, Ed., Commonwealth Secretariat, London, 1991, pp. 177–188.

Morgan, R.P.C. *Soil Erosion and Conservation.* Longmans, Essex, UK, 1986.

Morgan, R.P.C., D.D.V. Morgan, and H.J. Finney. A Predictive Model for the Assessment of Soil Erosion Risk. *J. Agric. Engineering Res.*, 30: 245–253, 1984.

Morgan, R.P.C., J.N. Quinton, and R.J. Rickson. Modelling Methodology for Soil Erosion Assessment and Soil Conservation Design: The EUROSEM Approach. *Outlook Agric.*, 23(1): 5–10, 1994.

Mrema, G.C. and E.U. Odigboh. Agricultural Development and Mechanization in Africa: Policy Perspective. *NAMA Newsletter,* 1(3): 11–49, 1993.

Murray, C.A. Summary of Annual Report of the Department of Conservation and Extension (Jan-Dec 1948). *Rhodesia Agric. J.*, 46(4): 234–236, 1949.

Netting, R. Kofyar Variations—When It Comes to Sheer Adaptability, Northern Nigeria Farmers Have Few Peers. *Ceres*, 148: 22–24, 1994.

Pereira, H.C. *Land Use and Water Resources.* Cambridge Univ. Press, London, 1973.

Pieters, A.J. and R. McKee. The Use of Cover and Green Manure Crops. In *Soils and Men*, USDA, Yearbook of Agriculture, 1938, pp. 431–444.

Rapp, A. Soil Erosion and Sedimentation in Tanzania and Lesotho. *AMBIO*, 4: 154–163, 1975.

Rapp, A., L. Berry, and P. Temple. Studies of Soil Erosion and Sedimentation in Tanzania. *Geogr. Ann.*, 54A (3/4): 105–379, 1972.

Reijutjes, C. Low Input Farming: Is It Worth the Work ? *Ceres*, 148: 20–22, 1994.

Robinson, D.A. Land Use Planning in Native Reserves in Southern Rhodesia. *Rhodesia Agric. J.*, 50(4): 327–337, 1953.

Roose, E.J. Dix-sept Annees Des Mesures Experimentales de l'erosion et du Ruissellement Sur un Sol Ferrallitique Sableux de Basse Cote d'Ivoire. Contribution a l'etude de l'erosion Hydrique en Milieu Intertropical. These Doct. Ing., Fac. Sci. Abidjan, 1973.

Roose, E.J. Use of the Universal Soil Loss Equation to Predict Erosion in West Africa. In: Soil Erosion: Prediction and Control. Soil Conservation Society of America, Special Publication No.21, pp. 60–74, 1977.

Rowland, J.W. The Need for Modernization of Farming Methods. *Rhodesia Agric. J.*, 54(5):395–402, 1957.

Rowland, J.W. The Conservation Ideal—Being the SARCUSS Record for the Period 1952–70. SARCUSS, Private Bag X116, Pretoria, 1974.

Rowland, J.W. The Conservation Ideal: The Third Decade: Being the SARCUSS Record for the Period 1970–80. SARCUSS, Private Bag X116, Pretoria, 1994.

Samad, M., Merrey, D., Vermillion, D., Fuchs-Carsch, M., Khalid, M. and R. Lenton. Irrigation Management Strategies for Improving the Performance of Irrigated Agriculture. *Outlook Agric.*, 21: 279–286, 1992.

Schertz, D.L. The Basis for Soil Loss Tolerance. *J. Soil Water Conserv.*, 38: 10–14, 1983.

Singh, G., Bohra, C.P. and B.K. Yadav. Energy Input in Production Agriculture of India During the Last Four Decades 1951–1991. *Agric. Eng. J. (AIT)*, 3: 1–20, 1994.

Starkey, P. *Animal Traction Directory: Africa.* Vieweg for German Appropriate Technology Exchange, GTZ, Eschborn, Germany, 1988.

Stehlik, O. Potencialni eroze pudy proudici vodou na uzemi CSR. *Studia Geographica* 42, Brno, 1975.

Stenhouse, A.S. Agriculture in the Matengo Highlands. *East African Agric. J.*, 10:22–27, 1944.

Stocking, M.A. The Cost of Soil Erosion in Zimbabwe in Terms of the Loss of Three Major Nu-

trients. Consultants' Working Paper No.3, Soil Conservation Programme, Land and Water Development Division, FAO, Rome, 1986.
Stocking, M.A. Assessing Vegetative Cover and Management Effects. In *Soil Erosion Research Methods*, R. Lal, Ed., International Society of Soil Science, Wageningen, 1988, pp. 163–186.
Stocking, M.A. and H.A. Elwell. Prediction of Sub-Tropical Storm Losses from Field Plot Studies. *Agric. Met.*, 12: 193–201, 1973a.
Stocking, M.A. and H.A. Elwell. Soil Erosion Hazard in Rhodesia. *Rhodesia Agric. J.*, 70: 93–101, 1973b.
Tiffen, M., Mortimore, M. and F. Gichuki. More People Less Erosion. *Environmental Recovery in Kenya*. J. Wiley and Sons, Chichester, UK, 1994.
Troeh, F.R., Hobbs, J.A. and R.L. Donahue. *Soil and Water Conservation for Productivity and Environmental Protection*. Prentice-Hall, Englewood Cliffs, NJ, 1980.
Wischmeier, W.H. and D.D. Smith. Predicting Rainfall Erosion Losses from Cropland East of the Rocky Mountains. USDA Agric. Handbook No. 282, 1965.
Wischmeier, W.H. and D.D. Smith. Predicting Rainfall Erosion Loss—A Guide to Conservation Planning. USDA Agric. Handbook No. 537, 1978.
WRI. World Resources 1986: An Assessment of the Resource Base that Supports the World Economy. World Resources Institute, New York, 1986.

THIRTEEN

Consequences of Soil Erosion on Soil Productivity and Its Restoration by Soil Management in Tanzania

Jens B. Aune, Ignance K. Kullaya, Method Kilasara, Fedilis S.B. Kaihura, Bal Ram Singh, and Rattan Lal

INTRODUCTION

Vast areas of once fertile lands have been rendered unproductive due to erosion-induced soil degradation in Tanzania. The global economic loss due to accelerated erosion is very high (Pimentel et al., 1995) and Tanzania is no exception. Accelerated soil erosion has raised both ecological and economic concerns. The ecological concerns are related to the problem of environmental quality in general and pollution of water resources in particular. Economic effects of erosion are of two types. The off-site effects are those related to the problem of sedimentation, siltation of waterways and reservoirs, and property damage by flooding. The on-site economic effects of accelerated soil erosion are due to reduction in crop yield, loss of water and plant nutrients, and overall degradation of the soil resources.

Agriculture is the mainstay of Tanzanian economy, employing more than 90% of its rural population, and accounting for over 58% of the gross domestic product (Kajumulo, 1992). Tanzania has an arable land area of about 3.4 million ha out of the total land area of 94.5 million ha. The land area of high agricultural potential is merely 1.6 million ha, including 0.15 million ha under irrigation. With total population of about 30 million increasing at the rate of 3.46%/yr, per capita arable land is merely 0.11 ha (Lal, 1995a; Muchena and Kiome, 1995; USDA, 1992). Input required for improving soil fertility and increasing productivity are either not available or are prohibitively expensive. Average rate of fertilizer use is low, about 15 kg/ha (USDA, 1992), consequently soil fertility is being depleted. Mining soil fertility and resultant soil degradation are responsible for low yields, agricultural unsustainability, and poor environmental quality.

Soil research in Eastern Africa dates back to the early 1920s. Most early research, however, concentrated on soil survey, soil classification, and mapping (Anderson, 1963; Hathout, 1975). Soil fertility and fertilizer use research began in the 1970s, leading to development of fertilizer recommendations (Samki and Harop, 1984; Mowo et al., 1993). Agronomic aspects of manure and compost use (Semoka et al., 1983; Ngaiza, 1983), biological nitrogen fixation (Chowdhury et al., 1983) and low-input agriculture (Rodale, 1983) have been discussed and reviewed. However, improving productivity of degraded soils with low nutrient reserves will require drastic measures to improve nutrient status by adding off-farm inputs.

In Tanzania, soil degradation caused by erosion is a widespread problem which takes place in different forms and intensities affecting nearly all major land uses. Degraded or partially degraded soils with low output potential occupy a considerable proportion of the cultivated lands, but there is a lack of appropriate technology that can solve the problem of soil degradation. Yet, restoration of degraded and depleted soils is crucial for bringing about drastic increase in agricultural productivity to feed the ever-increasing population in the country. Measures required to restore productivity cannot be adopted without experimental data on the extent of the problem and its economic consequences.

Neither the economic consequences of soil erosion nor the impact on crop production are sufficiently known in Tanzania and in the rest of Africa. There is a disturbing degree of correspondence between the regions affected by severe soil erosion and those prone to gross food deficit, notably in Africa (Lal, 1995b). Therefore, it is necessary that soil erosion and its effects be expressed in monetary terms related to loss in productivity, environmental degradation, and damages to infrastructure (Lal, 1985).

This chapter presents a synthesis of different studies that have been undertaken as a research collaboration between the Agricultural University of Norway and Tanzania research institution to assess the effect of soil erosion on productivity in different parts of Tanzania. Two different approaches were used to address this issue:

1. field study on the effect of topsoil depth on crop yields
2. modeling the effect of soil erosion on crop productivity.

SOIL EROSION/CROP PRODUCTIVITY FIELD STUDY

Objectives and Methodology

The need to relate erosion-induced soil degradation to productivity has been recognized by many researchers (Lal, 1995b; Pimental et al., 1995). The present two-phase study was initiated in 1992 in three contrasting ecoregions (semiarid , subhumid, and humid) of Tanzania to investigate the impact of soil erosion on crop yield and soil quality, and the effect of soil fertility management practices on restoration of soil productivity. Objectives of the Phase 1 (1992–1994) study were to:

(i) establish criteria for assessment of degree of past erosion from soil properties,
(ii) relate the degree of past erosion as measured by topsoil depth to yield of maize (Zea mays) and cowpea (Vigna unguiculata), and
(iii) identify the effect of soil erosion on soil chemical and physical properties.

Three agroecological zones (located in the Kilimanjaro, Tanga, and Morogoro geographical regions) were selected representing the humid, subhumid and subhumid - semiarid transition zone, respectively (Table 13.1). A detailed characterization of these agroecological zones was carried out by De Pauw (1984). The zones differ markedly in rainfall, geology of the parent material, and dominant soil type or soil characteristics. The typical soils of each zone were taken into account in selecting the sites for the study. Accordingly, two sites were selected in both Kilimanjaro and Tanga regions while sites were chosen in Morogoro. The study was conducted on agricultural lands with known history of man-

Table 13.1. Agroecological Characteristics of the Sites

Ecoregion	Altitude and Climate	Geology	Soils	Vegetation and Land Use
Kilimanjaro	altitude of 900 to 2300 meters above sea level, rainfall of 1500 mm per annum, temperature varies with altitude: 15 to 28°C, dependable growing season of 3 to 5 months	quartzo-feldspathetic gneiss and granulites, horneblende and biotite gneisses, localized thick crystalline limestone masses	deep well-drained dark reddish brown friable or firm clay loams and clays with strong structure; well-developed profile; pH: 6 to 7.5. CEC: 20-45 cmol/kg; moderate to high fertility; Nitosols common	coffee and bananas plantations; forest patches remaining on steep nonagricultural lands; wide range of annual crops, legumes and vegetables including maize, french beans, sunflower
Tanga	altitude of 150 to 500 meters, rainfall of 1100 mm per annum, temperature of 20 to 30°C, dependable growing season of 4 to 4.5 months.	danded gneiss with crystalline aluminous gneiss on the ridge summits and limestone gneisse in the plains	deep-drained reddish sandy clays to clays, weak structure, low fertility Ferralsols (occupy 60%) and moderate fertility Luvisols (occupy 40%)	open tree cover without interlaced canopies with grass and herb cover, main crops are sisal, maize, coconut, and citrus
Morogoro	altitude of 200 to 500 meters to rainfall of 600 to 1200 nnper annum, temperatures of 15 to 30°C, dependable growing season of 3 to 3.5 months	hornblende and biotite gneisses with abundant garnet in the Uluguru mountain range, unidentified sediments of Karoo age and Neogene deposits on plains	yellowish or reddish sandy clays to clays with subsoil acidity (Acrisol-Ferralsols type), together with reddish friable clays (Luvisols)	wooded grassland with canopy cover of less than 50%; maize, sisal, and sunflower are the main crops

agement practices. Sheet and rill erosion are the dominant processes, and gullies and landslides are rare on these sites.

An area of about two hectares was selected, each site was topographically surveyed, and divided into 5-m grid points at which minipits (30 by 60 cm wide) were dug to the depth of 50 cm. Soil morphological characteristics were systematically observed with increase in soil depth from the ground surface to the 50 cm depth in order to determine the boundary between the topsoil and the underlying subsoil. The morphological characteristics considered were: soil consistence both under field moisture and wet soil conditions, soil color, and soil texture. Each site was characterized into four topsoil depth (TSD) classes: (1) < 15 cm, (2) 16–20 cm, (3) 21–25 cm, and (4) >25 cm. The TSD classes 1, 2, 3, and 4 represent severely eroded, moderately eroded, slightly eroded, and least eroded phases, respectively (Kilasara et al., 1994a; b).

Twelve plots of 5 × 5 m were selected for each erosion class, and were used for soil sampling, field measurements, and growing crops. Crop yields were measured beginning with the 1993 season. Cowpeas was used as a test crop in the short rainy season (October-December) and maize (variety MTV-1) as test crop in the long rainy season (March to June). Cowpeas was sown at a spacing of 50 × 15 cm and maize at 75 × 30 cm spacing. The cowpeas crop was successful at Kilimanjaro and Tanga but failed in Morogoro due to drought. No fertilizer was applied to cowpeas, but for maize both nitrogen and phosphorous were applied each at a rate of 40 kg ha^{-1}. This amount was equivalent to half of the recommended rate of plant nutrients for maize in Tanzania.

The Phase 2 was initiated in 1994 with the objective to determine the extent to which soil management could offset the effect of loss in TSD on productivity and restore soil quality. One new soil type was added at Morogoro in the second phase and one new TSD class was identified at Misufini 1. The soil erosion classes established in Phase 1 constituted a block. Soil management practices were tested by superimposing treatments on different TSDs. In each block, treatments were randomly distributed and replicated three times. A real replication of the TSD was not possible because each TSD class was found at a distinct position in the surveyed area.

The following soil management treatments were tested:

1. Farmers practice: manual shallow cultivation just before planting in Tanga and Morogoro, and hand hoeing up to 10-15 cm depth before planting in Kilimanjaro,
2. Tie-ridging (TR): contour ridges cross-tied to conserve runoff,
3. Tie-ridging plus 20 Mg farmyard manure (FYM)/ha (air dry),
4. Tie-ridging plus nitrogen and phosphorous fertilizer at local recommendation rates, 100 kg N/ha and 13 kg P/ha at Kilimanjaro sites, 50 kg N/ha and 18 kg P/ha at Tanga sites, and 80 kg N/ha and 36 kg P/ha at the Morogoro sites.

Nitrogen and phosphorus equivalents applied in FYM were 59 kg N and 16 kg P/ha in Tanga, and 15kg N and 2.5 kg P/ha per hectare in Kilimanjaro.

Effect of Soil Erosion on Soil Chemical Properties

The sand content was the highest in Morogoro soils (Table 13.2), and clay content was higher in the subsoil than in the topsoil for all sites, suggesting a preferential movement of clay within the soil profile. Soil organic carbon (SOC) and soil organic nitrogen (SON) were much higher in the topsoil than in the subsoil. The only exception was the Kilimanjaro site at Xeno Helena, where SOC content was high in both topsoil and subsoil. Soil pH was not a limiting factor for productivity in any of these soils. Phosphorous

Table 13.2. Selected Surface and Subsurface Soil Properties at the Beginning of the Study

Site		Sand %	Silt %	Clay %	pH H_2O	O.C. %	N %	Av-P mg/kg
K. Boro	Topsoil	30	26	44	7.0	2.4	0.21	38
	Subsoil	22	22	56	6.6	0.8	0.11	29
X. Helena	Topsoil	44	22	34	6.0	2.0	0.14	7.6
	Subsoil	36	22	42	6.3	1.6	0.13	14
ML1	Topsoil	26	8	66	6.6	2.7	0.22	4.0
	Subsoil	24	4	72	6.5	0.9	0.07	1.0
ML2	Topsoil	48	8	44	6.5	2.3	0.19	4.0
	Subsoil	34	5	61	6.4	0.5	0.07	4.0
MS1	Topsoil	75	6	19	6.5	1.2	0.11	<1
	Subsoil	68	7	25	7.0	0.5	0.07	<1
MS3	Topsoil	76	3	21	6.3	1.0	0.12	1.0
	Subsoil	67	3	30	6.6	0.6	0.07	<1
Mindu	Topsoil	70	5	25	5.8	1.2	0.11	<1
	Subsoil	60	5	35	5.8	0.5	0.08	<1

K-boro = Kirima-Boro; X.He = Xeno Helena; ML1 = Mlingano 1; ML2= Mlingano 2; MS1 and MS2 = Misufini 1 and 2, respectively; ND = Not determined.

content was extremely low in all but the Kilimanjaro soil, which was relatively less sensitive to erosion effects on yield.

The effect of TSD on soil chemical properties was most apparent at the Morogoro and Mlingano sites (Table 13.3). The SOC and total soil N (TSN) increased with increasing TSD at all sites, but the effect was only significant at the Morogoro and Mlingano sites. The reduction in these properties with decrease in TSD is also in accord with the results presented by Lal (1976). Available phosphorous also increased with increasing TSD. However, there was no clear relationship between soil pH and TSD.

Available water-holding capacity (AWC) per cm soil decreased with reduction in TSD (Table 13.4), especially for the Morogoro soils. Reduced AWC may have seriously limited crop yields in Morogoro, as water deficit was a major problem at this site. In general, soil bulk density increased with decrease in TSD. However, a marked effect was observed only at the Mlingano 1 site where soil bulk density increased by 0.18 Mg/m^3 from the deepest to the shallowest TSD. Accelerated erosion had, therefore, degraded soil physical properties by increasing its bulk density and reducing AWC.

Soil erosion is, in many cases, a selective process which preferentially removes the fine soil particles and organic fraction. However, a clear selective removal of the clay was only observed at the Mindu site in Morogoro. At the Kirima Boro site in Kilimanjaro, the clay content decreased with increasing depth of the topsoil. The same trend was apparent at the Xeno Helena site. These contrasting results might indicate two different forms of soil erosion. In the Kilimanjaro region it is likely that severe soil erosion led to the removal of the entire surface soil rather than the selective removal of the fine soil particles. At the Morogoro sites selective removal of soil particles was the most likely process. Soil bulk density increased significantly with decreasing TSD at the Kirima Boro site, Mlingano 1 and 2 sites, and Misufini 3 site.

Effects of Restorative Management Practices on Soil Properties

Data for the effect of soil management on soil properties are available for Misufini 1, Kirima Boro, Xeno Helena, and Mlingano 1 and 2 (Table 13.5). The results were generally

Table 13.3. Impact of Past Erosion on Selected Chemical Properties in Established Erosion Classes for Each Major Agricultural Soil in the Three Ecoregions

Site	Erosion Class	pH H_2O	SOC	Total N	Av-P mg/kg	CEC Cmol/kg
			—— % ——			
Kirima	1	5.2	2.0	0.18	26.8	25.3
Boro	2	5.3	2.0	0.19	27.7	25.0
	3	5.3	2.2	0.20	38.4	32.9
	4	5.5	2.2	0.20	41.2	28.1
	LSD(0.05)	NS	NS	NS	6.3	NS
Xeno	1	4.9	1.8	0.21	35	26.4
Helena	2	4.9	1.8	0.21	34	26.7
	3	4.8	1.9	0.23	36	28.4
	LSD(0.05)	NS	NS	NS	NS	NS
Mlingano 1	1	6.0	1.9	0.18	1.3	14.8
	2	6.1	2.0	0.19	1.0	12.1
	3	6.4	2.3	0.20	1.7	15.0
	4	6.1	2.3	0.21	1.6	14.2
	LSD(0.05)	0.2	0.03	0.02	0.4	0.9*
Mlingano 2	1	6.2	2.1	0.18	1.0	13.5
	2	6.2	2.2	0.18	2.3	14.0
	3	6.5	2.6	0.21	3.8	14.6
	LSD(0.05)	0.2	0.17	0.02	1.0	0.9
Misufini 1	1	7.4	1.0	0.09	7	21.8
	2	7.3	1.5	0.12	11	21.3
	LSD(0.05)	NS	0.25	0.03	NS	NS
Misufini 3	2	6.0	1.3	0.10	3	17.3
	3	6.2	1.4	0.11	6	21.8
	4	5.9	1.5	0.11	3	22.4
	LSD(0.05)	0.6	0.15	NS	2	3.4
Mindu	1	6.8	1.2	0.18	2	20.4
	2	6.8	1.2	0.12	3	20.8
	3	6.6	1.4	0.14	3	23.3
	LSD(0.05)	NS	0.12	NS	0.58	NS

the same for all sites. As expected, the ridges did not change soil chemical properties as compared to the control. Nor was there any clear trend with regard to the effect of ridges on soil bulk density.

Application of FYM had a very marked effect on soil chemical properties. Compared to the control treatment, pH increased by 0.35, 0.23, 0.57, 0.34, and 0.48 units, respectively, for Misufini 1, Kirima Boro, Xeno Helena, Mlingano 1, and Mlingano 2. FYM also had a marked effect on SOC and TSN contents in all sites. In comparison with control, the SOC content increased by 0.39, 0.92, 0.06, 0.31, and 0.35% for Misufini 1, Kirima Boro, Xeno Helana, Mlingano 1, and Mlingano 2, respectively. This increase in SOC content indicates a general improvement of soil fertility. The effect was even more pronounced for available P. For example, for Misufini 1, Bray P increased from 2.68 to 19.84 mg/kg. When available P is above 10 mg/kg, it is generally not a limiting factor (Aune and Lal, 1995). Even for the Kilimanjaro soils, with relatively high levels of P, there was an effect of adding FYM on available P. FYM also increased the level of K; however, this increase probably did not affect productivity because K is generally not a limiting factor for the maize crop in Tanzania (Mnkeni, 1992).

Applications of N and P fertilizers also affected soil chemical properties favorably, but not as consistently as that of the FYM. The SOC was increased in the two Kilimanjoro

Table 13.4. Impact of Past Erosion on Selected Physical Properties for the Seven Soils in Different Ecozones

Site	Erosion Class	Sand	Silt	Clay	BD	IR	AWC
			%		Mg/m^3	cm/h	m m
Kirima	<15	33	18	49	1.2	48	106.5
Boro	15–20	31	19	50	1.2	29	123.6
	20–25	35	20	45	1.0	38	123.9
	>25	38	18	44	1.1	30	129.2
	LSD(0.05)		NS	2.7	0.08	13	15.1
Xeno	<15	51	14	35	0.97	36	117.5
Helena	15–20	50	13	37	0.87	45	124.7
	20–25	54	14	32	0.97	42	127.6
	LSD(0.05)		NS	NS	NS	13	NS
Mlingano 1	<15	54	8	38	1.39	5.7	
	15–20	57	9	34	1.29	10.3	
	20–25	53	11	36	1.30	16.3	
	>25	57	11	32	1.21	9.2	
	LSD(0.05)	1	1		1.52		
Mlingano 2	<15	44	09	47	1.46	6.4	8.0
	15–20	43	09	48	1.42	10.1	12.0
	20–25	45	07	48	1.41	3.2	20.0
	LSD(0.05)	NS	NS	1.8	0.03	4.36	1.9
Misufini 1	<15	52.5	16.1	31.4	1.47	ND	103.4
	15–20	51.0	14.8	32.5	1.48	ND	97.7
	LSD(0.05)	NS	1.18	0.92	NS		NS
Misufini 3	15–20	50.2	12.5	37.2	1.40	ND	89.7
	20–25	53.8	12.4	33.7	1.42	ND	121.9
	>25	50.1	13.2	36.7	1.35	ND	125.5
	LSD(0.05)	2.85	NS	2.42	0.06	–	23.5
Mindu	<15	41.3	16.3	41.6	1.39	ND	93.2
	15–20	41.5	11.8	36.1	1.41	ND	93.3
	20–25	36.6	13.3	49.8	1.35	ND	126.9
	LSD(0.05)	2.85	2.32	4.2	NS	–	20.5

BD = bulk density, IR = infiltration rate, AWC =available water capacity

sites, but not as much as by the FYM treatment. Fertilizer application can increase SOC content by increasing the amount of roots and biomass produced. Available P also increased when fertilizer was added. At the Kilimanjaro sites, fertilizer application increased available P as much as or slightly more than by FYM application, but for the other sites, application of FYM was more efficient in increasing available P. The FYM application increased available P more than that by use of fertilizer P although less P was added through FYM than through fertilizer application.

Effect of Soil Erosion on Crop Yield for Different Soil Management Practices

A regression analysis between TSD and yield was conducted for the results from the 1993 season (Figures 13.1 to 13.5). The effect of TSD on yield was linear. At Morogoro sites, TSD explained 81%, 68% and 57% of the variations in yield for Mindu, Misufini 3, and Misufini 1 locations, respectively. The magnitude of correlation coefficient was low for other sites. The average decrease in productivity per cm reduction in TSD was 88, 55, and 39 kg/ha for Morogoro, Mlingano, and Kilimanjaro sites, respectively. There was a significant effect of TSD in five of the seven sites (Table 13.6). The average loss in grain yield for the seven sites was 64 kg/ha/cm reduction in TSD.

The effect of TSD on yield was stronger for Morogoro than for other sites because of

Table 13.5. Effect of Soil Management on Soil Chemical and Physical Properties

Site	Treatment	pH	SOC (%)	Total N (%)	Av-P (mg/kg)	BD (Mg/m^3)
Misuf. 1	FP	6.28	0.94	0.15	2.68	1.46
	TR	6.51	0.94	0.15	4.40	1.38
	TR+FYM	6.63	1.33	0.18	19.84	1.32
	TR+NP	6.59	0.94	0.14	6.70	1.39
	LSD (5%)	NS	NS	0.00	4.35	0.06
Kirima Boro	FP	5.93	2.07	0.16	27.83	1.22
	TR	6.12	2.24	0.16	30.25	1.21
	TR+FYM	6.27	2.99	0.18	44.33	1.18
	TR+NP	5.81	2.63	0.18	48.92	1.26
	LSD (5%)	0.23	0.20	0.03	5.18	0.03
Xeno Helena	FP	5.62	1.86	0.20	23.22	1.11
	TR	5.76	1.97	0.18	31.56	1.08
	TR+FYM	6.19	2.63	0.26	40.44	1.08
	TR+NP	5.83	2.31	0.20	44.89	1.10
	LSD (5%)	0.23	0.23	0.05	4.15	NS
Mlin-gano 1	FP	5.88	2.01	0.20	1.07	
	TR	6.00	2.07	0.20	1.48	
	TR+FYM	6.22	2.32	0.22	6.45	
	TR+NP	5.95	1.93	0.19	2.71	
	LSD (5%)	NS	0.03	NS	1.17	
Mlin-gano 2	FP	6.26	2.27	0.18	0.86	
	TR	6.24	2.40	0.19	2.69	
	TR+FYM	6.74	2.62	0.22	14.80	
	TR+NP	6.19	2.30	0.19	7.56	
	LSD (5%)	0.07	0.24	0.03	4.05	

several factors. The topsoil at Morogoro had the lowest inherent fertility with SOC content of slightly > 1% while soils from Mlingano and Kilimanjaro had a SOC content > 2 %. In addition, the SOC content in the subsoil was about 0.5 % in the Morogoro soils. Thus, these soils had a very limited capacity to supply nitrogen once the topsoil was removed. The effect of decrease in TSD on soil chemical properties was also most pronounced in the Morogoro soils (Table 13.3). The strong reduction in yield with decreasing TSD at Morogoro might also be related to changes in soil physical properties. There

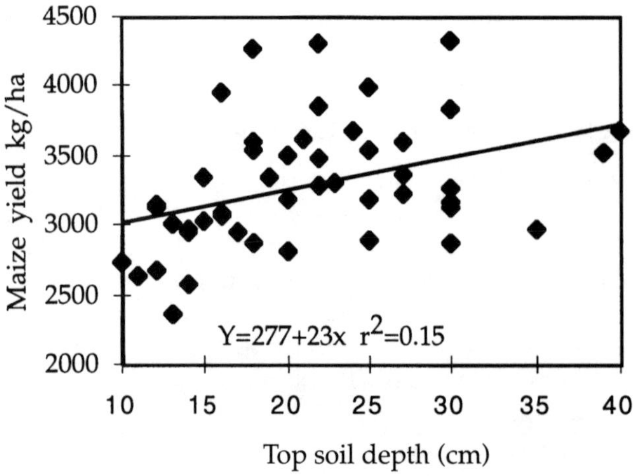

Figure 13.1. Impact of topsoil depth on maize yield at Kirima-Boro, Kilimanjaro ecoregion.

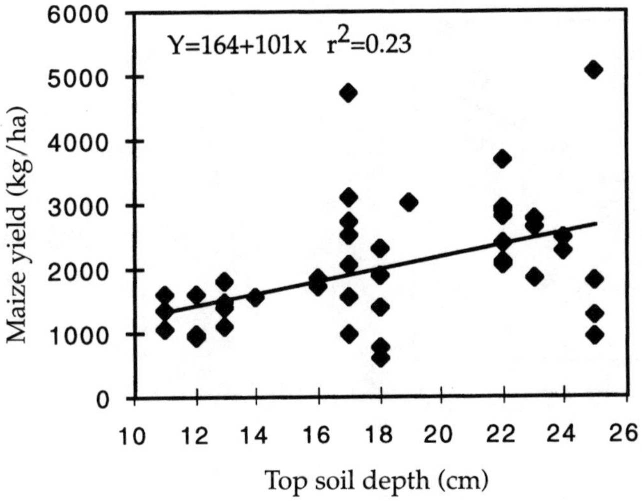

Figure 13.2. Impact of topsoil depth at Mlingano 2-Tanga ecoregion.

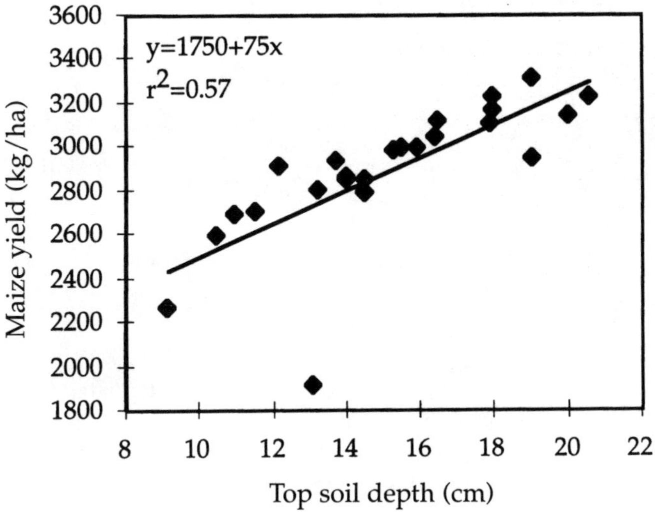

Figure 13.3. Impact of soil depth on maize yield at Misufini 1-Morogoro ecoregion.

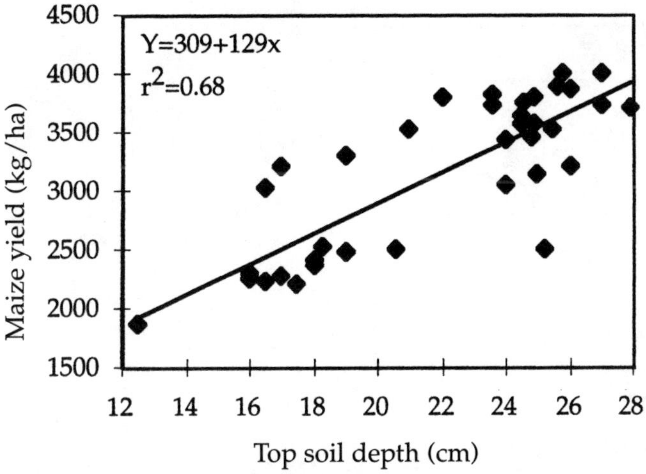

Figure 13.4. Impact of topsoil depth on maize yield at Misufini 3-Morogoro ecoregion.

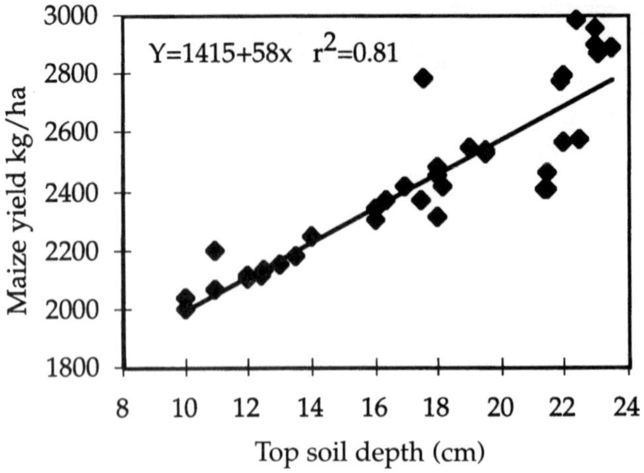

Figure 13.5. Impact of topsoil depth on maize yield at Mindu-Morogoro ecoregion.

Table 13.6. Maize Grain Loss (kg/ha) per Centimeter Reduction in Topsoil Depth for the Seven Major Agricultural Soils in Tanzania

Site	Reduction in Maize Yield in kg/ha per cm Reduction in TSD	R^2	Level of Significance
Kirima-Boro	23	0.14	P<0.01
Xeno-Helena	54	0.06	NS
Mlingano 1	9	0.01	NS
Mlingano 2	101	0.23	P<0.001
Misufini 1	75	0.57	P<0.001
Misufini 3	130	0.68	P<0.001
Mindu	58	0.81	P<0.001
Avg. yield loss per cm topsoil	64		

was a reduction in the AWC in the upper 20 cm with decreasing TSD in most sites. This reduction might have affected yields in semiarid Morogoro more drastically than in other sites with subhumid and humid climates. Reduction in TSD also significantly reduced cowpea yield (Figure 13.6), which was similarly affected at Xeno Helena and Kirima Boro, respectively (Table 13.7). Average reduction in cowpeas grain yield was 34

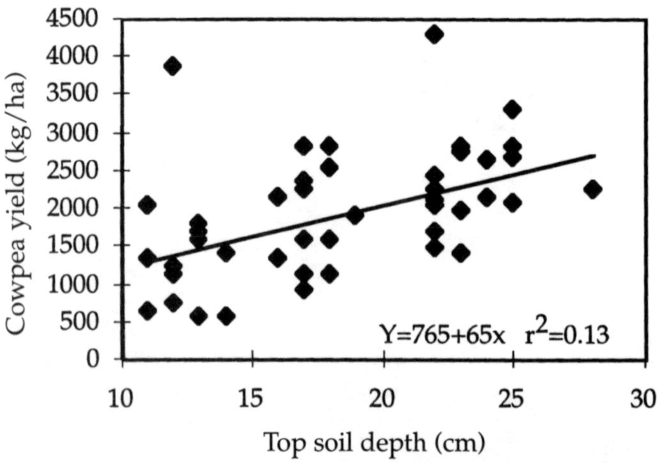

Figure 13.6. Impact of topsoil depth on cowpea yield at Mlingano 2-Tanga ecoregion.

kg/ha/cm of TSD loss. These results, however, indicate that cowpeas is less sensitive to changes in TSD compared to maize. Cowpeas is more drought tolerant than maize. Since decrease in TSD is associated with reduction in AWC, low sensitivity for cowpeas to TSD could be accounted for by its drought tolerance.

Crop yields in the second phase were higher for the thick TSD at each site as compared to the shallow TSD in six of the eight locations (Table 13.8). The most pronounced effect of TSD was observed at the Morogoro sites, while the effect at the Kilimanjaro sites was less apparent. A regression analysis between TSD and grain yield at each site showed that the grain yield of maize was on average reduced by 43 kg/ha/cm reduction in TSD with a variation between the sites from -4.9 to 186.5 kg/ha. This reduction in crop yield is lower than the effect observed in the first phase of the study where the average grain loss per cm of TSD was 64 kg/ha. Experiments in Phases 1 and 2 are not directly comparable as the soil management treatments were not the same in the two phases and the growing seasons are different.

Crop yields were consistently higher for those treatments that received NP fertilizers or FYM than in the control (Table 13.9). Ridging, however, had only a minor effect on

Table 13.7. Effect on Reduction in Topsoil Depth on Cowpea Grain Yield on Lixisols and Nitisols in Tanga and Kilimanjaro Ecoregions

Site	Cowpea Yield Reduction kg/ha per cm Reduction in TSD	R^2	Level of Significance
Mlingano 2	66	0.13	$P<0.05$
Xeno-Helena	29	0.17	$P<0.05$
Kirima-Boro	7	0.09	$P<0.05$
Avg. yield loss	34		

Table 13.8. Effect of Soil Depth on Maize Grain Yield Productivity for Eight Locations in Tanzania

TSD Class	Misu. 1	Misu. 2	Misu. 3	Mindu	Kiri. Boro	Xeno Hel.	Mli. 1	Mli. 2
< 15 cm	3457			3784	3142	4507	2582	3202
15-20 cm	4517	4478	4420	3550	3536	4894	2206	2741
20-25 cm	5322	4844	5122	4136	3577	4461	2776	3283
> 25 cm			4372		3598		2720	
Yield loss per cm soil reduction in TSD	186.5	73.2	-4.8	35.2	28.1	-4.6	19.7	8.1

Table 13.9. Effect of Soil Management on Maize Grain Yield for Eight Locations in Tanzania

TSD Class	Misu. 1	Misu. 2	Misu. 3	Mindu	Kiri. Boro	Xeno Hel.	Mli. 1	Mli. 2
Control	3613	4361	3947	3103	1974	3499	1993	2411
Ridges	3925	4024	4241	3584	2170	2954	1915	2933
Ridges + FYM	5454	5317	5759	4840	4804	5448	3384	3361
Ridges +NP	4736	4942	4605	3766	4905	6185	2993	3598
SE	180	271	130	206	80	170	171	248
$LSD_{5\%}$	630	949	455	721	280	595	599	

Table 13.10. Effect of Soil Management on Cowpea Yield at Kirima Boro

Treatment	Grain	Stover
Control	296	2032
Ridges	440	2475
Ridges + FYM	825	5485
Ridges + NP	651	3966
SE	83.5	497.7

Table 13.11. Effect on Maize Grain of Soil Depth and Treatments Across the Six Sites Which Had the Erosion Classes 1, 2, and 3

Treatment	< 15 cm	15-20 cm	20-25 cm	Mean
Control	2458	2673	3102	2744
Ridges	2697	2761	3186	2881
Ridges +FYM	4457	4530	4772	4586
Ridges + NP	4172	4330	4643	4382
Mean	3446	3574	3926	

$LSD_{5\%}$-depth=627; $LSD_{5\%}$-traitment=734; Interaction depth*traitment=n.s.

crop yield. For FYM and fertilizer treatments, average yields were 1732 kg/ha and 1353 kg/ha higher than in the control. The effect of FYM was most pronounced at the Kilimanjaro region, where yield was 2830 and 2346 kg/ha higher than that of the control for the Kirima Boro and Xeno Helena sites, respectively. The result of cowpeas yield also showed a positive effect of nutrient applications yields (Table 13.10).

A combined analyses across sites for six sites having the TSD classes 1, 2, and 3 showed that crop yield decreased with decreasing TSD and increased with addition of fertilizer or FYM (Table 13.11). The interaction between the TSD class and fertilizer treatment was not significant. However, there existed a tendency that the effect of TSD was slightly more pronounced when no fertilizer was applied. For treatments where fertilizer was applied, yields were 471 kg /ha higher in the TSD class 20-25 cm than in the TSD class < 15 cm. The corresponding difference was 644 kg/ha in the control treatment. The data show that it was not possible to fully compensate for the loss of TSD by adding fertilizers.

The effect of reduction in TSD on maize yield was less pronounced in this study than has been observed by others. This less sensitivity to TSD may partly be due to differences in the methodology and agroecological conditions. In Cameroon, a 50% reduction in maize yield was observed when 2.5 cm of the topsoil was artificially removed (Rehm, 1978). In Nigeria, maize yield decreased by 23% when 2.5 cm of soil was artificially removed (Lal, 1976).

MODELING THE EFFECT OF SOIL EROSION ON CROP YIELDS

A model for predicting yield for tropical conditions was developed by Aune and Lal (1995) and Aune (1995). The objective of this model was to predict how the choice of soil management practice affects long-term productivity and changes in soil properties. This model can also be used to identify the soil-degrading processes which contribute most to yield reductions.

The model is based on the principle of computing the fertility index, but considers the dynamic aspect of temporal changes in soil properties (Aune and Lal, 1995; Aune, 1995). The model consists of several submodels. Yields for the various crops are calculated according to the Mitcherlich and Baule equation (Baule, 1918) as follows:

$$Y_t = \text{Potential yield} * N_t * P_t * pH_t * SOC_t * Pest_t * M \quad (1)$$

where Y^t is expected yield, N^t is the nitrogen factor, P^t is phosphorous factor, pHt is the factor for soil acidity, SOC^t is the soil organic carbon factor, $Pest^t$ is pest control measures, and M is management factor. Potential yield is the maximum yield obtained under experimental conditions when all agronomic factors (except the climatic) are maintained at the optimum level. The N^t , P^t , pH^t, SOC^t, $Pest^t$ and M are crop-specific and time-dependent factors (time step one year) ranging from zero to one. These factors are based on the relationship between supply or the status of a variable and relative yield (Aune and Lal, 1995). For an optimum level, the growth factor has a rating of one. Potential yield is obtained when all factors are rated at 1, and no yield is obtained when one or all factors are rated at 0. The factors change according to management practice and with change in soil property. The management factor is a factor that can be used to calibrate yield to observed yield levels in the initial year and can therefore account for factors that are difficult to estimate such as efficiency of harvesting, incidence of pests and weeds, micronutrient deficiency, etc.

Calculation of Growth Factors

The nitrogen factor N_t expresses the relationship between available nitrogen in kg/ha and relative yield (Figure 13.7). The relationship was obtained by analyzing data of maize experiments in Tanzania (Mowo et al., 1993).

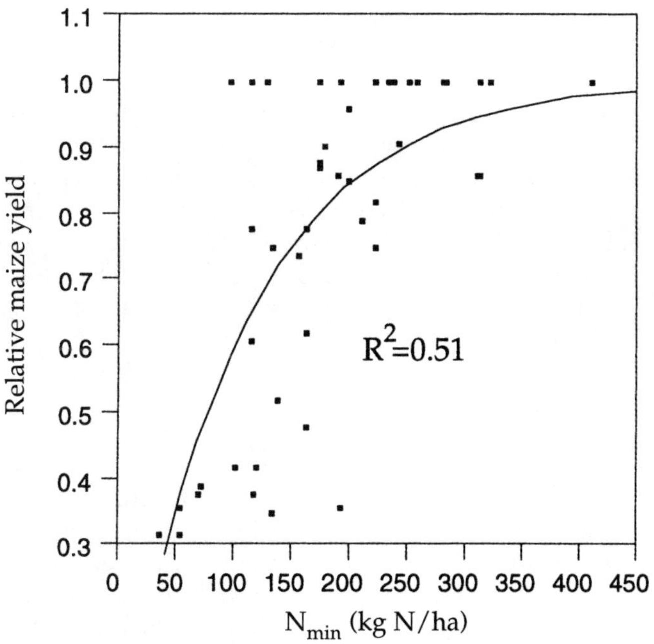

Figure 13.7. Relationship between available nitrogen and relative maize yields (N_t factor).

Available nitrogen is calculated as the sum of addition from decomposing SOC, mineralization of crop residues and manures, fertilizer use, and from atmospheric deposition (Aune, 1995). The change in SOC and SON contents are calculated according to the Nye and Greenland (1960) decomposition model:

$$N_{s(t)} = N_{s(t-1)}\, e^{-rt} + N_{r(t-1)}\, (\Omega) - N_{e(t-1)} \quad (2)$$

where $N_{s(t-1)}$ is nitrogen in SOC, r is a decomposition constant, $N_{r(t-1)}$ is the amount of nitrogen added by residues, W is the proportion of nitrogen in residues entering SON, and $N_{e(t-1)}$ is the annual loss of nitrogen through soil erosion. Soil erosion in this model is calculated according to the Universal Soil Loss Equation (FAO, 1979), but the factors are calibrated to get results in accordance with the observed levels of soil erosion in runoff plots in Tanzania (Aune, 1995). Soil erosion affects capacity of soils to release nitrogen by removing N from the SON pool. Change in pH is a function of the antecedent level, the quantity and type of nitrogen fertilizers used, additions of lime, and duration the field has been under cultivation.

A complete mathematical description of the model is presented elsewhere (Aune and Lal, 1995; Aune, 1995). The model is programmed in the Microsoft Excel spreadsheet, which has several advantages over using more basic programming languages, the most important advantage being the ease of replacing a function with locally adopted functions. The model can, therefore, be easily adopted to site-specific conditions. Where feasible, soil- and site-specific functions should be used. Recently, the model was programmed into an optimization model, which can now identify which combination of soil conservation methods and soil management practices gives the best economic returns. The economic return is calculated by using the net present value method.

Model Predictions

The model was used to assess how different soil degradation processes contribute to yield reductions in the Southern Highlands of Tanzania. Figure 13.8 shows the effect of soil erosion and nitrogen fertilizers on maize yields in this region. The effect of soil erosion on crop yield was more pronounced when no fertilizer was added than when 100 kg N/ha was applied annually.

A more detailed examination of the causes of soil degradation according to model predictions is presented in Figure 13.9 under a situation where no fertilizer was used. The change in SON concentration is the important factor causing most of the decline in yield, while change in SOC and pH contribute relatively less to the yield decline. Because antecedent soil pH was high (5.75), pH change did not affect maize yield. The effect of pH on productivity was significant when it was below 5.0. Soil erosion reduced maize yield by 92 kg/ha/cm reduction in TSD. This reduction is equivalent to the effect of soil erosion on productivity observed in the Morogoro region.

CONCLUSIONS

The data presented show that maize grain yield was strongly affected by the TSD. The average rate of reduction in grain yield was 64 and 43 kg/ha/cm of TSD for the first and second phase, respectively. Although not fully restored, grain yield on eroded soil increased with use of fertilizer and FYM. However, resource-poor farmers in Tanzania cannot purchase fertilizer or apply quantities of manure necessary to restore soil fertility. Therefore, subsistence farmers must have access to low-cost technologies which can pro-

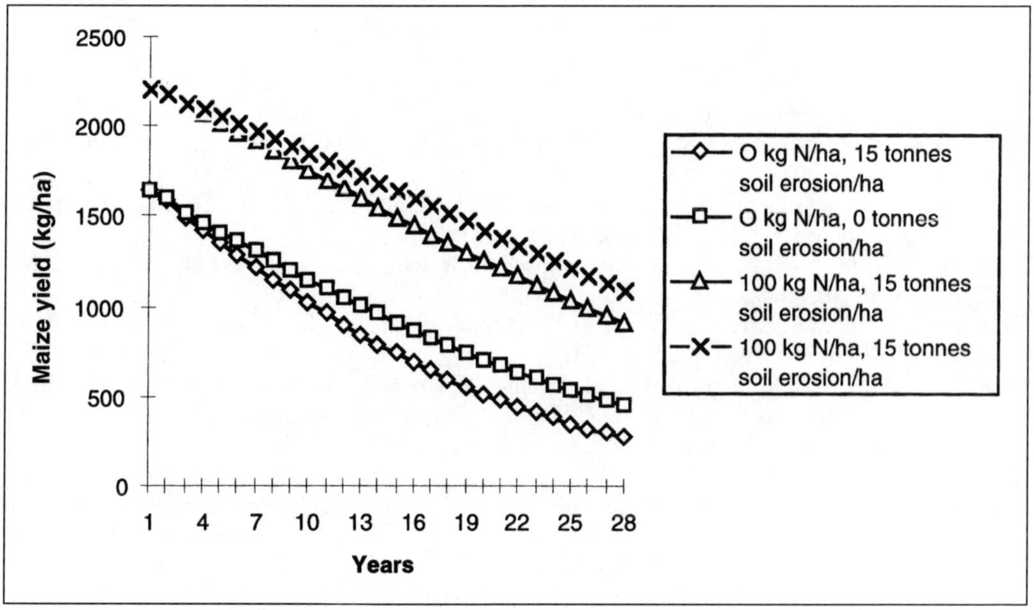

Figure 13.8. Predictions of the effect of soil erosion and nitrogen fertilizers on long-term maize productivity in the Southern Highlands of Tanzania.

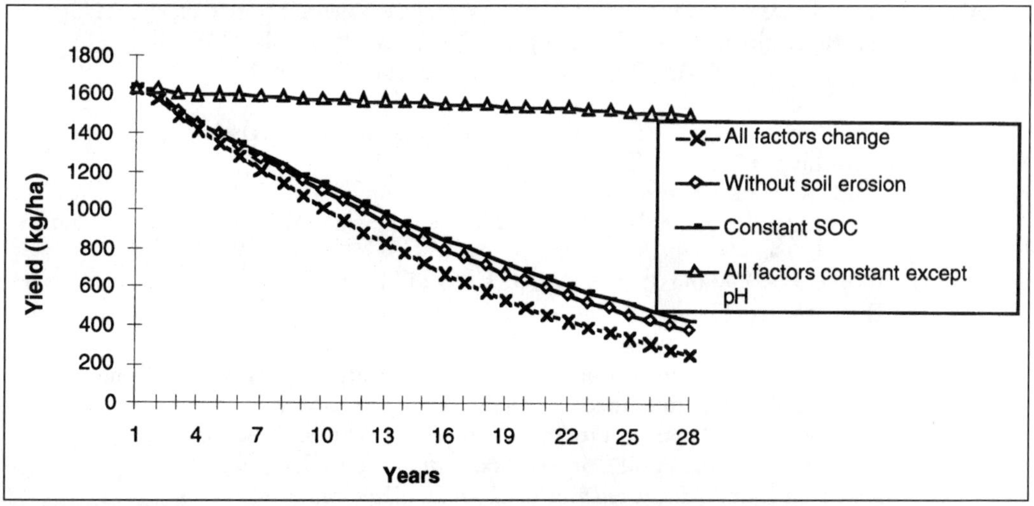

Figure 13.9. Causes of soil degradation in the Southern Highlands of Tanzania when no nitrogen is applied.

tect their soils against soil erosion. It is also important to develop policies that integrate these resource-poor farmers into a market economy. Access to agricultural credit and a well-functioning market will enable farmers to purchase the necessary inputs and sustain yields.

REFERENCES

Anderson, B. Soils of Tanganyika. Bulletin No. 16. Research Division. Ministry of Agriculture. Govt. Printer. Dar es Salaam, 1963.

Aune, J.B. Predicting Soil Degradation in Tanzania—A System Analysis Approach, Supplement 21. *Nor. J. Agric. Sci.,* 1995, pp. 47–60.

Aune, J.B. and R. Lal. The Tropical Soil Productivity Calculator—A Model for Assessing Effects of Soil Management on Productivity. In *Soil Management, Experimental Basis for Sustainability and Environmental Quality.* R. Lal and B.A. Stewart, Eds. Lewis Publishers, Boca Raton, FL, 1995, pp. 499–520.

Bie, S.W., I.C. MacGillivray, P.A. Sanchez, H. Zandstra, and C. Zulberti. Task Force Report: CGIAR Response to UNCED's Agenda 21, 1993.

Baule, B. Zu Mitcherlich gezetz der physiologischen Beziehungen. *Landwirtschaftliche Jahrbucher* 51, 1918, pp. 363–385.

Chapman, H.D. Cation Exchange Capacity. In *Methods of Soil Analysis.* Part 2.1, Black et al., Ed., ASA, Madison, WI, 1965, pp. 891–901.

Chowdhury, M.S., B.J. Ndunguru, and L.I. Lulandalla. Contribution of Biological Nitrogen Fixation to the Nitrogen Supply of Soils. Workshop on "Resource-Efficient Farming Methods For Tanzania." 16–20 May, 1983. Morogoro, 1983, pp. 86–94.

De Pauw, E.F. Soils, Physiography and Agroecological Zones of Tanzania. Crop Monitoring and Early Warning Systems. Min. of Agriculture, Dar es Salaam/FAO, Rome, Consultance Final Report, 1984.

FAO. A Provisional Methodology for Soil Degradation Assessment. Rome, FAO, 1979.

Greenland, D.J., A. Wild, and D. Adams Organic Matter Dynamics in Soils of the Tropics—From Myth to Complex Reality. In R.Lal and P.A. Sanchez, *Myth and Science of Soils of the Tropics.* SSSA Special publication No. 29. Soil Sc. Soc. of America, Madison, WI, 1992, pp. 17–31.

Greenland, D.J., G. Bowen, H. Eswaran, R. Rhoades, and C. Valentin. Soil Water and Nutrient Management Research IBSRAM Position paper. IBSRAM, Bangkok, Thailand, 1994.

Hathout, S.A. Soil Map of Tanzania. In "Atlas of Tanzania." 5th Edition, 1975.

Kajumulo, D.A. Marketing Intelligence Information Systems for Crop and Price Monitoring. Proc. Regional Workshop on "Effect of Structural Adjustment and Trade Liberalization on Marketing Institutions." Arusha, Tanzania, 1992.

Kilasara, M., F.S.B. Kaihura, I.K. Kullaya, J.B. Aune, B.R. Singh, and R. Lal. Establishment of Criteria for Distinguishing Levels of Past Erosion in Tanzania. *Nor. J. Agr. Sci.*, Supplement No. 21: 61–70, 1994a.

Kilasara, M., I. Kullaya, F. Kaihura, B.R. Singh, R. Lal, and J.B. Aune. Impact of Past Erosion on Crop Yield in Selected Ecological Regions of Tanzania. *Nor. J. Agric. Sci.,* Supplement No. 21:71–79, 1994b.

Lal, R. Soil Erosion Problems on an Alfisol in Western Nigeria and Their Control. Monograph 1, Institute for Tropical Agriculture, Ibadan, Nigeria, 1976.

Lal, R. Soil Erosion and its Relation to Productivity in Tropical Soils. In *Soil Erosion and Conservation*, S.A. Swaify et al., Eds. Soil Conserv. Soc. Am., Ankeny, IA, 1985, pp. 237–244.

Lal, R. Development in East Africa. *Geoderma* 67: 159–164, 1995a.

Lal, R. Erosion-Crop Productivity Relationships for Soils of Africa. *Soil Sci. Soc. Amer. J.* 59, 661–667, 1995b.

Liebhardt, W.C. Nitrogen Self-Sufficiency. Workshop on "Resource Efficient Farming Methods for Tanzania." 16–20 May, 1983. Morogoro, 1993, pp. 75–85.

Mnkeni, P.N.S. Role of Soil Management in Enhancing Sustainability of Small-Holder Cropping System in Some Agro-Ecosystems of Tanzania: A Review. Ecology and Development Paper No. 3. NLH. Norway, 1992.

Mowo, J.G., J. Floor, F.B.S. Kaihura, and J.P. Magoggo. Review of Fertilizer Recommendations in Tanzania. National Soil Service. Ministry of Agriculture. Tanga, Tanzania, 1993.

Muchena, F.N. and R.M. Kiome. The Role of Soil Science in Agricultural Development in East Africa. *Geoderma* 67: 141–157, 1995.

Ngaiza, E.M. Manure and Compost Handliing in Tanzania. Workshop on Resource-Efficient Farming Methods for Tanzania. 16–20 May, 1983, Morogoro, 1983, pp. 20–23.

Nye, P.H. and D.J. Greenland. The Soils under Shifting Cultivation. Technical communication No. 51. Commonwealth Bureau of Soils. Harpenden, 1960.

Pimental, D.C., Harvey, P. Resosudarmo, K. Sinclair, D. Kurz, M. McNair, S. Crist, L. Shpritz, L. Fitto, R. Saffouri, and R. Blair. Environmental and Economic Costs of Soil Erosion and Conservation Benefits. *Science* 267: 1117–1123, 1995.

Rehm, S. Land Development in the Humid Tropics. In *Proceedings International Agra-Mechanic.* GTZ, Frankfurt, West Germany, 1978.

Rodale, R. Importance of Resource Regeneration. Proc. Workshop on "Resource-Efficient Farming Methods For Tanzania." 16–20 May, 1983, Morogoro, 1983.

Samki, J.K., and J.F. Harop. General Problems in Preparing Fertilizer Use Recommendation in Tanzania. Proc. 6th Annual General Meeting of the Soil Sci. Soc. East Africa, 30 July–1 August, 1984, Nyeri, Kenya, 1984.

Semoka, J.M.R. B.J. Ndunguru, and W.C. Liebhardt. Agronomic Aspects of Manure and Compost Use. Workshop on Resource-Efficient Farming Methods for Tanzania." 16–20 May, 1983, Morogoro, 1983, pp. 64–71.

USDA. World Agriculture: Trends and Indicators. USDA-ERS. Washington. D.C., 1992.

FOURTEEN

The Role of Ruminant Livestock in Soil Fertility Management in Sub-Saharan Africa

J.W. Smith and A. Naazie

INTRODUCTION

The human population in sub-Saharan Africa (SSA) was estimated to be 498 million in 1990, and is projected to reach 1.2 billion by the year 2025 (Winrock, 1992). It is also estimated that by that time the rural to urban population ratio will be 0. 5 6: 1, reflecting a major demographic shift from the current ratio of 2.7:1. This rapid increase in population has major implications for agriculture in all regions of SSA. Because urban, as well as rural demand for food will have to be met largely from rural supply, growing rural populations subsisting on a fixed land base will likely lead to an overall intensification of agriculture (Boserup, 1981; Pingali et al., 1987). All of the land on the continent classified as very suitable for cultivation is already cultivated (FAO, 1986). Consequently, if deleterious environmental effects of agricultural expansion onto marginal lands are to be minimized, the increasing demand for food must be met largely by productivity gains rather than at the margin. Thus, the imperatives of prudent and efficient resource use are likely to drive crop and animal agriculture toward a fuller integration.

Pressures on land resources and the corresponding impetus toward intensification of agriculture will be brought about not only by expanding human population, but expanding ruminant populations as well. The total number of cattle in SSA will increase from 162 million in 1988 to 239 million by 2025, while sheep and goats will increase from 270 million to 562 million for the corresponding periods (Winrock, 1992). Under human and livestock population pressure, traditional farming systems are breaking down. In the highlands, for example, there is little or no opportunity to fallow lands. This is rapidly becoming the case in subhumid ecozones where traditional rangelands are rapidly being converted to crop lands, (McIntire et al., 1992), and as a consequence, traditional livestock grazing areas are diminishing. As agriculture in SSA savanna ecozones is progressively transformed from the traditional extensive fallow systems to continuous cropping, yields of crops and land productivity will decline (Kesseba, 1995) and threaten sustainability (Smith et al, 1995). The soils of sub-Saharan Africa have inherently low fertility and are fragile, highly weathered, and characterized by low activity clays with low cation exchange capacity (Brader, 1993). They show nutrient deficiency problems after only a short period of cultivation, with nitrogen being rapidly depleted (Padwick, 1983). Ruminant livestock have often been portrayed as degraders of the environment and soil condition but their role in soil enhancement can be tremendous when properly managed in an integrated resource management context. This chapter discusses the utility of ruminants in arresting resource, particularly soil, degradation in the context of (a) fertility management and (b) integrated resource management.

SOIL FERTILITY MAINTENANCE

The most important constraint to sustaining yields under short fallow or continuous cropping systems in SSA is that of soil fertility maintenance, although significant problems with pest, weed control, and erosion also arise. Kang (1993), for example, reported that soil pH, organic C, and exchangeable Ca and Mg levels declined markedly under continuous cropping. In plots where inorganic fertilizer was applied, soil pH declined faster than in unfertilized plots. Similar results were also reported by Adepetu et al. (1979) and Kang and Balasubramanian (1990). The use of inorganic fertilizers to restore soil fertility is an expensive and unsustainable option for resource-poor farmers in SSA.

It is known that the addition of organic matter improves soil physical properties, water retention, and reduces runoff and even in small quantities improves soil cation exchange capacity (Padwick, 1983). In this context, manure (feces and urine) produced by ruminants in large quantities in SSA could help in the restoration of soil fertility under continuous cropping or short fallows. Apart from the direct contribution of soil nutrients, manuring promotes organic matter accumulation, enhances cation exchange capacity (Powell, 1986; Bationo and Mokwunye, 1991), and seems to have beneficial effects on biological processes especially through the activity of earthworms and microbes (Tarawali and Ikwuegbu, 1993). Manure, as a contributor to maintenance of soil fertility and to meeting energy needs has long been recognized by resource-poor farmers in sub-saharan Africa, particularly those in the highlands. In other ecozones of SSA, manure remains an underutilized resource about which research is hardly conducted. However, those studies which have been done show promising results. For example, studies by Bationo and Mokwunye (1991) and Powell (1986) indicate some of the beneficial effects (Table 14.1) on soil pH, organic matter, total nitrogen, and available phosphorus from samples taken at first crop harvest after manure application. Another study by Powell (1986) in which

Table 14.1. Effect of Manure on Soil Surface Chemical Properties in West Africa[g]

Manure Application (t/ha)[g]	pH	Organic Matter (%)	Total Nitrogen (mg/kg)	Available Phosphorus (mg/kg)
5[a]	5.37 (4.98)[d]	0.39 (0.29)	202 (153)	10.3 (5.3)
20.0[a]	6.21 (4.98)	0.58 (0.29)	285 (153)	22.9 (5.3)
NR[b,e]	5.8 (5.1)	3.25 (2.55)	164 (131)	9.6 (4.6)
3.6[c]	6.51 (6.18)	0.18 (0.15)	NR (4.3)	5.7
2.1[c]	5.04[f] (4.86)	0.31 (0.28)	NR	NR
3.1	5.45 (5.20)	0.31 (0.28)	150 (138)	11.0 (6.1)

[a]From Bationo and Mokwunye, 1991.
[b]From Powell, 1986.
[c]From Powell, unpublished, from 40 farmers' fields in Niger.
[d]Values in parentheses are soil properties in nonmanured plots.
[e]Not recorded.
[f]pH in KCl, otherwise pH in water.
[g]Samples taken at first crop harvest after manure application.

Table 14.2. Effects of Overnight Manuring on Soil Properties, and Maize Grain Yields[a]

Soil Properties and Maize Yield	Non-Manured Plots		Manured Plots	
	Mean	sd	Mean	sd
pH	5.1	0.4	5.8	0.3
Soil organic carbon (%)	1.50	0.48	1.91	0.66
Bray-1 P (ppm)	4.6	1.8	9.6	2.5
Total N (%)	0.131	0.04	0.164	0.059
CEC (C mol(+)/kg)	4.34	1.77	6.15	1.88
Grain yield (kg/ha)	2800	1310	5280	2090

[a]Modified from Powell, 1986.

plots were manured overnight showed similar effects on the same soil parameters (Table 14.2).

Options for using manure on farms could include management routines which allow the deposition of manure and urine in situ, or collection and storage for use in the succeeding cropping season. In the latter case the contribution of urine is ignored. In situ deposition is usually accomplished by corralling animals on the designated land to be cropped or by integrating feeding strategies into land use patterns such as fodder banks (Tarawali and Ikwuegbu, 1993; see next section). Manure represents a very considerable resource of important livestock ecozones. It is estimated that one tonne of cattle manure contains 8 kg nitrogen, 4 kg phosphorus, and 16 kg potash (CTA, 1995). Depending on intake, approximately 100 g and 140 g of nitrogen per tropical livestock unit (TLU, equivalent to 250kg liveweight) are produced daily during the dry and wet seasons, respectively, with faecal and urinary nitrogen being produced in roughly a 60:40 ratio (Schlecht et al., 1993). When these figures are used to estimate the manure output from current estimates of ruminant livestock numbers (Jahnke, 1982), approximately 1.2 million, 74 and 106 thousand tonnes of nitrogen are produced annually by cattle, sheep, and goats, respectively, for the subhumid zone alone. These figures are projected to increase to 1.8 million, 108 and 154 thousand tonnes, respectively, by the year 2025 (Fig. 14.1; ILRI unpublished data). This is a considerable amount, and even if the manure available is not enough to completely restore soil fertility, the marginal requirement for inorganic fertilizer may be rendered affordable.

INTEGRATED CROP-LIVESTOCK MANAGEMENT

In those areas of SSA where both ruminant livestock and crop production can be practised, integrating the two units represents not only a reasonable choice from a resource management perspective, but may be the only option, and the synergies that could be exploited are tremendous. As indicated earlier, pressure on land will result in less range area for livestock. Hence ruminant livestock will depend more and more on crop residues as feed resources. For example, in Kaduna state (in the Nigerian subhumid zone), even if only 10% of crop residues are utilized as feed and cropped area expands at 2% per year (Winrock, 1992), crop residues will still be more important than range as a feed resource before the year 2010 (Fig. 14.2; ILRI, unpublished data). In addition to the feed issue, shortening fallow periods or continuous cultivation will result in rapid decline in soil condition and fertility as indicated earlier. Thus, resource-poor farmers will find manure from ruminants to be, perhaps, their best soil amendment measure. The mutual need for some form of symbiotic existence between crops and ruminant livestock in SSA is obvious.

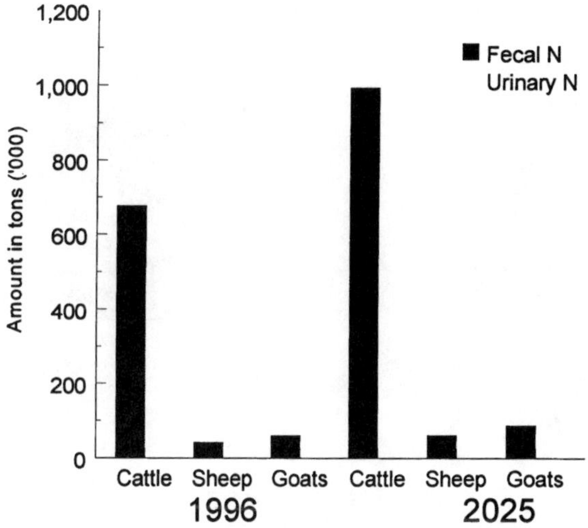

Figure 14.1. Estimated fecal and urine nitrogen from ruminant livestock populations in the subhumid zone of tropical Africa. From ILRI (1996).

Strategies that incorporate feed production with soil fertility management considerations will become paramount as continuous cropping replaces long fallows.

An analytical scheme showing phases of the transformation and the performance during a hypothetical ten year period is shown in a model (Fig. 14.3). In this schematic model, under the traditional long fallow cropping system, an aggregate of 3 years of cropping in 10 years was calculated, and an aggregate of 5 years of cropping out of 10 years was calculated, for the traditional crop-livestock system. The traditional crop-livestock system is largely supported by formal and informal relationships between crop farmers and transhuman livestock producers. The latter derive access to seasonal feed resources, while the former benefit from the manure deposited. A longer aggregate period of 7 years of cropping out of 10 years is calculated for the improved legume fallow system which uses such fallow/cover crops as *Mucuna puriens* and *Pueraria phaseoloides*. Even in this system, however, a fallow period of a minimum of three years is deemed to be necessary. Studies are currently being conducted to evaluate the potential of an alternative indicated in the model as the improved crop-livestock system. In that system the main cereal crop is relayed with a forage legume and plots are grazed between two cropping seasons. An important challenge for this research is to develop technological options which promote

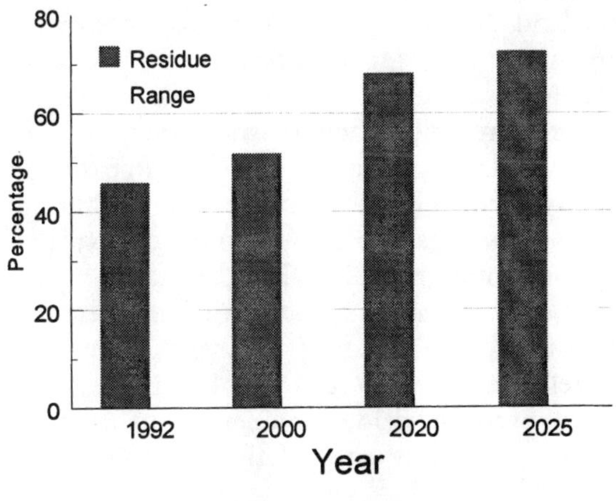

Figure 14.2. Relative importance of crop residues and range as feed sources, Kaduna state, if only 10% of the crop residues were utilized as feed. 28% of the land in Kaduna state is cropped and it is assumed that cropping is expanding at 2% per year.

FALLOW SYSTEMS

BENEFIT	TRADITIONAL (NATIVE SAVANNAS): CROP	TRADITIONAL (NATIVE SAVANNAS): CROP + LIVESTOCK	IMPROVED (PLANTED LEGUME): CROP	IMPROVED (PLANTED LEGUME): CROP + LIVESTOCK
	EX-POST	EX-POST	EX-POST	EX-ANTE
• SOIL IMPROVEMENT	++	++	+++	+++
• FEED	—	+	+++	+++
• ANIMAL PRODUCTS				
• MILK / MEAT	—	+	—	+++
• TRACTION	—	—	—	++
• MANURE	—	+	—	+++
• AVG. FALLOW DURATION (yrs)	7	5	3	< 1
• CROPPING YRS	3	5	7	10

\+ = MODERATE
++ = SIGNIFICANT
+++ = VERY SIGNIFICANT
_ = NO CONTRIBUTION

Figure 14.3. Analytical scheme showing contribution of livestock in fallow systems during a 10-yr cycle. From ILRI (1996).

crop-livestock synergies in a manner which facilitates expansion of food production and can sustain or even increase land and labor productivity under conditions of severe resource competition.

Figure 14.4 represents a conceptual model of the possible synergies or complementarities between ruminants and crops that could be exploited to ameliorate soil fertility problems of short fallow or continuous cropping systems, as well as generate feed to improve livestock productivity. The symbols used are explained in Figure 14.5. The basic degradable resource is the soil; its fertility can decline due to losses through many factors including uptake by the crop. The residue produced by the crop could serve as valuable ruminant livestock feed to improve their productivity, while the manure (faeces and urine) from the ruminants could be recycled to the land. Animal traction offers the potential to expand crop production, thereby increasing labour and land productivity and providing options for increasing feed and manure production as well.

Crop-livestock systems can include planting forage legume relays which are grazed between succeeding cropping seasons (Fig. 14.3). Relaying main crops with grain or forage legumes could substantially improve the quality of the residues as livestock feed while enhancing soil nitrogen content through fixation as well as improving manure quality. Preliminary results from a long-term multilocational study by ILRI, IITA, and NARS partners on the effects of no-grazing, grazing with sheep, or grazing with cattle on continuous seasonal cropping of maize, does not suggest that grazing residues in situ has any detrimental effect on crop yield. In fact, maize yields increased in legume relayed and grazed plots, and grazed plots showed higher increases of maize yield over a period of three years of cropping (Fig. 14.6). These results, though preliminary, show that farmers could po-

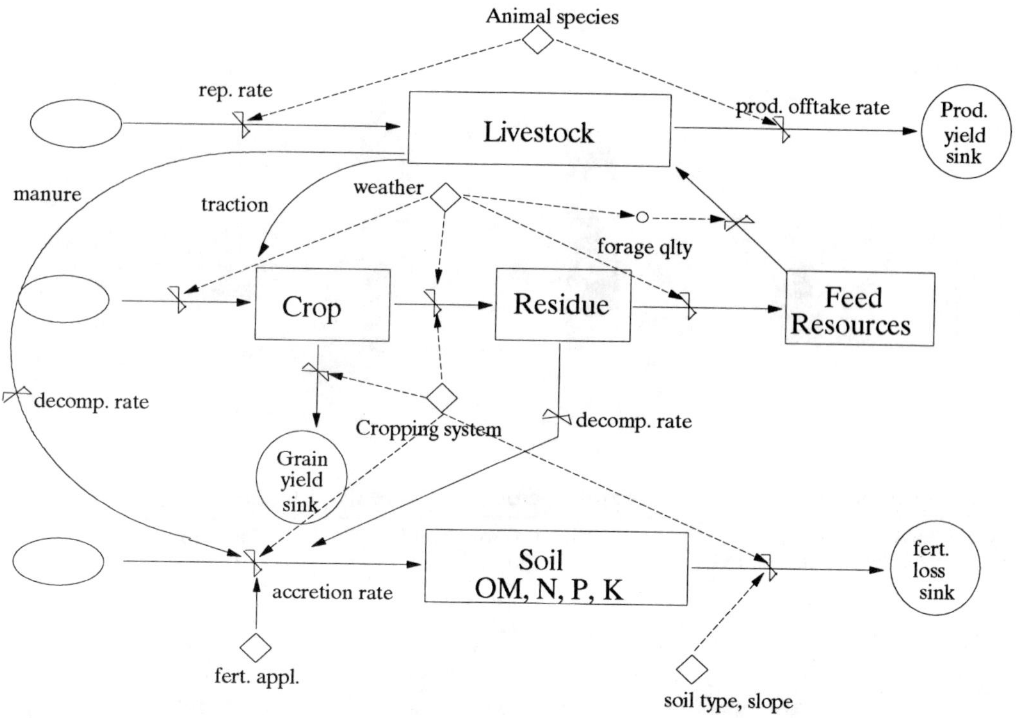

Figure 14.4. Conceptual model of the role of ruminants in crop–livestock systems.

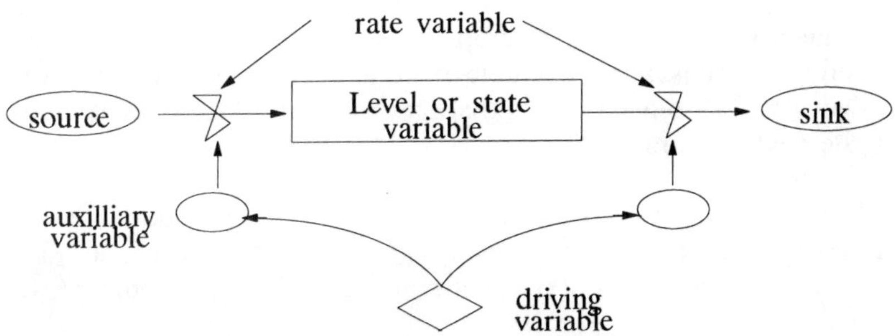

Explanation of terms

1. - Sources and sinks denote start or end conditions
2. - Rate variables or control valves determine rate of material transfer e.g., growth rate
3. - Auxilliary variables influence rates of material transfer
4. - State variable (levels) represent points where material accumulates, e.g., body weight
5. - Driving variables affect but are not affected by the rest of the system
6. - Arrows represent material tranfers. When used in dotted form they usually represent information transfer

Figure 14.5. Schematic representation and explanation of variables, relationships, and symbols used in the model (Fig. 14.4).

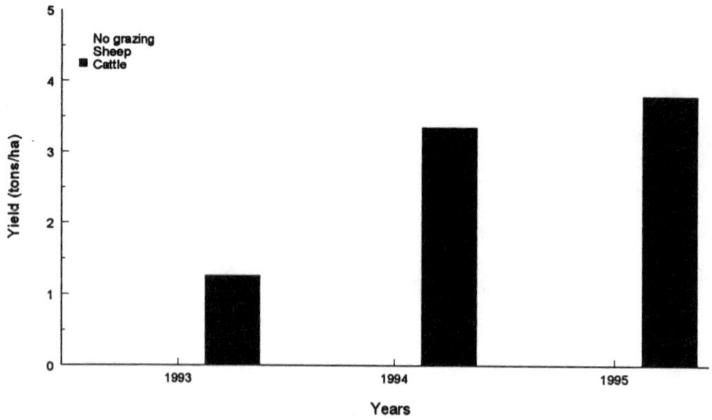

Figure 14.6. Effect of grazing livestock on maize grain yield under a modified land use system (improved fallow) in which maize is relayed with a forage legume. From ILRI (1996).

Table 14.3. Contribution of Stylo-N to Succeeding Maize Yields[a]

Previous Land Use	Maize Grain Yield (kg/ha)	N-Contribution[b] (kg/ha)
Cropped 3 years	460	—
Fallow	1270	30
S. hamata, 2 years	1330	32
S. hamata, 3 years	2510	90
S. guianensis, 1 year	1640	44
S. guianensis, 2 years	2700	110

[a]From Mohamed-Saleem and Otsyina (1986).
[b]The amount of N required by the 3-yr cropped area to attain similar grain yields as those obtained by other land uses.

tentially combine growing cereals/grain crop with provision of dry season grazing/feeding of livestock and derive considerable benefits by m aintaining or increasing crop yield and land productivity. It has also been shown that growing plots of legumes as fodder banks to supplement the feed of ruminants improves yield of subsequent cereal crops (Tarawali and lkwuegbu, 1993). Mohamed-Saleem and Otsyina (1986) demonstrated a considerable effect of *Stylosanthes hamata* and *Stylosanthes guianensis*, two fodder bank species, on maize yield over that of a traditional fallow (Table 14.3). These results also help to illustrate some of the potential reciprocal benefits to be derived from integrated crop-livestock systems.

Thus technologies which can integrate ruminant feed production into crop farming systems are currently and will progressively become more important. Utilization of crop residues as livestock feed is, however, not without implications for crop production since such residues when not used to feed livestock contribute to maintenance of soil structure and fertility. Studies (Kang, 1993) have shown that crop residue management could affect cowpea seed yield. Much research is still needed to ascertain the response surface of crop residues used as mulch versus being used as feed/manure.

REFERENCES

Adepetu, J.A., Obi, A.O. and Aduayi, E.A. Changes in Soil Fertility under Continuous Cultivation and Fertilization in South Western Nigeria. *Nigerian J. Agric. Sci.* 1: 15–20, 1979.

Bationo, A. and Mokwunye, A.U. Role of Manures and Crop Residue in Alleviating Soil Fertility Constraints to Crop Production: With Special Reference to the Sahelian and Sudanian Zones

of West Africa. In *Alleviating Soil Fertility Constraints to Increased Crop Production in West Africa*. A. U. Mokwunye, Ed., Kluwer Academic Publishers, Netherlands, 1991, pp. 217–225.
Bosenip, E. Population and Technological Change: A Study in Long-Term Trends, University of Chicago Press, Chicago, 1981.
Brader, L. Foreword. In *Soil Organic Matter Dynamics and Sustainability of Agriculture.* Mulongoy, K and Merckx, R., Eds. Proc. Int. Symp.Soil Organic Matter Dynamics and Sustainability of Tropical Agriculture: John Wiley and Sons, Chichester-New York, 1993, pp. 297–305.
CTA (Technical Centre for Agriculture and Rural Cooperation) Are Beasts a Burden? *Spore* 58: 1–4, 1995.
FAO African Agriculture: The Next 25 Years. Annex 11. The Land Resource Base. ARC/86/3. FAO, Rome, Italy, 1986.
Jahnke, H. E. *Livestock Production Systems and Livestock Development in Tropical Africa.* Kielers Wissechoftoverlag Vank, Kiel, Germany, 1982.
Kang, B.T. and Balasubramanian, V. Long Term Fertilizer Trials on Alfisols in West Africa. In *Transactions of XIV International Soil Science Society Congress*, Vol. 4. ISSS, Kyoto, Japan, 1990.
Kang, B.T. Changes in Soil Chemical Properties and Crop Performance with Continous Cropping on an Entisol in the Humid Tropics. In *Soil Organic Matter Dynamics and Sustainability of Agriculture.* Mulongoy, K. and Merckx, R., Eds. Proc. Int. Symp.Soil Organic Matter Dynamics and Sustainability of Tropical Agriculture, John Wiley and Sons, Chichester-New York, 1993.
Kesseba, A.M. Strategies for Sustainable Agriculture in Sub-Saharan Africa:Some Issues and Options. *Alley Farming Research and Development.* Proc. Int.Conf. Alley Farming. Ibadan, 14–18 September, 1992. Produced by the Alley Farming Network for Tropical Africa in collaboration with the International Institute of Tropical Agriculture, the International Centre for Research on Agroforestry and the International Livestock Research Institute, IITA, Ibadan, 1995.
Mohamed-Saleem, M.A. and Otsyna, R.M. 1986. Fodder Banks: For Pastoralists or Farmers. In *Potentials of Forage Legumes in Farming Systems of Sub-Saharan Africa.* Haque, I., Jutzi, S. and Neate P.J.H., Eds., Proc. Workshop ILCA, Addis Ababa, Ethiopia, 16–19 Sept., 1985, pp. 212–231.
McIntire, J., Bourzat, D. and Pingali, O. Crop-Livestock Interactions in Sub-Saharan Africa. World Bank, Washington D.C., 1992.
Padwick, G. W. Fifty Years of Experimental Agriculture. II. The Maintenance of Soil Fertility in Tropical Africa: A Review. *Expl. Agric.* 19: 293–310, 1983.
Pingali, P., Bigot. Y. and Binswanger, H.P. *Agricultural Mechanization and the Evolution of Farming Systems in Sub-Saharan Africa.* World Bank, Washington D.C., 1987.
Powell, J.M., Femandez-Rivera, S., Williams, T.O. and Renard, C., Eds. *Livestock and Sustainable Nutrient Cycling in Mixed Farming Systems of Sub-Saharan Africa.* Volume II: Technical Papers. Proceedings of an International Conference held in Addis Ababa, Ethiopia, 22–26 November, 1993, International Livestock Centre for Africa, Addis Ababa, Ethiopia, 1995.
Powell, J.M. Manure for Cropping: A Case Study from Central Nigeria. *Experimental Agriculture* 22:15–24, 1986.
Ruthenberg, H. *Farming Systems in the Tropics*, 3rd ed. Clarendon Press, Oxford, 1980.
Schlecht, E., Mahler, F., Sangari, M., Susenbeth, A. and Becker, K. Quantative and Qualitative Estimation of Nutrient Intake and Fecal Excretion of Zebu Cattle Grazing Natural Pastures in Semi-Arid Mali. In *Livestock and Sustainable Nutrient Cycling in Mixed Farming Systems of Sub-Saharan Africa,* Volume II: Technical Papers. Proc. Int. Conf. Addis Ababa, Ethiopia, 22–26 November, 1993, ILCA, Addis Ababa, Ethiopia, 1995.
Smith, J., Weber, G., Manyong, M.V., and Fakorede, M.A.B. Systems Dynamics and Heterogeneity—The Key to Unlocking West Africa's Potential: The Case of Maize in Nigeria. Paper presented at the Maize Workshop July 9–12, 1995, Kellogg Center, Michigan State University, East Lansing, MI, USA, 1995.
Tarawali, G. and Lkwuegbu, O.A., The Potential of Forage Legumes in Soil Management for Sustainable Livestock and Crop Production in the Subhumid Zone of Central Nigeria. In *Soil*

Management in Sustainable Agriculture. Cook, N. F. and Lee, N. C., Eds. Proc. III Int. Conf. Sustainable Agric. Wye College, London. Wye College Press, 1993, pp. 435–443.

Winrock International. Assessment of Animal Agriculture in Sub-Saharan Africa. Winrock International Institute for Agricultural Development, Morrilton, AK, 1992.

FIFTEEN

Maintenance of Soil Organic Matter and Nitrogen Supply in the Moist Savanna Zone of West Africa

R.J. Carsky, S.S. Jagtap, G. Tian, N. Sanginga, and B. Vanlauwe

IMPORTANCE OF SOIL ORGANIC MATTER IN THE MOIST SAVANNA ZONE OF WEST AFRICA

The moist savanna zone of West Africa is an important area for food production in sub-Saharan Africa. Rainfall (900–1500 mm yr^{-1} [Fig. 15.1]) and solar radiation (11.3–14.7 MJ m^{-2} day^{-1}) are generally adequate during the growing season (Jagtap, 1995). The length of growing period (Fig. 15.2) varies from five to nine months, with one five- or

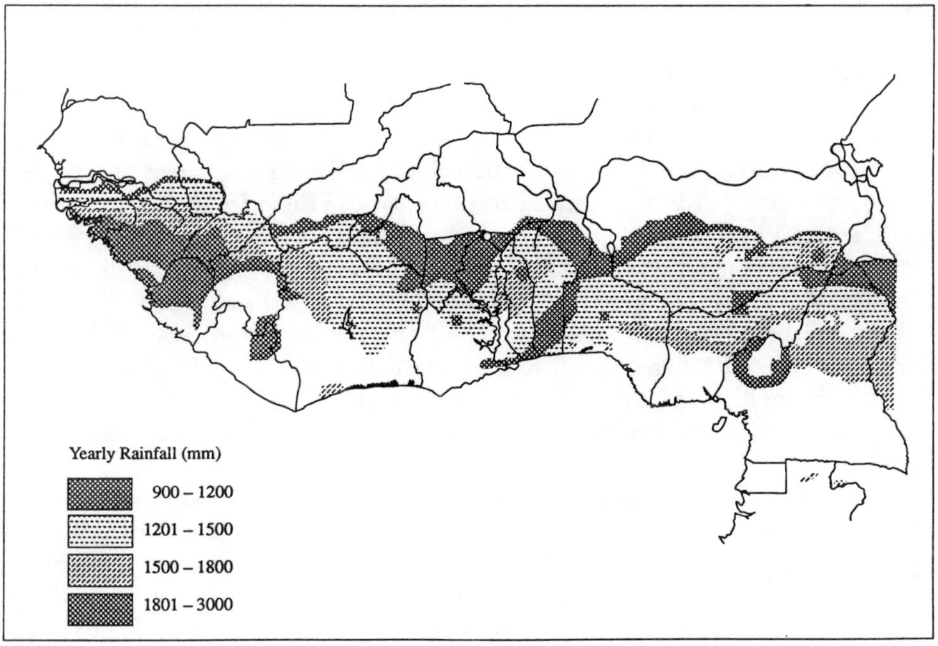

Figure 15.1. Total annual rainfall in the lowland (<800masl) moist savanna zone of West Africa.

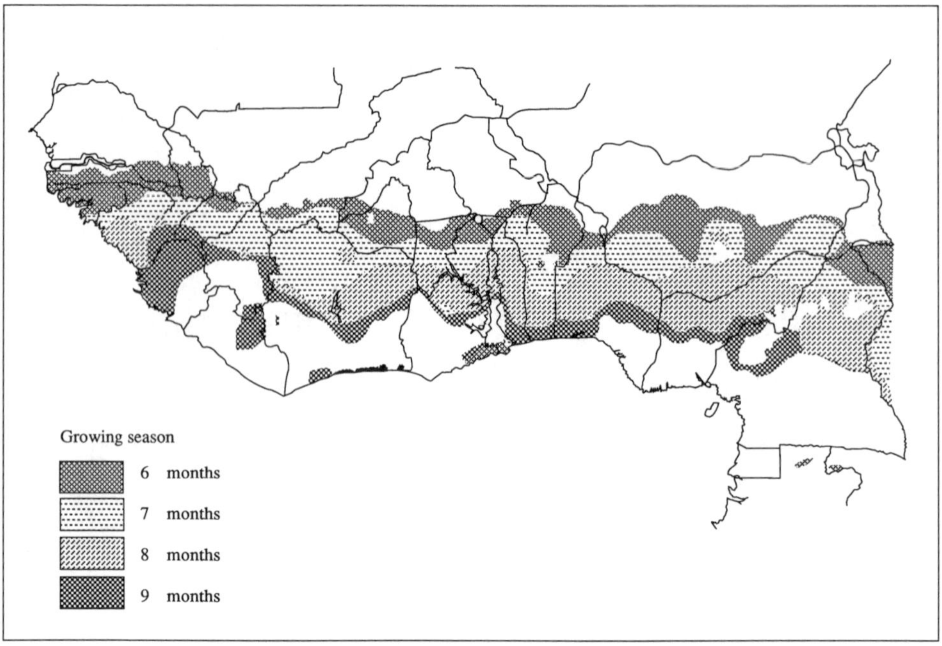

Figure 15.2. Length of growing period in the lowland moist savanna zone of West Africa.

six-month growing season in the 'Guinea' savanna (monomodal rainfall distribution) and two three- or four-month growing seasons in the 'derived' and 'coastal' savannas (bimodal). In terms of rainfall distribution, solar radiation, and temperature, much of the moist savanna zone is climatically more suitable than the semiarid or humid forest zones for grain crop production (Jagtap, 1995). Disease and pest pressure is generally lower than in the more humid zones. Thus the moist savanna zone is an area of high potential productivity.

The human population of the moist savannas of West Africa was estimated to be 70 million in 1980 (Fig. 15.3). The bulk of the area has low population density with higher densities only in parts of Nigeria and coastal areas. The population is expected to increase to approximately 220 million by the year 2020 (Fig. 15.4) with large areas of high density. Soils are dominated by Alfisols, Ultisols, and Inceptisols (Table 15.1). Soil constraints are very common and include sandy surface texture, shallow depth to an impenetrable layer, and low cation exchange capacity. Soil organic matter contributes a large proportion of soil cation exchange capacity (Juo and Adams, 1986). Most soils have low inherent fertility and soil organic matter is a major source of nutrients for crop growth. At the same time, fertilizer use in sub-Saharan Africa is generally low, averaging 9 kg ha^{-1} (Sombroek, 1995). Average application rates in the moist savanna zone of West Africa are similarly low with the highest rates going to cash crops, such as vegetables and cotton, and maize in some areas.

Two major roles of soil organic matter in the savanna zone of West Africa are nutrient supply and maintenance of soil structure. The level of soil organic matter varies with management and climatic factors in broadly understood ways. Soil organic matter was observed to be positively correlated to mean annual rainfall in West Africa by Jones (1973). This is clearly due to the balance between the amount of vegetation produced and its uses for agriculture and domestic needs. In the bimodal rainfall zone with a longer

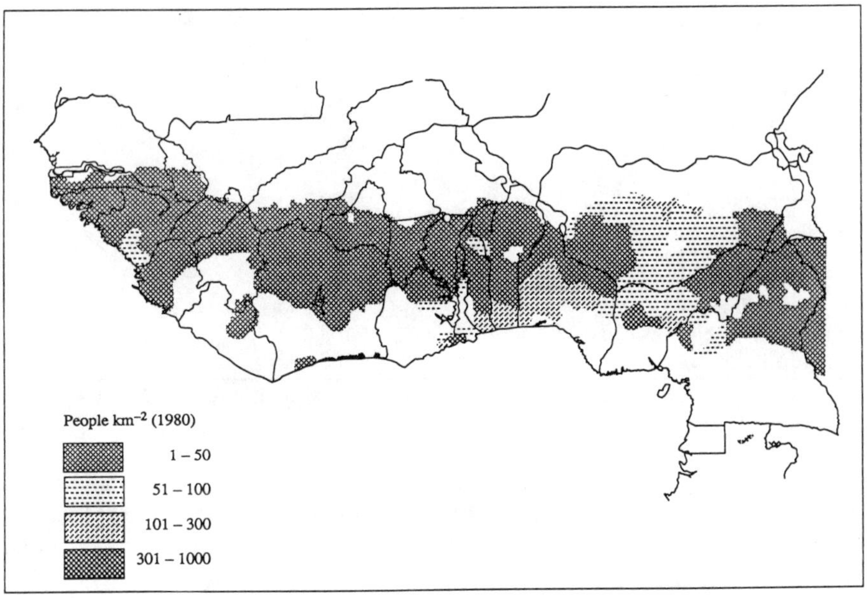

Figure 15.3. Population density in the moist savanna zone of West Africa in 1980.

growing period, vegetation is plentiful. Soil organic matter content is relatively high and weed competition is a major constraint along with poor nutrient supply. In the Guinea savanna zone with a shorter growing period, vegetation is sparse because of demands for animal feed, fuelwood, construction materials, and other domestic and agricultural uses. Soil organic matter content tends to be lower and related constraints are poor soil nutrient supply and water retention, and structural instability.

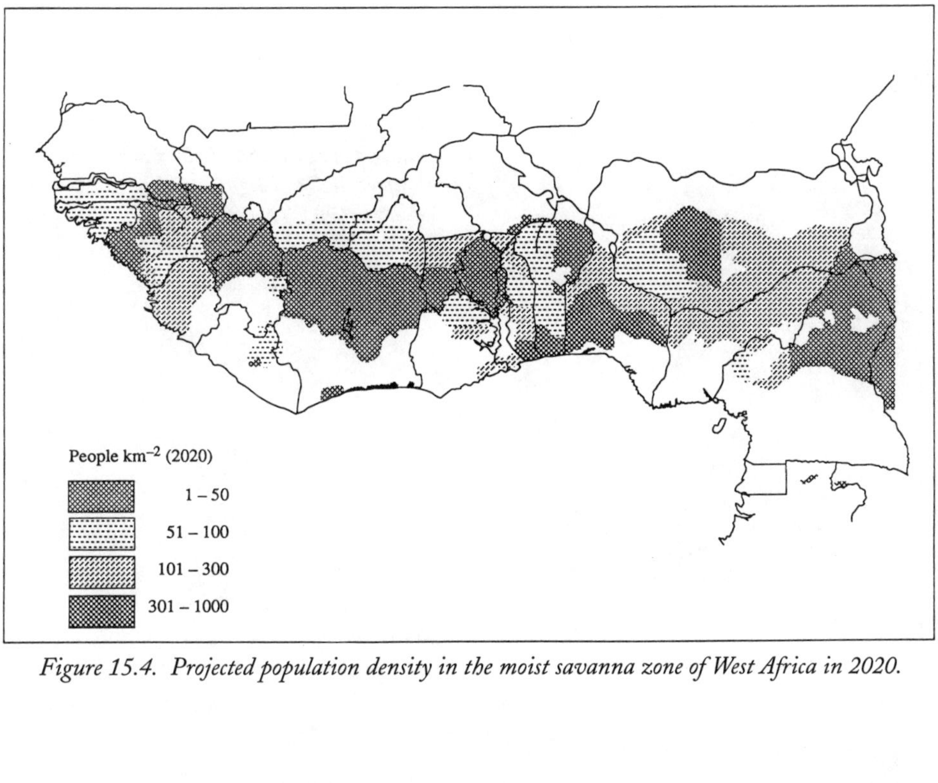

Figure 15.4. Projected population density in the moist savanna zone of West Africa in 2020.

Table 15.1. Biophysical Characteristics of West African Moist Savanna Subzones[a]

	Derived/Coastal Zone	Guinea Zone
Growing period (days)	210–270	150–210
Annual precipitation (mm)	1100–1800	900–1500
Rainfall pattern	Bimodal	Monomodal
Area covered (Senegal-Cameroon, km^2)	820,000	960,000
Major soils (km^2):		
Lithosols (Inceptisols)	124,000	200,000
Ferralsols (Oxisols)	141,000	16,000
Luvisols (Alfisols)	250,000	400,000
Nitosols (Alfisols/Ultisols)	108,000	86,000
Acrisols (Ultisols)	132,000	120,000

[a]From Agroecological Studies Unit, Resource and Crop Management Division, IITA, Ibadan, Nigeria.

In traditional agriculture, soil organic matter is the major source of nutrients, especially N, for crop production. Onyeanuforo (1994) provided a good example of this when he measured nitrate-N content of topsoil (0–15cm) in fields with different soil organic carbon (SOC) contents. Average nitrate-N content during the first eight weeks after planting was 65 kg ha^{-1} for a field with 1.8% SOC and 19 kg ha^{-1} for a field with 0.5% SOC (Fig. 15.5). Similarly, SOC was found to be a good predictor of maize grain yield on 34 farmers' fields in the northern Guinea savanna, provided cases of exceptional farmer management (late planting, late weeding, above average fertilizer application) were taken into account (Fig. 15.6). Certainly N supply was partially responsible for the strong relation-

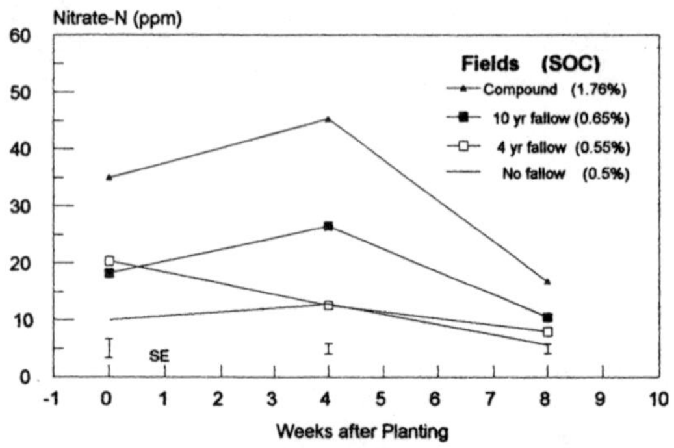

Figure 15.5. Nitrate-N in soil (0–30 cm depth) under a maize crop in fields with different fallow histories and soil organic carbon (SOC) content, Bauchi, 1993 (Onyeanuforo, 1994).

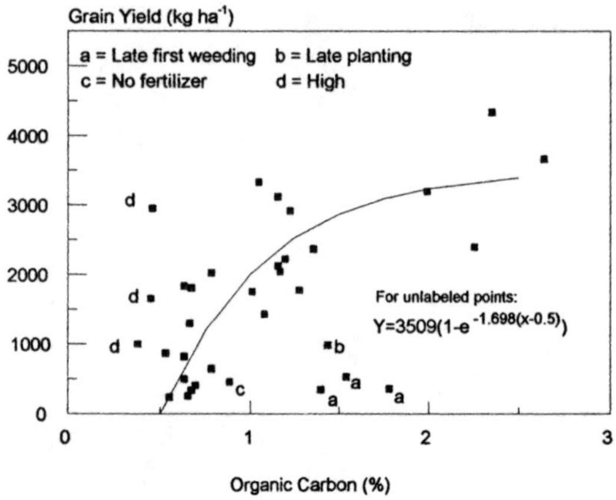

Figure 15.6. Relationship between maize grain yield and soil organic carbon (0–10 cm depth) in 34 farmer-managed trials, Bauchi, 1994 (Carsky and Lagoke, unpublished, 1996).

ship observed. However, at the low levels of SOC often observed in intensively cultivated savanna soils, the relationship between SOC and crop yield is less strong. In fact, in a study on 57 farmers' fields in an intensified zone of the Nigerian northern Guinea savanna, maize yield was not associated with SOC although it was related to soil nitrate content during the early part of the growing season (Weber et al., 1995). However, SOC was uniformly very low in the fields studied with a mean of 0.49% (SE=0.02) in the 0-15 cm depth.

Greenland (1994) argues that the level of soil organic matter may be taken as an indicator of the sustainability of a soil management system. Feller (1995) agrees that SOC should be used as an indicator of soil fertility, especially when a broad set of conditions are considered but that it should be modified by soil texture. He proposes the use of the equation: SOC (mg g^{-1}) = 0.87 + 0.32 (A + L), where SOC is organic carbon and A + L is the sum of the clay and fine silt contents (%). For example, the threshold level for a soil with 80% sand would be estimated at 7.3 mg g^{-1} or 0.73 %. Values below this threshold level were generally found on continuously cropped fields (and therefore at risk of physical degradation), and values above the threshold were found where fallow was practiced. Pieri (1989) analyzed 495 soil samples in the northern Guinea and Sudan savanna zones of West Africa and proposed a similar relationship between soil organic matter and the risk of physical degradation. This relationship should give a preliminary indication of the minimum acceptable SOC level for situations in which fertilizer is available. It should be differentiated according to soil type and rainfall regime rather than combined for all West African soils. The role of different SOM fractions should also be elucidated.

MAINTAINING OR INCREASING SOIL ORGANIC MATTER

Organic matter is rapidly depleted from native levels toward a low equilibrium under cultivation. In the northern Guinea savanna, Jones (1973) found topsoil SOC of 23 long-term uncultivated sites to average 1.03% and 19 cultivated sites to average 0.58%. The level under natural vegetation indicates the upper equilibrium which will be difficult to surpass with organic inputs. The work of Fauck et al. (1969) showed that native SOC depends on soil type but after many years of cultivation, SOC approaches approximately 0.5% or less. The equilibrium SOC values (upper and lower) appear to be most related to soil clay content (Jones, 1973). Jones and Wild (1975) synthesized results of several long-term trials. They found that the annual loss of SOC under cultivation ranges from 5 to 10% and that after many years of cultivation SOC is 25 to 45% of what it is under natural vegetation.

Fallowing is not an effective way to increase soil organic matter. Onyeanuforo (1994) sampled fields with different histories near Bauchi, in the northern Guinea savanna of Nigeria. The SOC content in a long-term cropped field was 0.5%, a four-year fallow had 0.55%, and a ten-year fallow had 0.65%. At Samaru, Jones (1971) reported SOC increases from long-term fallow rotations of 9% for 2/5 fallow (2 years fallow and 3 years cropping), 36% for 3/6 fallow, and 86% for 6/9 fallow, compared with continuous cropping which resulted in an SOC content of 0.22%.

Clearly, even small increases in soil organic matter content require repeated organic additions over many years. The results of several long-term trials were synthesized by de Ridder and van Keulen (1990) in which manure or compost were returned to the soil annually with and without fertilizer additions. Assuming SOC decomposition rate of 6% yr^{-1}, buildup of SOC from 0.3 to 0.4% would require approximately 15 t ha^{-1} of cereal straw and 31 t ha^{-1} of animal manure. These amounts of organic material are only available if fertilizer is used or if organic material from a large area is concentrated in a small area. Even to maintain these low SOC levels will require 2-3 t ha^{-1} of cereal straw and 5-6 t ha^{-1} of animal manure (de Ridder and van Keulen, 1990). Given enough time, farmers in many areas do develop 'compound' fields, small fields surrounding the homestead, which have higher SOC levels than the equilibrium normally found under natural vegetation (Prudencio, 1993). Study of these fields may help to design sustainable field management systems.

In savanna zones with a longer growing period than the above studies, similar soil organic matter depletion occurs. In the eight-month (bimodal) growing period zone Kang (1993) found that all cropping treatments caused a reduction of SOC from an initial level of approximately 1% to a range of 0.3 to 0.6 % after 10 years. The highest level under cropping resulted from fertilizer application and crop residue return. Only a natural fallow treatment maintained SOC at initial levels. Again in the eight-month growing period zone, Hauser and Kang (1993) reported that SOC was maintained at 1.08% with five years of alley cropping (1.23% in the hedgerows and 0.94% in the alleys) compared with 0.59% in a control without hedgerows.

MAINTAINING OR INCREASING N SUPPLY

In a socioeconomic environment where fertilizer is not available, alternative sources of N will be needed. The major alternatives are grain legume crop residues, forage legume pastures or fodder banks, cover crops, woody legumes, and animal manure. Each of these options has advantages and disadvantages which can be considered by producers. Two or more options can also be combined into an integrated nutrient supply system. Literature on N contribution of legume-based fallow systems which is synthesized below gives variable information. Often the differences are due to the ecological adaptation of the legume, the yield potential of the test crop, or variable legume residue management. Variable results are also due to the methodology used to estimate the N contribution. The contribution of biologically fixed N depends on the accuracy of methods used for estimation of fixation (Toomsan et al., 1995). Leguminous N contribution may be overestimated if the legume exploits less soil N than a cereal or fallow control, but this 'sparing effect' should also be of interest to farmers who are looking for ways to reduce N fertilizer to nonlegumes.

Grain Legumes

Grain legumes should generally be the first 'improved fallow' system to be tested in an area since they are much more likely to be adopted by farmers in the short to medium term. Traditional cowpea varieties which produce grain and ample hay have been improved with resistance to major stresses. It is good to include these dual purpose cowpeas in any test of improved fallow systems. High-yielding soybean varieties have been developed which do not respond to inoculation with Rhizobium. Their feed value is inferior to cowpea and groundnut.

A major advantage of grain legumes is the production of human food which makes them much more acceptable to farmers. A disadvantage may be the resulting small amount of N contributed to soil and subsequent crop in a rotation. Eaglesham et al. (1982) calculated negative N balance from soybean which derived about 60% of its N from fixation and had an N harvest index greater than 80%. Sanginga et al. (unpublished, 1996) have calculated N balance estimates from five soybean cultivars varying from -8 to 43 kg N ha^{-1}. Other estimates of fertilizer equivalence from soybean are 20-40 kg ha^{-1}

Table 15.2. Estimates from West Africa of N Contribution of Soybean and Cowpea to Maize in the Subsequent Year

Crop	Estimate (kg ha^{-1})	Source	Comments
Soybeans	45	a	Residues incorporated; previous yam control
Soybeans	20–40	b	Two varieties; residues exported, maize control
Cowpeas	40	a	Residues incorporated; yam control
Cowpeas	60	c	Residues incorporated; maize control
Cowpeas	80	d	Two cowpea crops in first year, residues left on surface; maize control

[a]Kaleem, 1993; [b]Carsky et al., in preparation; [c]Dakora et al., 1987; [d]Horst and Hardter, 1994.

and 40-60 kg ha^{-1} for cowpea from work in West Africa (Table 15.2). In each case the lower estimates are probably more realistic for current cropping systems where residue incorporation is rarely done. Indeed, residue export is more common in the Guinea savanna because it facilitates harvest (soybean) or is a popular animal feed (cowpea, groundnut).

Legume residue management is important, and similarly, management of the legume crop can influence the subsequent N contribution. For example, the average N fertilizer equivalent value of soybean was only 7 kg ha^{-1} in 10 trials conducted in Malawi (MacColl, 1989) in which soybean showed erratic nodulation, especially in the early years of the trial. Choice of legume crop variety is also important. Stoop and van Staveren (1982) demonstated that the impact on subsequent millet grain yield increases with the increasing maturity cycle of preceding cowpea. The same was observed for two soybean varieties (Carsky et al., in preparation).

Herbaceous Cover Legumes

The advantage of herbaceous legumes is that a much larger N contribution can be expected, although generally only in the first growing season after the herbaceous legume crop. The obvious disadvantage is the loss of the land to a direct food or cash generating crop. In the bimodal rainfall zone, one season can be used for a food crop and the second season for a soil-improving herbaceous legume. A long-term trial of continuous maize/cassava cropping at IITA Ibadan, in the bimodal rainfall zone comparing bush fallow with *Pueraria phaseoloides* regrowth during the second season has produced higher maize grain yields (1.46 vs. 1.28 t ha^{-1}) and lower cassava tuber yields (4.2 vs. 5.3 t ha^{-1}) in the *Pueraria* treatment (G. Tian, unpublished data). Sanginga et al. (1996) reported that a yearly rotation of *Mucuna* with maize in the bimodal rainfall savanna zone contributed approximately 100 kg N ha^{-1}.

A 'live mulch' system has been developed for the bimodal rainfall zone (Mulongoy and Akobundu, 1985). It is based on low-growing perennial legumes which contribute to weed suppression, N supply, and reduced evaporation without competing too much with the food crop. They are allowed to regrow before and after the food crop. In a study conducted from 1979 to 1981, the use of a live mulch cover with *Psophocarpus palustris* and *Centrosema pubescens* was compared with no tillage and conventional tillage systems without live mulch. Maize yield, which averaged 2 t ha^{-1} over five crops without N fertilizer, was at least twice that of the other systems. In the live mulch plots, maize showed little or no response to N fertilizer. It will be an interesting system for producers who learn to manage it but it is not expected to perform as well in the monomodal rainfall zone.

In the monomodal rainfall zone, there is less research conducted on soil-improving herbaceous legumes. Recent data from the monomodal zone such as those in Table 15.3

Table 15.3. Effect of Previous Annual Legumes on Yield of Maize at Kasuwan Magani, Central Kaduna State, 1995

Previous Crop	N Content at 6 WAP (kg ha^{-1})	Grain (kg ha^{-1})	Stalk (kg ha^{-1})
Maize	5.4	221	1430
Lablab purpureus	5.1	506	2530
Cajanus cajan	6.2	434	1550
Crotalaria ochroleuca	7.7	784	2140
Mucuna pruriens (white)	9.8	661	2600
Mucuna pruriens (black)	11.2	990	3490
SE	2.5	121.1	301

demonstrate the possibility of substantial response of maize to legumes grown in the previous year. A previous crop of *Mucuna* resulted in 100 % higher N content of six-week-old maize and much higher maize grain and stover yields compared with previous maize. Interestingly, the black-seeded *Mucuna* variety, which had a longer growth cycle than a white-seeded variety, also had more effect on subsequent maize yield than the white-seeded variety. The N fertilizer equivalence value of two years of forage or cover legumes was estimated at 20–40 kg ha^{-1} at Samaru (Fig. 15.7). In both trials, all plant residues were left on the soil surface. Part of the problem with herbaceous legume systems in the Guinea savanna zone is the long dry season during which much of the aboveground material either blows away or is eaten by animals. Only belowground material and aboveground material that has been incorporated by termites or other means are available to contribute nutrients in the subsequent rainy season. This may reduce the expected advantage of herbaceous legumes in the Guinea savanna zone. Recent results of Oikeh et al. (unpublished, 1996) from trials in the area around Samaru confirm that when most remaining aboveground residues were removed, subsequent maize yield was still 20–24% higher when maize was grown after *Stylosanthes hamata* and soybean, compared with previous maize. Total inorganic N in the 0–30 cm soil depth at the beginning of the subsequent growing season was 75, 52, and 44 kg ha^{-1} following previous soybean, stylo, and maize, respectively.

Substantial increases in use of herbaceous legumes are unlikely unless economic conditions become conducive to adoption of green manuring (Becker et al., 1995) or if direct economic products were to be realized such as human food (from large-seeded legumes

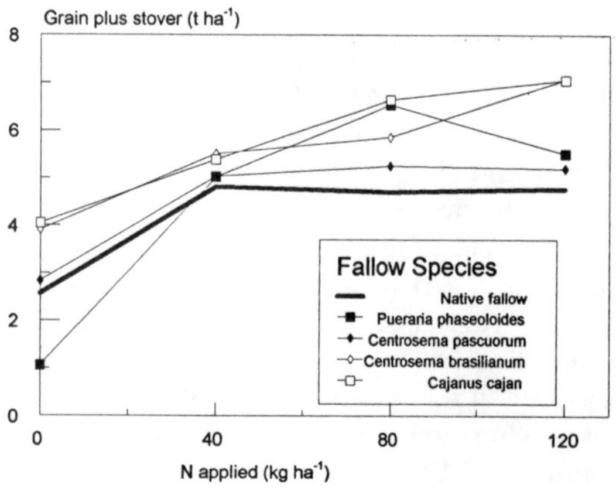

Figure 15.7. Maize aboveground dry matter response to N fertilizer after two years of herbaceous legume fallow, Samaru, 1995 (Iwuafor, unpublished data).

such as *Mucuna* and *Canavalia* spp.). Indirect benefits such as reduction of parasitism by nematodes or parasitic weeds would also help to encourage cultivation of herbaceous legumes. As long as these additional benefits are not apparently available, the use of herbaceous legumes for soil improvement will be confined mainly to research stations. Therefore, it is all the more important to test legume-based systems in collaboration with farmers to learn what systems are acceptable to them, while keeping in mind collateral benefits of legumes. Since 1987, *Mucuna* fallows have been tested with farmers in southern Benin Republic along with other technologies to enhance soil fertility. Researchers were open to the opinions of farmers and soon learned from them that *Mucuna* was often more prized for its ability to suppress a particularly noxious weed, *Imperata cylindrica*, than for soil fertility improvement. Thus *Mucuna* short fallows are being promoted on thousands of farmers' fields by the government extension services and several agricultural development NGOs in the Benin Republic. The first areas of adoption were high population density areas with little or no fallow period, nearly exhausted soils, and high *Imperata* infestation. *Mucuna* fallows are now being tested throughout the country and an extension bulletin exists for *Mucuna* fallow management.

An additional expected benefit of legumes is animal feed, although use of forage legumes in West Africa has been extremely limited (Thomas and Sumberg, 1995). Tarawali (1991) estimated an N fertilizer equivalence value of approximately 45 kg ha^{-1} after two to four year fodder banks composed of *Stylosanthes guianensis* or *S. hamata* in four sites in the Guinea savanna of Nigeria (monomodal). It should be noted that P fertilizer was applied yearly to the fodder bank and not to the control plots. The effect of the fodder bank on maize yield varied from approximately 20 kg N ha^{-1} to more than 60 kg N ha^{-1} and was related to forage biomass at the sites. The fodder bank package was extended in the subhumid zone of Nigeria during the 1980s, reaching more than 600 farmers by 1991 (Ajileye et al., 1994). Stylo forage banks are also being diffused by development agents in Mali where a technical bulletin has also been developed (E. Dembele, *Compagnie Malienne de Development de Textiles*, personal communication, 1996).

Because of the multitude of herbaceous legumes with multiple benefits, a user-friendly database of herbaceous legumes (called LEXSYS) has been developed at IITA with the help of NARS researchers of the Collaborative Group for Maize-Based Systems Research (COMBS, 1994). The database currently contains more than 100 species/accessions.

Woody Legumes

There have been cases in the bimodal zone in southwestern Nigeria where alley cropping with *Gliricidia sepium* and *Leucaena leucocephala* has had an N equivalence value of 45-90 kg N ha^{-1} (Atta-Krah and Kang, 1993; Sanginga et al., 1986; Tian et al., 1993), although lower N contribution is also common. Alley cropping with the same trees can maintain yields at an acceptable level if plant available P is sufficient, but a current trial at IITA-Ibadan, in the bimodal rainfall zone, suggests that without adding P, yields still decline. The same trial shows that for the last five years, average maize yields are 1.40 t ha^{-1} in alley cropping and 1.28 t ha^{-1} in conventional cropping (G. Tian, unpublished data). Another trial at the IITA station in southern Benin Republic (bimodal rainfall zone) has produced a combined grain yield of 2.9 t ha^{-1} over the last six years compared with 1.6 t ha^{-1} without alley cropping (M. Versteeg, personal communication, 1996).

There is much less information available from the monomodal savanna zone. Iwuafor and Kumar (1995) applied fresh *Leucaena leucocephala* prunings at a rate of 25 t ha^{-1} (containing 98 kg N ha^{-1}) and maize yield increase was equivalent to approximately 30 kg N ha^{-1} as split applied N fertilizer. Assuming 2% N in pruning dry matter, it will require 5 t ha^{-1} of dry prunings to obtain 100 kg N ha^{-1}. This quantity is obtainable in the monomodal zone from a limited number of species. In northern Ghana, *Cassia siamea* gave 4.6 t ha^{-1} containing 83 kg N ha^{-1} and *Acacia auriculiformis* gave 3.7 t ha^{-1} containing 74 kg N ha^{-1} (Nyamekye, 1992). Later in the season the hedgerow may compete with the crop for nutrients and moisture in the monomodal zone. In similar climatic zones in Zambia, Akyeampong et al. (1995) reported that *Cassia siamea* was responsible for a 37% benefit to maize and a simultaneous 17% yield loss due to competition. In the drier Machakos, Kenya area, competition from *Cassia siamea* was much more important than the benefit from the prunings.

Analysis of adoptability of alley cropping shows that it may be most appropriate where population density is high, maize is the preferred crop, and rainfall exceeds 1200 mm yr^{-1} (Whittome et al., 1995; Dvorak, 1996). This preliminary model should be validated periodically as more case studies become available using different species and different methods of introduction to farmers. This model cannot be assumed valid for agroforestry systems in general.

Animal Manure

There is no doubt about the value of animal manure for crop production. Not only does it supply nutrients, but its use can allow buildup of soil organic matter and helps counteract the acidifying effect of chemical fertilizer (de Ridder and van Keulen, 1990; Jones and Wild, 1975). High responses have been observed, the causes of which are increased nutrient supply and improved soil physical properties. Significant yield responses are observed to applications of 2.5-5 t ha^{-1} of manure (Jones and Wild, 1975) and, as mentioned above, maintaining SOC requires more than 5 t ha^{-1} (de Ridder and van Keulen, 1990).

The question remains as to the area of land which can be supplied with adequate animal manure every year. Modeling the supply and demand for manure indicated that the potential of manure to continuously sustain crop production in semiarid West Africa is limited by livestock population, spatial location of livestock at manuring time, manure excretion per animal, efficiency of manure collection, and amounts of feed available (Fernandez-Rivera et al., 1995). For the Guinea savanna of southern Mali, Bosma et al. (1995) agreed that herd size needs to be increased, forages need to be grown for dry-season grazing, and stable litter has to be transported to fields to maintain soil organic matter. They recommend a crop association of maize and lablab (*Dolichos lablab*) to improve the supply of good quality feed in combination with late dry-season pen rearing. Increased human labor will be necessary to improve the system where feed is transported to stabled animals and manure with litter is transported to crop fields. This will be realized when human population density increases with resulting reduction in the cost of labor. The use of animal manure for maintaining soil fertility can be improved by appropriate feeding and mixing with litter to manipulate the quality (Zoumana and César, 1994). Future research along these lines will be appropriate.

A Preliminary Decision Support System for Soil Fertility Maintenance in West Africa

The N contribution results presented above are synthesized in Table 15.4 along with additional factors which may influence adoption of legume-based soil fertility maintenance. Major legume benefits to be considered are human food, animal feed, and weed suppression. Locally important benefits, especially of woody species, are fuelwood, construction

Table 15.4. Elements Influencing Adoption of Legume-Based N Contribution in the Savanna Zone of West Africa

Legume Type	N Contribution (kg ha–1)	Other Benefits	Comments
Grain	0–60	human food	N contribution often inversely related to grain yield and directly related to legume growth duration.
Cover	30–100	weed suppression animal feed	Annuals will die during dry season. Mulch may persist until subsequent season in derived savanna but not in Guinea savanna.
Forage	20–70	animal feed weed suppression	Perennial species allow 2–4 years improved pasture for early rainy season and intensive dry season grazing.
Woody	30–90	fodder fuelwood medicine construction material	No benefit during tree establishment period. Greatest benefit at base of tree, diminishing with distance and depending on transport of prunings.

material, and medicine. Recommendations depend very much on the rainfall zone. In the monomodal zone, there is one long (5-6 month) growing season, enough for a sole crop or a relay intercrop but rarely for two crops. In the bimodal rainfall zone, the second cropping season can be used for the grain or herbaceous legume. Regrowth of the legume in the subsequent year may increase N contribution to nonlegume but may be difficult to manage. N contribution also increases with decreasing time interval between legume and subsequent nonlegume. Given the same amount of leguminous residue, a greater positive effect might be expected in the bimodal zone with a short dry season than in the monomodal zone with its longer dry season. The ideal system for the bimodal zone might be one in which the legume is grown during the first season followed immediately by the nonlegume crop. However, the second season is very often the shorter season which is not favorable for crop production. Therefore, a relay crop of legume into the first season nonlegume crops is more likely to be acceptable to producers.

Woody species will also accumulate more biomass in the bimodal than in the monomodal rainfall zone. Potential benefit is therefore greater, but on the other hand, pruning labor requirement will be greater. Belowground competition for moisture will be greater in the monomodal rainfall zone. Spatial arrangements which avoid competition with crops and which afford flexible management must be sought for both zones.

Legumes require a minimum level of available P to favor biological N fixation. Therefore N contribution of legumes can be expected to increase with P availability. The minimum level of available P should be defined for major soil types in the zone.

While many legume-based technologies are labor-demanding and may take the land out of food production, there have been and there will be instances of adoption. Matching production constraints with solutions and collateral benefits will provide possible answers to be tested. With soil-related production constraints increasing in West Africa, legume-based soil management systems are more likely to be adopted.

REFERENCES

Ajileye, E.O., D.V. Uza, and M.A. Farooqi. An Assessment of Farmers' Adoption Rate and Potential Impacts of Stylo-Based Feed Production Systems. In *Stylosanthes as a Forage and Fal-*

low Crop. P.N. de Leeuw, M.A. Mohamed-Saleem, and A.M. Nyamu, Eds. ILCA, Addis Ababa, Ethiopia, 1994, pp. 311–316

Akyeampong, E., B. Duguma, A.M. Heineman, C.S. Kamara, P. Kiepe, F. Kwesiga, C.K. Ong, H.J. Otieno, and M.R. Rao. A Synthesis of ICRAF's Research on Alley Cropping. In *Alley Farming Research and Development, Proceedings of an International Conference on Alley Farming*, B.T. Kang, O.O. Osiname, and A.Larbi, Eds. Ibadan, Nigeria, 14–18 September 1992, IITA, Ibadan, 1995, pp. 40–51.

Atta-Krah, A.N. and B.T. Kang. Alley Farming as a Potential Agricultural Production System for the Humid and Sub-Humid Tropics. In *Technologies for Sustainable Agriculture in the Tropics*. J. Ragland and R. Lal, Eds. ASA Special Publication Number 56, ASA, Madison, WI, USA, 1993, pp. 67–76.

Becker, M., J.K. Ladha, and M. Ali. Green Manure Technology: Potential, Usage, and Limitations. A Case Study for Lowland Rice. *Plant and Soil* 174: 181–194, 1995.

Bosma, R., M. Bengaly, and T. Defoer. Pour un système durable de production au Mali-Sud: accoître le rôle des ruminants dans le maintien de la matière orgaincique des sols. In *Livestock and Sustainable Nutrient Cycling in Mixed Farming Systems of Sub-Saharan Africa*. J.M. Powell, S. Fernandez-Rivera, T.O. Williams, and C. Renard, Eds. Proceedings of International Conference. Addis Ababa, Ethiopia, November, 1993, ILCA, Addis Ababa, Ethiopia. 1995, pp. 171–182.

Carsky, R.J., R. Abaidoo, K. Dashiell, and N. Sanginga. (in preparation) Nitrogen Contribution of Soybean Crop on Subsequent Maize in the Guinea Savanna Zone.

Carsky, R.J. and S.T.O. Lagoke. (in preparation). Determinants of Maize Yield in Farmer-Managed Trials in the West African Northern Guinea Savanna.

COMBS. Handbook for Use of LEXSYS (Legume Expert System): Decision Support for Integrating Herbaceous Legumes into Farming Systems. International Institute of Tropical Agriculture, Ibadan, Nigeria, 1994.

Dakora, F.D., R.A. Aboyinga, Y. Mahama, and J. Apaseku. Assessment of N Fixation in Groundnut (*Arachis hypogaea* L.) and Cowpea (*Vigna unguiculata* L. Walp) and Their Relative N Contribution to a Succeeding Maize Crop in Northern Ghana. *MIRCEN Journal*. 3:389–399, 1987.

de Ridder, N. and H. van Keulen. Some Aspects of the Role of Organic Matter in Sustainable Intensified Arable Farming Systems in the West-African Semi-Arid Tropics (SAT). *Fertilizer Research* 26: 299–310, 1990.

Dvorak, K.A. Adoption Potential of Alley Cropping: Final Project Report. Resource and Crop Managment Research Monograph No. 23, International Institute of Tropical Agriculture, Ibadan, Nigeria, 1996.

Eaglesham, A.R.J., A. Ayanaba, V. Ranga Rao, and D.L. Eskew. Mineral N Effects on Cowpea and Soybean Crops in a Nigerian Soil. II. Amounts of N Fixed and Accrual to the Soil. *Plant and Soil* 68: 183–192, 1982.

Fauck, R., Moureaux, C. and Thomann, C. Bilans de l'evolution des sols de Sefa (Casamance, Sénégal) après quinze années de culture continue. *L'Agronomie Tropicale* 24: 263–301, 1969.

Feller, C. La matière organique du sol: un indicateur de la fertilité. Application aux zones sahélienne et soudanienne. *Agriculture et développement*. 8:35–41, 1995.

Fernandez-Rivera, S., T.O. Williams, P. Hiernaux, and J.M. Powell. Faecal Excretion by Ruminants and Manure Availability for Crop Production in Semi-Arid West Africa. In *Livestock and Sustainable Nutrient Cycling in Mixed Farming Systems of Sub-Saharan Africa*. J.M. Powell, S. Fernandez-Rivera, T.O. Williams, and C. Renard, Eds. Proceedings of International Conference. Addis Ababa, Ethiopia, November, 1993, ILCA, Addis Ababa, Ethiopia, 1995, pp. 149–169.

Greenland, D.J. Soil Science and Sustainable Land Management. In *Soil Science and Sustainable Land Management*. J.K. Syers and D.L. Rimmer, Eds. CAB International, Wallingford, UK, 1994, pp.1–15.

Hauser, S. And B.T. Kang. Nutrient Dynamics, Maize Yield and Soil Organic Matter in Alley Cropping with *Leucaena leucocephala*. In *Soil Organic Matter Dynamics and Sustainability of*

Tropical Agriculture. K. Mulongoy and R. Merckx, Eds. John Wiley and Sons, New York, 1993, pp. 215–222.

Horst, W.J. and R. Hardter. Rotation of Maize with Cowpea Improves Yield and Nutrient Use of Maize Compared to Maize Monocropping in an Alfisol in the Northern Guinea Savanna of Ghana. *Plant and Soil.* 160:171–183, 1994.

Iwuafor, E.N.O. and V. Kumar. Alley Cropping of Maize and *L. Leucocephala* in the Northern Guinea Savanna of Nigeria. In *Alley Farming Research and Development, Proceedings of an International Conference on Alley Farming*, Ibadan, Nigeria, 14–18 September, 1992, IITA, Ibadan, 1995, pp. 304–310.

Jagtap, S.S. Environmental Characterization of the Moist Lowland Savanna of Africa. In *Moist Savannas of Africa: Potentials and Constraints for Crop Production.* B.T. Kang et al., Eds. Proceedings of an International Workshop at Cotonou, 19–23 September 1994. IITA, Ibadan, Nigeria, 1995, pp. 9–30.

Jones, M.J. The Maintenance of Soil Organic Matter Under Continuous Cultivation at Samaru, Nigeria. *J. Agric. Sci., Cambridge* 77: 473–482, 1971.

Jones, M.J. The organic matter content of the savanna soils of West Africa. *J. Soil Sci.* 24: 42–53, 1973.

Jones, M.J. and A. Wild. Soils of the West African Savanna. Commonwealth Agricultural Bureaux, Wallingford, England, 1975.

Juo, A.S.R. and F. Adams. Chemistry of LAC soils. In *Proceedings of the Symposium on Low Activity Clay (LAC) Soils*, SMSS Tech. Monograph No. 14, Soil Conservation Service, Washington, DC, USA, 1986, pp. 37–62.

Kaleem, F.Z. Assessment of N Benefit from Legumes to Following Maize Crop. In *1989 Annual Report of Nyankpala Agricultural Experiment Station*. Tamale, Ghana. GTZ, Eschborn, Germany, 1993, pp. 109–113.

Kang, B.T. Changes in Soil Chemical Properties and Crop Performance with Continuous Cropping on an Entisol in the Humid Tropics. In *Soil Organic Matter Dynamics and Sustainability of Tropical Agriculture.* K. Mulongoy and R. Merckx, Eds. John Wiley and Sons, New York, 1993, pp. 297–305.

MacColl, D. Studies on Maize (*Zea mays*) at Bunda, Malawi. 2. Yield in Short Rotations with Legumes. *Experimental Agric.* 25: 367–374, 1989.

Mulongoy, K. and I.O. Akobundu. Nitrogen Uptake of Maize in Live Mulch Systems. In *Nitrogen Management in Farming Systems of the Humid and Subhumid Tropics.* B.T. Kang and J. van der Heide, Eds. IB Haren, The Netherlands, 1985, pp. 285–290.

Nyamekye, A.L. Determination of Site Adaptability, Growth, and Yield of Various Tree Species under Alley Farming Management in the Sudan Savanna Zone of Ghana. In Nyankpala Agricultural Experiment Station (NAES) Annual Report, 1992. NAES, Nyankpala, Ghana, 1992.

Oikeh, S., V.O. Chude, R.J. Carsky, G.K. Weber, and W. Horst. (in preparation) Legume Rotation in the Moist Tropical Savanna: Managing Soil Dynamics and Cereal Yield in Farmers' Fields.

Onyeanuforo, C.C. Net Mineralization and N Release Patterns in the Savanna Ecology of Nigeria. Unpublished M.Sc. thesis, University of Ibadan, Ibadan, Nigeria, 1994.

Pieri, C. Fertilité des terres de savanes. Bilan de trente ans de recherche et de développement au Sud du Sahara. CIRAD, Paris, 1989.

Prudencio, C.Y. Ring Management of Soils and Crops in the West African Semi-Arid Tropics: The Case of the Mossi Farming System in Burkina Faso. *Agric. Ecosys. Environ.* 47:237–264, 1993.

Sanginga, N., K. Mulongoy, and A. Ayanaba. Inoculation of *Leucaena leucocephala* (Lam.) de Wit with *Rhizobium* and its Nitrogen Contribution to a Subsequent Intercrop. *Biol. Agric. Hortic.* 3: 347–352, 1986.

Sanginga, N., B. Ibewiro, P. Houngnandan, B. VanLauwe, J.A. Okogun, I.O. Akobundu, and M. Versteeg. Evaluation of Symbiotic Properties and Nitrogen Contribution of Mucuna to Maize Grown in the Derived Savanna of West Africa. *Plant and Soil* 179: 119–129, 1996.

Sanginga, N., K. Dashiell, J.A. Okogun, and G. Thottappilly. (in preparation) Nitrogen Fixation in Promiscuous Soybeans in the Guinea Savanna: II. Time Course of Nitrogen Fixation, Partitioning and Contribution to Nitrogen Balance.

Sombroek, W.G. Sustainable Agriculture: Land and Water Use and the Role of Local and External Inputs for Rural Development. In *Integrated Plant Nutrition Systems*, R. Dudal and R.N. Roy, Eds. FAO Fertilizer and Plant Nutrition Bulletin No. 12, FAO, Rome, 1995, pp. 15–18.

Stoop, W.A. and J.P. van Staveren. Effect of Cowpeas in Cereal Rotations on Subsequent Crop Yields Under Semi-Arid Conditions in Upper Volta. In *Biological Nitrogen Fixation Technology for Tropical Agriculture*, P.H. Graham and S.C. Harris, Eds. CIAT, Cali, Colombia, 1982, pp. 653–657.

Tarawali, G. The Residual Effect of *Stylosanthes* Fodder Banks on Maize Yield at Several Locations in Nigeria. *Tropical Grasslands* 25: 26–31, 1991.

Thomas, D. and J.E. Sumberg. A Review of the Evaluation and Use of Tropical Forage Legumes in Sub-Saharan Africa. *Agric., Ecosys. Environ.* 54: 151–163, 1995.

Tian, G. B.T. Kang, and L. Brussaard. Mulching Effect of Plant Residues with Chemically Contrasting Compositions on Maize Growth and Nutrient Accumulation. *Plant and Soil* 153: 179–187, 1993.

Toomsan, B., J.F. McDonagh, V. Limpinuntana, and K.E. Giller. Nitrogen Fixation by Groundnut and Soyabean and Residual Nitrogen Benefits to Rice in Farmers' Fields in Northeast Thailand. *Plant and Soil* 175: 45–56, 1995.

Weber, G., V. Chude, J. Pleysier, and S. Oikeh. On-Farm Evaluation of Nitrate-Nitrogen Dynamics Under Maize in the Northern Guinea Savanna of Nigeria. *Experimental Agric.* 31: 333–344, 1995.

Whittome, M.P.B., D.S.C. Spencer, and T. Bayliss-Smith. IITA and ILCA On-Farm Alley Farming Research: Lessons for Extension Workers. In *Alley Farming Research and Development, Proceedings of an International Conference on Alley Farming*, B.T. Kang, O.O. Osiname, and A.Larbi, Eds. Ibadan, Nigeria, 14–18 September, 1992, IITA, Ibadan, 1995, pp. 423–435.

Zoumana, C. and J. César. Production et utilisation du fumier dans la zone dense de Korhogo en Côte d'Ivoire. *Agric. Sys. Africa* 4(2): 20–27, 1994.

SIXTEEN

Grow the Soils to Grow the Crops in Africa

G.D. Anderson

INTRODUCTION

All soils result from the action of the five factors of soil formation: parent material acted on by climate and organisms (microbes to elephants, including vegetation and man) as conditioned by relief over periods of time. The influence of these on the nature of the soil has probably been best described by Jenny (1941). But soil must be understood not as an immutable product of nature, but something that is dynamic, constantly changing, with its course of development greatly influenced by man and animals, through their effect on vegetation, including crop plants.

The concept of "growing the soil" was introduced to the writer by Crompton, who applied it to growing grass on the soils of northwest England (1953). He emphasized the need to observe the soil profile and to try to understand the causes of various features, thereby learning to modify the conditions to develop the kind of profile desirable for crops. For instance, the mottled or gleyed colors associated with waterlogging may indicate not only the need for drainage, but the most effective placement depth in the profile and the spacing for tile or mole drains to remove excess water. Drainage can thus encourage a deeper root system and greater drought tolerance.

Crompton (1960) further showed that the weathering/leaching ratio could be used to explain the nature of soils, and this concept was used to interpret the soils on Mt. Kilimanjaro (Anderson, 1968a). Man and animals can often tip the balance in soil development, and Milne (1947) was one of the first to emphasize the need for man to put himself in control of the soil to prevent soil degradation processes taking their toll of the greatest natural resource given to mankind. Before considering remedial measures to promote better crop growth, it is necessary to consider briefly the soils physical factors which influence this crop.

Soil Physical Factors Affecting Crop Growth

Soil depth influences crop growth greatly, because it is the major factor giving support and determining moisture availability to crops in most environments. However, the depth of soil overlying hard rocks is not the only determinant of effective root depth. Both surface and subsoil waterlogging, either temporary or permanent, can affect the ability of roots to survive at depth. Lack of oxygen may be as much a cause of poor root penetration as excessive water or soil compaction. Soil acidity and nutrient deficiencies can also prevent good root development in both temperate (Anderson, 1977) and tropical regions (Sumner and Hylton, 1994). Salinity and alkalinity can affect both grassland species and their effective root depth, as well as the migration patterns of the wild animals grazing them (Anderson and Talbot, 1965; Anderson and Herlocker, 1973).

Soil texture primarily influences soil permeability to air and water, moisture retentivity and availability to crops, but also ease of tilth, nutrient reserve, and frequency of fertilizer inputs. The structure of subsoils is much influenced by the nature of the parent material and the texture, but surface structure is more prone to the hazards of the climate and cropping systems, particularly in the tropics, as Milne (1940) has emphasized. Porosity is an assessment of the pore space within and cracks between solid aggregates which together determine the permeability. This influences the exchange of gases and water and the flow of nutrients to plant roots. Stoniness causes damage and wear to cultivation implements and a dilution of the moisture and nutrient reserve for plants. If stones are small and not too numerous, these tend to help aeration, drainage, ease of cultivation, and trafficability.

Surface-capping occurs most commonly on soils high in silt and fine sand, with low organic matter, where the surface structure is weak. The production of a cap reduces the rainfall infiltration, and increases runoff, with major consequences of water loss to the crop, and erosion of the soil. Surface-capping can also prevent or delay seedling emergence and can cause crop failure. Preventing capping is very much dependent on the cropping system, particularly including crops with fibrous root systems in the rotation, such as grasses and fodder legumes, and applying crop residues to the surface. Hardpans occur naturally by in-filling and cementing of soil pores by clay or oxides of iron and aluminium, or avoidance by roots due to some deficiency or toxicity. Compact subsoils can also occur through tractor wheelings or the use of implements on moist or wet soil. Whatever the cause, subsoiling, often together with placement of phosphate, gypsum, lime or other nutrients, can alleviate subsoil compaction. Vigorous rooting crops such as lucern can also help, but for this correction of soil, acidity may be necessary (Munns, 1965). The use of winged subsoilers is recommended, and timing when the subsoil is dry is critical if adequate shatter is to be achieved (Spoor and Godwin, 1978).

Slope position, degree, and shape, by influencing the amount of water received and lost, affects the availability of water and nutrients for the crop. For example, convex upper-slope soils are prone to droughtiness and nutrient deficiencies, while concave lower-slope soils tend to receive excess water and nutrients, and in dry areas these may be prone to salinity. Waterlogging, wherever it occurs in the catena (Milne, 1935a), usually means that drainage is desirable to facilitate water removal and oxygen entry to the subsoils, and hence deeper rooting. Drainage, however, is not always economic and crops more adapted to waterlogging or salinity may be the answer.

All these factors affect moisture availability and aeration, and consequently plant-rooting and nutrient-uptake. While some cannot easily be altered by man, farming techniques and land use can modify others. As Batey (1988) has emphasized, it is important to examine both the profile by digging holes to assess the subsoil conditions and the soil surface, before deciding on the most effective measures to correct defects in the soil's physical condition.

Chemical Factors Affecting Crop Growth

Chemical factors with major influences on crop growth include lime, phosphorus, potassium, and nitrogen. In many humid regions, once forest has been cleared, calcium tends to be leached fairly rapidly and periodic lime application is essential for the main crop plants, including productive grasses and legumes. The exceptions to this may occur where acid-tolerant crops such as tea or potatoes are grown. In semiarid regions, lime responses on many crops are rarely expected in the first year of cropping, but may become more important in subsequent years (Anderson, 1970a; Foster, 1970).

Phosphorus application is often vital to maintain maize yields (Le Mare, 1959; Evans, 1963; Anderson, 1970b) and was widely beneficial when applied with nitrogen at a low rate on farmers' fields (Anderson, 1969). In many mature soils phosphorus is most depleted in the subsurface horizons. As a result, subsoil placement of phosphorus has been found beneficial on cocoa in aiding establishment, early growth, and yields (Anderson, 1968b). Plowing down of super phosphate prior to planting 'blueprint' grown potatoes is also recommended (Anderson and Hewgill, 1978). Pastures frequently give response to phosphorus, especially the legume contents of them, which can greatly increase, even with surface applications (Anderson 1968c). The contribution of mycorrhizae to the phosphorus nutrition of crops on a phosphorus-deficient Oxisol has been demonstrated by Vost and Fox (1979).

Potassium shortage, like lime, is most prevalent in areas of moderate to high rainfall, especially on soils with low clay content. In other situations potassium deficiency is due to potassium removal by the crop. Deficiency can occur where there is heavy crop removal of potassium-demanding crops such as bananas, sugar cane, conserved fodder, or potatoes. In the latter case, 'blueprint' grown potatoes have been shown to take up over 300 kg / ha K_2O even in dry years, when the maximum yield of 90 t/ha was reduced to below 60 t/ha (Anderson and Hewgill, 1978). The time since first cultivated or since under pasture or bush fallow, as well as the intensity and nature of the cropping, influence crop response to potassium. Optimum responses occur only when there is an adequate supply of other nutrients, particularly nitrogen, phosphorus, calcium, magnesium, and sulfur. A review of the potassium responses of various crops in East Africa (Anderson,1973) suggests that the demand for potassium fertilizers is likely to increase with more continuous and heavier cropping. Pasture legumes are more responsive to potassium than the grasses, as Dovrat (1966) has shown.

Available nitrogen is often the governor of plant growth where other nutrients are adequate, but is very transient in the soil due to losses as ammonia from the surface, denitrification, and leaching. Denitrification under wet anaerobic conditions has been shown to be highly correlated with the supply of readily decomposable organic matter, mineralizable carbon and water-soluble carbon providing good indices of this (Burford and Bremner, 1975). The cycles of wetting and drying of soils have been shown to be the means of nitrogen mineralization for crops in the field (Birch, 1960). Responses of wheat to nitrogen fertilizers have been related to soils, cultivation history, and the weather in northern Tanzania (Anderson et al., 1966). Maize yield response in western Tanzania has been well correlated with the increase in soil mineral nitrogen upon incubation (Robinson, 1968). However precise yield increases due to nitrogen fertilizer on any crop are always dependent on the rainfall pattern prior to and during crop growth. With transport costs of mineral fertilizers to outlying districts and the precarious yields in many areas of marginal and sporadic rainfall, growing grain and pasture legumes is recommended to supply at least part of the nitrogen requirement of maize and other nonleguminous crops. Some indication of the nitrogen that can be fixed and the percentage of total crop nitrogen fixed and harvested in leguminous food crops, is given in Table 16.1 (after Giller et al., 1994). Nitrogen fixation by tropical pasture legumes is also given by the same authors (Table 16.2). This shows the outstanding potential of *Desmodium* species, also found in Northern Tanzania (Anderson and Naveh, 1968). Here, however, hairy Peruvian and

Table 16.1. N_2-Fixation by Grain Legumes Grown as Monocrops in the Tropics[a]

Legume Food Crop		N Fixed as % of N in Crop	N Fixed (kg / ha^{-1})	N Harvest Index %
Groundnut	*(Arachis hypogaea)*	47–92	68–206	35–64
Pigeonpea	*(Cajanus cajan)*	65–88	68–88	25–62
Chickpea	*(Cicer arietinum)*	60–80	60–84	70
Soya bean	*(Glycine maximum)*	70–87	26–188	60–90
Common bean	*(Phaseolus vulgaris)*	15–72	3–91	50 -70
Black gram	*(Vigna mungo)*	95–98	119–140	52
Mung bean	*(Vigna radiata)*	89–90	58–107	49
Cowpea	*(Vigna unguiculata)*	32–76	47–201	31–70

[a]From Giller, K.E., J.F. McDonagh, and A. Cadisch, 1994, p. 176, with permission.

Table 16.2. N_2-Fixation in Tropical Pasture Legumes[a]

Species	N Fixed as % of N in Crop	N Fixed (kg / ha^{-1})
Calopognium muncnoides	—	136–182
Centrosema spp.	82–83	41–280
Desmodium spp.	70	25–380
Macroptilium atropurpureum	—	126
Puereria phaseoloides	88	115
Stylosanthes spp.	71–92	1–263
Zornia glabra	88	61

[a]From Giller, K.E., J.F. McDonagh, and A. Cadisch, 1994, p. 186, with permission.

Hunter River lucernes, *Dolichos formosus, Rychosia sennaarensis,* and *Glycine javanica* produced similar yields to *Desmodium uncinatum* on fertile volcanic soils, all in the range of 12,500–18,500 lb/ac dry matter in pure stands on nine sites, over about two years of cutting.

Biological Effects on Soils

As long ago as Ovid, it was recognized that "soil becomes exhausted by constant tillage" and though in the temperate regions adherents of Jethro Tull may have believed for a while the finer that one tilled the soil, the better crops one could achieve, those adherents lived largely in the days when implements had less destructive power than they do today. The close association of soil and vegetation were recognized early in Africa by Milne (1935b, 1940). Nye (1960) published data on the nutrients cycling under mature forest in Ghana (Table 16.3). Notable features are that 16% of the annual rainfall was intercepted by the canopy and never reached the ground. At the same time, large amounts of potassium and significant amounts of phosphorus and magnesium were washed out of the canopy. Most of the available nitrogen and calcium added to the soil reached it in the leaf fall. Cunningham (1963) studied the effect of clearing a tropical forest soil at Tafo in Ghana. He showed that after three years of exposure, the soil produced less mineral nitrogen, had smaller exchange capacity, lower exchangeable potassium and pH. Exposure also compacted the soil, impeded drainage through the surface layer, and made the soil erode more easily—clear indications of the process of degradation that ensues when tropical forest is cleared. Cunningham and Smith (1961) showed that one-third of the or-

Table 16.3. The Nutrient Cycle Under Mature Forest (lb/acre/annum)[a]

Additions to Soil Surface	Weight of Material (Oven Dry)	N	P	K	Ca	Mg
Litter fall	9,400	178	6.5	61	184	40
Timber fall	ca. 10,000	32	2.6	5	73	7
Rain wash		11	3.5	196	26	16
Total addition to soil		221	12.4	262	283	63

[a]From Nye, 1960, p. 339, with permission.

ganic matter was lost in under two years after clear felling, unless cover crops were established to protect the soil. In semiarid regions, Russell (1973) has pointed out that sorghum and millet can recover from severe wilt at any time up to ear emergence, while maize recovers from wilt up to the 3–4 leaf stage only. Early-planted maize, through making the most of nitrogen flushes and subsoil moisture, generally yields better than late-planted in most situations, but can run into periods of water stress and wilt beyond recovery, as Akehurst and Shreedharan (1965) have shown. From the soil conservation point of view, an earlier planted crop would seem to further the objective of keeping the soil covered as much as possible.

Apart from protecting the soil surface, another benefit of crop plants is the penetration of the subsoil. While wheat and groundnuts may have root systems to over 2 meters, lucerne is an outstanding example of a crop plant with an ability to extend its tap root to establish contact with deep water supplies (Russell, 1977), while Buffle grass, *Chenchrus ciliaris*, has been shown to root to a depth of 6 meters at Kongwa in Tanzania. In humid climates, particularly under tropical forest, roots tend to be very superficial, mostly in the top 15 cm, whereas in semiarid areas grass roots may be more abundant at 30 cm depth than near the surface. Deep-rooting crops in the rotation have the advantage of not only utilizing soil moisture better, but improving the structure and organic content of the subsoil. Superficial root systems also occur in temperate acid-matted grasslands, where nutrient cycling and microbial activity are weak. In both this and the tropical forest situation the soil surface horizons tend to be acid, and low in available phosphate, while a high proportion of the total phosphorus is in organic form at the surface. There is need to encourage deeper rooting, and a more active flora and fauna in such soils by using lime and in some cases, subsoil fertilization.

Beside cover crops of various kinds in the humid environments (Cunningham and Smith, 1961), organic mulch has been shown to be beneficial on coffee (Robinson and Hosegood, 1965). Grass mulch significantly raised the yield of groundnuts in The Gambia (Ashrif and Thornton, 1965), but in the long-term, continuous mulching, while aiding rainfall infiltration and microbial activity, may lead to potassium imbalance, causing reduced magnesium or phosphorus uptake. While mulch from one crop, e.g., bananas, may be very beneficial if transported to or used on intercropped coffee, there is need to watch for deficiencies occurring in the bananas and imbalanced uptake in the coffee.

In both subhumid and semiarid areas the need for soil stability and nitrogen to be introduced to the rotation means that legume cover crops, which may be fed to cattle or sheep or legume-rich ley pastures may be the key to facilitate some arable cropping. Such legume-rich pastures have been established in Uganda (Horrell, 1958); Kenya (Strange, 1958); Nigeria (Moore, 1965); northern Tanzania (Anderson and Naveh, 1968); and elsewhere. The effect of plowing out a three-year stand of lucerne meant that nitrogen fertilizer was not needed on two subsequent maize crops (Boawn et al., 1963).

Development of Sustainable Cropping Systems to Suit the Nature of the Soils and Environment

Greenland (1994) quotes a range of definitions of sustainable agricultural systems, his own being "A sustainable land management system is one that does not degrade the soil, or significantly contaminate the environment, while providing necessary support for human life."

In every environment the pattern or catena of soils must be looked at as a whole. What is a sustainable pattern of land use in one area is not necessarily so in another. Furthermore, what is a desirable farming system for long-term sustainability is not necessarily acceptable to the people. Because of human or animal population pressure they may wish to cultivate more frequently, or graze their animals more densely than is sustainable. Greenland (1994) suggests that the criteria of sustainability should be the maintenance or improvement of the physical, chemical or biological condition of the soil. Long-term experiments are needed to quantify processes in different soils and environments, so that soil deterioration can be measured in relation to productivity. Such experiments may be an ideal long-term approach but time is not on the side of neat answers of what precisely is sustainable. Population growth of 3.2% per annum in sub-Saharan Africa (Lal, 1990) and increasingly obvious erosion mean that conservation measures to protect and grow the soil must be implemented now. Sustainable land management systems have to preserve the productive state of the soil. Lal (1990) has suggested different steps in resource use strategy for arable, and perennial crops and pastures (Table 16.4), but the writer is not convinced that the steps are always in the right order. For instance, fertility management and soil surface management may be necessary before the development of a sustainable cropping system can be achieved. Also, cover crops may be a necessary prerequisite to the

Table 16.4. Flow Chart Outlining Different Steps in Resource Use Strategy[a]

What is appropriate land use of catena?

Arable Land	Perennial Crops	Pastures
|	|	|
Cropping Systems	Agroforestry Systems	Improved Species
|	|	|
Soil Surface Management	Cover Crops	Appropriate Ley Farming Systems
|	|	|
Fertility Management	Management of Biomass	Recycling Animal and Crop Wastes
|	|	|
	Water Management Engineering Structures	

[a]From Lal, 1990, with permission.

Table 16.5. Some Near-Sustainable Cropping Systems of the Tropics

Cropping System	Location and Duration	Example
Pasture - wheat	Waite Institute, Adelaide, Australia 69 years	Greenland, 1971
Irrigated rice	Los Banos, Philippines 24 years	Greenland, 1994
Shifting cultivation	Africa, south of Sahara 2 - 3 years	Nye and Greenland, 1960
Cotton with compost, lime, and fertilizers	Western Tanzania 9 years	Le Mare, 1972
Continuous cropping of maize, groundnuts, and cassava with N, P, K, fertilizers, mulch, and kraal manure	Different zones of Ghana 9 years	Djokoto and Stephens, 1961
Coffee and bananas with banana mulch	High altitude in Kenya, Tanzania, and Cameroon Many years	Robinson and Hosegood, 1965
Sisal with leguminous cover crops, with and without grazing	Tanzania Many years	Osborne, 1965
Coconuts with fertilizer, legumes, and cattle grazing underneath	Coastal sands of Tanzania Many years	Anderson, 1967 and 1968c
Intercropped sorghum or maize, with groundnuts	Tanzania and elsewhere	Evan, 1960
Lowland rice, sugar cane and pasture	Lowland areas of Southern Africa, Many years	Various
Fertilized pastures of grass /legume mixtures in different environments	Parts of Kenya, Uganda, Tanzania, and Nigeria Many years	Strange, 1958 Moore, 1965

successful establishment of perennial crops like coffee or tea after forest clearance. A list of some 'near sustainable' systems of cropping in the tropics is given in Table 16.5.

GROWING THE SOIL

Growing a soil involves three things: protecting the soil, feeding the soil, and encouraging a deep and prolific root system. It may be likened to a child growing up and needing protection, good food, and an environment of love, exercise, and safe exploration in which it can put down roots into the community, and develop morally and spiritually into a worthwhile citizen, able to withstand the hazards and setbacks of life. In a similar way, a soil that has been well-treated physically, well-nourished chemically and biologically, with a good biologically active subsoil, in which plant roots are able to penetrate deeply and explore, is better able to support crops in times of drought or temporary nutrient stress.

Protecting the Soil

In most parts of Africa, erosion by water or wind is a very destructive process, preventing soils from producing the crop yields for which they once had the capability. Soil erosion by water is generally the greatest hazard, but in the arid zone south of the Sahara and in other arid and semidesert areas, wind erosion can cause great soil losses. Erosion is usually an irreversible degradation process and the urgent need to produce food crops must not be allowed to cause such destruction. An example of the effect of different cropping systems on steep slopes in Rwanda is given by Roose (1987) (Table 16.6). In this situation it is clear that only pine afforestation, coffee with mulch, or casava with mulch, and grass strips, can prevent serious erosion. Even costly bench terraces are inadequate.

On a lesser slope of 18% at Tengeru in Tanzania, soil and water losses were unacceptable on clean weeded coffee and maize without conservation measures (Table 16.7) (Anderson, 1962). While Rhodes grass pasture had negligible soil loss, the only acceptable maize growing system was with trash bunds at 7-foot vertical intervals. An example of different land preparations prior to planting maize, and their effect on soil loss by wind erosion is given by Lal (1990) (Table 16.8). It is clear that in this situation, some form of prior crop cover such as maize residue, drilled oats, or rye is needed to protect the soil before maize planting. It is well known, even from temperate regions, that hedges and tree breaks can markedly reduce the amount of wind erosion.

In open landscapes, grass strips or plants like *Leucaena* or *Glyricidia* (Swift et al., 1994), may be important in preventing both water and wind erosion. If strips of grass contour-planted, bananas or legume cover crops are grown between strips of arable crops such as maize, sorghum, or groundnuts, then organic matter from these can be used to mulch the

Table 16.6. Effects of Cropping Systems on Runoff and Erosion in Rwanda[a]

Cropping System	Conservation	Slope %	Runoff Rate (% of Rainfall)	Erosion (t/ha/year)
Bare	None	40	7.2	440–880
Pine *Kesiya*	Weeding	50	0.2	0–1
Coffee	Mulch	50	0.1	0–0.4
Cassava on bunds	Mulch + 2 strips	49	0.4	0–0.4
Cassava	Trashline bunds	49	10–17	72–84
Cassava on bunds	Bench terraces	49	2.4	5–11
Maize and beans	Traditional bunds	49	–	150

[a]From Roose, E. J., 1987, quoted by Lal, 1990, with permission.

Table 16.7. Average Annual Loss of Soil and Water from the Runoff Plots at Tengeru, Tanzania on a Slope of 18%, Over the Years 1958–1960[a]

Treatment	Eroded Soil (Tons/Acre)	Water Loss (% of Annual Rainfall)
Bananas	0.28	1.8
Coffee, clean weeded	13.58	5.0
Rhodes grass	0.02	1.4
Maize with elephant grass bunds at 9-foot vertical intervals	4.35	2.3
Maize with elephant grass bunds at 7-foot vertical intervals	3.03	2.2
Maize with trash bunds at 9-foot vertical intervals	2.38	2.1
Maize with trash bunds at 7-foot vertical intervals	0.60	2.0
Maize without soil conservation measures	7.27	3.4

[a]From Anderson, 1962.

Table 16.8. Effect of Land Preparation on Soil Erosion by Wind on Newly Prepared and Planted Maize Fields, on Loamy Sands in Central Wisconsin[a]

Land Preparation	Soil Loss (t / ha)
Plowed and planted	188.0
Disced and planted	62.7
Plowed and planted with crust broken	44.8
No tillage and planted	33.6
Untilled maize planted	6.7
Disced after winter drilled oats	1.8
Standing chemically killed rye	0.09

[a]From Lal, 1990, with permission.

arable crop (e.g., Lawes 1966). In areas where livestock are kept, planting pure stands of legumes or legumes with grasses such as Rhodes grass, Buffel grass, *Panicum maximum, Cynodon plectostacyus,* or *Setaria splendida* (Anderson and Naveh, 1968) as a break between arable crops may have a stabilizing influence on the soil and provide some nitrogen for subsequent cropping. Growing a heavier crop by using optimum fertilizer rates can also protect soil from rainbeat and high soil temperatures, both of which encourage erosion (Cunningham, 1963).

Though cultivation often causes soil loss, it can also aid rainfall infiltration and prevent runoff and erosion. Water management, too, can facilitate tilth rejuvenation on vertisols (Hussein and Adey, 1994). Mechanized plowing or cultivation should be done along the contour, wherever possible. Chisel plowing at the beginning of the dry season can help trap any rainfall before the main planting rains. While tine cultivation may help to break up a subsurface layer, a winged subsoiler (Spoor and Godwin, 1978) can shatter compact deeper layers, facilitating greater moisture storage and oxygen supply, and promoting more prolific root systems. Similar or better effects are possible on smaller plots or gardens by deep digging, and when planting perennial crops like coffee, deep incorporation of lime or phosphate can be done simultaneously. Planting crops on contour ridges can greatly increase rainfall infiltration and water supply to the crop, but it is important that outfalls do not erode during heavy downpours.

One way of preventing water buildup in furrow bottoms, causing bursts of the ridges with consequent erosion, is to tie the ridges. This can be done by hoe or mechanically (Dagg and McCartney, 1968). Water-infiltration into the furrow bottoms can also be improved by vertical-mulching (Lawes, 1966). This also helps biological activity. Early planting to use rainfall more efficiently can extend the growing season in regions where the wet season is marginal (Tabi Owono et al., 1994). On steep slopes, making bunds and terraces is generally necessary to protect the soil, and where large-scale mechanized wheat or maize growing is practiced, as in northern Tanzania, Kenya, Zimbabwe and elsewhere, the erection of terraces by mechanical graders is desirable. However, to be sustainable, such systems should include grass—legume leys to aid structural stability, prevent erosion, and provide nitrogen to the arable crops.

Some of the most eroded areas of the tropics are those used for grazing cattle, sheep, or goats. These animals can cause erosion along their tracks where vegetation becomes sparse, especially in semiarid conditions. However, at the height of the dry season it is close grazing by sheep and even root-eating by goats that causes much damage to the grass cover. Overstocking of grazings nearly always occurs in the dry season, but in years of very poor rains catastrophic erosion often happens, with consequent silting-up of watercourses and dam sites. Such erosion is widespread in in Kongwa district, Tanzania (Payton and Shishira, 1994). The problem in such areas is the control of both stock numbers and common grazing. It is difficult to see how erosion of these semiarid areas can be prevented without controls on stock movement, by fences, and legally binding agreements. Once stock control measures are taken, there may be a case for overseeding such pastures with legumes, like *Stylosanthus gracilis*, possibly with some superphosphate in less fertile areas. This could increase plant cover and stocking capacity considerably. However, fencing, bush clearing, and grazing control are essentials to protect such areas (Pereira et al., 1961). To ensure that soils are adequately protected against erosion, both surface and subsoil conditions need regular examination. This will aid in the best choice of ameliorative measures.

Feeding the Soil

Stocking (1994) has listed the soil limitations resulting from erosion which inhibit productivity. Besides physical defects, the loss of fine-textured material and organic matter from the soil surface causes reduced cation-exchange capacity, deficiencies in nitrogen, phosphorus, and potassium, and often aluminium and manganese toxicities. Wherever these conditions develop, crop yields fall and the replenishment of lost nutrients is vital to reestablish good crop cover. In mature tropical forest soils, where most nitrogen, phosphorus, and potassium are in the surface few centimeters, loss of topsoil may be disastrous. On more fertile volcanic soils, where phosphorus and potassium can be considerable in the subsoil, loss of surface organic matter and available nitrogen can still reduce yields. Even where erosion is not a serious problem, there are widespread deficiencies of the major elements throughout Africa. With the cereals, nitrogen fertilizer is almost universally required for optimum yields of grain; for example, with wheat (Anderson et al., 1966) and maize (Le Mare, 1959; Djokoto and Stephens, 1961).

This is also true with grasses grown without legumes (Brockington, 1964). With tree crops like tea, coffee, and cocoa, nitrogen fertilizer is generally beneficial unless these are heavily mulched. Full response to nitrogen is rarely achieved unless supplies of phosphorus and potassium are adequate. As fertilizer, nitrogen is frequently lost from the soil surface or leached; wherever possible, some of it should be provided by interplanted or previously cropped grain or fodder legumes.

Next to nitrogen, phosphorus is probably the most widespread deficiency in Africa. Responses to super phosphate have been shown on cereals, groundnuts, and soya and on many tree crops, particularly in the planting and early growth stages. Phosphorus has a major role in encouraging vigorous root development, in grain filling, and early ripening in cereals. It promotes nodulation and nitrogen fixation in legumes. While potassium is not widely deficient in grazed grasslands, except on very sandy soils (e.g., Anderson, 1968c), wherever fodder conservation and continuous crop off-take occurs, soils tend to become deficient. So the need for potassium application under heavy cropping is likely to increase for a wide range of crops (Anderson, 1973). Where mulch or organic manures are applied to the soil, potassium fertilizers are less likely to be needed.

In all cases of fertilizer application the law of the minimum needs to be remembered, so it is no use supplying one nutrient to the soil if another is acutely deficient. For instance, the need for lime to correct acidity may need to be recognized before the optimum response to phosphate is achievable. The need to assess, by soil and plant analysis and diagnostic approaches, which are the most critical nutrient deficiencies (Sumner and Hylton, 1994) is therefore of paramount importance in achieving economic crop responses to fertilizer applications.

Encouraging Deep and Prolific Root Systems

In many soils, particularly deeply weathered Oxisols and Ultisols, most available nutrients occur in the top few centimeters and this has led to most roots growing in the same layer.

While a few roots in such soils may penetrate to a depth of several meters, there are often very few in subsurface horizons. Generally, this is caused by plants extracting phosphorus from such zones over many years and depositing it on the soil surface, largely in organic form. The removal of bases from subsurface horizons also means that acidity has brought about phosphate fixation as insoluble iron and aluminium phosphates. Aluminium and manganese toxicities may also occur. It is hardly surprising that roots avoid these zones and the lack of flora and faunal activity often leads to a compact soil, which is weakly constructed and hard to dig. Similar compact subsoils can occur in semiarid regions, even where there is much less organic topsoil. Shattering such zones by subsoiling can encourage roots to proliferate in them. Even greater root penetration can often be achieved by subsoil placement of phosphate or compound fertilizer. With tree crops, nutrients can be placed in planting holes, as done for cocoa (Anderson, 1968b). Fertilizers can be applied behind a subsoiler or deep plowed down for arable crops or prior to ley establishment. Subsoiling and tine cultivation are also means of breaking up compaction caused by machinery wheelings or cattle poaching. Such cultivation can enhance rainfall infiltration and improve permeability to air and water. Where fine-textured clays or low-lying situations cause poor drainage and lack of root penetration because of waterlogging, whether temporary or permanent, drainage with ditches, pipes, or mole plow is desirable. However the cost is often prohibitive unless high value crops such as sugarcane or lucerne are to be grown. Improving drainage may certainly inhibit salinity buildup under irrigation, though amendments such as gypsum may also be necessary.

Crops have a variety of rooting habits, though stoniness, compaction, drainage, texture, and structure in soils can often alter the norm of these. For example, wheat and perennial grasses will usually root to well over two meters depth (Batey, 1988), whereas potatoes, soya, and maize have much shallower root systems. However, in order to maintain or improve root distribution in soils, it is helpful to grow a variety of crops, even if there is no fixed rotation. Often it may be beneficial to grow a deep rooting crop, such as lucerne or legume-rich pasture in order to develop both structure and a more active flora and fauna, with nitrogen supply for subsequent crops, e.g., maize, as Boaum et al. (1963) showed. Though less research has been done, it is thought likely that other, less nutrient-demanding legumes such as Stylo may have similar benefits. Whatever is needed to develop a better root system under any crop requires appraisal of the soil profile by digging holes at least beyond the present rooting depth.

CONCLUSION

Growing the Soil Involves:

1. ***Regular examination of soil in the field,*** both on the surface and by digging holes to expose the subsoil features and rooting patterns. This will help decisions about remedial techniques to improve rainfall infiltration, prevent runoff, and encourage deeper and more prolific root systems.
2. ***Means of protecting the soil from erosion and soil degradation***, including grazing controls for livestock, cultivations, tied-ridge cropping, subsoiling, contour-furrowing, bunding and terracing, strip and intercropping, mulching, and cover crops, including legumes in pure stands and in grass mixtures.
3. ***Feeding the soil*** involves applications of nitrogen, phosphorus, potassium, lime, other nutrients and organic manures, as and when these are most required to achieve optimum crop yields in the rotation.
4. ***Encouraging deeper and more prolific root systems*** may require both physical improve-

ments of the subsoil, nutrient enhancement of deficient and compact zones, organic inputs, and growing deep-rooted legume-rich crops.

In all this, the farmer must take control of the soil to protect, feed, and nurture it, so that he may reap the benefits of profitable crops. To achieve this, he needs the help of the research worker, the teacher, and the soils advisor.

REFERENCES

Akehurst, B.C. and A. Sreedharan. Time of Planting—A Brief Review of Experimental Work in Tanzania. *E. Africa Agric. For. J.* 30, 189, 1965.

Anderson, G.D. The Effect of Various Cropping Systems on Soil Fertility. *Proc. E. Afric. Soil Fertility Spec. Comm.* Tengeru, Tanzania, 1962.

Anderson, G.D., B.G. Houston, and P.J. Northwood. Effects of Soil Cultivation History and Weather on Responses of Wheat to Fertiliser in Northern Tanzania. *Expl. Agric.* 2, 183–200, 1966.

Anderson, G.D. Increasing Coconut Yields and Income on the Sandy Soils of the Tanganyika Coast. *E. Afric. Agric. For. J.* 32, 310–314, 1967.

Anderson, G.D. *A Survey of the Soils and Land Use Potential of the Southern and Eastern Slopes of Mt. Kilimanjaro, Tanzania*, with Soil, Land Use Capability and Natural Vegetation Maps. Min. of Agric. and Co-op, Dar es Salaam, 1968a, pp. 1–73.

Anderson, G.D. Early Results from a Fertiliser Experiment on Cocoa at Mangubu, Muheza, Tanga Region, Tanzania. *Tengeru Report No 126*, Min. of Agric. and Co-ops, Tanzania, 1968b.

Anderson, G.D. Effects of Fertilizers on Botanical Composition and Productivity of Pasture on the Sandy Soils of the Tanganyika Coast. *E. Africa Agric. For. J.* 34, 207–216, 1968c.

Anderson, G.D. and Z. Naveh. Promising Pasture Plants for Northern Tanzania: V Overall Comparison of Promising Plants. *E. Africa Agric. For. J.* 34 (1), 84–105, 1968.

Anderson, G.D. Response of Maize to Application of Compound Fertilisers on Farmers' Fields in Ten Districts of Tanzania. *East Africa Agric. For. J.* 34 (3), 382, 1969.

Anderson, G.D. Fertility Studies on a Sandy Loam in Semi-Arid Tanzania 2. Effects of Phosphorus, Potassium and Lime on Yields of Groundnuts. *Expl. Agric.* 6, 213–222, 1970a.

Anderson, G.D. Fertility Studies on a Sandy Loam in Semi-Arid Tanzania 1. Effects of Nitrogen, Phosphorus and Potassium Fertilisers on Yields of Maize, and Soil Nutrient Status. *Exptl. Agric.* 6, 1–12, 1970b.

Anderson, G.D. Potassium Responses of Various Crops in East Africa. *Proc. 10th Coll. Int. Potash Institute,* Abidjan, 1973, pp. 413–437.

Anderson, G.D. and D.J. Herlocker. Soil Factors Affecting the Distribution of the Vegetation Types, and Their Utilisation by Wild Animals in Ngorongoro Crater, Tanzania. *J. Ecol.* 61, 627–651, 1973.

Anderson, G.D. *A Survey of the Soils, and Land Use Potential of Darnbrook Development Farm,* (with Soil, Land Capability, and Potential Use Suitability Maps) A.D.A.S. Leeds, UK, 1977, pp. 1–115.

Anderson, G.D. and D. Hewgill. Effects of Nitrogen, Phosphorus and Potassium on 'Blueprint' Grown Potatoes. Proc. A.D.A.S. / A.R.C. *Symposium on Maximising Yields of Crops.* Harrogate, UK, 1978, pp. 139–150.

Anderson, G.D. and L.M. Talbot. Soil Factors Affecting the Distribution of the Grassland Types, and Their Utilisation by Wild Animals on the Serengeti Plains, Tanganyika. *J. Ecol.* 53, 33–56, 1965.

Ashrif, M.I. and I. Thornton. Effects of Grass Mulch on Groundnuts in Gambia. *Exptl. Agric.* 1, No. 2, 1965, pp. 145–152.

Batey, T. *Soil Husbandry.* Pub. Soil and Land Use Consultants Ltd., Aberdeen, 1988, p. 157.

Birch, H.F. Nitrification in Soils after Different Periods of Dryness. *Plant and Soil* 12 (1), 81–96, 1960.

Boawn, L.C., J.L. Nelson and C.L. Crawford. Residual Nitrogen from Ammonium Nitrate Fertiliser and from Alfalfa Ploughed Under. *Agron. J.* 55 (3), 231, 1963.
Brocklington, N.R. Fertiliser Response of Giant Rhodes Grass in Northern Rhodesia. *Emp. J. Exptl. Agric.* 32, 76, 1964.
Burford, J.R. and J.M. Bremner. Relationship Between the Denitrification Capacities of Soils and Total Water Soluble and Readily Decomposable Soil Organic Matter. *Soil Biol. Biochem.* 1, 389–394, 1975.
Crompton, E. Grow the Soil to Grow the Grass. *J. Min. Agric.* 60, 301, 1953.
Crompton, E. The Significance of the Weathering / Leaching Ratio on the Differentiation of Major Soil Groups. *Trans. 7th Int. Cong. Soil Sci.* 4, 406–412, 1960.
Cunningham, R.K. The Effect of Clearing Tropical Forest Soil. *J. Soil Sci.* 14, 334, 1963.
Cunningham, R.K. and R.W. Smith. Comparison of Soil Covers During Cocoa Establishment on Clear Felled Land. *Trop. Agric.* 36 (1),13, 1961.
Dagg, M. and J.C. McCartney. A System of Tied-Ridge Cultivation for Semi-Arid Areas of Tanzania. *Exptl. Agric.* 4, 279, 1968.
Djokoto, R.K. and D. Stephens. Longterm Fertiliser Experiments under Continuous Cropping in Ghana. *Emp. J. Exptl. Agric.* 29, 181, 1961.
Dovrat, A. Responses of Rhodes Grass Overseeded with Legumes to N and K Fertilisers, and to the Availability of Soil K in Israel. *Exptl. Agric.* 2, 255–263, 1966.
Evans, A.C. Studies of Intercropping: 1. Maize and Sorghum with Groundnuts. *East Africa Agric. For. J.* 26, 1–10, 1960.
Evans, A.C. Soil Fertility Studies in Tanganyika. *East Africa Agric. For. J.* 28, 228-239, 1963.
Foster, H. Liming Continuously Cultivated Soils in Uganda. *East Africa Agric. For. J.* 36, 58, 1970.
Giller, K.E., J.F. McDonagh, and A. Cadisch. Can Biological Nitrogen Fixation Sustain Agriculture in the Tropics? In *Soil Science and Sustainable Land Management in the Tropics*. Syers, J.K. and D.L. Rimmer, Eds. C.A.B. Oxford, 1994, pp. 173–191.
Greenland, D.J. Soil Science and Sustainable Land Management. In *Soil Science and Sustainable Land Management in the Tropics*. Syers, J.K. and D.L. Rimmer, Eds. C.A.B. Oxford, 1994, pp. 173–191.
Greenland, D.J. Changes in Nitrogen and Physical Condition of Soils Under Pasture. *Soils and Fertilisers* 34, 237–251, 1971.
Horrell, C.R. Herbage Plants at Serere Experimental Station, Uganda. *East Africa Agric. For. J.* 24, 41 and 24, 133, 1958.
Hussein, J. and M.A. Adey. Sustainable Regeneration of Surface Tilth in Zimbabwean Vertisols Through Water Management. In *Soil Science and Sustainable Land Management in the Tropics*. Syers, J.K. and D.L. Rimmer, Eds. C.A.B. Oxford, 1994, p. 120.
Jenny, H. *Factors of Soil Formation*. McGraw-Hill, New York, 1941.
Lal, R. *Soil Erosion in the Tropics*. McGraw-Hill, New York, 1990.
Lawes, D.A. Rainfall Conservation and the Yields of Sorghum and Groundnuts in Northern Nigeria. *Exptl. Agric.* 2, 139–146, 1966.
Le Mare, P.H. Soil Fertility Studies in Three Areas of Tanganyika. *Emp. J. Exptl. Agric.* 27, 197, 1959.
Le Mare, P.H. A Long Term Experiment on Soil Fertility and Cotton Yield in Tanzania. *Exptl. Agric.* 8, 299–310, 1972.
Milne, G. Composite Units for the Mapping of Complex Soil Associations. *Trans. 3rd Int. Cong. Soil Sci.* 1, 345–347, 1935a.
Milne, G. Essays in Applied Pedology. 1. Soil Type and Soil Management in Relation to Plantation Management in East Usambara. *East Africa For. J.* 3, 1935b.
Milne, G. Soil and Vegetation. *East Africa Agric. For. J.* 8, 1940.
Milne, G. A Soil Reconnaissance Journey Through Parts of Tanganyika Territory, Dec. 1933–Feb. 1936 *J. Ecol.* 35, Nos. 1 and 2, 1947.
Moore, A.W. The Influence of Fertilisation, and Cutting, on a Tropical Grass-Legume Pasture. *Exptl. Agric.*1, No. 3, 13, 1965.

Munns, D.N. Interactions of Lime with Nitrogen and Phosphate on Growth of *Medicago sativa* and *Trifolium subterraineum. Aust. J. Agric. Res.* 16, 733–741, 1965.

Nye, P. Organic Matter and Nutrient Cycles Under Moist Tropical Forrest. *Plant and Soil* 13, 333–346, 1960.

Nye, D.J. and D.J. Greenland. *The Soil Under Shifting Cultivation.* Tech. Comm. No. 51, Com. Bur. Soil, Farnham Royal, Bucks, C.A.B. Oxford, 1960.

Osborne, J.F. *Handbook For Sisal Planters.* Tanganyika Sisal Growers Assoc., Tanga, Tanganyika, 1965.

Payton, R.W. and E.K. Shishira. Effects of Soil Erosion and Sedimentation on Land Quality. In *Soil Science and Sustainable Land Management in the Tropics.* Syers, J.K. and D.L. Rimmer, Eds. C.A.B. Oxford, 1994, pp. 88–119.

Pereira, H.C., P.H. Hosegood and D.B. Thomas. The Productivity of Tropical Semi-Arid Thorn Scrub Country Under Intensive Management. *Emp. J. Exptl. Agric.* 29, 269, 1961.

Robinson, J.B.D. A Simple Available Soil N Index, Pt. 2 Field Crop Evaluation. *J. Soil Sci.* 19, 280, 1968.

Robinson, J.B.D. and P.H. Hosegood. Effects of Organic Mulch on Fertility of a Latosolic Coffee Soil in Kenya. *Exptl. Agric.* 1, 67–80, 1965.

Roose, E.J. Soil Conservation on Steep Lands in Africa. *Proc. Soil and Water Conservation on Steep Lands.* San Juan, Puerto Rico, 1987. Quoted by Lal, 1990, with permission.

Russell, E.W. Water Conservation in Semi Arid Regions. In *Soil Conditions and Plant Growth,* 10th edition, Longman, London, 1973, p. 786.

Russell, R. S. *Plant Root Systems.* McGraw-Hill, Maidenhead, UK, 1977.

Spoor, G. and R.J. Godwin An Experimental Investigation into the Deep Loosening of the Soil by Rigid Tines. *J. Agric. Eng. Res.* 23, 243–258, 1978.

Stocking, M. Soil Erosion and Conservation—A Place for Soil Science? In *Soil Science and Sustainable Land Management in the Tropics.* Syers, J.K. and D.L. Rimmer, Eds. C.A.B. Oxford, 1994, pp. 40–58.

Strange, R. Preliminary Trials of Grasses and Legumes under Grazing. *East Africa Agric. For. J.* 24, 92–102, 1958.

Sumner, M.E. and K. Hylton. A Diagnostic Approach to Solving Soil Fertility Problems in the Tropics. In *Soil Science and Sustainable Land Management in the Tropics.* Syers, J.K. and D.L. Rimmer, Eds. C.A.B. Oxford, 1994, pp. 215–134.

Swift, M.D., K.A. Dvorak, K. Mulongoy, M. Musoko, N. Sanginga and A. Tian. The Role of Soil Organisms in the Sustainability of Tropical Cropping Systems. In *Soil Science and Sustainable Land Management in the Tropics.* Syers, J.K. and D.L. Rimmer, Eds. C.A.B. Oxford, 1994, pp. 155–172.

Tabi Owono, J., D.A. Rose and R.S. Shiel. Extending the Growing Season by Efficient Use of Rainfall. In *Soil Science and Sustainable Land Management in the Tropics.* Syers, J.K. and D.L. Rimmer, Eds. C.A.B. Oxford, 1994, pp. 145–154.

Vost, R.S. and R.L. Fox Contribution of Mycorrhizae to Phosphorus Nutrition of Crops Growing on an Oxisol. *J. Agron.* 71, 903–909, 1979.

SEVENTEEN

No Till and Mulching Effects on Soil Physical Quality of a Tropical Alfisol in Western Nigeria

R. Lal

INTRODUCTION

In contrast to other regions of the world, per capita food production has not kept pace with the increase in population in sub-Saharan Africa, which has been declining at about 2% per year since 1960 (World Bank, 1996). About 37% of the total population or 175 million people remain chronically underfed in the region, and the percentage has remained unchanged for the two decades ending in 1990. The absolute number of chronically underfed population doubled from 94 million in 1970 to 175 million in 1990 (Young, 1997; Uvin, 1994).

One of the numerous reasons of agrarian stagnation and perpetuation of hunger and malnutrition is the soil degradation. Depletion of soil fertility, accelerated soil erosion, drought and desertification have taken a severe toll on soil quality and its productivity. Sanchez et al. (1996) observed that the magnitude of nutrient mining in Africa is huge, with a net per hectare loss during the 30-year period ending in 1996 estimated at 700 kg N, 100 kg P, and 450 kg K in about 100 million hectares of cultivated land. Similar observations were made by Van der Pol, (1992); Smaling, (1993); and Stoorvogel et al. (1993). The problem of nutrient depletion is exacerbated by accelerated erosion and drought.

Adoption of conservation tillage and use of crop residue mulch have proved useful in minimizing soil erosion risks, conserving soil water and reducing drought, recycling plant nutrients, and regulating soil temperature. The beneficial effects of conservation tillage and mulch farming techniques, however, are soil and ecoregion specific (Lal, 1985; 1989; 1993; Lal et al., 1980; Aina, 1979; Dunham, 1988; Roose, 1976; 1987). Plowing, in combination with use of heavy equipment for deforestation and/or harvest traffic, can cause severe problems of compaction and soil degradation (Lal, 1984; 1985; 1996a, b, c).

No till system of crop production is a specific variant of conservation tillage, and its success depends on the availability of crop residue mulch, judicious use of chemical fertilizers, and adequate weed control measures (Lal, 1985). No till system may not be successful on soils prone to poor drainage and compaction, and where residue mulch is not available in enough quantity to minimize soil erosion, decrease evaporation losses, and regulate soil temperature.

No till affects agricultural sustainability and environmental quality through its impact on soil properties and processes. Notable among these is soil physical quality as determined by soil structure. Soil structural impacts of no till and mulch farming may be related to water stable aggregation, mean weight diameter of aggregates, and water retention and transmission properties. Tillage effects on soil physical properties, however, are not widely studied for soils of tropical Africa. Therefore, the objective of this experiment was to evaluate long-term effects of no till and soil management practices on soil physical properties.

MATERIAL AND METHODS

The experiment was conducted from 1979 through 1987 at the research farm of the International Institute of Tropical Agriculture (IITA), Ibadan, Nigeria. These experiments were conducted as a part of the long-term study of tillage effects on soil properties and crop yield. Results of these experiments have been described in detail elsewhere (Lal, 1974; 1976; 1978; 1982; 1995). The soil of the experimental site is derived from fine-grained biotite gneiss and schist parent material, and is classified as Paleustalf according to Soil Taxonomy (Soil Survey Staff, 1975), and Luvisol according to the FAO classification system (FAO, 1977). Two tillage systems were maintained for the duration of the experiment. Tillage treatments were: (a) no till in which crops were sown without mechanical seedbed preparation and with use of herbicides to control weeds, and (b) plow till in which soil was disced to about 20 cm depth and then rotovated to prepare a fine seed bed. Two tillage treatments were not randomized because they were applied in two parallel contour strips, one for each tillage treatment. There were 12 subplots within each strip which comprised four combinations of fertilizer and mulch, with three replications. The subplot size within each strip was 20 x 10 m. However, three replications of four treatments within each tillage strip were completely randomized. The residue mulch was removed from some treatments (M0), and added to the residue produced in others (M1). The fertilizer variable included none (F0) and 120 kg N/ha applied 1/3 at planting and 2/3 after 4 weeks, 26 kg P/ha and 30 kg K/ha (F1). Nitrogen was applied as urea, P as single super phosphate, and K as muriate of potash.

Soil physical properties were measured for 0–5 cm and 5–10 cm depths. Soil samples were obtained during the dry season (December through March) every year beginning in 1980. Soil samples were air dried, gently ground, and passed through 2-mm sieve. Soil texture was determined by the hydrometer method (Gee and Bauder, 1986). Soil moisture retention characteristics were determined on unsieved samples using a combination of tension table and pressure plate extractors (Lal, 1979; Klute, 1986). Double ring infiltrometers were used to measure infiltration rate. The oxygen diffusion rate was measured using platinum microelectrodes.

All data were statistically analyzed using years as fixed variables for computing the analysis of variance and the least significant difference (Steel and Torrie, 1980).

RESULTS AND DISCUSSION

The data are discussed to highlight tillage and management effects on temporal variations in texture, infiltration, and soil moisture retention characteristics.

Table 17.1. Tillage and Management Effects on Clay Contents

		0–5 cm depth					5–10 cm depth				
Tillage	Treatment	1981	1982	1983	1984	1987	1981	1982	1983	1984	1987
		% by weight									
No till	M_0F_0	16.0	20.0	18.4	20.0	12.1	16.0	22.5	20.4	22.0	12.8
	M_0F_1	16.7	20.5	20.4	21.3	13.5	16.0	19.9	21.1	22.7	14.8
	M_1F_0	16.5	19.2	17.7	18.7	14.1	17.2	19.2	21.1	19.3	12.1
	M_1F_1	16.1	19.9	19.1	19.1	10.8	15.5	19.2	19.7	17.3	11.5
	Mean	16.3	19.9	18.9	19.8	12.6	16.2	20.2	20.6	20.3	12.8
Plow till	M_0F_0	18.3	19.9	19.7	20.7	13.5	18.3	19.9	19.7	21.3	14.8
	M_0F_1	17.3	17.2	18.4	19.3	15.5	18.3	23.2	19.7	19.3	15.5
	M_1F_0	16.9	18.5	17.7	18.7	14.1	18.3	19.2	19.1	20.0	13.5
	M_1F_1	18.7	19.3	18.4	22.0	18.5	19.3	21.2	21.7	20.0	18.7
	Mean	17.8	18.7	18.6	20.2	15.4	18.6	20.9	20.1	20.2	15.6
LSD (.05)											
Year (Y)		1.4					2.1				
Tillage (T)		1.2					1.4				
Mulch (M)		0.8					1.0				
Fertilizer (F)		0.8					1.0				
M × Y		1.7					2.2				
T × Y		2.8					3.1				
T × M		1.1					1.4				
T × F		1.1					1.4				

F_0 = no fertilizer, F_1 = recommended fertilizer rate, M_0 = crop residue removed, M_1 = crop residue retrieved as surface mulch.

Table 17.2. Tillage and Management Effects on Silt Content

		0–5 cm depth					5–10 cm depth				
Tillage	Treatment	1981	1982	1983	1984	1987	1981	1982	1983	1984	1987
		% by weight									
No till	M_0F_0	10.3	10.5	11.3	11.2	11.5	10.3	9.9	10.7	10.1	11.2
	M_0F_1	11.6	9.2	11.3	12.0	13.5	10.9	8.5	9.3	8.0	12.8
	M_1F_0	10.3	9.9	10.0	11.3	17.0	9.6	8.5	8.7	10.7	12.8
	M_1F_1	9.6	11.2	10.7	10.7	14.8	8.3	8.5	10.0	10.7	12.1
	Mean	10.5	10.0	10.8	11.3	14.2	9.8	8.9	9.7	9.9	12.2
Plow till	M_0F_0	8.9	9.9	11.3	9.2	9.7	8.9	10.5	10.0	9.1	11.7
	M_0F_1	9.1	9.5	8.7	6.8	10.7	8.3	10.1	9.3	9.5	10.0
	M_1F_0	8.9	9.9	10.7	9.9	11.2	6.9	10.5	10.0	8.5	9.2
	M_1F_1	8.9	9.9	10.7	9.1	12.1	9.6	9.2	11.3	9.7	11.8
	Mean	9.0	9.8	10.4	8.8	10.9	8.4	10.1	10.2	9.2	10.7
LSD (.05)											
Year (Y)		1.1					1.5				
Tillage (T)		0.9					0.9				
Mulch (M)		0.6					0.7				
Fertilizer (F)		0.6					0.7				
M × Y		1.3					1.5				
T × Y		1.9					2.0				
T × M		0.8					0.9				
T × F		0.8					0.9				

F_0 = no fertilizer, F_1 = recommended fertilizer rate, M_0 = crop residue removed, M_1 = crop residue retrieved as surface mulch.

Figure 17.1. Plow till treatments had higher erosion losses than no till treatments.

Textural Properties

Tillage and management effects on clay and silt contents are shown in Tables 17.1 and 17.2, respectively. There were no consistent temporal trends in the clay content with regard to tillage methods (Table 17.1). However, the overall mean clay content for no till compared with plow till treatments was 17.5% vs. 18.1% for 0–5 cm depth, and 18.0% vs. 19.1% for 5–10 cm depth. Somewhat more clay content for the plow till treatment may be due to the higher erosion rate (Fig. 17.1) and mixing of the clayey subsoil with the surface soil. In contrast to the clay, silt content was generally more for the no till compared with the plow till treatment (Table 17.2). Differences in silt content in favor of the no till treatment were consistent for 0–5 cm depth. The overall mean silt content for no till compared with plow till treatment was 11.4% vs. 9.8% for 0–5 cm depth and 10.1% vs. 9.7% for 5–10 cm depth. High silt content in the surface 0–5 cm layer of the no till treatment is consistent with less soil erosion. Being mostly in an unaggregated state, silt fraction is preferentially removed by surface runoff (Fig. 17.2) during the erosional process.

In accord with the data on clay and silt fractions, the surface layer of the plow till treatment had visibly more gravel content than that of the no till treatment. At any given time, soil in the plow till treatment also contained lower soil moisture content compared with the no till plot. Therefore, all other factors remaining the same, the aeration porosity might have been more for plow till than no till soil. Consequently, the oxygen diffusion rate (ODR) was generally more for plow till compared with the no till treatment (Table 17.3). Irrespective of the tillage methods, the ODR was generally more for treatment with than without crop residue mulch. Presence of crop residue mulch on the soil surface minimized crusting and surface seal formation, and therefore increased aeration and ODR.

Figure 17.2. Silt fractions preferentially removed by water runoff.

Water Infiltration

The data on cumulative infiltration are shown in Table 17.4. There are no definite temporal trends in cumulative infiltration. However, cumulative infiltration was significantly more for no till compared with the plow till treatment. The ratio of cumulative infiltration for no till:plow till treatment was 2.7 for March 1981, 2.6 for October 1981, 1.3 for

Table 17.3. Tillage and Mulching Effects on Oxygen Diffusion Rate at 15 cm Depth Measured During the First Growing Season in 1983

Tillage	Treatment	5 May 1983	19 May 1983	24 May 1983
			μmols/cm²/sec	
No till	M_0F_0	0.37	0.58	0.38
	M_0F_1	0.42	0.58	0.37
	M_1F_0	0.41	0.69	0.33
	M_1F_1	0.47	0.71	0.43
	Mean	0.42	0.64	0.38
Plow till	M_0F_0	0.58	0.73	0.50
	M_0F_1	0.22	0.75	0.51
	M_1F_0	0.71	0.70	0.42
	M_1F_1	0.69	0.67	0.51
	Mean	0.55	0.71	0.49
LSD (.05)				
Tillage (T)		0.01	0.13	0.11
Mulch (M)		0.14	0.03	0.14
Fertilizer (F)		0.08	0.06	0.10
T × M		0.20	0.05	0.20
T × F		0.11	0.09	0.14
F × M		0.11	0.09	0.14

F_0 = no fertilizer, F_1 = recommended fertilizer rate, M_0 = crop residue removed, M_1 = crop residue retrieved as surface mulch.

Table 17.4. Tillage and Management Effects on Accumulative Infiltration at 2 Hours After Surface Ponding

Tillage	Residue	Fertilizer	March '81	Oct. '81	Feb. '82	Feb. '83	Feb. '84	Jan. '85	Feb. '86
						cm			
No till	Removed	None	95.5	84.1	76.6	116.1	107.7	85.6	84.6
	Returned	None	93.7	93.7	115.4	122.5	63.8	117.1	68.6
	Removed	Applied	59.0	59.0	117.3	111.5	59.4	69.1	79.2
	Returned	Applied	80.7	81.3	134.8	114.4	79.7	112.2	68.3
	Mean		82.2	79.5	111.0	116.1	77.7	96.0	75.2
Plow till	Removed	None	31.8	31.8	70.3	33.5	49.4	34.9	30.9
	Returned	None	36.0	36.0	83.4	47.5	60.5	29.0	20.5
	Removed	Applied	43.8	43.8	130.9	79.8	26.4	31.4	24.3
	Returned	Applied	8.9	8.9	50.5	51.9	50.5	62.5	34.8
	Mean		30.1	30.1	83.8	53.2	46.7	39.5	27.6
LSD (.05)									
Year (Y)			22.6						
Tillage (T)			18.7						
Residue (R)			18.2						
Fertilizer (F)			18.2						
R x Y			48.2						
Y x T			49.3						
T x M			25.8						
T x F			48.2						

February 1982, 2.2 for February 1983, 1.7 for February 1984, 2.4 for January 1985, and 2.7 for February 1986. The infiltration capacity, or the equilibrium infiltration rate, was two to three times more in no till compared with the plow till method of seedbed preparation. Further, cumulative infiltration in no till treatment was more when crop residue

Figure 17.3. Earthworm activity was high in no till plots with residue mulch.

Figure 17.4. There was a considerable concentration of gravels on the surface layer of the plow till treatment.

was returned as mulch than when it was removed. The beneficial effect of mulch on cumulative infiltration was more in no till than plow till methods.

Significantly more infiltration in no till compared with the plow till treatment was due to differences in soil structure in relation to the continuity, size, and relative proportion of the macropores. Soil managed with no till, especially in the absence of any tractor traffic because all farm operations were performed manually, had more earthworm activity (Fig. 17.3) and higher proportion of biopores (worm channels) than plowed soil. Worm channels are large, stable, continuous, and very effective in conducting water from surface into the subsoil. It was because of the high infiltration rate that there was a little or no runoff from no till plots, and there was minimal soil erosion. Even if there were a small runoff, there was little or no soil splash or soil erosion because of the high strength of soil aggregates and the protective effect of crop residue mulch (Lal, 1989). High runoff and erosion caused visibly more gravels on the surface of plow till plots (Fig. 17.4).

Soil Moisture Retention Characteristics

High Energy Soil Moisture Retention

The data on soil moisture retention characteristics for 0–5 cm and 5–10 cm depths are shown in Tables 17.5 and 17.6, respectively. The data on soil moisture retention at low suctions, determined on the tension table, show that soil moisture retention was consistently more for no till compared with the plow till soil. The ratio of soil moisture retention was consistently more for no till compared with the plow till soil. The ratio of soil moisture retention at 0 suction (saturation point) for no till:plow till treatment was 1.365 for 1981, 1.188 for 1982, 1.390 for 1983, 1.314 for 1984, and 1.328 for 1985 (Table 17.5). Therefore, the saturation percent was 19% to 40% more in no till than plow till treatment. Similarly, soil moisture retention at 30 cm suction (0.003 MPa) was consistently more for no till than plow till treatment, and the difference in favor of the no till treatment was about 20%. A significant point of interest is the volume of pores drained at 30 cm suction, which was drastically more in no till compared with the plow till treatment (Table 17.5). The higher drainable porosity in no till soil was in accord with the favorable soil structure and more cumulative infiltration in the no till treatment.

The data on high energy soil moisture retention for 5–10 cm depth are shown in Table 17.6. Trends in soil moisture retention with regard to the tillage methods were similar to that of the 0–5 cm depth shown in Table 17.5. Soil moisture retention at 0 suction (the saturation point) was more for no till than the plow till treatment for all years of observation. The ratio of saturation point for no till:plow till treatment was 1.22 for 1981, 1.16 for 1982, 1.21 for 1983, 1.19 for 1984, and 1.23 for 1987. There were no temporal trends in the saturation point except that the least value was observed for 1987 in both tillage treatments. Similar trends with regard to the tillage methods were observed for the 30 cm suction, except that differences due to tillage treatments were rather small. The ratio of soil moisture retention at 30 cm suction for no till: plow till treatments was 1.02 for 1981, 1.04 for 1982, 1.12 for 1983, 1.02 for 1984, and 1.10 for 1987 (Table 17.6). Residue mulch had no consistent effect on soil moisture retention for either of the tillage treatments.

Field Capacity

Tillage methods and soil management effects on soil moisture retention at field capacity are shown in Tables 17.7 and 17.8 for 0–5 cm and 5–10 cm depths, respectively. Field capacity for these soils is more likely to be in the vicinity of 0.01 MPa suction than at 0.03 MPa suction (Lal, 1979). Similar to the data of the high energy soil moisture retention, soil moisture retention for 0–5 cm depth at 0.01 and 0.05 MPa suctions was consistently more for no till compared with the plow till treatment (Table 17.7). The ratio of soil moisture content at 0.01 MPa suction for no till:plow till treatment was 1.08 for 1981, 1.30 for 1982, 1.38 for 1983, 1.46 for 1984, and 1.51 for 1987. Similar ratio for soil moisture content at 0.05 MPa suction was 1.23 for 1981, 1.23 for 1982, 1.30 for 1983, 1.20 for 1984, and 1.31 for 1987.

Table 17.5. Tillage and Management Effects on Soil Moisture Retention at Low Suctions (0–5 cm Depths)

		0 suction					0.003 MPa suction				
Tillage	**Treatment**	**1981**	**1982**	**1983**	**1984**	**1987**	**1981**	**1982**	**1983**	**1984**	**1987**
		% by weight									
No till	M_0F_0	43.1	40.0	42.6	38.0	35.4	19.5	28.9	20.4	22.8	21.5
	M_0F_1	35.9	34.1	39.0	39.7	36.7	22.0	25.3	21.4	27.5	19.8
	M_1F_0	40.9	35.7	40.1	38.0	43.3	19.9	24.2	18.5	20.7	22.2
	M_1F_1	40.2	39.2	42.6	39.9	39.7	19.7	26.8	17.1	19.8	21.1
	Mean	40.0	37.3	41.0	38.9	38.8	20.3	26.3	19.4	22.7	21.2
Plow till	M_0F_0	28.9	31.7	31.0	28.9	28.6	18.0	24.5	19.7	18.0	18.7
	M_0F_1	28.6	30.5	25.6	27.3	26.2	14.4	19.0	14.5	18.1	16.2
	M_1F_0	29.5	32.2	30.5	30.9	29.9	16.0	19.7	16.9	19.2	18.9
	M_1F_1	30.1	31.1	30.9	31.2	31.9	18.3	22.9	18.3	21.4	19.8
	Mean	29.3	31.4	29.5	29.6	29.2	16.7	21.5	17.4	19.2	18.4
LSD (.05)											
Year (Y)		1.2					2.3				
Tillage (T)		1.9					2.2				
Mulch (M)		1.0					1.3				
Fertilizer (F)		1.7					2.0				
M × Y		1.5					1.8				
T × Y		2.7					3.1				
T × M		1.5					1.8				
T × F		2.3					2.8				

F_0 = no fertilizer, F_1 = recommended fertilizer rate, M_0 = crop residue removed, M_1 = crop residue retrieved as surface mulch.

Table 17.6. Tillage and Management Effects on Soil Moisture Retention at Low Suctions (5–10 cm Depth)

		0 suction					0.003 MPa suction				
Tillage	**Treatment**	**1981**	**1982**	**1983**	**1984**	**1987**	**1981**	**1982**	**1983**	**1984**	**1987**
		% by weight									
No till	M_0F_0	36.1	37.7	35.3	37.1	32.7	17.6	18.6	18.6	21.9	18.5
	M_0F_1	32.6	32.4	32.2	34.4	33.8	19.8	21.8	19.1	20.5	19.0
	M_1F_0	33.9	33.2	34.0	36.7	33.6	20.1	22.8	21.7	24.0	22.3
	M_1F_1	33.9	33.3	34.3	33.2	33.5	14.8	23.9	20.0	16.0	19.8
	Mean	34.1	34.2	34.0	35.4	33.4	18.1	21.8	19.9	20.6	19.9
Plow till	M_0F_0	29.0	29.3	27.8	32.0	28.7	18.5	23.3	19.9	22.7	19.0
	M_0F_1	27.3	28.3	29.2	30.2	27.5	16.5	19.2	16.2	17.6	15.9
	M_1F_0	26.9	28.5	26.6	26.7	24.8	16.2	18.7	16.9	18.9	17.6
	M_1F_1	28.6	31.4	29.0	29.9	27.9	20.4	22.7	17.8	21.2	19.7
	Mean	28.0	29.4	28.2	29.7	27.2	17.9	21.0	17.7	20.1	18.1
LSD (.05)											
Year (Y)		0.5					2.6				
Tillage (T)		2.2					2.9				
Mulch (M)		0.9					1.3				
Fertilizer (F)		1.4					2.0				
Y × M		1.3					1.8				
Y × T		3.1					4.2				
T × M		1.3					1.8				
T × F		2.0					2.8				

F_0 = no fertilizer, F_1 = recommended fertilizer rate, M_0 = crop residue removed, M_1 = crop residue retrieved as surface mulch.

Table 17.7. Tillage and Management Effects on Soil Moisture Retention at Medium Suctions (0–5 cm Depth)

		0 suction					0.003 MPa suction				
Tillage	Treatment	1981	1982	1983	1984	1987	1981	1982	1983	1984	1987
		% by weight									
No till	M_0F_0	12.5	11.5	14.9	17.8	12.5	8.7	8.4	10.1	8.0	8.1
	M_0F_1	12.9	10.9	15.5	18.4	13.7	8.9	7.8	9.9	9.3	8.2
	M_1F_0	14.7	11.2	12.8	16.2	18.1	9.2	7.7	9.9	8.6	13.2
	M_1F_1	12.9	11.1	14.3	16.5	12.3	8.0	7.5	9.7	7.5	8.2
	Mean	13.3	11.2	14.4	17.2	14.2	8.7	7.9	9.9	8.4	9.4
Plow till	M_0F_0	12.3	8.9	9.8	11.6	8.9	7.6	6.8	7.7	7.8	6.7
	M_0F_1	9.9	7.4	9.9	10.7	8.0	5.6	5.3	6.5	5.0	6.5
	M_1F_0	13.5	8.1	9.0	11.4	9.1	6.5	6.5	7.6	7.1	6.9
	M_1F_1	13.6	9.9	12.8	13.5	11.5	8.8	6.8	8.5	8.0	8.8
	Mean	12.3	8.6	10.4	11.8	9.4	7.1	6.4	7.6	7.0	7.2
LSD (0-5)											
Year (Y)		0.8					1.2				
Tillage (T)		1.5					0.9				
Mulch (M)		0.7					0.5				
Fertilizer (F)		1.1					0.8				
M × Y		0.7					0.7				
T × Y		2.1					1.3				
T × M		0.9					0.7				
T × F		1.5					1.1				

F_0 = no fertilizer, F_1 = recommended fertilizer rate, M_0 = crop residue removed, M_1 = crop residue retrieved as surface mulch.

Table 17.8. Tillage and Management Effects on Soil Moisture Retention at 0.01 and 0.05 MPa Suctions (5–10 cm Depth)

		0.01 MPa suction					0.05 MPa suction				
Tillage	Treatment	1981	1982	1983	1984	1987	1981	1982	1983	1984	1987
		% by weight									
No till	M_0F_0	10.7	11.9	13.8	16.6	12.8	6.7	8.2	8.4	7.9	7.9
	M_0F_1	11.5	10.4	12.3	15.6	11.2	7.7	7.1	8.2	8.1	7.9
	M_1F_0	10.7	10.7	13.6	17.1	10.6	7.4	7.2	9.1	8.4	7.0
	M_1F_1	13.5	9.4	12.7	13.2	10.6	7.1	6.8	8.4	6.9	7.4
	Mean	11.6	10.6	13.1	15.6	11.3	7.2	7.3	8.5	7.8	7.6
Plow till	M_0F_0	14.5	12.9	10.1	12.3	9.0	7.8	7.2	8.1	8.0	7.2
	M_0F_1	12.4	8.4	9.1	11.0	9.3	6.8	6.6	7.3	7.6	6.9
	M_1F_0	.8	7.7	9.7	10.6	7.9	6.6	5.7	7.0	6.0	6.0
	M_1F_1	14.8	9.4	12.0	14.2	10.4	9.0	7.6	8.4	7.6	8.7
	Mean	12.9	9.6	10.2	12.0	9.2	7.6	6.8	7.7	7.3	7.2
LSD (.05)											
Year (Y)		2.8					1.2				
Tillage (T)		2.0					1.0				
Mulch (M)		0.8					0.4				
Fertilizer (F)		1.3					0.7				
M × Y		1.2					0.6				
T × Y		2.8					1.4				
T × M		1.2					0.6				
T × F		1.9					0.9				

F_0 = no fertilizer, F_1 = recommended fertilizer rate, M_0 = crop residue removed, M_1 = crop residue retrieved as surface mulch.

The data in Table 17.8 for 5–10 cm depth did not observe a consistent trend with regard to the tillage methods. For four out of five years, however, moisture contents at 0.01 and 0.05 MPa suctions were more for no till compared with plow till treatments. The overall mean moisture retention, average of all years, for 0.01 MPa suction was 12.4% for no till compared with 10.8% for plow till treatments. Similarly, the overall mean moisture content for 0.05 MPa suction was 7.7% for no till compared with 7.3% for plow till treatment (Table 17.8).

Low Energy Moisture Retention and the Permanent Wilting Point

Just as the field capacity for these soils is at 0.006 to 0.01 MPa suction, the permanent wilting point for these soils may also be at 0.1 to 0.5 MPa suction. In fact, the soil moisture characteristics show little, if any, difference in moisture retention at 0.1 and 1.5 MPa suctions.

The data in Table 17.9 for moisture retention at 0.1 MPa and 0.5 MPa suctions consistently show higher moisture content for no till compared with the plow till treatment. The overall mean moisture content at 0.1 MPa suction was 7.8% for no till compared with 5.9% for plow till, with a ratio of 1.32. Similarly, the overall mean moisture content at 0.5 MPa suction was 6.8% for no till compared with 5.2% for plow till treatment, with a ratio of 1.31 (Table 17.9). The data in Table 17.10 for 5–10 cm depth followed a trend similar to those for 0–5 cm depth. Soil from the no till treatments consistently retained more moisture at 0.1 MPa and 0.5 MPa suctions than that from the plow till treatment. The overall mean moisture content for 0.1 MPa suction was 6.7% for no till compared with 6.0% for plow till, with a ratio of 1.12. The overall mean moisture content for 0.5 MPa suction was 5.8% for no till compared with 5.4% for plow till, with a ratio of 1.07 (Table 17.10).

Tillage method effects on soil moisture content at 1.5 MPa suction are shown in Table 17.11. Soil moisture content at 1.5 MPa suction was consistently more for no till compared with the plow till treatment for both depths. The overall mean moisture content for 0–5 cm depth was 5.9% for no till compared with 4.6% for plow till, with a ratio of 1.28. Similarly, the mean moisture content at 1.5 MPa suction for 5–10 cm depth was 5.2% for no till compared with 4.6% for plow till, with a ratio of 1.13.

GENERAL DISCUSSION AND CONCLUSIONS

Soil physical quality was superior in no till compared with the plow till method of seed bed preparation, because for this soil it is a function of infiltration capacity, plant available water reserve, and the soil structure (Eq. 1).

$$S_q \text{ (physical)} = f(\text{AWC}, i_c, \text{WSA}) \tag{1}$$

where S_q refers to soil quality, AWC to available water capacity, i_c to the equilibrium infiltration rate, and WSA to water stable aggregation.

The AWC in no till vs. plow till treatment is shown by the data in Table 17.12. The AWC was calculated as the difference in moisture content at 0.01 MPa and 1.5 MPa suctions (Eq. 2).

$$\text{AWC} = (W_{0.1\text{MPa}} - W_{1.5\text{MPa}}) \tag{2}$$

where W refers to the gravimetric moisture content. Ideally, the gravimetric moisture content should be converted to the volumetric moisture content by multiplication with

Table 17.9. Tillage and Management Effects on Soil Moisture Retention at High Suctions (0-5 cm Depth)

Tillage	Treatment	0.1 MPa suction					0.5 MPa suction				
		1981	1982	1983	1984	1987	1981	1982	1983	1984	1987
		% by weight									
No till	M_0F_0	8.6	7.4	7.5	7.8	7.4	7.4	6.6	8.0	6.5	6.2
	M_0F_1	7.8	6.9	7.2	8.3	7.5	6.7	5.7	7.6	7.3	6.2
	M_1F_0	8.6	6.6	7.9	7.9	10.7	6.8	5.9	7.6	6.4	6.2
	M_1F_1	8.1	6.8	7.6	7.2	7.6	6.8	5.7	8.3	6.1	6.9
	Mean	8.3	6.9	7.6	7.8	8.3	6.9	6.0	7.9	6.6	6.4
Plow till	M_0F_0	5.9	6.1	6.1	5.9	5.8	5.6	5.9	6.1	4.5	4.6
	M_0F_1	5.4	5.0	4.7	4.5	5.8	4.7	4.7	4.7	4.0	4.5
	M_1F_0	6.1	5.9	5.5	5.9	6.0	5.1	5.0	6.0	4.5	4.9
	M_1F_1	6.8	6.3	6.3	6.6	7.7	6.0	5.7	5.9	5.4	6.7
	Mean	6.1	5.8	5.7	5.7	6.3	5.4	5.3	5.7	4.6	5.2
LSD (.05)											
Year (Y)		0.8					0.2				
Tillage (T)		0.6					0.8				
Mulch (M)		0.4					0.4				
Fertilizer (F)		0.7					0.6				
M × Y		0.6					0.5				
T × Y		0.9					1.1				
T × M		0.6					0.5				
T × F		0.9					0.9				

F_0 = no fertilizer, F_1 = recommended fertilizer rate, M_0 = crop residue removed, M_1 = crop residue retrieved as surface mulch.

Table 17.10. Tillage and Management Effects on Soil Moisture Retention at 0.1 and 0.5 MPa Suctions (5-10 cm Depth)

Tillage	Treatment	0.1 MPa suction					0.5 MPa suction				
		1981	1982	1983	1984	1987	1981	1982	1983	1984	1987
		% by weight									
No till	M_0F_0	6.7	7.3	6.5	7.7	6.9	5.8	6.0	7.4	6.3	5.6
	M_0F_1	7.1	6.1	6.1	6.8	6.9	5.9	5.4	6.0	5.9	5.6
	M_1F_0	6.6	6.1	7.0	7.6	6.6	5.7	4.6	6.7	6.7	5.4
	M_1F_1	6.4	6.1	6.4	5.9	6.7	5.5	5.4	5.9	4.9	5.3
	Mean	6.7	6.3	6.5	7.0	6.8	5.7	5.4	6.5	6.0	5.5
Plow till	M_0F_0	6.4	6.4	6.4	6.4	6.0	5.8	5.7	6.7	5.1	5.0
	M_0F_1	6.3	5.6	5.5	5.8	5.5	5.7	5.5	6.0	4.8	4.6
	M_1F_0	5.5	5.1	5.0	5.1	5.5	5.1	5.1	4.7	4.3	4.5
	M_1F_1	7.1	6.2	6.1	7.0	7.8	5.5	5.7	5.4	5.7	6.2
	Mean	6.3	5.8	5.8	6.1	6.2	5.5	5.5	5.7	5.0	5.1
LSD (.05)											
Year (Y)		0.7					0.7				
Tillage (T)		0.7					0.8				
Mulch (M)		0.3					0.3				
Fertilizer (F)		0.4					0.5				
M × Y		0.4					0.5				
T × Y		1.0					1.1				
T × M		0.4					0.5				
T × F		0.6					0.7				

F_0 = no fertilizer, F_1 = recommended fertilizer rate, M_0 = crop residue removed, M_1 = crop residue retrieved as surface mulch.

Table 17.11. Tillage and Management Effects on Soil Moisture Retention at 1.5 MP Suction

		0–5 cm depth					5–10 cm depth				
Tillage	**Treatment**	**1981**	**1982**	**1983**	**1984**	**1987**	**1981**	**1982**	**1983**	**1984**	**1987**
						% by weight					
No till	M_0F_0	7.1	6.5	5.3	5.2	5.4	5.5	5.9	5.6	4.5	5.0
	M_0F_1	7.1	5.4	6.2	5.1	7.2	5.9	5.0	5.0	4.6	4.8
	M_1F_0	7.0	5.3	5.7	5.7	4.9	5.9	4.5	6.7	4.9	5.0
	M_1F_1	7.1	5.3	5.2	4.6	5.8	5.8	4.7	5.9	3.8	4.6
	Mean	7.1	5.6	5.6	5.2	5.8	5.8	5.0	5.8	4.5	4.9
Plow till	M_0F_0	5.6	4.7	4.8	4.3	4.0	5.2	4.4	4.9	4.9	4.3
	M_0F_1	5.0	4.2	4.8	4.1	4.3	5.1	4.5	4.7	4.2	4.0
	M_1F_0	5.2	3.7	3.8	3.7	4.0	4.9	4.1	3.7	4.0	3.8
	M_1F_1	5.2	5.1	4.8	4.9	5.7	5.6	5.3	4.5	4.0	5.5
	Mean	5.3	4.4	4.6	4.3	4.5	5.2	4.6	4.5	4.3	4.4
LSD (.05)											
Year (Y)		0.1					0.8				
Tillage (T)		0.6					0.7				
Mulch (M)		0.3					0.3				
Fertilizer (F)		0.5					0.5				
M × Y		0.5					0.4				
T × Y		0.9					1.0				
T × M		0.5					0.4				
T × F		0.7					0.7				

F_0 = no fertilizer, F_1 = recommended fertilizer rate, M_0 = crop residue removed, M_1 = crop residue retrieved as surface mulch.

Table 17.12. The AWC for No Till Compared with the Plow Till Treatments

	0–5 cm depth					5–10 cm depth				
Tillage	**1981**	**1982**	**1983**	**1984**	**1987**	**1981**	**1982**	**1983**	**1984**	**1987**
					% by weight					
No till	6.2	5.6	8.8	12.0	8.4	5.8	5.6	7.3	11.1	6.4
Plow till	7.0	4.2	5.8	7.5	4.9	7.7	5.0	5.7	7.7	4.8

the specific gravity. The data in Table 17.12 show generally more AWC for no till compared with the plow till treatments. The overall mean AWC for 0–5 cm depth was 8.2% for no till compared with 5.9% for plow till treatment, with a ratio of 1.39. Similarly, the mean AWC for 5–10 cm depth was 7.2% for no till compared with 6.2% for plow till treatment, with a ratio of 1.16. Inevitably, therefore, crops in no till were less susceptible to drought than those under the plow till method of seedbed preparation.

Principal reasons of favorable soil structure under no till compared to plow till system are: (a) higher soil organic carbon content, (b) higher activity of earthworms and other soil fauna, and (c) more microbial biomass carbon. Less susceptibility to drought under no till compared to plow till method of seedbed preparation is due to: (a) low runoff, (b) low soil evaporation, (c) high soil moisture retention due to favorable pore size distribution, and (d) better root system development into the subsoil horizon.

A prerequisite for exploitation of the benefits of no till farming, however, is the development and validation of a package of cultural practices needed specifically for the no till system. This package is very soil and ecoregion specific, and varies among cropping/farming systems. The package includes specific provisions for crop residue mulch, weed control measures, appropriate seeding equipment, suitable techniques for fertilizer application, and effective and economic measures for pest control.

There is an urgent need to develop a soil guide to tillage methods. Development of such a guide requires establishment of long-term tillage experiments on principal soils of the major ecoregions of the tropics.

REFERENCES

Aina, P.O. Soil Changes Resulting from Long-Term Management Practices in Western Nigeria. *Soil Sci. Soc. Am. J.* 43: 272–277, 1979.

Dunham, R.J. Diverse Tillage Systems in Semi-Arid West Africa. Proc. Int'l Soil Tillage Res. Organization, 11th Conference, Edinburgh, Scotland, 1988, pp. 631–636.

FAO-UNESCO. *Soil Map of the World.* Vol. 6 Africa, UNESCO, Paris, 1977.

Gee, G.W. and Bauder, J.W. Particle-Size Analysis. In *Method of Soil Analysis,* A. Klute, Ed. Part 1, 2nd ed., Agronomy Monograph No. 9, ASA and SSSA, Madison, WI, 1986, pp. 383–411.

Klute, A. Water Retention: Laboratory Methods. In *Method of Soil Analysis,* A. Klute, Ed. Part 1, Physical and Mechanical Methods, 2nd ed., Soil Science Society of America, Vol. 9, Madison, WI, 1986.

Lal, R. No Till Effects on Soil Properties and Maize (*Zea mays*) Production in Western Nigeria. *Plant Soil* 40: 321–331, 1974.

Lal, R. No Tillage Effects on Soil Properties under Different Crops in Western Nigeria. *Soil Sci. Soc. Am. J.* 40: 726–768, 1976.

Lal, R. Influence of Six Years of No Tillage and Conventional Plowing on Fertilizer Response of Maize (*Zea mays*) on an Alfisol in the Tropics. *Soil Sci. Soc. Am. J.* 43: 399–403, 1978.

Lal, R. Soil Physical Properties: Measurements and Prediction. In *Soil Physical Properties and Crop Production in the Tropics,* R. Lal and D.J. Greenland, Eds. John Wiley & Sons, Chichester, 1979.

Lal, R. Effects of 10 Years of No Till and Conventional Plowing on Maize Yield and Properties of a Tropical Soil. Proc. IX ISTRO Conference, Osijek, Yugoslavia, 1982, pp. 111–117.

Lal, R. Mechanized Tillage Systems Effects on Soil Erosion from an Alfisol in Watershed Cropped to Maize. *Soil & Tillage Res.* 4: 349–360, 1984.

Lal, R. Mechanized Tillage Systems Effects on Properties of a Tropical Alfisol in Watershed Cropped to Maize. *Soil & Tillage Res.* 6: 149–161, 1985.

Lal, R. Conservation Tillage for Sustainable Agriculture. *Adv. Agron.* 42: 85–197, 1989.

Lal, R. Tillage Effects on Soil Degradation, Soil Resilience, Soil Quality and Sustainability. *Soil & Tillage Res.* 27: 1–7, 1993.

Lal, R. Tillage and Mulching Effects on Maize Yield for Seventeen Consecutive Seasons on a Tropical Alfisol. *J. Sustainable Agric.* 5: 79–93, 1995.

Lal, R, Deforestation and Land Use Effects on Soil Degradation and Rehabilitation in Western Nigeria. I. Soil Physical and Hydrological Properties. *Land Degradation & Development* 7: 19–45, 1996a.

Lal, R. Deforestation and Land Use Effects on Soil Degradation and Rehabilitation in Western Nigeria. II. Soil Chemical Properties. *Land Degradation & Development* 7: 87–98, 1996b.

Lal, R. Deforestation and Land Use Effects on Soil Degradation and Rehabilitation in Western Nigeria. III. Runoff, Soil Erosion and Nutrient Loss. *Land Degradation & Development* 7: 99–119, 1996c.

Lal, R., De Vleeschauwer, D. and Malafa Nganje, R. Changes in Properties of a Newly Cleared Tropical Alfisol as Affected by Mulching. *Soil Sci. Soc. Am. J.* 44: 827–837, 1980.

Roose, E.J. Use of USLE to Predict Erosion in West Africa. SWCS Spec. Publ. 21: 60–74, 1976.

Roose, E.J. Soil Conservation on Steeplands in Africa. In *Soil and Water Conservation on Steeplands,* W.C. Moldenhauer and N.W. Hudson, Eds. SWCS, Ankeny, IA, 1987.
Sanchez, P.A. and R.R.B. Leakey. Land Use Transformations in Africa: Three Determinants for Balancing Food Security with Natural Resource Utilization. ICRAF, Nairobi, Kenya, 1996.
Sanchez, P.A., A-M. Izac, I. Valencia and C. Pieri. Soil Fertility Replenishment in Africa: A Concept Note. In *Achieving Greater Impact from Research Investments in Africa*, S.A. Breth, Ed. ICRAF, Nairobi, Kenya, 200–207, 1996.
Smaling, E. An Agro-Ecologic Framework for Integrated Nutrient Management with Special Reference to Kenya. Ph.D. thesis, Agric. University, Wageningen, The Netherlands, 1993.
Soil Survey Staff. Soil Taxonomy: A Basic System of Soil Classification for Making and Interpreting Soil Surveys. Agricultural Handbook No. 436, SCS, USDA, Washington, D.C., 1975.
Steel, R.G. and Torrie, J.H. *Principles and Procedures of Statistics*, 2nd edition, McGraw Hill, New York, 1980.
Stoorvogel, J.J., E.M.A. Smaling and B.H. Janssen. Calculating Soil Nutrient Balances in Africa at Different Scales. *Fertilizer Res.* 35: 227–235, 1993.
Uvin, P. The Hunger Report: 1993. Brown University, USA, 1994.
Van der Pol, F. Soil Mining. An Unseen Contributor to Farm Income in Southern Mali. Royal Tropical Institute Bulletin 325, Royal Tropical Institute, Amsterdam, 1992.
World Bank. African Development Indicators 1996. World Bank, Washington, D.C., 1996.
Young, E.M. World Hunger. Routledge, London, 1997.

FIVE

Soil Quality Management in Tropical America

EIGHTEEN

Research Strategies for Sustainable Agriculture in Acid Savannas of Latin America

Raúl R. Vera

INTRODUCTION

The world's savannas cover approximately 20% of the planet's land surface. In South America, temperate and tropical savannas occupy 45% of the surface area; of this area, approximately 240 million hectares are covered by neotropical savannas distributed among Brazil (*Cerrados*), Colombia and Venezuela (*Llanos*), Bolivia (*Sabana*), and Guyana (*Intermediate savannhas*), as shown in Figure 18.1.

The majority of the soils underlying the neotropical savannas are old and highly weathered. Thus, 68% of the *Llanos* and 96% of the *Cerrados* are covered by Oxisols, and an additional 2–4% by Ultisols (Cochrane et al., 1985). Soil chemical characteristics are well known and have been amply discussed in the world literature (for detailed profile analyses see Cochrane et al., 1985). In essence, neotropical Oxisols have Al saturation levels ranging between 80 and 90%, and correspondingly very low levels of mono and divalent cations; similarly, they seldom exceed 3 ppm P. Physical properties of these soils in their natural state have generally been recognized as adequate, but recent progress in understanding physical processes and constraints in these soils are leading to a reevaluation of these older concepts (Amezquita, pers. comm.).

In this context, the objective of this chapter is to summarize the research strategies pursued by Centro Internacional de Agricultura Tropical's (CIAT's) Tropical Lowlands Program in the well-drained savannas of South America, and also to briefly note some of the advances made.

CIAT's Tropical Lowland Program resulted from the merging in late 1993 of the former Savannas and Forest Margins programs originally proposed in CIAT's Strategic Plan and implemented in mid-1992.

The Savannas initiative had its origin in the long presence of CIAT in the neotropical savannas through research carried out by the Rice Program, the former Tropical Pastures Program, and to a lesser extent the Cassava Program. Savanna land resources, as well as those of the rest of the South American tropical lowlands, had been extensively charac-

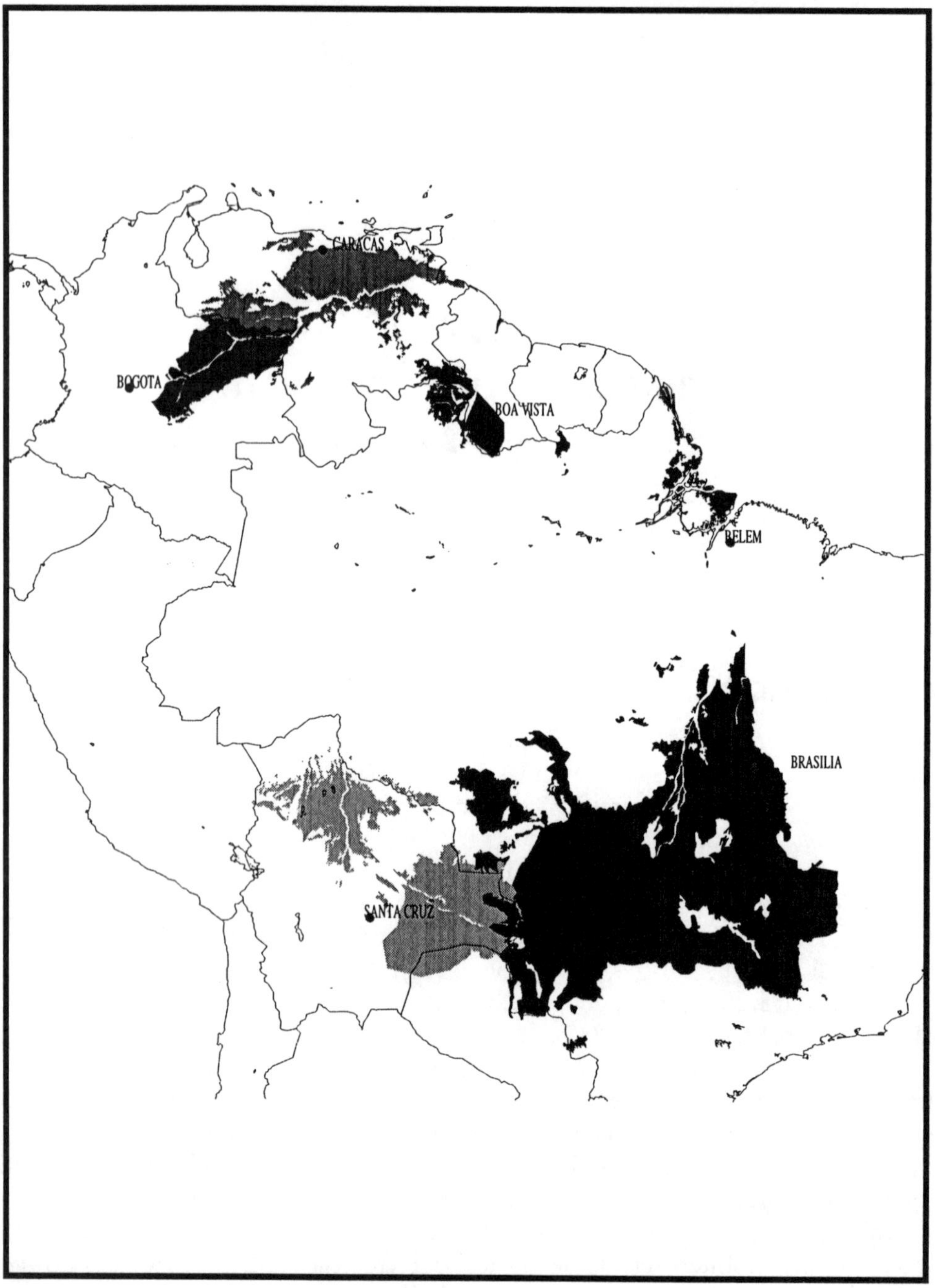

Figure 18.1. Savannas and grasslands of tropical South America. From Land Management Unit, CIAT.

terized from a biophysical point of view by work carried out by the former Agroecological Studies Unit (AES) in the late 1970s and early 1980s (Cochrane et al., 1985).

The stock of knowledge on the Forest Margins area was less, and was primarily based on research undertaken by the former Tropical Pastures Program in the Peruvian Amazon (Ucayali department), and to a lesser extent, the Napo department in Ecuador and Caqueta department in Colombia, all of which used intensively the studies developed by the AES mentioned above.

Merging research in those two agroecosystems was a pragmatic decision associated with financial constraints but it made sense strategically in that both agroecosystems share a number of geographic and biophysical characteristics, most notably acid soils of low fertility. Also there is spatial continuity between them. In effect, all of the savanna countries, namely Bolivia, Brazil, Colombia, Guyana, and Venezuela, are also Amazonian countries. This has implied that throughout much of the present century, both agroecosystems have interacted closely as a consequence of the movement of large segments of population along the savanna-humid forest transition zone. In countries such as Bolivia, Brazil, and Guyana this transitional area is currently subjected to rapid rural and urban development.

RATIONALE

In the view of some like Borlaug and Dowswell (1994), the neotropical savannas constitute the last significant agricultural frontier in the world, whereas for others it is a fragile and precious natural resource, rich in flora and fauna, and located in the basin of major American rivers. The savannas of tropical South America extend over 240 million hectares, and include the *Cerrado* (180-205 million ha), the *Llanos* of Colombia (17 million ha), and of Venezuela (28 million ha) and large areas of Bolivia (14 million ha) and Guyana. Biodiversity in the savannas has been less well studied than that of the Amazonian rain forest, and has certainly attracted much less attention in the world press and scientific literature. Nevertheless, and as an example, the Brazilian savannas or Cerrados include well over 160,000 plant species (approximately 5% of the known plant species worldwide). Similarly, recent studies in the gallery forests of the savannas have shown that the richness and diversity of the small fauna in them is as large or larger on a per hectare basis than that of the Amazon rain forest.

Over the last 40 years or so, the neotropical savannas have been settled and therefore have been heavily intervened by human action. This has led the Cerrados, for example, to become a major supplier of cereals, oilseeds and beef. In 1992, 40% of Brazil's rice and soybean acreage was located in the Cerrado, as well as 23% of its maize area. Similarly, 45% of Brazil's cattle herd was located in the region. Current estimates of sown grass pastures in the neotropical savannas amount to 4050 million hectares, while annual cropping varies between 10 and 12 million hectares per year. In some countries, sugarcane, tree plantations, and various other annual crops constitute major land uses.

Despite the large and continually increasing contribution of the neotropical savannas to agricultural production, food security, and income, it should be remembered that the Latin American population is highly urbanized, in contrast to that of Africa and Asia. This is particularly true in the case of the savanna countries; 75% of the population is urban, and there are areas in the savannas where cropping and cattle operations are particularly successful, where urbanization already exceeded 85% in the late 1980s. In these regions, the issue of the linkages between agriculture, agroindustry, and the rest of the economy in terms of employment generation and overall economic activity has not been investigated, although there is every reason to believe that these linkages are strong and have major multiplicative effects (see Trigo, 1990 for the issue of urbanization and agricultural intensification, and Schetjman, 1994 for linkages).

Land distribution in the neotropical savannas is highly skewed as elsewhere in Latin America. Nevertheless, more than 50% of the farms in the Cerrados are less than 50 hectares in size. Also, in Bolivia, parts of Colombia and of Venezuela, relatively large areas are home to a variety of indigenous peoples.

Although the situation is rapidly changing, inappropriate technology has been used in much of the above areas. Erosion, soil chemical and physical degradation, buildup of pests in monocropped areas, and others are common problems. The impact of land use on flora and fauna has been scarcely assessed, but constitutes a major problem area, as shown by recent studies conducted by the National University of Colombia under the auspices of CIAT (Mejia, 1995).

An aspect that has received very little attention is the fact that the cerrados of Bolivia and Brazil are the origin of all of the major rivers draining south to the River Plate (for example, the Paraná river can be traced back to an area near Brasilia). Similarly, the Bolivian and Brazilian Cerrados also give rise to major effluents of the Amazon river. Other savanna areas, such as those of the Colombia portion of the Guianas and Venezuela constitute the watershed of the Orinoco river. It is then obvious that land-use changes in the neotropical savannas are bound to have very major effects downstream. In fact, there is abundant anecdotal evidence, and also some quantitative estimates, of significant contamination and siltation of major rivers (e.g., the Paraná, San Francisco, and Orinoco rivers), as well as point estimates of unacceptably high levels of organochlorides and organophosphates in fish of some of these rivers.

The savannas of Bolivia, Colombia, and Venezuela are a major source of oil for these countries, with subsequent road and other infrastructure development closely associated with further and rapid settlement and agricultural intensification. The effects of road and infrastructure development on land-use change can be seen today most dramatically in the Bolivian portion of the Cerrados, about to be transversed by an interoceanic road that will link Sao Paulo in Brazil, with Iquique in Chile.

Rapid frontier expansion occurred also in the Forest Margins (FM) areas beginning in the 1960s. Between 1970 and 1985 the rate of deforestation in the Brazilian Amazon was estimated at 1.5 to 2.0 million ha per year. However, there is still considerable opportunity for influencing future land-use patterns, since only 6% of that area has been cleared. In the FM, land concentration is lower than in the savannas, but still high, with a Gini coefficient of 0.79 in 1985, with small farms coexisting with large landlords. Shifting cultivation by smallholders was estimated to account for 32% of deforestation in 1980. There are large differences in land-use patterns in cleared areas between countries and between regions within countries. As an example, 29% of the cleared areas was dedicated to agriculture and 57% to pasture in the states of Acre and Rondonia, Brazil, in 1985. The consequences of this type of intervention in the tropical humid forests have been amply discussed in the literature and do not need further elaboration here, but it is important to reiterate that many of these settlement areas are geographically and politically contiguous with savanna lands, and share many of the same soils and biophysical constraints.

RESEARCH STRATEGY IN THE NEOTROPICAL SAVANNAS

The overall goal of the program is ***to develop and test a diverse set of sustainable land use forms for the acid-soil savannas***. Ideally, these alternative land uses will combine increased productivity with preservation and even enhancement of the resource base, and with improved equity.

It is of utmost importance that diversification of land use in terms of native vegetation, annual crops, perennial crops and plantations, and livestock be considered as a major research thrust, among other reasons because of the increasing need of the region to become competitive in world markets and create for itself and for its products, unique market niches. CIAT and the CGIAR system have promoted a series of consultation meetings with policy makers of the region. These meetings, the last of which was held in Colombia in 1995, have consistently and repeatedly concluded with the plea that the research system should focus its efforts on how to increase agricultural competitiveness. This should not be understood as synonymous with the ability to compete in traditional crops such as corn or soybeans, though these are essential components of today's savanna farming systems, but rather, with the need to develop and adapt nontraditional tropical commodities also.

It is realized, of course, that this ideal combination of diversified and productive commodities with resource preservation and equity may not be attainable, or only partially so. If that is indeed the case, tradeoffs have to be identified and quantified, and alternative development paths or scenarios must be constructed, so that policy makers, and society, will be able to make decisions on an informed and scientific basis.

The strategies to achieve that goal revolve around three major interrelated project areas:

- The study of the dynamics and trends of land-use patterns, leading to: (a) biophysical and socioeconomic characterization of both agroecosystems; (b) identification of representative sites and extrapolation domains; (c) identification of causes and consequences of land-use change, and (d) modeling alternative development scenarios;
- Understanding of the biophysical and socioeconomic processes that affect resource management, following a holistic approach at several spatial scales (Fig. 18.2), depending upon the specific processes under study;
- Development and testing of alternative prototypes of agropastoral and agro-silvo-pastoral systems, based on the understanding developed above.

It was recognized from the outset that these objectives are too ambitious for a single program and even for CIAT as a whole. Therefore, implementation of research projects in each of the above areas has implied close collaboration with a wide range of government and nongovernment institutions, listed in various CIAT documents (CIAT, 1994; 1995).

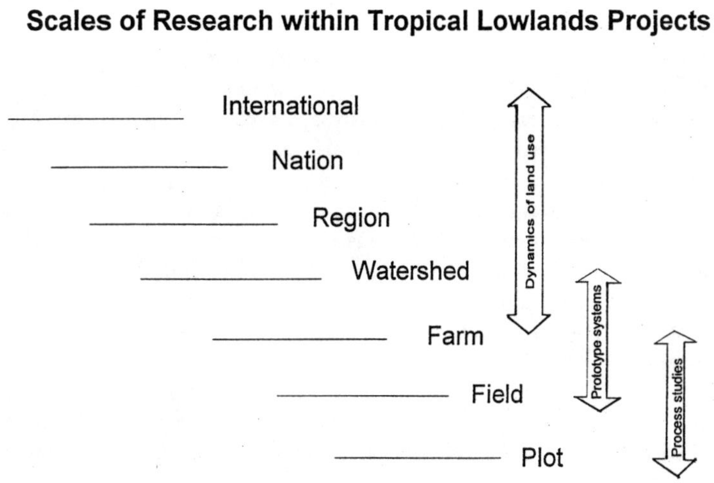

Figure 18.2. CIAT's Tropical Lowlands Program research projects in relation to scale of analysis and aggregation.

Figure 18.2 shows how our three main projects, presented below, target a relatively narrow range of scales, and how by overlapping, they attempt to cover a range that allows analysis and extrapolation across the spectrum of scales.

Dynamics of Land Use

The overall purpose is to provide the basis for developing technologies and policies that contribute to sustainable agriculture by studying the dynamic processes underlying patterns of land use. The project characterizes the determinants and consequences of land-use change in representative sites of savannas and forest margins via the study of secondary data, participatory field studies, analyses supported by geographical information systems, and modeling at the farm and landscape levels. In the final analysis, it aims to assess the consequences of past and present policies and derive technological recommendations and policy options.

The program has ongoing studies of land use across a range of sites in the tropical lowlands which are examining trends and consequences that may affect technology generation and adoption, potential policy interventions and their interactions, as well as possible tradeoffs incurred in terms of productivity, equity, and resource conservation. The range of sites includes:

a. extensive and semi-intensive pastoral lands in the savannas of Colombia;
b. intensive cropland and mixed farming in the Brazilian Cerrados, a transition area between the Brazilian Cerrados and the Amazon basin, and
c. the benchmark site of the Alternatives to Slash and Burn Global Program located between the states of Acre and Rondonia in Brazil.

Important highlights to date include:

a. The detailed characterization of approximately 60% of the Cerrados areas for which there is sufficient and consistent secondary information, based on over 30 biophysical and socioeconomic variables (Jones et al., 1992), complemented by a further socioeconomic analysis carried out by a contracted NGO. These studies were led by CIAT's Land Management Unit, and it involved a CIAT interprogram working group and Brazilian researchers from several different institutions.

   It led to identification of high priority representative sites in the Cerrados. Among these, CIAT and EMBRAPA-CPAC chose Uberlandia to begin initiation of field work, pending further, and ongoing, detailed analyses based on Rapid Rural Appraisals, field surveys and interviews with qualified informants, and collection and analyses of additional secondary data.
b. Detailed Rapid Rural Appraisals carried out in several candidate areas for research on the Brazilian forest margins area, namely, areas in the States of Para, Maranhao, Acre, and Rondonia, and in savannas areas of the Cerrados and Colombia's Llanos. These studies were carried out by interdisciplinary and interinstitutional teams, and were supported by analyses of secondary data, satellite images, and remote sensing.

   For example, it led to the selection of the stretch of land extending between Rio Branco, Acre, and Theobroma, Rondonia, as the initial site for research on the forest margins by both the TLP and the Alternatives to Slash and Burn Global Project (Avila, 1994), and to the selection of Uberlandia in the Brazilian cerrados, as our main benchmark site.
c. Highly advanced studies on the dynamics of deforestation and its underlying causes, and more generally on the dynamics of land use in the Acre-Rondonia site and more recently,

Table 18.1. Adoption of Improved Pastures in the Colombian llanos[a]

	1979	1995
Improved grass pasture, % of the area	2	20
Animal units/ha	0.09	0.28
Beef production (kg/ha)	16	50
Potential herd size (000 heads) on well-drained savannas	959	1726

[a]From Tropical Lowlands Program Annual Report, 1995.

a comparable site in the area of influence of Pucallpa, Perú. A comparable study in areas of the Bolivian Cerrados, still relatively untouched, would be highly desirable.

d. Adoption studies (Table 18.1) initially conducted in the savannas of Colombia have identified both opportunities and constraints to the adoption of grass-legume leys in relation to a number of variables of the socioeconomic context, thus contributing to better focusing research efforts in this area.

It should be noted that crops, and to a much lesser extent, cattle production, are already fairly intensive in much of the neotropical savannas. From a purely production and economic point of view, the adaptation of soybeans to the Cerrados has been a major success achieved by research conducted largely by the public sector of Brazil. During the last five years, 10 million hectares were sown yearly to soybeans. In fact, this crop originally introduced into the southern region of Brazil, rapidly expanded north into the Cerrados, and more recently south and west into the pampas of Argentina, the Paraguayan Chaco and the Bolivian savannas. Current estimates of areas sown yearly to soybeans in that group of countries average 20 million hectares, with a trend to continued horizontal expansion. As with any monocrop, problems of soil management rapidly arose, thus requiring second generation research, such as that exemplified by the following two projects.

Mechanistic Understanding of Soil-Plant-Animal Processes

The project investigates the biophysical processes that affect resource management, develops and uses land quality indicators, and analyzes the tradeoffs incurred within existing and alternative systems. The latter is achieved within the context of the findings from the land-use dynamics and prototype systems projects.

The aim of this project is to improve the biophysical efficiency of agricultural production such that the efficiency of use of both internal and external inputs is maximized, and environmental effects are minimized. Ensuring maximum efficiency requires quantitative knowledge of underlying processes such as nutrient cycling, and changes in physical and biological soil properties. The project therefore seeks to quantify soil and soil/plant processes associated with changes in primary biomass productivity in contrasting land-use systems that extend from no external input native savanna, to intensive crop systems supported by medium levels of external inputs. This project interacts most closely with the *Prototype Sustainable Cropping Systems* project described below.

Currently, the project is implemented via two major long-term experiments established in the Cerrados of Brazil and savannas of Colombia, respectively. They involve a variety of land-use systems ranging from grazed native vegetation to input-intensive cropping systems, and constitute an excellent platform for researchers and institutions interested in assessing specific variables. For example, the Cerrados experiment, and shortly also our

Colombian trial, is being used by Cornell University researchers to measure gas fluxes as related to land-use systems. Similarly, French researchers are monitoring the dynamics of soil meso fauna in an attempt to learn how to manage it and enhance soil biological activity.

Important highlights of this project to date include:

a. The finding that some deep-rooted introduced forage grasses can contribute substantial amounts of organic matter to the soil, particularly when associated with productive, well-adapted legumes (Fisher et al., 1995). Although many questions about this phenomenon remain unanswered, it does support the hypothesis that agropastoral systems which involve planned rotations of annual crops and perennial forage species contribute to soil enhancement and may have positive effects in terms of global warming.
b. A forage phase in the crop sequence has been shown to lead to increased soil biological activity, increased soil fauna biomass and maintenance of its species diversity, increased VMA infection potential, and beneficial effects on nutrient partitioning in soil organic matter.
c. Much-improved and sensitive methods for soil organic matter and P partitioning have been tested and adjusted for acid soils. These methods, though more time-consuming and expensive than traditional ones, provide a much better understanding of P dynamics and appear to have more predictive power.
d. A clearer and more quantitative understanding of N cycling in acid tropical soils, generally low in organic matter, has begun to emerge. The relative roles of N cycling via feces, urine, and litter has been quantified and a simple N cycling model with considerable predictive power has been developed.

As implied above, these research activities are long-term and are therefore expected to provide more and significant outputs in the near future such as models of nutrient cycling and of physical and biological soil degradation and rehabilitation. These models, and the improved understanding that they represent, will constitute the basis for the design of sound prototype land-use systems.

Prototype Sustainable Cropping Systems

In the present context, cropping systems include both pastoral and perennial plantations components as well as annual crops. Numerous alternative cropping systems are practiced by farmers throughout the lowland tropics; these systems are under rapid evolution in response to changing economic and environmental conditions, and rapidly changing technologies. As consequence of this diversity, only a carefully selected small number of very contrasting *prototypes* can be examined in any detail.

The project seeks to build on existing farm-based systems, in order to generate technologies, land management strategies, and policy options for the sustainable agricultural development of the target regions. Research activities and selected farming systems are based on the identification of the major problems, opportunities, and niches identified in the land-use dynamics projects, and are supported by the understanding of soil-plant-animal processes developed in the respective project.

The project includes biophysical and socioeconomic monitoring of extant land-use systems, and on farm participatory evaluation of alternative systems. These types of activities have been implemented with various degrees of intensity in three sites: the Uberlandia region in the Brazilian Cerrados representing input- and management-intense agropastoral systems; the area of influence of Puerto López in the Colombian Llanos representing more extensive, cattle-based land-use systems; the colonization projects of

Pedro Peixoto (Acre) in the Brazilian Amazon, and the area near Pucallpa in the Peruvian Amazon representing different situations of small settlers practicing shifting agriculture.

Some of the highlights are as follows:

a. A reliable method that combines use of SPOT images and ground measurements has been developed and tested in the Colombian Llanos to assess the extent and intensity of degradation of native savanna plant communities in response to various management practices. Long-term monitoring of fixed points is providing quantitative evidence on the relative resilience of different plant communities, and may lead to identify 'keystone" species as well.
b. The yield and economic advantages of the (low external input) rice-pasture prototype system has been amply demonstrated. Soil enhancement under this system has also been shown (Table 18.2).

   Nevertheless, the precise niche(s) for this type of system still needs to be defined. Derived prototypes using new acid soil tolerant varieties of maize and soybeans are being tested. The inclusion of systems based on tropical fruit plantations is being planned in cooperation with the Colombian national research organization, CORPOICA.
c. The positive effects in terms of land use and food production of integrating crops and cattle production have been documented for a small number of on-farm case studies. Nevertheless, it is clear that these systems are management-intensive.

This type of research project builds on earlier, and continuing, research on crop and forages germ plasm introduction and evaluation. For example, the systematic introduction since the early 1950s, intensified in the 1970s and 1980s, of grasses of African origin into tropical America has been a success comparable to that of soybeans mentioned above. These grasses, such as *Brachiaria* spp., *Andropogon gayanus, Panicum maximum*, and a few others, currently occupy about 50 million hectares, including approximately 40 million hectares in the savannas biome. Their introduction and widespread adoption has generally led to multiplying beef production per hectare by a factor of between 10 and 20 times

Table 18.2. Beneficial Effects of Grass/Legume Pastures in Agropastoral Systems[a]

Parameter	Benefit	Reference
Animal production		
Animal live weight gain	++	Vera et al., 1992; 1994
Reduction of age at first calving	+	Vera, 1991; Vera et al., 1993
Reduction in calving interval	+	Vera et al., 1993
Milk yields	+	Lascano and Avila, 1991; Ullrich, et al., 1994
Crop Yield		
Rice after pasture	++	Thomas et al., 1996
Soil Improvement		
Total soil carbon	+	Fisher et al., 1994
Soil carbon in sand fractions	++	Guggenberger et al., 1995
Potential N mineralization	+	Rao et al., 1994
Wet aggregate stability	+	Gijsman and Thomas, 1996
Water infiltration rate	+	Gijsman and Thomas, 1996
Microbial-P	+	CIAT, 1995
Nutrient cycling via litter	+	Thomas and Asakawa, 1993; Guggenberger et al., 1995
Earthworm populations	++	Decaens et al., 1994

[a]Modified from Thomas et al., 1996.

+ = an increase of < 100% compared with grass-only or native savanna pastures.

++ = an increase of > 100% compared with grass-only or native savanna pastures.

that of native vegetation. It is hypothesized that they have contributed to the maintenance, and even slow increase, of the rate of beef supply in the region despite cattle being increasingly displaced to poorer soils and more marginal conditions. Nevertheless, as is true for the case of monocropped soybeans, this development has created its own problems such as the widely perceived but poorly documented process of "pasture degradation," and an increase in the incidence of some pests such as the *spittlebug* in some of the *Brachiaria* species. It is in this setting that projects such as that described above, that attempt to deal with issues such as crop-pasture integration, pasture and savanna integration and biodiversity, and landscape diversity, can contribute to maintain increased levels of productivity in an environmentally benign way.

Institutional Relations

The program's strategies and research activities are only a small part of the set of actions required to achieve sustainable development of the tropical lowlands. The program is therefore actively developing collaborative studies with a diverse suite of national, international, and regional R&D institutions. Formal projects and networks (i.e., the Agropastoral Research Network, MAS, etc.) as well as numerous informal linkages have been, and continue to be established. Only a few major partners can be listed here, including CIAT-Sta. Cruz, CORPOICA, EMBRAPA, and FONAIAP as *national research institutions* in Bolivia, Colombia, Brazil, and Venezuela respectively; *universities* such as National University, U. Javeriana and Technological University of the Llanos in Colombia, UNELLEZ and U. San Bolivar in Venezuela, the Federal University of Uberlandia in Brazil, Hohenheim and Gottingen universities in Germany, Colorado SU, Cornell, and Ohio SU in the US and U.Complutense in Spain; various NGOs such as several cooperatives in the Cerrados, farmers and cattlemen associations in Colombia and Venezuela, ISPN in Brazil; international institutions like IFDC, TSBF and CIRAD; regional consortia like PROCITROPICOS and international consortia like ASB.

REFERENCES

Avila, M. , compiler. *Report of a Research-Site Selection in Acre and Rondonia States of Amazon Region, Brazil.* ICRAF, Nairobi, Kenya, 1994.

Borlaug, N. E. and C. R. Dowswell. *Feeding a Human Population that Increasingly Crowds a Fragile Planet.* Keynote lecture, 15th World Congress of Soil Science, Acapulco, Mexico, 1994, pp. 1–15.

CIAT. *Tropical Lowlands Program, Annual Report.* CIAT, Cali, Colombia, 1994.

CIAT. *External Program and Management Review*, October 2–9, 1994. CIAT, Cali, Colombia, 1994.

CIAT. *Tropical Lowlands Program, Annual Report.* CIAT, Cali, Colombia, 1995.

Cochrane, T. T., L. G. Sánchez, L. G. de Azevedo, J. A. Porras and C. L. Garver. *Land in Tropical America,* 3 volumes, CIAT-EMBRAPA-CPAC, Cali, Colombia, 1985.

Decaens, T., P. Lavelle, J. J. Jimenez Jaen, G. Escobar and G. Rippstein. 1994. Impact of Land Management on Soil Macrofauna in the Oriental Llanos of Colombia. *Eur. J. Soil Biol.* 30: 157–168, 1994.

Fisher, M. J. , L. M. Rao, R. J. Thomas, M. A. Ayarza, C. E. Lascano, J. L. Sanz and R. R. Vera. Carbon Storage Deep in the Soil by Introduced Pastures in the South American Savannas. *Nature* 371 (15 September 1994): 236–238, 1994.

Guggenberger, G., W. Zech and R. J. Thomas. Lignin and Carbohydrate Alteration in Particle-Size Separates of an Oxisol under Tropical Pastures Following Native Savanna. *Soil Biol. Biochem.* 27: 1629–1638, 1995.

Gijsman, A. G., Thomas, R. J. Evaluation of some physical properties of an oxisol after conversion of native savanna into legume-based or pure grass pastures. *Tropical Grasslands* 30: 237–248, 1996.

Jones, P. G., M. Rincón and L. A. Clavijo. Area Classification and Mapping for the Cerrados Region of Brazil, second draft. CIAT, Cali, Colombia, 1992.

Lascano, C. E. and P. Avila. Potencial de producción de leche en pasturas solas y asociadas con leguminosas adaptadas a suelos ácidos. *Pasturas Tropicales* 13: 2–10.

Mejía, M. A. Efecto del déficit hídrico sobre la densidad poblacional de la fauna mayor en un segmento del bosque de galeria del C./. Carimagua (Depto. Del Meta) Colombia. Universidad Nacional de Colombia, Departamento de Biologfa, Bogotá, Colombia, 1995.

Mueller, C., H. Torres and G. Martine. *An Analysis of Forest Margins and Savanna Agroecosystems in Brazil,* Institute for the Study of Society, Population and Nature (ISPN), Brasilia, Brazil, 1992.

Rao, I.M., Ayarza, M. A., and R. J. Thomas. 1994. The Use of Carbon Isotope Ratios to Evaluate Legume Contribution to Soil Enhancement in Tropical Pastures. *Plant* and Soil 162: 177–182, 1994.

Schetjman, A. Agroindustria y transformación productive de la pequeña agriculture. Revista de la CEPAL 53: 1467–157, 1994.

Smith, J. et al. "*The Forest Margins and Savannas of Latin America: A Unique Opportunity for Contributing to Natural Resource Management,*" paper presented at the 2020 Ecoregional Workshop, Airlie House, Virginia, USA, Nov. 7–10, 1994.

Smith, J., J. V. Cadavid, A. Rincon, and R. Vera. Land Speculation and Intensification at the Frontier. A Seeming Paradox in the Colombian Savanna. *Agricultural Systems* (in press), 1996.

Thomas, R. J. and N. M. Asakawa. Decomposition of Leaf Litter from Tropical Forage Grasses and Legumes. *Soil Biol. Biochem.* 25: 1351–1361, 1993.

Thomas, R. J., J. Smith, J. I. Sanz, and R. R. Vera. Approaches to the Sustainable Management of the Tropical Lowlands in Latin America. In *Agricultural Research and Sustainable Management of Natural Resources*, Hans-Joachim A. Preuss, Ed. Münster: Lit Verlag, pp. 185–195, 1996.

Trigo, E. J. *Agriculture, Technological Change and the Environment in Latin America: A* 2020 *Perspective* Food, Agriculture and the Environment Discussion Paper 9, IFPRI, Washington D.C., 1995.

Vera, R. R., R. Thomas, L. Sanint, and J.1. Sanz. *Development of Sustainable Ley-Farming* Systems *for the Acid-Soil Savannas of Tropical America.* (English). Simpósio lnternacional sobre Ecologia e Agricultura Sustentável nos Trópicos, Rio de Janeiro, Brasil, 3–6 de fevereiro de 1992, published in Anais da Academia Brasileira de Ciencias 64 (supl. 1):105–125, 1992.

Vera, R. R., J. 1. Sanz, P. Hoyos, D. L. Molina, M. Rivera and M. C. Moya. 1994. Pasture Establishment and Recuperation with Undersown Rice on the Acid Soil Savannas of South America (English). In *Biological Basis of Sustainable Animal Production,* Huisman, E. A., J. W. M. Osse, D. van der Heide, S. Tamminga, B. J. Tolkamp, W. G. P. Schouten, C. E. Hollingworth and G. L. van Winkel, Eds., Proceedings of the Zodiac symposium, Wageningen, The Netherlands, April 13–15, 1993. EAAP publication No. 67, 1994, Wageningen Pers, Wageningen, The Netherlands, 1994.

Vera, R. R. *Growth and Conception in Continuously Underfed Brahman Heifers. Animal Production* 53:45–50, 1991.

Vera, R. R., Ca. A. Ramirez and H. Ayala. *Reproduction in Continuously Underfed Brahman Cows. Animal Production* 57: 193–198, 1993.

Ullrich, C., R. R. Vera, and J. H. Weniger. Milk Production by Dual Purpose Cows on Grass-Alone and Grass-Legume Pastures. *Pasturas Tropicales* 16(3):27–30, 1994.

NINETEEN

Long-Term Research on Natural Resource Management: A "Commodity Center" Perspective

Jeffrey W. White, Olaf Erenstein, Larry Harrington, and Ken Sayre

INTRODUCTION

Having a global mandate to develop sustainable maize and wheat systems, CIMMYT views the coming decades as a period of tremendous challenges. Projected increases in food demand will require sustaining agricultural production growth rates at over 2% per annum (Rosegrant et al., 1995). However, this increase must be achieved while facing the dual constraints of rapid environmental degradation and of limited opportunity for increasing land under cultivation. A formal, global assessment of resource degradation problems for maize and wheat systems is lacking, but our informal review of major maize and wheat production environments suggests that such problems are widespread and threaten the majority of production areas in developing countries (Tables 19.1 and 19.2).

Recognizing that cultivar improvement alone cannot ensure needed maize and wheat production increases, CIMMYT and collaborating national programs have long complemented crop improvement efforts with research on agronomic practices. This work has encompassed not just matching of agronomic practices to new plant types but research to conserve or restore the natural resource base. Examples include assessments of the potential of minimum tillage wheat after rice in the Indian subcontinent (Majid et al., 1988), strategies of soil fertility management for smallholders on southern Africa (Kumwenda et al., 1996), and the potential of conservation tillage for maize-based systems in Mesoamerica (CIMMYT, 1991; Erenstein, 1996). Many such practices appear to have relevance in multiple regions and can be applied both to maize and wheat systems (Tables 19.1 and 19.2).

These research efforts have had notable successes. Over 80% of wheat growers in the Yaqui Valley of Mexico have shifted from planting in flat basins ("*melgas*") to planting on beds (Flores, et al., 1996). This has allowed major reductions in irrigation water and herbicide use and permits more flexible fertilizer use. Conservation tillage practices are now widely used for maize production in Brazil (Busscher et al., 1996) and adoption is increasing in Central America (Sain and Barreto, 1996).

Although pleased with these successes, CIMMYT concluded that greater impact

Table 19.1. Examples of Natural Resource Concerns and Possible Solutions for Maize Systems in Developing Countries

Natural Resource Concerns	Promising Technologies for Alleviating Resource Problems	Representative Regions
Soil erosion	Green manures Living barriers Conservation tillage Improved maize crop cover	Mexican highlands Central America Central and southern Brazil East African highlands
Soil fertility declines	Intercropping Improved fallows Phosphorus recapitalization Green manures	Sub-Saharan Africa Southern Mexico
Misuse of agrochemicals	Conservation tillage Integrated pest and disease management	Diverse irrigated regions
Salinization	Improved water management	Egypt
Loss of crop genetic diversity	In situ and ex situ germ plasm conservation Participatory crop improvements	Mexican highlands Peruvian highlands

Table 19.2. Examples of Natural Resource Concerns and Possible Solutions for Wheat Systems in Developing Countries

Natural Resource Concerns	Promising Technologies for Alleviating Resource Problems	Representative Regions
Excess fertilizer inputs	Split applications of fertilizer (possible with bed planting)	Diverse irrigated regions
Herbicide resistance in weeds	Integrated weed management, including mechanical cultivation (possible with bed planting)	Yaqui Valley of Mexico Indus and Gangetic Valleys
Soil erosion	Conservation tillage Living barriers Improved wheat crop cover	Mexican highlands Andean highlands
Decreased quality and availability of irrigation water	Replacing rice or other rotation crops that have high water requirements Improved water management (possible with bed planting)	India Pakistan Yaqui Valley of Mexico
Loss of crop genetic diversity	In situ and ex situ germ plasm conservation Participatory crop improvements	Iran Middle East Turkey

would come from uniting the center's natural resource management research under an umbrella structure. Thus, the Natural Resource Group (NRG) was created in early 1994 and formalized in late 1995. This chapter discusses research activities of the NRG with emphasis on the role of long-term experiments (LTEs), which have, and will continue to play, a key role in research on natural resource management. Many issues related to the design and management of LTEs illustrate the anticipated benefits of promoting a cross-commodity focus for natural resource management research.

CIMMYT'S NATURAL RESOURCE GROUP

The NRG was formed to help CIMMYT and its collaborators answer key questions such as:

- What are the untapped opportunities for sustainable increases in maize and wheat yields?
- How can maize and wheat technologies help slow or reverse resource degradation?
- How can the long-term consequences of technical change best be assessed?

The NRG structures its research in a problem-solving framework that includes phases including problem definition, developing an array of technology options to address priority problems, tailoring promising prototypes to farmers' needs, understanding and accelerating adoption, scaling adoption strategies up from specific sites to the regional level, and understanding the consequences of change—or of allowing resource degradation to continue (Table 19.3). This approach promotes efficient matching of research tools and methods to needed research functions. Emphasis is also given to working through partnerships such as provided by ecoregional initiatives and bilateral projects with NARS.

Among the research methods that the NRG uses or promotes, LTEs are tools of choice for research on the microlevel pace and consequences of resource degradation (Table 19.3). However, the utility of LTEs is much broader than this. They can serve as laboratories for understanding processes of resource degradation, as test-beds for evaluating and understanding promising technologies, as vehicles for refining prototypes, and as fora for promoting adoption among farmers. They also are valuable complements to simulation modeling since they can provide data for development and testing, giving greater confidence in even longer term predictions possible with models (Acock and Acock, 1991).

LONG-TERM EXPERIMENTS AT CIMMYT

CIMMYT's experience with LTEs includes both in-house experience at our experiment stations and collaborative experiments largely conducted by NARS with CIMMYT participation through our regional agronomists. A prominent example of an in-house experiment is the Long-Term Rotation and Tillage Experiment for Rainfed Highlands, which is discussed at length to provide background for the section on issues concerning LTEs. Examples of LTEs that CIMMYT regional staff participate in are listed in Table 19.4 to emphasize the importance CIMMYT ascribes to such work and to provide an entry point for a reader who wishes to obtain further information on specific LTEs.

The Long-Term Rotation and Tillage Experiment for Rainfed Highlands

The Wheat Program started the Long-Term Rotation and Tillage Experiment for Rainfed Highlands in 1991 under the leadership of R.A. Fischer (former Wheat Program Director). The experiment is located at CIMMYT's main station, El Batan (elev. 2250 m, 19°31′N, 98°50′W). Currently, there are 16 core treatments of wheat and maize rotations or continuous cropping, with conventional or zero till, and either residue removal, incorporation, or use as mulch. These treatments are intended to document basic effects of crop management on maize and wheat and will be maintained for the life of the experiment. An additional 16 treatments are used to examine management options that might allow farmers to overcome problems associated with introduction of reduced tillage or with leaving crop residues in the field. From 1991 to 1995, these additional treatments included fodder crops and green manures such as oats and vetch. In 1996, a new set of treatments were defined, primarily looking at options with other types of legumes.

Table 19.3. Research Phases of a Problem-Solving Framework for Natural Resource Management and Examples of Corresponding Data Sources and Research Methods

Phase	Data Sources or Research Methods
Understanding and defining problems and opportunities	
• Processes	strategic disciplinary research, crop growth or soil process models
• Consequences and pace of change	long-term trials, farmer monitoring, chronosequence studies, historical comparisons, retrospective information from farmers, watershed studies, dynamic modeling, gender analysis
• Incidence	agroecosystem mapping, community resource mapping, use of available remote sensing data, database development, GIS
• Causes and interactions among problems	diagnostic surveys, participatory appraisal, agroecosystem mapping, policy analysis, gender analysis
Expanding and understanding the array of options	
• Identifying options	understanding of problem causes, scientists' experience, scientific literature, experience of successful projects, farmers' innovations by subgroup (e.g., gender)
• Understanding the performance of prototypes	agronomy, strategic technical research, modeling
Tailoring prototypes to farming systems	
• Tailoring prototypes	participatory adaptive experimentation
• Assessing profitability	financial and economic analysis of adopted practices
Understanding and accelerating adoption	
• Understanding adoption	adoption studies, review of case studies, gender analysis
• Fostering appropriate policy and institutional changes	policy analysis, policy workshops
Scaling up	
• Synthesis across sites	GIS, modeling, site similarity studies
• Extrapolation of practices	GIS, modeling, extrapolation domain definition
• Links among system levels	GIS, modeling, watershed studies, decision support systems
Understanding the consequences of change	
• Consequences at different levels, over different time scales, and for different groups	long-term trials, farmer monitoring, chronosequence studies, historical comparisons, retrospective information from farmers, watershed studies, dynamic modeling, gender analysis

In contrast to many trials, this long-term experiment is envisioned as much as a research and training facility as a trial to address a specific research question. Therefore, the objectives are varied and are expected to evolve over time. The broadest research objective currently is to develop improved rotation and tillage systems relevant to environments similar to the high valleys of Mexico (representative of highland, low-latitude wheat and maize environments, ME2). Under the NRG's problem-solving framework (Table 19.3), the specific research issues that are addressed include:

Understanding and Defining Problems and Opportunities:

- characterizing processes underlying soil crusting, compaction, and organic matter dynamics evaluation or development of simulation models dealing with crop rotations.

Table 19.4. Examples of LTEs with Participation of CIMMYT Regional Staff

Location	Research Topic	Participants	CIMMYT Contact
Jalisco, Mexico	No-tillage and residue management in maize systems	INIFAP	Eric Scopel
Bolivia	No-tillage and residue management in wheat systems	NARS Various NGOs	Patrick Wall
Ethiopia	Crop rotations x N x P fertilization Stubble management	IAR	Doug Tanner
Zimbabwe Malawi	Soil fertility management for maize systems	Soil Fertility Research Network	Stephen Waddington
India Nepal Pakistan	Rice-wheat systems with major focus on nutrient management	NARS (Coordination by the Rice-Wheat Consortium)	Peter Hobbs

Expanding and Understanding the Array of Options:

- Effects of tillage options on maize and wheat production.
- Use of residues as mulch and soil ameliorants.
- Crop-livestock interactions, including use of residues as feed and strategies for producing alternative high quality forage.
- Carryover effects of legumes in crop rotations.

The experiment also serves as a training facility for demonstrating issues in sustainable cropping systems for trainees and visiting scientists, including to promote wider use of LTEs, and is a valuable platform for planning more relevant and sustainable agronomic management for CIMMYT experiment stations.

Sowing arrangements, populations, and fertilizer levels are selected to be representative of practices for rainfed highland regions. Since the focus of the experiment is on tillage effects and soil nutrient dynamics for rainfed environments, weeds, diseases, and pests are controlled using recommended practices. The experiment is only given emergency irrigations when lack of precipitation threatens crop establishment or survival of critical treatments. Cultivars are rotated to assure they represent the best available materials. This is particularly important for disease resistance since cultivars may become unusable if a race-specific resistance breaks down.

The experiment is arranged as two randomized complete blocks with 32 treatments (see Table 19.5 for details). The low number of replicates is partially compensated for by running treatments in paired sets in alternate years. For example, maize-wheat rotations over two years (treatments 13 to 16) is complementary to wheat-maize rotations (treatments 5 to 8) over the same period. Individual plots measure 7 m by 20 m.

The "minimum data set" recorded each year is grain yield, yield components, aboveground dry weight (at harvest), total aboveground nitrogen content, principal growth stages, and basic management practices. Specialized studies to date have measured soil moisture, nitrogen, and organic matter dynamics.

Data from 1991 to 1995 are still being analyzed. Prominent effects noted to date include:

- Reduced growth, especially of maize, under continuous maize or maize-wheat rotations with zero-tillage residue removal.
- Wheat withstands continuous cropping much better than maize.

Table 19.5. List of Treatments for the Long-Term Rotation and Tillage Experiment at El Batan

Treatment No.	Rotation[a]	Tillage	Straw Management
Core treatments			
1	W-W	zero	keep
2	W-W	zero	remove
3	M-M	zero	keep
4	M-M	zero	remove
5	W-M	conventional	keep
6	W-M	conventional	remove
7	W-M	zero	keep
8	W-M	zero	remove
9	W-W	conventional	keep
10	W-W	conventional	remove
11	M-M	conventional	keep
12	M-M	conventional	remove
13	M-W	zero	keep
14	M-W	zero	remove
15	M-W	conventional	keep
16	M-W	conventional	remove
Exploratory treatments			
17	W-M	ridge-till (beds)	keep
18	W-M	ridge-till (beds)	remove maize, keep wheat
19	M-W	ridge-till (beds)	keep
20	M-W	ridge-till (beds)	remove maize, keep wheat
21	W+B-M+B	ridge-till (beds)	keep maize, remove wheat
22	M+B-W+B	ridge-till (beds)	keep maize, remove wheat
23[b]	C+Md-W	conventional-zero	remove
24[b]	W-C+Md	zero-conventional	remove
25[c]	C+MD-M	ridge-till (beds)	remove
26[c]	M-C+Md	ridge-till (beds)	remove
27	W+FB-M+F	minimum	keep
28	M+F-W+FB	minimum	keep
29	W+B-M+V	minimum	remove
30	M+V-W+B	minimum	remove
31	W-F	minimum	keep
32	F-W	minimum	keep

[a]For 1996, all treatments are on the second rotation. W = Wheat (trigo); M = Maize; B = Berseem clover; C = Barley (cebada); Md = Medicago; FB = Faba Bean; F = Bean (frijol).
[b]Barley cut near heading to simulate early pasture-variety = hooded type.
[c]Barley cut for hay at milk-dough stage.

- Poor establishment and persistence of legumes are constraints to their use as green manures in this environment.

Issues in Long-Term Experiments for Natural Research Management

Long-term field experiments pose many management problems that are particular to their long durations (Frye and Thomas, 1991). CIMMYT's recent experiences with LTEs, combined with a broader view from research on farming systems and strategic agronomy, suggest several issues that affect the utility of such experiments:

- *Number of treatments*: Depending on the objective, there is justification for a wide range in the number of treatments. For exploratory research (such as the Long-Term Rotation and

Tillage Experiment for Rainfed Highlands) a wide range of treatments is required, but for more adaptive work, with farmer input, only a few main treatments should be used. Satellite trials are often preferable to handling a large number of treatments within a single LTE.

- *Flexibility of designs*: To assure the integrity of a LTE, a set of core treatments should be maintained over time. However, if additional exploratory treatments are included in the original design, the experiment can be adapted in view of previous results or new technologies.
- *Plot size*: We prefer plots that are large enough to observe tillage effects easily. This also permits future subdividing of plots for soil fertility or other treatments.
- *On-station vs. on-farm*: Both on-station and on-farm LTEs have valuable functions depending on the research objectives. Farmers are usually receptive to LTEs if their management doesn't complicate other farm activities and if the objectives are made clear.
- *Farmer participation*: Experience has shown that the value of farmer-input is often underestimated. We emphasize that farmers can participate even in on-station LTEs effectively, as illustrated by the University of California at Davis experience in comparing conventional, low-input, and organic farming systems (Temple et al., 1994).
- *Basic vs. applied focus*: This issue is easily overemphasized. LTEs should serve as research tools, allowing both basic and applied research in a single experimental context.
- *Minimum data sets*: A rigid standard for minimum data sets is unrealistic given the variable objectives and research infrastructure of LTEs. Nonetheless, a useful guideline is to ensure that sufficient data are recorded to permit simulating treatment effects over multiple years. For this purpose, the DSSAT standards provide a description of such requirements (IBSNAT, 1996).
- *Data ownership and warehousing*: To maximize the utility of a LTE, all data must be available in a format that permits ready access by scientists not directly involved in the experiment. International centers nominally are the owners of data recorded by their scientists, but this relation is seldom enforced. Ownership of data in collaborative LTEs is understandably problematic and merits more attention. Access to data could be vastly improved by use of databases. A promising example is the International Crop Information System (Fox and Skovmand, 1996), which is being developed cooperatively among several international centers to promote storage and retrieval of data in flexible but standardized formats.
- *Sustainable leadership and funding*: Assuring continuity in LTEs is problematic in the context of worldwide declining support for agricultural research. Beyond the primary need for LTEs to examine long-term trends, justifications for continued support to LTEs include that LTEs are relatively low-cost to manage once they are established, that they can serve as a valuable training and demonstration vehicle, and that they serve as laboratories where researchers from various disciplines can work in a single experiment.

CONCLUSIONS

LTEs are invaluable for examining problems related to sustainable wheat and maize production for developing countries. CIMMYT will continue to promote their use for measuring the rate of resource degradation, understanding underlying processes of degradation, and for developing and testing prototype solutions. Nonetheless, as with any research tool, they require careful management and an appreciation of their limitations. It should also be emphasized that for some objectives related to sustainable production, there may be complementary or more cost-effective methods. These include farmer monitoring, chronosequence studies, historical comparisons, retrospective information from farmers, watershed studies, and dynamic modeling.

REFERENCES

Acock, B. and M.C. Acock. Potential for Using Long-Term Field Research Data to Develop and Validate Crop Simulators, *Agron. J.* 83: 56–61, 1991.

Busscher, W.J., D.W. Reeves, R.A. Kochhann, P.J. Bauer, G.L. Mullins, W.M. Clapham, W.D. Kemper, and P.R. Galerani. Conservation Farming in Southern Brazil: Using Cover Crops to Decrease Erosion and Increase Infiltration. *J. Soil Water Conserv.* 51: 188–192, 1996.

CIMMYT. *Maize Conservation Tillage,* Training Working Document No. 7, CIMMYT, Mexico, D.F., 1991.

Dowswell, C.D., R.L. Paliwal, and R.P. Cantrell. *Maize in the Third World,* Westview Press, Inc., Boulder, CO, USA, 1996.

Erenstein, O., *Evaluating the Potential of Conservation Tillage in Maize-based Farming Systems in the Mexican Tropics.* CIMMYT Natural Resources Group Reprint Series #1, CIMMYT, Mexico, D.F., 1996.

Flores, D., B. Avalos, E. Rice, and J. Harris. *Summary of Farming Practices in the Yaqui Valley* (unpublished memorandum), CIMMYT, Mexico, D.F., 1996.

Fox, P.N. and B. Skovmand. The International Crop Information System (ICIS) – Connects Genebank to Breeder to Farmer's Field. In *Plant Adaptation and Crop Improvement*, Cooper, M. and Hammer, G.L., Eds., CAB International, Wallingford, U.K., (in press), 1996.

Frye, W.W. and G.W. Thomas. Management of Long-Term Field Experiments, *Agron. J.* 83: 38–44, 1991.

IBSNAT. *Technical Report 5. Documentation for IBSNAT Crop Model Input and Output Files, Version 1.1.* Dept. of Agronomy and Soil Science, Univ. of Hawaii, Honolulu, HI, 1996.

Kumwenda, J.D.T., S.R. Waddington, S.S. Snapp, R.B. Jones, and M.J. Blackie. *Soil Fertility Management for the Maize Cropping Systems of Smallholders in Southern Africa: A Review.* CIMMYT NRG Paper 96-02, CIMMYT, Mexico, D.F., 1996.

Majid, A., M. Aslam, N.I. Hasmi, and P.R. Hobbs. Potential Use of Minimum Tillage in Wheat after Rice. In *Wheat Production Constraints in Tropical Environments,* Klatt, A.R., Ed., CIMMYT, Mexico, D.F., 1988.

Rajaram, S. and M. van Ginkel. *A Guide to the CIMMYT Bread Wheat Section.* Wheat Special Report No. 5, CIMMYT, Mexico, D.F., Mexico, 1996.

Rosegrant, M.W., M. Agcaoili-Sombilla, and N.D. Perez. *Global Food Projections to 2020: Implications for Investment,* International Food Policy Research Institute, Washington, D.C., 1995.

Sain, G.E, and H.J. Barreto. The Adoption of Soil Conservation Technology in El Salvador: Linking Productivity and Conservation. *J. Soil Water Conserv.* 51: 313–321, 1996.

Temple, S.R., O.A. Somasco, M. Kirk, and D. Friedman. Conventional, Low-Input and Organic Farming Systems Compared, *California Agric.* 48:14–19, 1994.

TWENTY

Agricultural Intensification Systems and Their Impact on Soil and Water Quality in the Cerrados of Brazil

Dimas Vital Siqueira Resck

INTRODUCTION

The Cerrados region contributes significantly to the national agropastoral and forest production. However, crop yields are low despite the technological advances being made. Latosols and quartz sand are the most common soils of the region. Due to the inadequate management systems adopted, these soils are easily degraded, leading to decrease in crop yields, erosion, high runoff, sedimentation, and contamination of the rivers in the region.

Mankind is faced with the challenge of feeding a rapidly increasing population while natural resources are diminishing. Two important natural resources of fundamental importance are soil and water. The Cerrados region covers 207 million ha, and is the greatest agricultural frontier on the planet. The available research information has shown that it is feasible to produce 60% to 70% of the national grain production of about 80 million ton (in 1996) without clearing any new land. The Amazon region, which comprises 56% of the Brazilian territory, is protected by the Cerrados region. Considerable research is needed to use the Amazon region for production of food, fibers, and energy while preserving its richness in biodiversity.

This chapter demonstrates the global importance of the Cerrado region, and the impact of management systems on soil physical, chemical, and microbiological properties and on water balance and water quality.

THE CERRADOS ECOSYSTEM

The Cerrados region occupies an area of 207 million hectares (24.42% of the Brazilian territory). It is located mostly in the midwest region, but also northeast and north regions, covering a vast area from 5°50′S to 21°26′S.

The vegetation is characterized by a high physiognomical diversity and is called "Campo limpo" if grass is dominant; "Campo Sujo" if shrubs and trees are dispersed in the grass; "Cerrado" latu sensus when shrubs, trees, and grass are equally dominant; and

Table 20.1. Some Characteristics of Principal Ecoregions of the Cerrados[a]

Cerrados Subregions	Total Precipitation (mm)	Dry Season, # of Months	Rainfall Season	Average Temperature, (°C)	Dominant Soils
Core area	1400 to 1700	4–5	spring and summer	22–24	Red-Yellow Latosol and Litolics
Northeast influence	500 to 800	8–9	summer	24–26	Quartz Sand, Red-Yellow Latosol
Amazonic influence	2000 to 2300	4–5	spring and summer	24–26	Red-Yellow Latosol, Quartz Sand, and Laterites
Austral-Atlantic influence	1400 to 1700	4–5	spring and summer	18–20	Dark-Red and Red-Yellow Latosol, Podzolics, and Litolics
Austral-continental influence	1100 to 1400	2–3	spring and summer	20–22	Red-Yellow Podzolic

[a]Adapted from Azevedo and Caser (1980).

"Cerradão" when trees are the dominant vegetation, with some trees reaching 6 to 7 m high.

Azevedo and Caser (1980) described climatic features of these regions as shown in Table 20.1. These characteristics are briefly outlined below.

About 80% of the rainfall occurs from October to March with a short dry spell of two to three weeks called "veranico." The rainfall in 75% of the region ranges from 1000 to 1800 mm and the water deficit in 88% of the region lasts from 4 to 7 months.

This region comprises 5 out of 9 important watersheds of the country. These watersheds ranked in the order of area are: Tocantins-Araguaia (29%), Paraná (19%), São Francisco (19%), Paraguai (17%), and Amazônia (16%). Three of the most important hydroelectric plants depend on this region: Itaipu (Paraná river, 12,600 MW), Tocantins (Tocantins river, 7,745 MW), and Paulo Afonso (São Francisco river, 3,986 MW). The potential capacity of these rivers is at least four times greater than the current output. Soil erosion and sediment deposition are diminishing their capacity and affect the water quantity and quality (IBGE, 1989; Resck, 1991).

Latosols (Oxisols) occupy 46% of the region. Other soils include Quartz Sand (15.2%) and Podzolics (15.1%) (Adámoli et al., 1986). Latosols and Quartz Sand are developed on gently rolling and flat uplands and are managed by intensive mechanization.

Latosols have two main constraints: (1) These soils have low inherent fertility and contain predominantly low activity clays. Their mineral fraction is composed mainly of kaolinite, gibbsite and iron oxides, associated with mean organic matter content of 2.5% to 3.0%. Latosols also have high Al^{+3} content throughout the profile or at least to 40 cm depth. (2) Latosols are deep and very well structured with extremely high porosity (70%) with equal proportion of macropores and micropores. However, the available water capacity (between −6 kPa and −100 kPa water potential) is low at about 10% by volume. Thus, plants have available water of less than 1 mm of water per cm of soil horizon (Resck and Silva, 1990).

Quartz Sand soils also have some severe constraints including low inherent fertility, low activity clays, weak soil structure, and low organic matter content of about 1%. These constraints render them even more fragile than Latosols, especially when managed inappropriately. These soils usually have less than 15% clay and more than 80% of sand.

Population Migration into the Cerrados Region

The population of the midwest region of the Cerrados in 1980 was 5.1 million (68%) in urban centers and 2.4 million (32%) in the rural areas. The population of Brazil in 1980 was 80.4 million (68%) in urban centers and 38.6 million (32%) in rural areas. The population of the midwest in 1990 was 7.6 million (81%) in urban centers and 1.8 million (19%) in rural areas compared with 110.9 million (75%) in urban centers and 36.9 million (25%) in the rural areas of Brazil. In ten years, urban population of the midwest region increased by 2.5 million or by 26%, mostly through migration from rural areas. Out of this, 1.9 million or 74% of the total increase came from the south and southeast regions, supported by the federal government programs whose main objectives were to promote occupation of the region back in 1960 and to bring development to these almost desert places (Resck, 1996). The average demographic pressure increased from 8.12 inhabitant/km^2 in 1980 to 10.20 inhabitant/km^2 in 1991. The rate of increase in population for the same period for Brazil was 2.90% for urban centers and –0.63% for rural areas. The corresponding rate of increase for the Cerrados region was 3.34% for urban center and 0.97% for rural areas (Pereira et al., 1996, unpublished).

Because of the population migrations, the rural labor force decreased in proportion to that migration process, thus the degree of mechanization in the agricultural sector increased. Meanwhile the poverty also increased due to the high degree of mechanization, underemployment, and low salaries paid to the rural workers.

Region's Contribution to the National Grain, Beef Cattle and Wood Production

States of the Cerrados region, their corresponding areas, land use changes, production and yields of principal crops and the number of livestock from 1975 to 1993 are shown in Table 20.2.

Five states of the region represent 66.3% of the Cerrados region and 15.91% of Brazil territory. These states contribute significantly to the national grain and beef cattle production.

For the 18 year period from 1975 to 1993, area under soybean cultivation increased from 235,736 ha (4% of the total) to 3,930,734 ha (37% of the total). There was an increase of 1567% in area cultivated to soybean in the Cerrados region during this 18 year period. In addition to the area, soybean yield also increased by 48% from 1463 kg/ha to 2160 kg/ha. The contribution to national production of soybean from 3% in 1975 to 39% in 1993 therefore is a phenomenal increase of 2725%.

There was also an increase in the area under corn cultivation, which was 1,527,111 ha (14% of the land) in 1975 to 2,099,367 ha (17.7% of the land), leading to an increase of 37% in areas cultivated to corn in the Cerrados region. Corn grain yield was 1,542 kg/ha in 1975 and increased to 3,163 kg/ha in 1993; i.e., an increase of 105%. Contribution to the total national production of corn was 16% in 1975 (2,625,513 t) and increased to 22% in 1993 (6,611,293 t); i.e., an increase of 147%.

Table 20.2. Temporal Changes in Food Crops and Cattle Production in the Cerrados Region from 1975 to 1993

States	Area of the Cerrados Region	Crop or Cattle Herd	Area (ha)	Production (t) or Herd	Yield (kg/ha)	Area (ha)	Production (t) or Herd	Yield (kg/ha)
	%			1975			1993	
Minas Gerais	19	Soybean	68,578	80,073	1,168	520,338	1,055,279	2,028
		Corn	808,402	1,248,255	1,544	881,170	2,499,200	2,836
		Rice	532,793	433,090	813	242,724	375,149	1,546
		Beans	278,859	126,774	455	264,416	18,631	827
		Cattle		11,086,756			12,689,459	
Mato Grosso	20	Soybean	325	638	1,960	1,557,787	3,838,363	2,464
		Corn	59,176	91,705	1,550	237,779	692,202	2,911
		Rice	139,609	214,576	1,537	333,199	389,493	1,169
		Beans	16,448	13,647	830	14,766	10,952	742
		Cattle		1,868,830			6,583,993	
Mato Grosso do Sul	10	Soybean	111,233	156,135	1,404	824,966	1,769,672	2,145
		Corn	67,260	100,926	1,501	228,701	629,883	2,754
		Rice	443,486	528,092	1,191	80,074	147,797	1,846
		Beans	9,535	7,771	815	15,196	10,851	714
		Cattle		4,365,668			14,188,303	
Goiás	17	Soybean	55,600	73,392	1,320	982,883	2,004,277	2,039
		Corn	589,995	1,182,091	2,004	734,389	2,597,866	3,537
		Rice	806,917	707,086	876	304,510	384,142	1,262
		Beans	207,193	105,018	507	143,262	125,218	874
		Cattle		10,704,168			18,580,908	
Distrit 0 Federa 1	0.3	Soybean	0	0	0	44,760	95,160	2,126
		Corn	2,278	2,536	1,113	17,328	65,482	3,779
		Rice	1,003	1,044	1,041	3,016	3,438	1,140
		Beans	1,725	602	349	5,919	9,144	1,545
		Cattle		40,165			123,569	
Brazil	24	Soybean	5,824,492	9,893,008	1,699	10,635,330	22,574,762	2,123
		Corn	10,854,687	16,334,516	1,505	11,870,763	30,051,333	2,532
		Rice	5,306,270	7,781,538	1,466	4,411,315	10,108,434	2,291
		Beans	4,145,916	2,282,466	551	3,884,341	2,478,325	638
		Cattle		101,673,753			155,134,073	

The area sown to rice in 1975 was 1,923,808 ha (36% of the land), and decreased to 963,523 ha (22% of the land), in 1993, a 50% decrease in the area cultivated to rice. However, there was an increase of 28% in rice yield (1,092 kg/ha to 1,393 kg/ha) during the same period. Consequently, the region's contribution to rice production dropped from 24% (1,883,888 t) in 1975 to 13% (1,300,019 t) in 1993 with an overall decrease of 31%.

The land area cultivated to beans decreased from 513,760 ha (12% of the land) in 1975 to 443,559 ha (11%) in 1993 with an overall decrease of about 14% in 18 years. Crop yield was 551 kg/ha in 1975 and increased to 638 kg/ha (16% of increase). The total grain production in 1975 was 253,812 t (11% of the national production) and increased to 374,796 t in 1993 (15% of the national beans production), representing an overall increase of 48%.

Total livestock population of Cerrados region was 28,065,587 cattle (28% of the country herd) in 1975, which increased to 52,166,232 (34% of the country herd) in 1993, representing an increase of 86%.

Forest in the region has an important impact on national economy. Land area devoted to forestry activity is about 1.9 million ha and has been stable since 1982. Forestry products contribute more than 50% of the national production.

Anthropogenic perturbations for agricultural land use has affected about 18% of the land area of the Cerrados region. The present land use includes 27% under native pasture, 3% under reserve, and 52% is yet to be developed (EMBRAPA, 1994). A total of about 40 million ha have been planted to pasture (Spain et al., 1996), and about 15 million ha to annual crops and forestry leaving a balance of about 56.5% of the region yet to be developed. The undeveloped area is approximately 126 million ha of land comprising mainly Oxisols (Latosols) and Inceptisols (Quartz Sand).

Predominant Farming Systems

Soils of the region are low in phosphorus and because of mineralogical composition, have high P fixation capacity requiring high rates of P fertilization. Depending on soils and crops, the remedial rates of P fertilization can be as high as 240 kg P_2O_5/ha and maintenance rates of 60 to 80 kg of P_2O_5 ha. Lime and gypsum applications are also required to neutralize Al and to rapidly translocate Ca to subsoil. Some soils also need K and micronutrients applications for high yields.

Due to high costs of the fertilizers and lack of bank credits, recommended rates of fertilizers are rarely practiced by the farmers. Therefore, crop yields are lower than the potentially achievable yields. In the mid-1980s, the region used only 32% of the total fertilizers (1,080,000 t) consumption of the southeast region of Brazil. The southeast region includes the industrially developed states comprising São Paulo, Rio de Janeiro, and Minas Gerais (Resck, 1991). The fertilizer consumption of the region is still low.

Farm machinery and implements are important aspects of the farming system. The ratio of cultivated land area per tractor is 198 for developing countries compared with 34 for developed economies (IBGE, 1987). In the Cerrados region, this ratio was 66969 ha/tractor in 1920 and 85 ha/tractor in 1985. Since most large farmers migrated to the region from São Paulo, Rio de Janeiro and Minas Gerais, they prefer the heavy disk harrow because of its velocity (5 km/h to 10 km/h) and field efficiency (70 to 90%) (Folle and Brandini, 1995). The efficiency from an engineering point of view implies less force (kN/m) required to pull the implement and therefore less horsepower of the tractor needed. High engineering efficiency, however, is not necessarily suitable for judicious soil management.

SOIL AND WATER DEGRADATION PROCESS

Tillage operations are normally started during the dry season from July to September, because there is a large area to be prepared and there are few tractors and implements. The heavy disk harrow is the most common implement, with numerous adverse consequences. These soils, normally well aggregated, are easily destroyed by this implement which cuts and exposes organic matter for decomposition by microorganisms. The surface horizons in undisturbed (virgin area under Cerrados vegetation, Brachiaria pasture and Eucalyptus forest) Latosols with 45% of clay have more than 80% of the aggregates > 2.0 mm in diameter. The surface horizon in soils cultivated with heavy disk harrow for 10 years under soybean-maize rotation contained only 20% of aggregates > 2.0 mm in diameter. In addition to decrease in aggregation, there also was a corresponding loss of carbon estimated at 13.23 t/ha (0.66% of one hectare weighing 2,000 t). The loss of carbon in areas cultivated with green manure for ten years and prepared with rototiller was about 16.82 t/ha (1.18%) along with 42% of aggregates > 2.0 mm. These implements, therefore, destroy soil structure despite production of high biomass (about 10 t/ha in both cases). The organic carbon content of aggregates > 2.0 mm for these systems was 1.62% in Cerrados and pasture, 1.28% in forest and green manure, 1.33% in soybean-corn crop rotation, and 1.45% in rice-corn crop rotation (Resck and Silva, 1990). The failure to maintain a high percentage of aggregates > 2.0 mm in the disturbed area was attributed to decrease in the pool of physically protected organic matter (POM) (Duxbury et al., 1989) and the amount of organic matter which is several-fold less in the disturbed systems compared to the undisturbed systems.

Experiments have also been conducted to evaluate the effects of 11 years of cultivation on soil organic matter content of highly weathered clayey Red-Yellow Latosol (> 70% of clay). The soil organic carbon content of the surface horizon was 1.86% immediately after clearing the Cerrados. It increased to about 2.26% during the following two years of upland rice cultivation, presumably because of the decomposition of root residue from the natural vegetation. Under continuous cultivation of soybean using heavy disk harrow, soil organic carbon content decreased to less than 1.74% because of a decrease in total residue inputs and increase in microbial activities resulting from liming and fertilization. This relatively low rate of decrease in organic carbon was because of the high clay content which makes the organic carbon a physically protected pool. The soil surface horizon contained 90% of the soil aggregates > 2.0 mm under Cerrados vegetation and only 62% of aggregates > 2.0 mm after 11 years of cultivation despite the fact that the organic carbon content of aggregates was 2.09%. Consequently, aggregates were characterized by a wide range of sizes, i.e. > 1 mm, 0.5 mm, 0.25 mm, 0.1 mm and < 0.1 mm (Woomer et al., 1994). Proportion of the naturally dispersed clay increased from 10% under natural vegetation to 44% after 11 years of cultivation. The disk harrow implement pulverized soil structure exposing the organic matter to microbial decomposition. In addition, this implement formed a plow pan at 12-15 cm depth. The dense pan is subsequently coated by the illuviated clay. The surface soil is prone to crusting that decreases water infiltration. Latosols are normally very resistant to erosion, but become susceptible to soil erosion after these degradative processes. Soil sediments are laden with nutrients having an enrichment ratio of 14 for P, 2 for Ca and Mg and 5 for organic matter (Resck et al., 1991; Resck, 1981).

The rainfall erosivity (EI_{30}) index increases from July to January, and this increasing trend coincides with the time of soil tillage and relatively bare soil surface. Consequently, soil erosion and runoff losses are maximum during this period (Dedecek et al., 1986).

Another example of inappropriate soil management is the destructive effect of the heavy disk harrow for cultivating Quartz Sand and Latosols in Bahia. The rate of decline of soil organic matter (SOM) with time of cultivation (TC), and its effect on cation exchange capacity (CEC) were analyzed in a soil continuously cropped to soybeans in the west Cerrados region of Bahia. The soils were grouped accordingly to clay content in (Group I) soils with clay < than 15% and from 15 to 30% combined because of similar behavior, and (Group II) soils with clay > 30%. The amount of SOM lost after five years of cultivation was lower in Group I soils (15.74 t/ha) than in Group II soils (18.90 t/ha). However, those losses represented 69% and 46% of the initial SOM stock in Groups I and II soils, respectively. The SOM decay with TC was studied, fitting the following model to the data (Dalal, 1986):

$$SOM = SOM_E + (SOM_I - SOM_E)*EXP(-K*TC) \quad (1)$$

where SOM is the organic matter content at a given time, SOM_E is SOM at equilibrium, SOM_I is the initial stock of SOM (before cropping), and K is the SOM decomposition rate accordingly. K values estimated for Groups I and II were 0.300 and 0.244% yr^{-1}, and $t_{1/2}$ (half-life) values were 2.74 and 6.07 yr, respectively, showing that the rate of loss of SOM was higher in soils with low than high clay content. These decay constants in Quartz Sand soils were K = 0.299% yr^{-1} and $t_{1/2}$ = 21.5 yr. The SOM decay was also correlated with the decline in CEC with a correlation coefficient of over 0.80 for all soils. Stepwise regression analysis, introducing the clay content into the linear model, showed that the clay content was significantly important only in Group II soils. Decline in CEC per unit percent loss of SOM was higher in Group I than in Group II soils, probably due to relatively high contribution of the clay fraction to CEC in clayey soils (Silva et al., 1994).

A watershed experiment conducted on the Taquara Creek (Brasilia District) showed that the minimum, average, and maximum daily discharge were 0.5, 0.8, and 1.2 m^3/s, respectively. The nutrient analyses of stream flow showed that about 30 μg/L of total P and 180 μg/L of NO_3-N were lost during the month of January 1993 (Resck et al., unpublished). The turbidity factor (representing the sediment concentration) was about 15 in January compared to 2 during the dry season in August. The total precipitation was 250 mm and the maximum precipitation in 24 hours was 100 mm. Most farmers in the 4,300 ha-watershed use heavy disk harrow and rototiller to grow horticultural crops. This equipment leads to decline in soil structure, compaction, soil erosion, and runoff. Inappropriate use of disk plow, another common implement in the region, also adversely impacts soil physical and chemical properties but to a lesser intensity than heavy disk harrow and rototiller.

THE TROPICAL PARADOX

The clay fraction of Latosols and Quartz Sand predominantly contains 1:1 type clay minerals (with little or even no isomorphic substitution) and iron and aluminum oxides. Consequently, these soils have a low net negative surface charge density and CEC. Most of the CEC of these soils is contributed by SOM due to ionization of the carboxylic, enolic, and phenolic groups. Increase in soil pH increases charge on SOM and clays, but SOM has much greater effect on CEC than clay. Helling et al. (1964) observed that in the pH range of 2.5 to 8.0 the CEC of the inorganic fraction increased approximately twice, whereas that of the organic fraction increased seven times. Effects of SOM and clay fraction on CEC are additive because part of the clay charge is blocked by the SOM and it cannot be detected by the normal procedures of CEC determination.

Silva et al. (1994) showed that SOM is responsible for about 75% of the total CEC. Van Raij (1969) also showed that the average contribution of SOM to CEC in soils from the State of São Paulo was 74% for the surface layers compared with 35% for B horizons.

In uncultivated soils, therefore, charge can be increased by liming and fertilization. Resck (1986) found that liming increased CEC in a virgin Dark-Red Latosol (Typic Haplustox) by 93% compared to an increase of only 6% in a cultivated soil. The objective in cultivated soils is to improve aggregation and water storage capacity by increased nutrient supply to the plants. The more plant biomass is produced and returned to the soil, the better are the soil physical properties due to enhanced microbial activity. The addition of biomass strengthens nutrient recycling mechanisms. These interactive effects can be summarized as follows:

Better soil chemical condition → more CEC → more plant biomass → more microbial biomass → more charge and cementing by-products → more soil aggregation → more water storage → more nutrients availability to the plants.

Once the soil physical and chemical quality is restored, soil management practices adopted must maintain or sustain the good soil quality thus achieved.

It is a paradox because in order to do that, soil must initially be restored by improving its chemical, physical, physicochemical, and microbiological properties. Subsequently, the management systems adopted must maintain the soil status and good conditions achieved. Undisturbed systems such as forest, pasture, fruit trees, no-tillage or minimum tillage practices are better suited to maintain the desirable soil condition.

SOIL AND WATER MANAGEMENT AND CONSERVATION IN THE CERRADOS REGION

Soil management involves a range of operations and practices, e.g., clearing the natural vegetation, liming to neutralize surface and subsoil acidity and Al^{+3} (factor A), applying corrective and a maintenance fertilizers (factor B), alternating soil tillage systems (factor C) and crop rotation (factor D) which includes annual crops, pastures, and forest (Resck, 1996).

All these four factors must be synchronized because they interact among themselves like four corners of a tetrahedron (Fig. 20.1). The reason why high yields have not yet been achieved, despite the availability of improved technology, is that the application of these four factors has been made in a very disorganized way without creating a harmonious balance between them. For example, a lot of farmers apply 240 kg of P_2O_5/ha but do not apply lime, do not follow a crop rotation, and use only one type of implement such as a heavy disk harrow.

Interaction among these factors is very dynamic and their effect on productivity is exponential (Fig. 20.2). The productivity attains the maximum over time because natural resources impose that restriction; i.e., there is an upper limit to the productivity. Once the plateau is attained, it is important to consider another tetrahedron specific to the Natural Resources including factors such as climate (A), soil (B), vegetation (C), and hydrology (D). A third tetrahedron called Production Systems should also be considered. This tetrahedron is composed of climate (A), soil (B), agropastoral and forest systems (C), and social-economical aspects (D). These three tetrahedrons interact among themselves in determining productivity (Fig. 20.3).

Now those three tetrahedrons determine sustainability of agropastoral, silvicultural, and arable land-use systems under irrigated and rainfed conditions. The sustainable productivity is represented by an oval shape because that is the most stable of the geometric forms (Fig. 20.4). Being a dynamic function, productivity will evolve over time, each evolutionary trend will have its own plateau, because there is a limit for the energy (derived from the sun) that reaches the biosphere (Fig. 20.5).

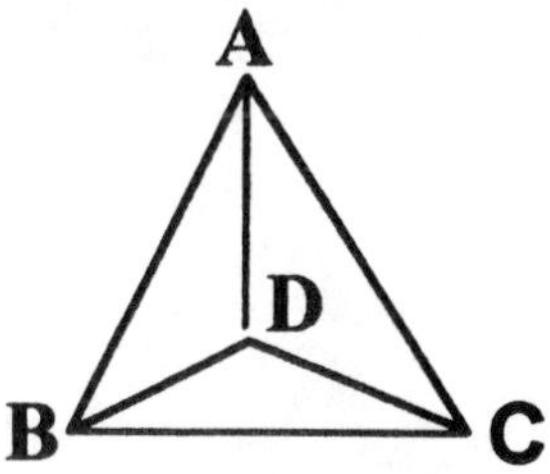

Figure 20.1. Harmonic representation among the soil management factors.

From the agronomic point of view, one of the most important factors that must be taken care of is the dynamics of soil tillage systems. It is the dynamics in a sense because each implement has its own time of operation and certain functional characteristics to accomplish specific objectives.

Since the first and foremost objective is to increase soil pH and base saturation besides increasing availability of phospho-

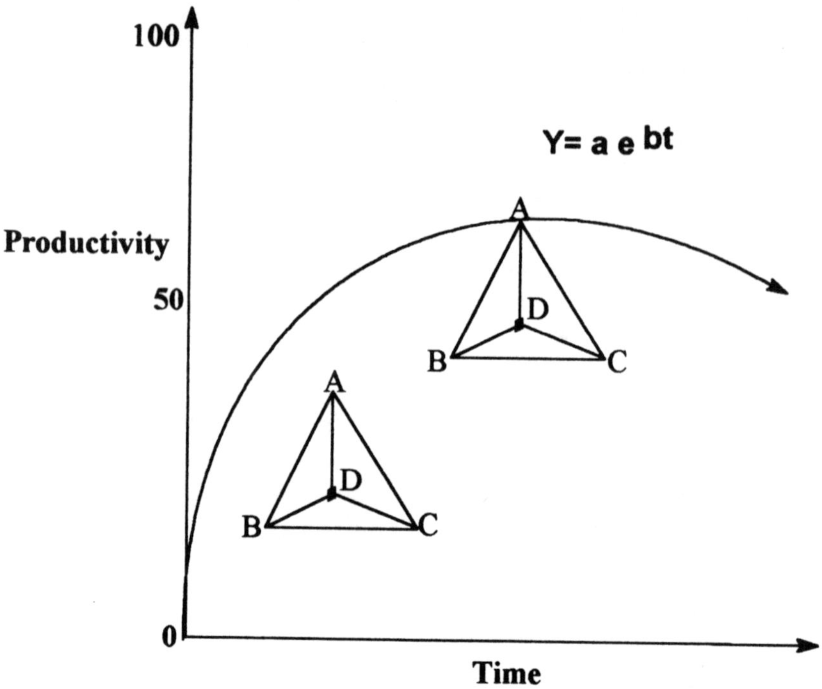

Figure 20.2. Efficiency of the soil management factors applied harmonically with time.

rus and potassium to alleviate soil fertility constraints, the best implement to achieve these through soil surface management is the disk plow. When pulled by the tractor, it mixes the soil and enhances soil chemical reaction. Furthermore, to apply lime in two split applications, one before plowing and another after but before harrowing with light disks, disk plow facilitates uniform distribution of lime into the soil profile because it can reach up to 25–30 cm depth under favorable soil moisture condition and friable tilth.

Once the soil chemical problems have been corrected, it is necessary to incorporate the maximum possible organic carbon into the soil profile to improve structure. Crop residues and green manure are important to achieve this objective. The best implement to incorporate residue is the moldboard plow which can bury the biomass up to 40 cm-depth for rapid decomposition and humification (Arai et al., 1994). Kumada (1987) fractionated soil organic matter based on solubility characteristics using soluble fractions in NaOh

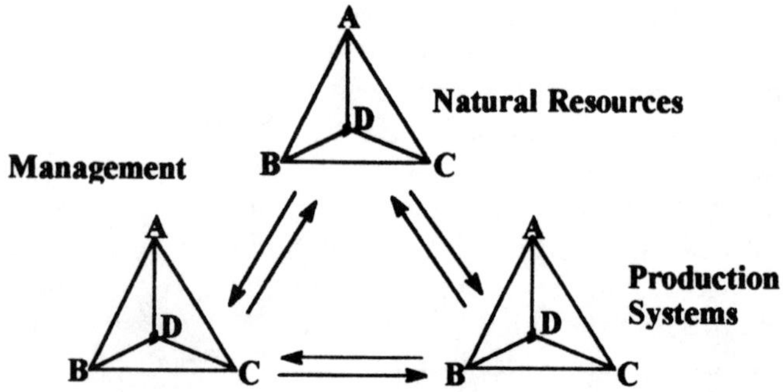

Figure 20.3. Schematic representation of the interrelationships among Natural Resources, Soil Management and Production Systems.

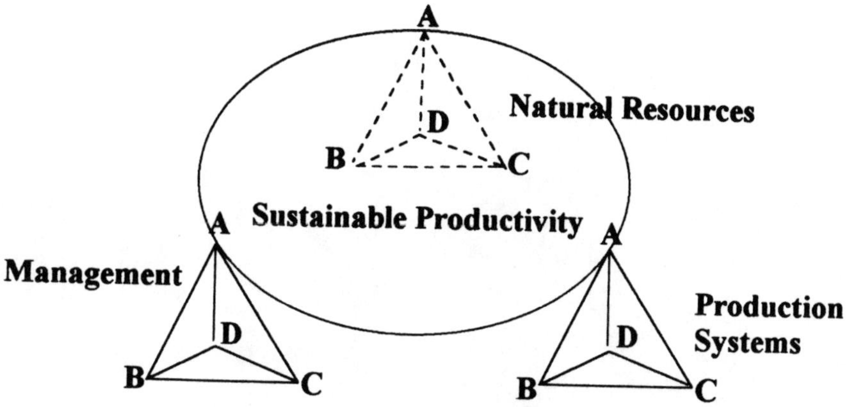

Figure 20.4. Representation of the sustainability by the Natural Resources, Management and Production Systems tetrahedrons.

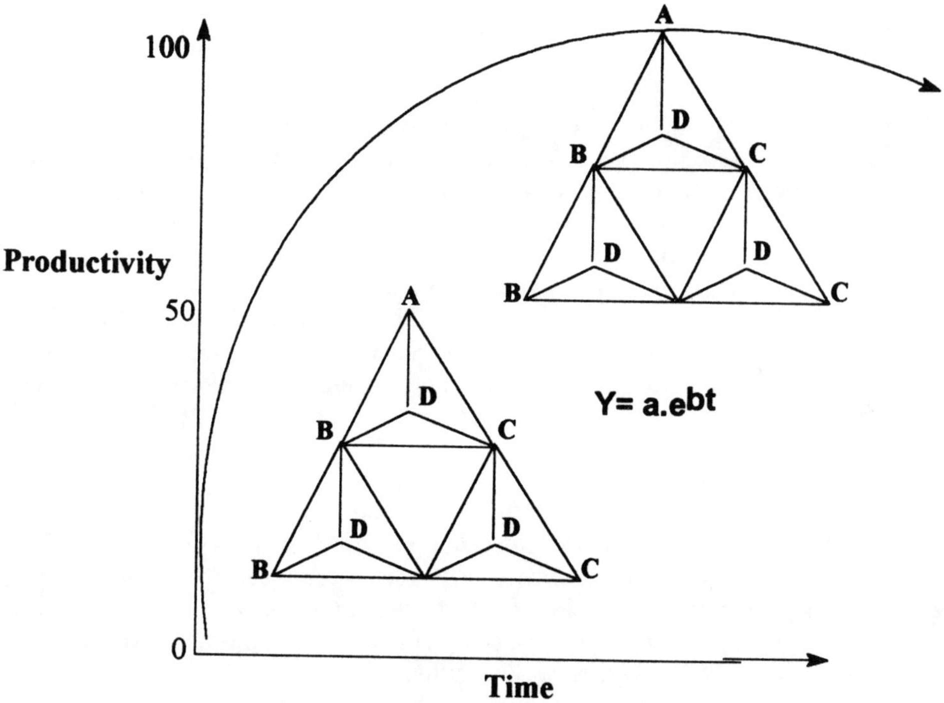

Figure 20.5. Sustainable productivity evolution with time.

(free organic matter and linked to Al), $Na_4P_2O_7$ (linked to Ca and Fe) and $NaBH_4$ (occlude inside the clay minerals) solutions and precipitated in H_2SO_4 which constitutes the Humic Acid (HA). Denaturation of humic acid was reflected in its optical properties (relative color intensity (K_{600}/C) and color index = log {(Absorption at 400 nm/Abs at 600 nm)}. K_{600} refers to optical density at 600 nm of a HA solution, and C is the mL of

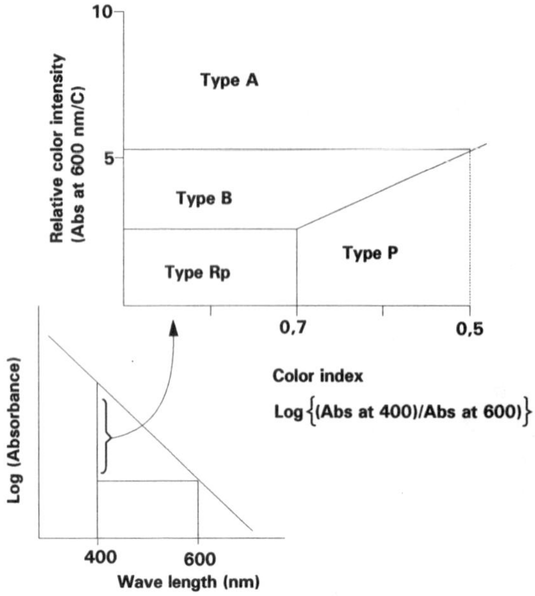

Figure 20.6. Classification of humic acid types.

$KMnO_4$ consumed by 30 mL of the HA solution to oxidize carbon. According to these properties, humic acid was classified into either Rp, B, P, or A types (Fig. 20.6). During the humification process, the color intensity increases and the value of color index decreases. HAs from disk and moldboard plow treatments were classified as type A reflecting high degree of humification. However, relative color intensity (K_{600}/C) and color index values for both tillage implements were 0.63 and 6.4 compared with 0.54 and 8.0, respectively. These results demonstrate the higher degree of humification obtained with moldboard plowing. It is likely that aluminum protects humic acid against decomposition in Cerrados soils leading to an increase in humin and a decrease in fulvic acids (FA) when soils are cultivated using these implements. The ratio of carbon in HA:FA was 0.68 for the moldboard plow and 0.26 for the disk plow (Resck et al., 1990; unpublished), indicating a higher degree of polymerization.

This process of soil conditioning can take at least four years to improve soil physical and chemical properties. Having accomplished the first stage, farmers proceed to the second stage by either choosing an implement that causes least soil disturbance such as chisel plow for minimum tillages or shift to no-tillages depending upon their technical skills and financial resources. Moreover, farmers can adopt pasture or forest systems to maintain and even improve soil properties over time.

Water storage in the soil profile is much greater because of the increase in the soil tortuosity factor, and as consequence there is higher flux of nutrients to plants, and high yield.

Finally, in order to attain sustainability, soil and water management should be planned on a watershed basis. Agropastoral and forest ecosystems are rational land uses within a watershed. From a practical point of view, it is better to consider a hydrologically simple watershed (1st, 2nd, or 3rd order) which covers about 1,000 to 10,000 ha (Resck, 1992). The goal is to double the production of grain, beef, wood, fibers, and energy without compromising the natural resources but improving the social-economical status of the population. Furthermore, the water discharge from the watershed should be of a much better quality; e.g., without any sediments, pesticides, or pathogens.

CONCLUSIONS

There are about 71.19 million ha of arable land yet to be developed in the Cerrados region, considering that only 43.5% of the land has been cultivated, contributing about 40% to the national production. The average grain yield for predominant crops is 1,939 ± 962 kg/ha and a lot remains to be accomplished in achieving high yields on cultivated lands prior to clearing any new land.

There is a need to develop a policy about the population migration from rural and/or other places to the cities which imposes additional stress on natural resources, and requires substantial increase in food production.

The four factors related to soil and water management and conservation (pH correction, fertilizer applications, crop rotation, and dynamics of soil tillage systems) are not being used systematically. Consequently, there is a serious problem of soil erosion, diseases, water pollution, and low crop yields.

Because of the mineralogy of most soils in the region, organic matter plays a very important role in CEC, soil structure, water, and nutrients availability.

Because of the low available water to the plants, these soils (mainly Latosols and Quartz Sand) must be plowed as deep as possible with disk plow (chemical correction) and moldboard plow (physical correction) to increase soil water storage and improve soil quality (right H and bases saturation plus P). Long dry spells of 40 days or more can occur during summer. Minimum or no-tillage, pasture and forest plantation are the best choices to achieve sustainable use.

Above all, watershed should be the planning unit for integrated soil and water management and conservation to increase food, fibers, energy, and good water supply.

REFERENCES

Adámoli, J., Macedo,J., Azevedo, L.G. and Madeira Netto, J. Caracterização da região dos Cerrados. In Solos dos Cerrodas: tecnologias e estratégias de manejo. Goedert, W.J., Ed. São Paulo: Nobel/Brasília: EMBRAPA-CPAC, 1986, pp. 33–74.

Arai, S., Resck, D.V.S., Cardoso, A.N. and Linhares, N.W. Quantitative and Qualitative Characteristics of Soil Organic Matter under Cerrados. In Relatório Técnico do Projecto Nipo-Brasileiro de Cooperação em Pesquisa Agrícola nos Cerrados 1987/1992. Brasília. Reports, EBMRAPA-JICA, 1994, pp. 477–487.

Azevedo, L.G. de and Caser, R.L. Regionalização do Cerrado. In Simpósio Sobre o Cerrado, 5, 1979, Brasília. Uso E Manejo. Anais. Planaltina: EMBRAPA-CPAC, 1980, pp. 211–229.

Dalal, R.C. and Mayer, R.J. Long Term Trends in Fertility of Soils under Continuous Cultivation and Cereal Cropping in Southern Queensland. II. Total Carbon and Its Rate of Loss from the Soil Profile. *Aust. J. Soil Res.*, 24, 281–292, 1986.

Dedecek, R.A., Resck, D.V.S., and Freitas Júnior, E. de. Perdas de solo, água e nutrientes por erosão em Latossolo Vermelho-Escuro dos Cerrados em diferentes cultivos sob chuva natural. *R. bras. Ci. Solo*, 10: 265–272, 1986.

Duxbury, J.M., Smith, M.S., and Doran, J.W. Soil Organic Matter as Source and Sink of Plant Nutrients. In *Dynamics of Soil Organic Matter in Tropical Ecosystems.* Coleman, D.C., Oades, J.M. and Uehara, G. Eds. Honolulu, University of Hawaii, 1989, pp. 33–67.

Empresa Brasileira de Pesquisa Agropecuária. Atlas do meio ambiente do Brasil. Brasília: EMBRAPA-SPI, 1994.

Folle, S.M. and Brandini, A. Uso de máquinas e implementos agrícolas no preparo de solos de Cerrado. In Simpósio Sobre o Cerrado, 7, 1989, Brasília. Estratégias de Utilização. Anais. Planaltina: EMBRAPA-CPAC, 1995, pp. 98–114.

Helling, C.S., Chesters, G. and Corey, R.B. Contribution of Organic Matter and Clay to Soil

Cation-Exchange Capacity as Affected by the pH of the Saturating Solution. *Soil Sci. Soc. Amer. Proc.*, 28, 517–520, 1964.

IBGE. Anuário Estatístico do Brasil—1975. Rio de Janeiro, IBGE, Vol. 36, 1989.

IBGE. Sinopse preliminar do censo agropecuário. Rio de Janeiro, IBGE, Vol. 4, 1987.

IBGE. Anuário Estatístico do Brasil—1989. Rio de Janeiro, IBGE, Vol. 49, 1989.

IBGE. Anuário Estatístico do Brasil—1994. Rio de Janeiro, IBGE, Vol. 54, pp. 3–35 to 3–38, 1994.

Kumada, K. *Chemistry of Soil Organic Matter.* Japan Scientific Societies Press, Tokyo, 1987.

Resck, D.V.S. Soil Water and Chemical Elements Losses by Using Rainfall Simulation on Soybean Cycle. Brasília, EMBRAPA-DID, 1981. (EMBRAPA-CPAC. Boletim de Pesquisa, 5).

Resck, D.V.S. Effect of Crop Residues and Liming on Soil Physical and Chemical Properties of Tropical and Temperate Soils. Ph.D. thesis, Purdue University, West Lafayette, IN, 1986.

Resck, D.V.S. and Silva, J.E. da. Soil Organic Matter Dynamics Under Different Tillage Systems in the Cerrados Region in Brazil. In International Congress of Soil Science, 14, 1990. Kyoto, Japão. Transactions. *Kyoto: International Society of Soil Science, 1990*, Vol. 6, pp. 325–326.

Resck, D.V.S. Uso e ocupação do solo no Brasil Central. Planaltina: EMBRAPA-CPAC, 1991. (EMBRAPA-CPAC. Documentos, 35).

Resck, D.V.S., Pereira, J. and Silva, J.E. da. Dinâmica da matéria orgânica na região dos Cerrados. Planaltina: EMBRAPA-CPAC, 1991. (EMBRAPA-CPAC. Documentos, 36).

Resck, D.V.S. Manejo e conservação do solo em microbacias hidrográficas na região dos Cerrados. Planaltina: EMBRAPA-CPAC, 1992. (EMBRAPA-CPAC. Documentos, 40).

Resck, D.V.S. Dinâmica da matéria orgânica e seus efeitos nas propriedades do solo. In *Encontro Nacional de Rotação de Culturas*, 2. 1992. Campo Mourão. Anais. Campo Mourão: AEACM, 1993, pp. 117–143.

Resck, D.V.S. Manejo de solos e sustentabilidade dos sistemas agrossilvipastoris na região dos Cerrados. In *International Symposium on Tropical Savannas: Biodiversity and Sustainable Production of Food and Fibers in Tropical Savannas*, 1. 1996. Brasília. Proceedings. Brasília: EMBRAPA-CPAC, 1996, pp. 81–89.

Spain, J.M., Ayarza, M.A., and Vilela. Crop Pasture Rotation in the Brazilian Cerrados. In *International Symposium on Tropical Savannas: Biodiversity and Sustainable Production of Food and Fibers in Tropical Savannas*, 1. 1996. Brasília. Proceedings. Brasília: EMBRAPA-CPAC, 1996, pp. 39–45.

Silva, J.E., Lemainski, J., and Resck, D.V.S. Soil Organic Matter Losses and Their Relationships with Cation Exchange Capacity in Soils of the Cerrados Region of Western Bahia, Brazil. *Revista Brasileira de Ciência do Solo, Campinas*, 18 (3), 541–547, 1994.

Van Raij, B. *A capacidade de troca de cátions das frações orgânica e mineral em solos*. Bragantia, Campinas, 28, (8), 85–112, 1969.

Woomer, P.L., Martin, A. Albrecht, A., Resck, D.V.S. and Sharpenseel, H.W. The Importance and Management of Soil Organic Matter in the Tropics. In *The Biological Management of Tropical Soil Fertility*. Woomer, P.L. and Swift, M.J., Eds. John Wiley and Sons, New York, 1994, pp. 47–80.

TWENTY-ONE

Production Systems in the Savannas of Brazil: Key Factors to Sustainability

Carlos R. Spehar

INTRODUCTION

The Brazilian savannas (Cerrados) occupy 2.1 million km^2, with a gradient of vegetation, from open fields to dense woods that resemble the African savannas, except for differences in soils. Acid soils (Ferralsols) are predominant, with low levels of nutrients, high available aluminium and low moisture holding capacity. Soil research has contributed to expand modern agriculture in its domain (Spehar and Souza, 1995). The concomitant genetic improvement of annual crops, like soybeans (*Glycine max* [*L.*] Merrill), has performed an important role in the modern agricultural settlement (Kiihl and Garcia, 1989; Souza and Goedert, 1989; Spehar, 1994a; Spehar, 1995a).

Soil pH in native areas ranges from 4.0 to 5.5, with high Al-saturation (up to 80% of the base saturation). Soil clay fraction consists mainly of kaolinite and iron and aluminium hydroxides, with low cation exchange capacity (CEC<13.0, most common around 4.0 cmolc/kg) and high adsorption capacity for phosphates and, in a minor scale, sulfates. (Spehar, 1994c). Availability of P, K, Ca, Mg, and Zn is low in virgin areas and highly dependent of organic matter (OM) content (Resck et al., 1991).

The climate of the Cerrado presents two moisture regimes: the rainy and the dry season. The average yearly precipitation varies between 1,000 to 2,000 mm and is concentrated between September and May, with changes according to the influences of neighboring physiographic regions, i.e., Amazon Rain Forest, Pantanal, Atlantic Rain Forest and Caatinga (Fig. 21.1). Only 15% of the rainfall occurs in the dry season and this is a limiting factor for cultivation of annual crops throughout the year. The excess water in the rainy season results in large water infiltration in the ground. This forms the streams and tributaries of three major river basins of South America which have been a source of irrigating water to extend the cultivation into the dry season. Although not high in the rainy season, the range between minimum and maximum temperature increases in the dry season in the core area of the Cerrados, at an average altitude of 1,000 m above the sea level. In the lowlands, the range for temperature diminishes.

Dry spells usually occur during the rainy season and are aggravated in the borders to the semiarid zone (Caatinga). They affect the performance of annual crops when they happen at the reproductive phase (Assad et al., 1994). Yield depression is variable accord-

Figure 21.1. Cerrado of Brazil: Its location in the country in relation to adjacent physiographic regions of the Amazon Rain Forest, Pantanal, Atlantic Rain Forest, and Caatinga (semiarid).

ing to crop characteristics. Rice (*Oryza sativa*) and maize (*Zea mays*) are more susceptible than soybeans, which escape moisture stress due to its longer flowering period. Soybean adaptation to the Cerrados has been a key factor for establishment of annual cropping systems (Spehar, 1994a; 1995a). The shallow rooting of cultivated plants is considered to be the main cause for susceptibility to dry spells. The benefits of fertility improvement will reach deeper layers of the soil and improve root growth conditions (Ritchey et al., 1980; 1982). Aluminum and low-Ca tolerance has also contributed to solving the problem (Fageria et al., 1989; Spehar, 1994c; 1994b; 1995b; 1995c; Spehar and Makita, 1994). The conjugation of soil improvement and crop breeding allows exploitation of leached down nutrients and subsoil water for yield stability.

The vast predominantly flat areas of the Cerrados are less exposed to runoff and soil erosion. This has facilitated the expansion of mechanized farming of crops. Careless land development and repeated tillage, with uncovered soil during four to six months of the dry season, have, however, increased vulnerability of the system, due mainly to OM loss and its consequences. Soil texture ranges from clayey to very sandy (Inceptisols), with even more reduced moisture-holding capacity and higher susceptibility to erosion (Resck et al., 1991). Soybean, maize, upland rice, and common beans have shown extraordinary performance in recently developed areas, with moderate disease, insect, and weed threat. This chapter is focused on the present farming systems, pointing out the problems of reduced alternatives of land and crop uses, with reduced income, aiming at the future exploitation on a sustainable basis.

CHARACTERISTICS AND LIMITATIONS OF PRODUCTION SYSTEMS

The Cerrados were exploited for subsistence agriculture and beef cattle ranching, in the rate of one head/5 ha of native grassland until the early 1970s. The local populations were confined to small villages scattered in a large area, and population increase is directly related to recent land development (Spehar and Souza, 1995). Earlier settlements in relatively fertile patches are marked by small farm holds and, in recent times, they cover medium to large areas in naturally low-fertile, improved soils. Reduced machinery, in all cases, has resulted in bad soil and crop husbandry practice (Urben Filho and Souza, 1993). Poor practices resulted in yield losses and in soil erosion, particularly in areas with high rain intensity. Soybeans and other subsequent annual crops that became cultivated in the region, partly or fully mechanized, have been affected by scarcity of and untrained local labor. All these factors compounded to increase farming problems in the Cerrados.

At the farm level, local differences in topography, soil texture, springs, and native vegetation associated with these environmental components have not been considered before land development, and integrated farming system concepts were not developed until the late 1980s. Decline of worldwide agricultural commodity prices also affected Cerrados cultivation, which concentrated mainly on soybean monocrop (Spehar and Souza, 1995).

Farming in the Cerrados has also been limited by reduced storage and transportation facilities. The rapid growth, a consequence of low interest credit, caused disturbance in new agricultural developments. The large farm holds have organized their own base, while the small ones have relied on cooperative associations to cut input costs. Recent cropping system diversification to further reduce costs incurred in pest and disease control has become necessary, although other long-term benefits have not been taken into account. The access to technology by farmers is directly related to farm size: the small farmers are, in general, less informed than medium to large farmers and liable to use bad technology. This is one of the reasons for differences in efficiency among them, since technology is adaptable and can be validated in farming systems.

Soil Fertilization and Crop Performance

Methods to overcome low fertility of soils have been defined. Lime is abundant and has been supplied to farmers. However, due to low buffer capacity of soils, overliming and naturally low soil reserve has led to micronutrient deficiency problems (Ritchey et al., 1982; Galrao, 1991). For clayey soils, 4 t/ha of lime (100% $CaCO_3$ equivalent) are needed to obtain 50% base saturation.

Cerrado soils show striking responses of cultivated plants to P application. In the past, farmers relied on government subsidies for P fertilization for economical grain production, but this is no longer available. One strategy to improve available P is the combination of partly-soluble rock phosphate and cultivation of perennial pasture before shifting into grain production. There are, however, other important lines of research that need study, such as efficient P use, ability to solubize, and/or extra P in plant species (Ak et al., 1995).

Potassium availability depends on soil CEC, liming produces an increase in negative charges that retain K. However, this can be enhanced by improving the OM content further to prevent its leaching. Bad soil and crop residue management causes reduction in OM, resulting in decrease in available K. In sandy areas the problem is more intensified, and best results are obtained by split application of K for annual crops: half in the furrow at sowing and half broadcast before bloom (Vilela et al., 1985).

Subsoil improvement can be achieved by the use of gypsum ($CaSO_4$) to increase Ca movement in the soil profile. A rate of 200 kg/ha $CaSO_4$/crop is needed for soils with high Al saturation rate (>30%) at depths of >35 cm and with a Ca level <0.2 cmolc/kg). Gypsum is recommended at rates of 0.5, 1.0, and 1.5 t/ha for sandy, mid-textured, and clayey soils, respectively.

Soil Organic Matter Management

The OM status of the Cerrado soils is in the range of 1.8% to 2.6% for areas with soybean monocrop. The level is even lower in sandy soils and this becomes critical, compromising long-term production. Following land clearing, soil liming, and fertilization, the OM content declines. Soybean cultivation initially increases soil OM and causes its decline on subsequent monocrops (Resck et al., 1991). Cultivation and burning of crop residues in the dry season, as weed control means, increases the soil susceptibility for erosion, further reduces OM and nutrient adsorption. The consequences of OM loss are: soil structure destruction, increased density, erosion, CEC reduction, nutrient loss, crop vulnerability to drought, reduction in water infiltration, intermittence in spring and river flow, destruction of long-term production perspective. Monitoring to assess the status of this and other key environmental attributes is necessary (Lal, 1994).

Crop rotation can contribute to maintenance of soil OM, resulting from differences in C/N ratios of the crop residues. Sole cultivation of soybeans causes OM reduction because of the low C/N ratio, with consequent high rates of decomposition (Resck et al., 1991). Ground cover during the dry season helps to reduce weed infestation, protects the soil against excess radiation, and contributes to maintain and/or increase OM, ensuring yield stability. Thus, the need for alternative, drought-tolerant crops with high fiber and large biomass production to be adapted to the production systems.

Impacts of Crop Rotation

Monocrop soybean, associated with bad tillage, has increased diseases, pests, nutrient disorders, soil erosion, and fertility loss, causing negative impact on the environment and on the economics of farming. Crop rotation and double cropping are the logical alternative, which has started since the late 1980s.

The advantages of soybean-maize rotation has already been reported (Araujo et al., 1989). In a crop rotation experiment, Vasconcelos et al. (1989) evaluated the dry matter production and nutrient yield of different cropping systems. Inclusion of legume green manure either sole or intercropped with maize, can recycle large amounts of nutrients and provide additional N by biological fixation. Higher yields for maize were obtained in rotation with soybeans, in addition to its positive nutrient recycling and nematode control.

Association of green manure with annual crops has been an alternative to maintain soil fertility. Pereira et al. (1992) found that grain crop yields following green manure increased up to 46.6% higher than sole crops. Similar results have been obtained by Resck and Pereira (1979) in experiments with soybeans as a sole crop compared with soybeans following green manure. Different species of plants were utilized as green manure and incorporated into the soil at blooming.

Another possible system which is becoming more successful in the Cerrados is the annual crop-pasture rotation. This is a very efficient nutrient recycling system that also contributes positively to enhance soil structure. The grass-legume association in rotation with annual crops offers the following advantages: soil fertility improvement by incorporating/cycling N, P, S, and other nutrients through crop residues, increased biological activity in the subsoil due to the deep root growth of perennial acid-tolerant plants and efficient nutrient recycling by plants and animals, improvement of soil physical conditions by better OM management and, consequently, reduced erosion, improvement of the soil microbiological activity ('biological tillage'), reduction of weed, pest and disease problems, elimination of some noxious weeds on pasture fields by plowing them for annual crop (Spain et al., 1996).

To keep the prospect for long-term soil productivity, it is necessary to establish production systems that maintain and, ideally, improve the soil structure. It is observed in farmers' fields that annual crop-pasture rotation enhanced soil aggregation. This has resulted in a more than two-fold increase in the animal/stocking of the pasture following annual crops (Ayarza et al., 1993).

Additional studies on annual grain-pasture cropping systems are being undertaken to refine information to farmers. For grain production, a more dynamic shift to grain-pasture is desirable. It is expected that with diversification of cropping systems, the partial soil fertility improvement in extensive pastures will shift into more intensive beef and dairy cattle production. The maximum economical yield of the system, instead of sole crops, shall be assessed, and this will result in more stability to the Cerrado agricultural production.

In addition to OM maintenance, other advantages of rotation are higher yields of crops, reduced risk in marketing, increased supply of food and feed for livestock, and environment preservation due to control of pests and diseases. These arguments can be used to validate technology and to convince the farmers of the need for crop diversification, although it should be backed by appropriate government policies.

CROP ADAPTATION; TRADITIONAL AND POTENTIAL USES

Present Situation

Upland, rainfed rice and artificially established pasture are Al-tolerant, relatively low input crops that pioneered cultivation, prior to modern grain production. Soybean cultivation in the low latitudes of the Brazilian Cerrados, mostly situated between the tropical and equatorial zones, was limited by its day-length response, causing reduced plant height and yield. Response to day length, climate, and soil conditions was taken into account in local cultivar acquisition. This consolidated soybean cultivation in the 1990s. With varying vegetative and reproductive phases, the Cerrado selected soybean varieties ensures suitable crop planning (Urben Filho and Souza, 1993; Spehar, 1995a).

In 20 years, the Brazilian Cerrados have already accounted for 50% of the national production of soybeans, with mostly higher yields than in the traditional temperate zone. The expansion in the cultivated area is a direct consequence of new high-yielding genotypes which incorporate genes for lateness and juvenility and aluminium tolerance (Kiihl and Garcia 1989; Spehar 1994a).

Maize has recently become a major crop in the Cerrado region. From a share of 17%, mostly concentrated in naturally fertile areas, it evolved to about one-third of the national production. It has been employed in rotation with soybeans, with yields as high as in the best areas where it is traditionally cultivated in the country. In the areas where the rainfall is prolonged, maize is also cultivated in double cropping, after the soybeans. Although fully mechanized like the soybeans, subsistence cultivation is a common practice that still contributes to low average yields.

Maize expansion has been carried out by farmers who have already had the infrastructure for soybeans. However, there are limiting factors such as the need for nitrogen fertilization (30 to 120 kg/ha), and additional care in storage. These are compensated for by the high-yielding ability of hybrids and varieties (EMBRAPA-CNPMS, 1992).

In modern cultivation, upland rice has been used in rotation with other crops and associated with pasture renovation. Reduced fertilization of the soil, bad crop husbandry, incidence of brusone disease, and the high frequency of dry spells in the growing season have been the main factors for low yield, as compared to other parts of Brazil. However, new high-yielding and disease resistant varieties and technologies have been developed to enable the farmers to obtain higher and more stable levels of productivity in rain-favored areas (Kluthcouski et al., 1991).

Common beans (*Phaseolus vulgaris*) were cultivated in small scale, to sustain the scattered local population until 20 years ago. It was a low-input crop, practiced by subsistence farmers. This mostly explains the low yields obtained by the farmers, in contrast with modern bean cultivation under a central pivot irrigation system. Yield records of 4,000 kg/ha have been obtained with new high-yielding and disease resistant varieties. Disease and pest management, appropriate fertilization, and suitable crop husbandry have contributed to increased bean production in the Cerrados, which already accounts for about 25% of the national production.

The productivity of the four main annual crops, i.e., beans, maize, soybeans, and rice, is shown to increase with the time after savanna occupation (Fig.21.2). This is a consequence of research contribution in crop breeding and husbandry and soil fertility improvement, despite long-term threats due to monocropping and bad soil management practices.

Wheat (*Triticum aestivam*) and brewery barley (*Hordeum vulgare*) have been cultivated in the Cerrados of Southern Mato Grosso State, in double cropping after soybeans. The region has contributed to 10% of the national wheat production. Due to climate requirements, there is little opportunity for this rainfed cropping system to expand. A more recent trend is to grow these cereals during the dry season, under irrigation, following full season soybeans. This, however, depends on market prices to compensate for irrigation costs.

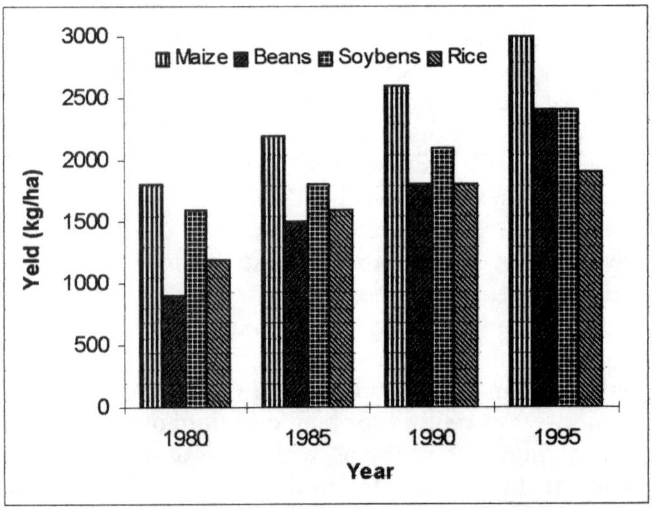

Figure 21.2. Grain yield (kg/ha) evolution in the Brazilian Cerrado, for beans, maize, soybeans, and rice.

Brachiaria decumbens has been the basic, long-term, pasture. It is, however, susceptible to insect (spittlebug), has low forage quality in the dry season and declines gradually over time. To overcome the scarcity of good quality forage, new species of exotic grasses have been selected, the promising are *Andropogon gayanus*, *Panicum maximum*, and *Paspalum* spp., the first two originated from Africa and the last from the fringe areas between the Cerrados and other physiographic regions such as the Pantanal to the southwest. Among the legumes, varieties of native *Stylosanthes guianensis* and *Arachis pinto* and *A. hypogea* (Pizarro et al., 1997) have shown potential for improving pasture quality. Studies on mineral element requirements of forage species have been conducted to achieve short-term large biomass production (Vilela et al., 1991).

There is, however, still a gap in what can be achieved and that obtained by farmers (Spehar and Souza, 1995). This is a direct consequence of input reduction and limited infrastructure, which results from insufficient or inappropriate K and P application, caused by increased fertilizer and reduced commodity prices, inappropriate crop diversification and husbandry (sowing pattern and date, plant population and variety), soil exposure during the dry season, disease, pest, and weed incidence.

The high specialization in modern production has led to a dangerously narrow genetic basis. The germ plasm collections have been used only to extract a few genes to disease and insect resistance. Repeated cultivation method leads to considerable losses in physical and chemical properties of soils. Inadequate weed management in the present system, without ground cover after the main crop is harvested, contributes to increase soil density and erosion problems, increasing production costs and negative impacts on the environment. The prevailing monocrop leads to a reduced use of production potential (energy harvest) on the one hand and bad plant and soil management on the other. This will lead into soil quality loss. There will be a trend to increase nutritional imbalance and reduced yield quality of final product. This increasingly high input-low output cycle, after repeated bad management, reaches unsustainability in short-medium term (Fig. 21.3).

Future Situation

The introduction, characterization, and conservation of germ plasm from species with economical potential, to follow or anticipate the main crop, is needed to provide cover

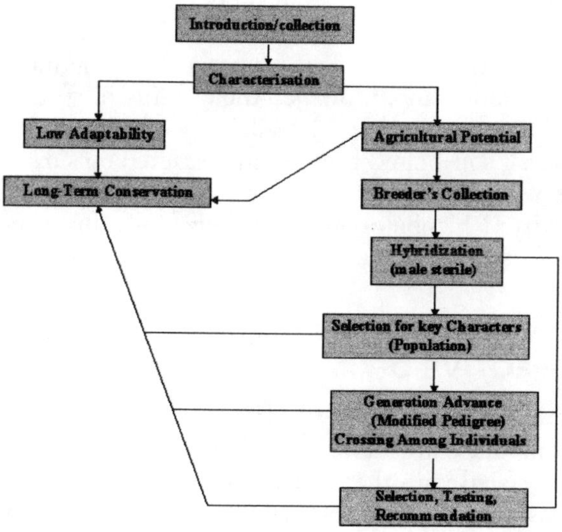

Figure 21.3. Schematic sequence to adapt potential crops to production systems.

and additional output. This will protect the soil in the dry season, help to increase nutrient availability and recycling, reduce weed infestation, maximize land utilization, and provide extra income to the farmers.

The Cerrado region has a rainy season ranging between six to eight months. Thus the amount of residual moisture is variable and its use by cover crops depends on their relative tolerance to drought. To overcome the climatic risks, small seed species like quinoa (*Chenopodium quinoa* Willd.), amaranthus (*Amaranthus* spp), and te'f (*Eragrostis tef)* can be relay sown. Moreover, if there are losses due to irregular rainfall at the beginning or at the end of the rainy season, the fields can easily be resown at low cost. The farmer would dedicate only a reduced area to produce seeds for next crop season. If 2-3 kg/ha are sown to cultivate 200 ha, only 400-600 kg will be needed.

The germ plasm for new species should be representative and characterized to meet the following targets: drought tolerance, speed of growth, biomass and grain production, competitiveness against weeds, and quality of final product. It is necessary to consider the biology and mode of reproduction of each species for effective conservation. The modern approach in adapting these nontraditional species should be applied (Marshall and Brown, 1995). The core collections should represent 10% of big collections and supply the user a sample of the species diversity. From these breeders collections will be developed genotypes to supply specific needs.

The rates of genetic changes in domesticated species is more intense when compared to the evolutionary process (Hintum, 1992). Without human interference, environmental factors influence gene frequency, whereas in domesticated species the variability tend to restrict to utility genes. These are relative to human needs, and, consequently, underexploited species may have been developed for multiple purposes. Quinoa and te'f are examples of these: farmers harvest seeds for direct consumption and the remains are used as fodder.

Thus, for cover and green manure crops, the level of adaptation of modern cultivation is low, even though utilized in local production systems (Vietmeyer, 1986). Their diversified use as grain and forage production, nutrient recycling and enhancement, and biomass are directly related to fitness. The main objectives of introducing these nontraditional plants are: to improve botanical diversity, characterize their respective germ plasm to derive breeders collections, assess diversity in desirable characters, identify gene sources for future breeding programs, supply demand for alternative, high-value plant products, and, finally, to turn the Savanna Research Centre into a reference in production systems to these underexploited plants.

Benefits of crop rotation have gradually been understood, and among them the most known is break in disease and pest cycles of monocrop. However, the benefits of species diversification still need to be validated. Among them are: (a) availability of nutrients—the effect of root exudates by *Cajanus cajan* on soil-bound P has been measured on subsequent crops in Vertisols, and opens the way to further (Ak et al., 1995) study other species, (b) increased microbiological activity (Rhizobium and Micorrhiza) as a function of cropping sequence. Other synergistic effects shall be identified to enhance the production systems.

NEW CROPS TO ENHANCE PERSPECTIVE OF NONCONVENTIONAL PRODUCTION SYSTEMS

Germ plasm introduction, characterization, and conservation for cover crops and green manure shall contribute to fill the gap associated with the main crop and turn the grain production viable. Weeds, pests, and diseases common to the main crops shall interrupt

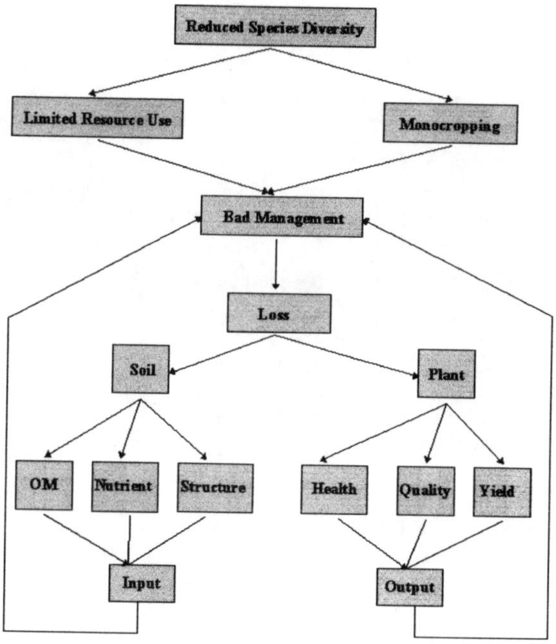

Figure 21.4. Consequences of reduced diversity of production systems in the long term.

their cycle. Biomass production that protects and fertilizes the soil with nutrient recycling will intensify, additional income to the farmer will turn real (Spehar, 1996). How can a crop fit into the system? The schematic sequence in Figure 21.4 illustrates the improved steps for participation of nonconventional crops in the system, with production on a competitive basis.

Their adaptation depends on finding alternative uses for products (Spehar and Souza, 1993). Thus, the more applications of plant products, the greater the possibility of its participation in the system. Moreover, nonconventional uses for grain crop products shall increase the possibilities to achieve ground cover that yields net income to farmers. Most important in this dynamics of biomass/grain production is the maximum use of land with least soil disturbance and this includes a combination of tillage methods and cropping sequences, with constant monitoring (Lal, 1994). Some information is available and can be validated at farmers' levels, others need to be generated. Table 21.1 summarizes the traditional and alternative crops with present and potential uses. It can be seen that nonconventional grain crops and new uses for traditional ones can be found when compared to other parts of Brazil and the world.

This list is far from being comprehensive, and illustrates the possible integration of crops with different uses to improve the farming system. The approach should be based on biomass with the prospect of transforming grain and other plant parts into added value products and OM maintenance. The presence of perennial grasses and legumes, grown as annual/biennial species, turns the biomass production feasible with respective ground cover in large parts of Cerrados, with emphasis on the ones where sandy soil texture is associated with reduced rainfall for double cropping. New alternative management such as associated and oversowing of subsequent crops to save rainfed moisture, shall be developed to increase the possibilities of maximum production per unit of capital.

If crops are considered from the perspective of different uses, there are several new combinations which have not yet been exploited. In Table 21.2 there are some traditional and potential grain crop sequences that may become part of the farming system. The

Table 21.1. Present and Potential Participation and Uses of Annual/Biennial Crops in the Farming Systems of the Cerrado Region[a]

	Present												Potential											
	Cerrado				Brazil				World				Cerrado				Brazil				WorldCrop			
	G	M	F	O	G	M	F	O	G	M	F	O	G	M	F	O	G	M	F	O	G	M	F	O
Amaranths									*		*	*	*		*	*	*		*	*	*		*	*
Barley					*				*	*	*	*	*	*		*	*	*	*	*	*	*	*	*
Beans	*				*				*	*		*	*	*		*	*	*		*	*	*		*
Chickpeas									*	*		*	*						*		*	*		*
Crotalaria						*								*	*	*		*	*	*		*	*	*
Ground nut					*				*	*	*		*	*	*		*	*	*		*	*	*	
Guar										*	*	*		*	*	*		*	*			*	*	*
Forest turnip		*				*	*			*	*	*		*	*	*		*	*	*		*	*	*
Maize	*	*			*		*		*	*	*	*	*	*	*	*	*	*	*	*	*	*	*	*
Millet		*				*	*		*	*	*	*	*	*	*	*	*	*	*	*	*	*	*	*
Oats					*	*	*		*	*	*	*	*	*	*		*	*	*		*	*	*	*
Pasture			*			*	*			*	*	*		*	*	*		*	*	*		*	*	*
Peas	*				*				*		*		*		*		*		*		*		*	
Quinoa									*		*		*	*	*	*	*	*	*	*	*	*	*	*
Rice	*				*	*			*	*		*	*	*		*	*	*		*	*	*		*
Pigeon pea							*		*	*	*	*	*	*	*	*	*	*	*	*	*	*	*	*
Sesame									*		*		*	*	*		*	*	*		*	*	*	
Sorghum	*	*			*	*	*		*	*	*	*	*	*	*	*	*	*	*	*	*	*	*	*
Soybeans	*				*				*		*	*	*		*	*	*		*	*	*		*	*
Sunflower					*				*	*		*	*	*		*	*	*		*	*	*		*
Te'f									*	*	*		*	*	*		*	*	*		*	*	*	
Wheat	*				*				*	*	*	*	*	*	*	*	*	*	*	*	*	*	*	*

[a]Grain = G; Forage = F; Mulch = M; Other = O.

Table 21.2. Traditional (T) and Potential (P) Grain Cropping Sequences, in Relation to Component Characteristics of Disease and Pest Reaction (P = positive; N = negative), Nutrient Improvements/Recycling (H = high; L = low), Growth-Yield Reduction (GR = high; LR = low), Cost (E = effective; I = ineffective), and Organic Matter (S = stable; U = unstable)

Cropping System	Component Characteristics						Remarks
Soybean-soybean	T	N	H	LR	E	U	Vulnerable in short run; disease threat
Bean-bean	T	N	H	LR	I	U	Highly vulnerable; disease and pest threat
Maize-maize	T	N	L	LR	I	S	Vulnerable in short run, disease and pest threat
Rice-rice	T	N	L	GR	I	S	Vulnerable in short run, disease, growth inhibition
Soybean-maize	T	P	H	LR	E	S	Vulnerable economically, biologically viable
Soybean-maize (sequence)	T	P	H	LR	E	S	Biologically viable in extended rainy season
Soybean-maize (forage)-beans	P	P	H	LR	E	U	Viable in some areas; applicable in mixed farming
Beans-beans-maize (sequence)	T	N	H	LR	I	U	Vulnerable in the short run; disease and pest threat
Maize-stylosanthes (associated)	P	P	H	LR	E	S	Applicable in large areas, mixed farming
Maize-grass (associated)	P	P	H	LR	U	S	Short-term, dynamic cycling, mixed farming
Soybean-millet (sequence)	P	P	H	LR	E	S	Applicable in extended rainy season; mixed farming
Soybean-millet (oversown)	P	P	H	LR	E	S	Applicable in large area; mixed farming
Soybean-quinoa (oversown)	P	P	H	LR	E	S	Applicable in large area; mixed farming
Soybean-sunflower (sequence)	P	P	H	LR	E	E	Viable in extended rainy season
Soybean-wheat	P	P	H	LR	E	S	Viable in irrigated systems

comparative advantages of their use stimulates farmers to start at a small scale, for technology validation. Some of them rely on appropriate seed bed preparation, and the integration of tillage methods will follow crop diversification. The farmers shall look at the farming system as a whole, considering the factors that affect long-term land use and only after careful planning, move on into new systems. Monitoring shall be employed by farmers and training shall be provided to them. By this means, the causes of yield reduction are identified for adjustment in the cropping system.

TILLAGE METHODS IN RELATION TO GRAIN CROPPING SYSTEMS

Maximum economical yield of grain crops depends on sowing at the right time and at the suitable population, given other yield components have been taken care of. Day-length sensitive soybeans show a clear effect of delayed planting (Urben Filho and Souza, 1993). It also affects yield of long-cycle species and part of the available rain may be wasted, causing soil erosion problems.

Soil structure maintenance depends on seedbed preparation, organic matter maintenance, and drainage. Plowing under dry or excessively wet conditions destroys soil structure, and recovery may take several years of suitable management. Repetitive disk harrowing, commonly used, is the main cause of soil structure destruction (Resck et al., 1991).

Erosivity in the Cerrado region can be very high, with 70% precipitation occurring between November and March. This coincides with early stages of annual crop growth, which, in most cases, have not fully covered the ground. The amount of soil and nutrient loss following heavy rain with different cropping systems has already been demonstrated.

Erosion problems can be prevented when, before land clearing, farmers identify the soil type, topography, vegetation distribution, and water springs in the area. Ideally, cleared areas shall be large enough to handle with the available machinery, to avoid uncovered soil. Disk harrowing after harvest and during the dry season, carried out to reduce weeds, exposes the soil to severe wind erosion, particularly in the plateau region. The key to success is to increase efficiency of sowing before rainfall intensifies, for maximum crop yield, optimization of growing season, and soil protection.

Repeated tillage for every crop in seedbed preparation creates two distinct soil layers: one loose superficial layer and a compact sublayer. Different plowing depths have been shown to be effective in reducing density and, in turn, soil erosion and nutrient loss (Spehar and Souza, 1995).

Another alternative is using double cropping, in which the second cultivation is done by no-tillage sowing. This provides ground cover and utilizes residual soil moisture. At the beginning of the next rainy season, plant material is cut by shallow disk, prior to sowing. This method is only suitable in the areas with a long rainy season that can accommodate two crops.

An additional advantage of the no-tillage is rapid operation, suitable for double cropping and crop rotation, time and money-saving durable machinery. For the no-tillage method to be successful requires fertile soil, efficient weed control, adequate mulch cover, and suitable machinery and farming operations. For compact soil it is necessary to ameliorate the compact layer. The best root growth with depth is the mouldboard treatment preceded by no-tillage, when soybean and maize yields are increased.

No-tillage is of particular interest in intensive cultivation systems with irrigation in the dry season. This is the recent trend in farming which enables three crops in the same year, e.g., maize, soybeans or field beans, and wheat, soybeans, maize, and peas (*Pisum sativum*). These sequences apply whenever commodities have attractive prices and different potential applications. Before deciding the cropping systems, the farmers should consider that mulch reduces weed control cost, soil and water loss, and maintains/improves soil physical and chemical attributes.

The main limitation besides increasing soil density due to excessive seed bed preparation is the decline in soil structure with OM loss, soil and nutrient loss, and compromise of the future exploitation of agriculture. This contrasts with no-tillage, which, in addition, enhances the efficiency of sowing in the Cerrados. The decision on what tillage system to use depends, however, on appropriate monitoring of soil chemical and physical characteristics along the cropping history.

ECONOMICAL PROSPECTS FOR GRAIN PRODUCTION SYSTEMS

Maximum Economic Yields

Considering that soil fertility is not limiting and high-yielding varieties are grown, pests and diseases increase and compromise potential yields in sole crops; and production costs tend to reduce in rotation-double cropping systems. Analysis of grain production indicates the economical advantages of crop diversification.

Input saving crops such as legume should be employed to maximize other components of the system. Thus, if P level is high for cultivation, the limiting element would be N. It can be partly supplied to the system by legume and other species sequence.

Output should be measured over input. This ratio reflects the income of cropping sequences. Economical analysis should be carried out to assure that the maximum yield is attained in the sequence, with nutrient recycling. Projected net income per crop sequence should be computed beforehand, and guide farmers' planning.

Table 21.3. Economical Analysis of Cropping Systems, Including Soybeans, Maize and Pasture (Average over Three Years)[a]

Cropping Sequence	Production Cost/ha (R $)	Output/ha (R $)	Net Income/ha (R $)
Soybeans Monocrop	400	500	100
Maize monocrop	450	600	150
Maize-soybeans	430	550	120
Maize-soy-pasture[b]	330	650	220

[a]Soybeans yield = 3.0 t/ha; maize yield = 7.2 t/ha; meat production = 360 kg/ha.
[b]Three years soybean monocrop, after land development, followed by maize-pasture, pasture and soybeans; 20 months pasture, 3 heads/has, before shifting to grain production.

In Table 21.3, some cropping sequences, total input for the system, total output, and net income are listed. If soil management and disease and pest problems are considered, the monocrops should show high income reduction. If new uses for traditional crops and introduction of new crops are considered, it is expected that the system would be more advantageous economically than only grain production. The approach is to maximize output per capital unit, with preservation of environmental attributes.

ENVIRONMENTAL IMPACTS OF GRAIN PRODUCTION

Several species of plants, typical of the Brazilian savanna's unique ecological environment, and still unexploited by mankind, are in danger of extinction. The presently available law determines the procedures for land occupation such that the vegetation along river banks, lakes, and natural or artificial water reservoirs, in catchment areas, on top of hills and mountains, and on fringes of plateau and slopes be preserved (Spehar and Souza, 1995). If the areas are flat and there are differences in native species distribution, ideally, 20% of the virgin area in the property should be kept as reserve, in slopes, virgin land clearing should be done in contour belts, with terracing at the preservation areas and to prevent erosion. Considerable destruction of virgin areas, however, has been carried out, in a complete environmental abuse. Law enforcement should be first priority to avoid further destruction of valuable genetic resources.

The approach to make the environment productive, taking advantage of abundant heat and moisture, was, in the past, achieved mostly by soil fertility improvement and plant adaptation. Liming and fertilization, although now routinely employed, were attained by experiments where tentative levels were tested against economical productivity of potential crops. More attention must be devoted to the genotype x environment interaction in a cooperative effort to direct breeding programs for the Cerrado needs. This will enhance the opportunity for profit and, in the long run, farmers will capitalize in more intense use of land and preserve remaining native areas.

More complex problems such as stress tolerance (moisture, toxic elements, nutrients), nutritional balance, crop rotation and nutrient recycling, disease and pest integrated control and resistance shall be tackled by team effort. The results will contribute not only to rationalize solutions to these problems but also to prevent the occurrence of unfavorable environmental impact.

The integration of a research team to all sectors and organizations involved in agriculture and environment will result in overall planning for taking the right decision at the right time. It is necessary to learn from the troublesome beginning of Cerrado occupation. Farmers could, then, artificially afford to pay for mistakes. Society as a whole had a burden to carry as a consequence, by paying new and higher taxes, thus draining resources from one sector to another. Producers of agricultural machinery, lime, fertilizers, and those involved in trading and shipment of agricultural goods benefited the most.

The massive incentives in extensive agricultural countries during the three decades after World War II increased food supply and are the probable origin of present low commodity prices. This scheme worked to buffer the overheated and/or war-disrupted economies. However, the trends indicate little possibility to artificially support agricultural production. Farmers need to be guided about the consequences of bad farming management for themselves and for society as a whole (Spehar and Souza, 1995). In this area, only the good managers with real inclination for agricultural enterprises will stay active. The government, however, should provide the opportunity for management training.

The establishment of cooperatives and associations among farmers should be included in government policies at a high priority so that all the people involved in agricultural production will project a target of sustainability. The most recent approach to agricultural exploitation in an environmentally balanced system is related to the microbasin or catchment integrated planning. In this system farmers share and utilize the common natural resources. Even though they are complex, it has been suggested that they should be the basis for occupation planning (Resck et al., 1991). From the main stream, up to the catchment areas, the potential environmental attributes should be identified according to soil type, climate, relief, natural vegetation, hydrology, and geology. These will determine the farming systems that should successfully create permanent settlements.

There are pilot projects from which further studies will be carried out to systematically determine the pattern of settlement/exploitation that preserves the environment and still suits the needs of an already growing regional population. It is expected that these projects will be conducted by research organizations, state extension services, cooperatives, and communities to solve the present problems and to prevent future problems. It is also expected that appropriate government policies, taking integrated planning as a priority for incentive, will stimulate the communities to rapidly reach the stage for more rational use of the Cerrado environment.

Monitoring soil OM, nutrient, and water dynamics can be done by special techniques, surface erosion is determined by monitoring soil level after crop season, OM is a function of cropping sequence management to maintain and/or raise its levels, nutrient cycle depends on OM in the soil and soil preservation, water dynamics depends on suitable management that maintains water flow in the system (Lal, 1994). Springs, streams, and main rivers will change their volume and flow as a consequence of reduced water infiltration. This will cause further destruction of native species in reserves and compromise irrigation projects, turning them unfeasible (Spehar, 1996).

The Brazilian Cerrado occupation phases are contrasted with trends for declining yields and environmental deterioration. Integrated crop management should stress the farming patterns that reduce risks, conserve and improve soil fertility, recycle nutrients, increase crop productivity and value added at the farmer's level. A system approach, to use already available technology and the formation of multidisciplinary research teams to tackle complex problems, should be aimed at sustainability, defined by maximum income per unit of investment with least negative environmental impact.

In Figure 21.5 are the main components for successful establishment of sustainable grain crop activity. This depends on comprehensive analysis of production components in which the respective negative impact is contrasted with environmental attributes. It shall be used as a guideline for farmers, extentionists, researchers, and policy makers.

Quantifying the effects of environmental components on commodities allows definition of infrastructure needs and the input levels. The integrated farming system shall be

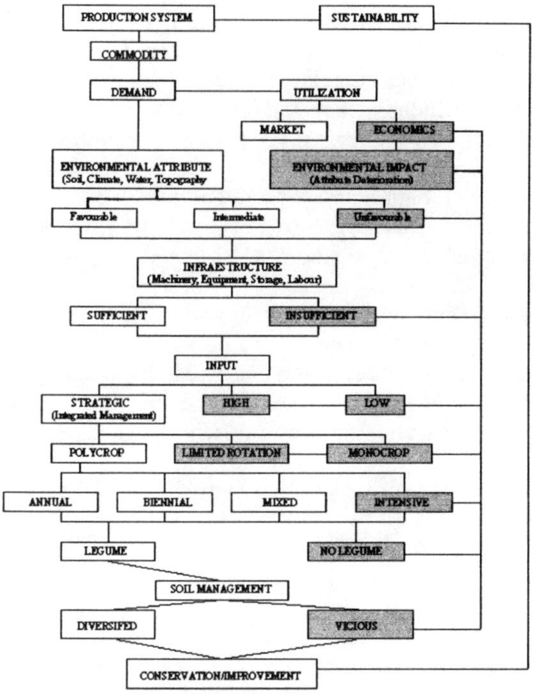

Figure 21.5. Factors affecting the establishment of sustainable agriculture

conducted to reduce negative impact which is monitored during cultivation practice; this allows determining relative importance of the production component.

The prerequisites for a given commodity are presented in Figure 21.6. When the environment attribute is contrasted with species characteristics, it becomes possible to come up with production systems that may bring about maximum yield, and cause the least negative impact. These attributes should direct research planning which, to be effective, influences political decisions. Technology should be available to farmers, who decide what to fit into their system, based on demand and production cost. The following basic points should be taken into account to achieve sustainability of the system:

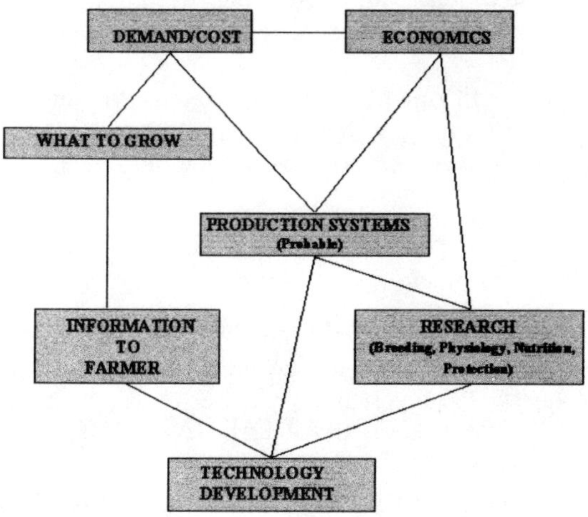

Figure 21.6. Participation of commodity in production systems.

1. Additional investment and formation of a critical mass of researchers in genetics, breeding, crop physiology, microbiology, production modeling, soil management and fertility improvement, water and natural resources management and conservation,
2. Greater integration among researchers and institutions in a real, dynamic, multidisciplinary approach to improve system undertaking of complex problems. Some important lines of research are selection of species and varieties stress tolerant (water, low-nutrient, toxic elements), efficient in nutrient solubilization, high biomass production, high quality food, and feed, nutrient recycling, microbiological efficiency, nutritional requirements for economical yields of crops, crop rotation dynamics, and resistance to major diseases and insects,
3. Hierarchical classification of problems for experimentation, research, and development, to prevent occurrence of negative impacts to the environment,
4. Integration of researchers and extensionists for demand definition and research planning,
5. Holistic approach that integrates all sectors involved in the production chain, such that the overall planning may influence definition of realistic, long-term policies for land use,
6. Planning agricultural exploitation that takes into account environment attributes in catchment areas, with a community-based concept to exploit the common natural resources, to be extended to the basin, taking into account the climate, soil, hydrology, natural vegetation, relief, and geology. These attributes will determine the probable farming systems,
7. Establishment of sustainability indices for the Cerrados, comparing available information for other parts of the world and in the Cerrados to quantify relative importance and means to stop the bad agricultural practices before natural resources and potential production of farming systems are irreversibly destroyed. These indices should be obtained after constant monitoring.

CONCLUSIONS

1. Agricultural expansion in the Cerrados has been favored by environmental conditions such as sufficient rainfall, definite wet and dry seasons, temperature, mechanized extensive flat areas, good soil drainage and low price of land, as opposed to low fertile acid soils, reduced water holding capacity, distance, and lack of infrastructure.
2. Lack of suitable planning and technology transfer, reduced and low-qualified labor, and little knowledge of the new environment by the settlers explained most of the failure at the beginning of Cerrado settlement.
3. Balanced fertilization and suitable OM management in cropping systems is of paramount importance for yield stability and can be achieved through alternate cropping and tillage systems.
4. Tillage method should be based on soil and crop monitoring; no till is beneficial to moisture saving, reduced erosion, OM and soil aggregation maintenance, and machinery and energy saving.
5. Development of a community-based concept for utilizing common natural resources and realistic government policies on transportation, education, welfare, and infrastructure will help to define farming systems for permanent settlements to turn the Cerrado into a major world food supplier.

REFERENCES

Ak, N., Otani, T., Arihara, J. Effects of Specific Compounds Exuded from Roots on Phosphorus Nutrition. In *Proceedings of Genetic Manipulation of Crop Plants to Enhance Integrated Nutrient Management in Cropping Systems*—1. Phosphorus. 1994. Johansen, C., Lee, K.K., Sharma, K.K., Subbarao, G.V., and Kueneman, E., Eds. Patancheru, India: FAO/IAEA/ICRISAT. 1995, pp. 117–128.

Araujo, N.B., Rushell, R., Eleuterio, A., Silva, N.C., and Santos, G. Rotacao anual das culturas de milho e soja. Goiania, Go, EMGOPA-DDT. 12p. (EMGOPA, Boletim de Pesquisa 15), 1989.

Assad, E.D., Sano, E.E., Matsumoto, R., Castro, L.H.R., and Silva, F.A.M. Veranicos na regiao dos Cerrados Brasileiros: Frequencia e probabilidade de ocorrencia. In Chuva nos Cerrados: Analise e espacializacao. Empresa Brasileira de Pesquisa Agropecuaria, Centro de Pesquisa Agropecuaria dos Cerrados, Brasilia, DF, EMBRAPA-CPAC: EMBRAPA-SPI, Assad, E., Coord., 1994, pp. 43–48.

Ayarza, M., Vilela, L., and Rauscher, F. Rotacao de culturas e pastagens em um solo de cerrados: estudo de cave. Congresso Brasileiro de Ciencia do Solo, Goiania, Go, Brazil, 1993.

EMBRAPA-CNPMS. Empresa Brasileira de Pesquisa Agropecuaria, EMBRAPA. Centro Nacional de Pesquisa de Milho e Sorgo, CNPMS. Relatorio Tecnico Anual—Periodo 1988–1991, Sete Lagoas, MG, Brazil: EMBRAPA-CNPMS, 1992.

Fageria, N.K., Baligar, V.C., and Wright R.J. The Effects of Aluminum on Growth and Uptake of Al and P by Rice. *Pesquisa Agropecuaria Brasileira*, v. 24, pp. 677–682, 1989.

Galrao, E.Z. Micronutrients and Cobalt on Soybean Yield in a Cerrado Soil. R *bras. Ci. Solo* 15:117–120, 1991.

Hintum, Th. J.L. van. The Structure of Genetic Diversity. In *International Workshop on Core Collection of Plant Germplasm*. Brasilia, DF: EMBRAPA-CENARGEN-SEMAMIBAMA/IBPGR-CGN. mimeo, 1992.

Kiihl, R.A.S. and Garcia, A. The Use of the Long-Juvenile Trait in Breeding Soybean Cultivate. In *Proceedings of the World Soybean Research Conference*, 4, 1989. Buenos Aires: AASOJA, v. 2, pp. 994–1000, 1989.

Kluthcouski, J., Pacheco, A.R., Teixeira, S.M., and Oliveira, E.T. Renovacao de pastagens de cerrado com arroz. I. Sistema Barreirao. Goiania, Go, EMBRAPA-CNPAF. (EMBRAPA/CNPAF. Documento No. 33), 1991.

Lal, R. Methods and Guidelines for Assessing Sustainable Use of Soil and Water Resources in the Tropics. Soil Management Support Services/Soil Conservation Service/USDA/The Ohio University/Department of Agronomy. (SMSS Technical Monograph No. 21), 1994.

Marshall, D.R. and Brown, A.H.D. Basic Sampling Strategy: Theory and Practice. In Guarino, L., Rao, V.R., and Reid, R. *Collecting Plant Genetic Diversity*. London: CAB International, 1995, pp.75–92.

Pereira, J., Burle, M.L., and Resck, D.V.S. Adubos verdes e sue utilizacao nos Cerrados. In *Simposio Sobre O Manejo E Conservacao Do Solo No Cerrado, 1992*, Campinas, SP, Brazil: Fundacao Cargill, 1992, pp. 140–154.

Pizarro, E.A., Valls, J.F.M., Ramos, A.K.B., Godoy, I.J., Carvalho, M.A., and Singh, A.K. Potencial forrajero de Arachis hypogaea en el Cerrado brasileno. *Pasturas Tropicales,* 18 (3):17–24, 1997.

Resck, D.V.J., Pereira, J., and Silva, J.E. Dinamica da materia organica na Regiao dos Cerrados. Planaltina, DF, Brazil: EMBRAPA-CPAC (Documento 36), 1991.

Resck, D.V.S., and Pereira, J. Efeito da incorporacao de restos culturais e adubo verde nas propriedades fisicas de um Latossolo Vermelho Amarelo, fase Cerrado. In *Proceedings of Congresso Brasileiro de Ciencia do Solo*, 17. Manaus, AM, Brazil, p.1. Rio de Janeiro, RJ, Brazil, SBCS. (Abstract), 1979.

Ritchey, K.D., Sousa, D.M.G., Lobato, E., and Correa, O. Calcium Leaching to Increase Rooting Depth in a Brazilian Savannah Oxisol. *Agron. J.*, 72:40–43, 1980.

Ritchey, K.D., Silva, J.E., and Costa, A.U.F. Calcium Deficiency in Clayey B Horizon of Savanna Oxisols. *Soil Sci.*, 133:378–382, 1982.

Ritchey, K.D., Urben Filho, G., and Spehar, C.R. Manganese Deficiency Induced by Excessive Liming in a *latossolo vermelho-escuro* Cerrado Soil In *Proceedings of National Soybean Research Seminar*, 2, 1981, Brasilia, DF. Londrina, PR, Brazil: EMBRAPA-CNPS, 1982, pp. 541–544.

Souza, P.I.M., and Goedert, W.J. Soybeans in the Brazilian Cerrados: Soil Fertility and Management. In *Proceedings of the World Soybean Research Conference*, 4, 1989. Buenos Aires. Buenos Aires: AASOJA, 2, 1989, pp. 2231–2244.

Spain, J.M., Ayarza, M.A., and Vilela, L. Crop Pasture Rotations in the Brazilian Cerrados. In

Simposio Sobre o Cerrado, 8, 1996, Brasilia, Anais do 8O Simposio Sobre o Cerrado: biodiversidade e producao sustentavel de alimentos e fibras nos Cerrados e Proceedings of the 1st International Symposium on Tropical Savannas: Biodiversity and Sustainable Production of Food and Fibers in the Tropical Savannas. Planaltina: EMBRAPACPAC, 1996, pp.139–151.

Spehar, C.R., Monteiro, P.M.F.O., and Zuffo, N.L. Soybean Breeding in Central Brazil. In *Proceedings of the Simposio Sobre A Cultura Da Soja Nos Cerrados*. Arantes, N.E., and Souza, P.I.M., Eds. Uberaba, MG, 1992. Piracicaba, SP: EMBRAPA-CNPS/CPAC. 1993, pp. 229–251.

Spehar, C.R., and Souza P.I.M. Adaptacao da quinoa (*Chenopodium quinoa* Willd.) ao cultivo nos cerrados do Planalto Central: Resultados preliminares. *Pesquisa Agropecuaria Brasileira*, 28 (5):635–639, 1993.

Spehar, C.R. Breeding Soybeans to the Low Latitudes of Brazilian Cerrados (Savannahs). *Pesquisa Agropecuaria Brasileira*, 29 (8):1167–1180, 1994a.

Spehar, C.R. Aluminum Tolerance of Soybean Genotypes in Short Term Experiments. *Euphytica*, 76:73–80, 1994b.

Spehar, C.R. Field Screening of Soya Bean [(*Glycine max* (L.) Merrill] Germplasm for Aluminium Tolerance by the Use of Augmented Design. *Euphytica*, 76(3):203–213, 1994c.

Spehar, C.R. Impact of Strategic Genes in the Soybean [*Glycine max* (L.) Merrill] on Agricultural Development in the Brazilian Tropical Savannahs. *Field Crops Res.*: 41:141–146, 1995a.

Spehar, C.R. Diallel Analysis for Mineral Element Absorption in Tropical Adapted Soybeans (*Glycine max* [L] Merrill). *Theoretical and Applied Genetics*, 90(5):707–713. 1995b.

Spehar, C.R. Genetic Differences in the Accumulation of Mineral Elements in Seeds of Tropical Soybeans *Glycine Max* (L.) Merrill. *Pesquisa Agropecuaria Brasileira*, 30(1):89–94, 1995c.

Spehar, C.R., and Makita, M. Aluminum Tolerance in Soybean Seedlings and its Utilization. *Pesquisa Agropecuaria Brasileira*, 29(12):1927–1932, 1994.

Spehar, C.R. Prospects for Sustainable Grain Production Systems in the Cerrados (Brazilian Savannahs). In: *Simposio Sobre o Cerrado*, 8, 1996, Brasilia, Anais do 8O Simposio Sobre o Cerrado: biodiversidade e producao sustentavel de alimentos e fibras nos Cerrados e Proceedings of the 1st International Symposium on Tropical Savannas: Biodiversity and Sustainable Production of Food and Fibers in the Tropical Savannas. Planaltina: EMBRAPACPAC, 1996, pp.139–151.

Spehar, C.R., and Souza, P.I.M. Developing Sustainable Cropping Systems to the Brazilian Cerrados (Savannas). In *Proceedings of the Workshop on Moist Savanna Zone*, 13 a 26-091994. Benin, Africa, 1995, pp. 325–356.

Urben Filho, G., Souza, P.I.M. Soybean Crop Management in the Cerrados In *Proceedings of the Simposio Sobre a Cultura da Soja nos Cerrados*. Arantes, N.E., and Souza, P.I.M., Eds. Uberaba, MG, 1992. Piracicaba, SP: EMBRAPA-CNPS/CPAC, 1993, pp. 299–332.

Vasconcelos, C.A., Marriel, I.E., and Pinto, N.F.J.A. *Rota, cao de culturas e produtividade de milho em solo sob vegetacao de cerrado*. Sete Lagoas, MG, Brazil: EMBRAPA-CNPMS (Comunicado Tecnico No. 5), 1989.

Vietmeyer, N.D. Lesser Known Plants of Potential Use in Agriculture and Forestry. *Science*, 232(4756):1379–1384, 1986.

Vilela, L., Silva, J.E., Ritchey, K.D., and Sousa, D.M.G. Potassio In *Solos dos Cerrados: tecnologias e estrategias de manejo*. Goedert, W.J., Ed., Planaltina, DF, EMBRAPA-CPAC and Sao Paulo, SP, Brazil. Nobel, 1985, pp. 203–222.

Vilela, L., Barcellos, A.O., Sanzonowicz. C., and Spain, J.M. Recuperacao de pastagem de *Brachiaria ruziziensis* atraves do uso de grade aradora nitrogenio e introducao de leguminosas In *Relatorio Tecnico Anual do Centro de Pesquisa Agropecuaria dos Cerrados 1985–1987*. Planaltina, Planaltina, DF, Brazil: EMBRAPA-CPAC, 1991, pp. 239–241

TWENTY-TWO

PHYSICAL AND ECONOMIC MODEL INTEGRATION FOR THE MEASUREMENT OF ENVIRONMENTAL IMPACTS OF FARMING

Charles C. Crissman and John M. Antle

GOALS FOR SUSTAINABLE AGRICULTURE AND CHANGES IN AGRICULTURAL RESEARCH

Stimulated in part by the Rio Conference of 1992, governments at all levels have incorporated sustainability objectives into agricultural, environmental, and other policies. These changes have profoundly affected the mandates of agricultural research institutions in both developed and developing countries. The incorporation of environmental considerations stimulated fundamental changes in agricultural research that resulted in alliance building, research program reorganization, and changes in research paradigms.

First, the range of problems raised as part of the concept of sustainability increased the complexity of the research problems faced to the point where many agricultural research organizations could not treat the entire range of environmental and agricultural problems in-house. These organizations sought to overcome human resource and financial constraints by building new collaborative links to other centers with complementary strengths. An array of new organizational structures has arisen to facilitate local, regional, and global research networks. Second, the addition of sustainability goals pushed the logic of systems several additional steps beyond farming systems and forced reorganizations in many institutions that resulted in adding ecoregional programs to the mix of commodity programs normally found there. Third, the complexities and heterogeneity of addressing problems in an ecoregional context makes the research model of trial plot experimentation less efficient. This has further stimulated interest in the use of systems models which can more efficiently address the numerous environmental and agricultural variables and extrapolate those results to larger areas.

Like many other organizations, the International Potato Center (CIP) has adapted to the changed setting by building new strategic alliances, developing new programs, and increasing its capacity for systems analysis by adding new staff and creating strategic alliances with advanced research organizations. Reflecting the new global environment for organization of agricultural and environmental research, the Consultative Group for International Agricultural Research (CGIAR) has approved the development of numerous system-wide and ecoregional programs among its member institutions. Making use of its quarter century of experience in the Andes, CIP is the convening center for the CGIAR program on Sustainable Mountain Agricultural Development. This program works in the tropical mountains of the Andes, the East African Highlands, and the Himalayas. The International Center for Research on Agroforestry (ICRAF) leads the effort in Africa while the non-CGIAR center, the International Center for Integrated Mountain Research (ICIMOD) is responsible for the Himalayas (TAC, 1996). CIP has primary responsibility for the high Andean ecoregion (see Fig. 22.1) and works through the Consortium for the Sustainable Development of the Andean Ecoregion (CONDESAN).

In addition to its CGIAR alliances, CIP has associated with another distinct set of collaborators via CONDESAN. CONDESAN is an open consortium consisting of over 90 governmental and nongovernmental organizations, universities, and advanced research organizations in the Andean and industrialized countries. The research priorities for CONDESAN and the CGIAR Mountain Program include: (1) characterizing agroecosystems, (2) intensifying land use and natural resource management, (3) exploiting biological diversity in a sustainable manner for economic growth, and (4) stimulating productivity and enhancing resource conservation through policies and public sector investment (CIP, 1995).

There exists a large knowledge base about Andean agricultural and environmental systems; however, due to the piecemeal approach to development and research, this knowledge has limited usefulness because of a lack of integration. CONDESAN is employing several strategies to overcome this limitation. They include the integration of available knowledge at different hierarchical levels, efficient communication between people and institutions within and outside the ecoregion, field studies that evaluate the performance of land use and commodity systems, and policy studies (Quiroz et al., 1995). Long-term work is programmed in benchmark sites in representative agroecoregions in Ecuador, Peru, and Bolivia. At present there are four sites: Carchi in the humid, nothern Andes, Cajamarca in the transition zone between the northern and southern Andes, Puno in the Lake Titicaca region, and Patacamaya in the dry altiplano (Fig. 22.1). In the future, additional sites may be added in Colombia, Chile, and Argentina.

The extreme heterogeneity of the agroecoregions in the Andes emphasizes the value of systems analysis. In CIP there are several efforts underway that explore innovative methods to overcome limitations in systems modeling in the complex environments in which they are applied. Quiroz et al. (1995) describe an effort between CIP and CONDESAN collaborators that quantifies many years of diagnostic and research experience in the Andes in simulation models. The effort is focused at the integration of simulation models, describing systems at different hierarchical levels with farm and regional models. Data from traditional sources as well as geo-referenced data obtained from remote sensing are being utilized.

In collaboration with IFDC, CIP has also begun work to calibrate for the tropical mountainous conditions of the Andes, various of the crop and soil models in the decision support system for agrotechnology transfer (DSSAT). This brings CIP into contact with the collaborating institutions in the International Consortium for Agricultural Systems Applications (ICASA).

A third effort in systems modeling involves the exploration of the integration of economic and physical models to assess the impact of policies and technology on agriculture and the environment. This work is entering a second phase as part of the recently restarted Soil Management CRSP (Anonymous, 1996).

Dent, Edwards-Jones, and McGregor (1995) state that though agricultural scientists have been working with a systems perspective in agricultural models for at least 20 years,

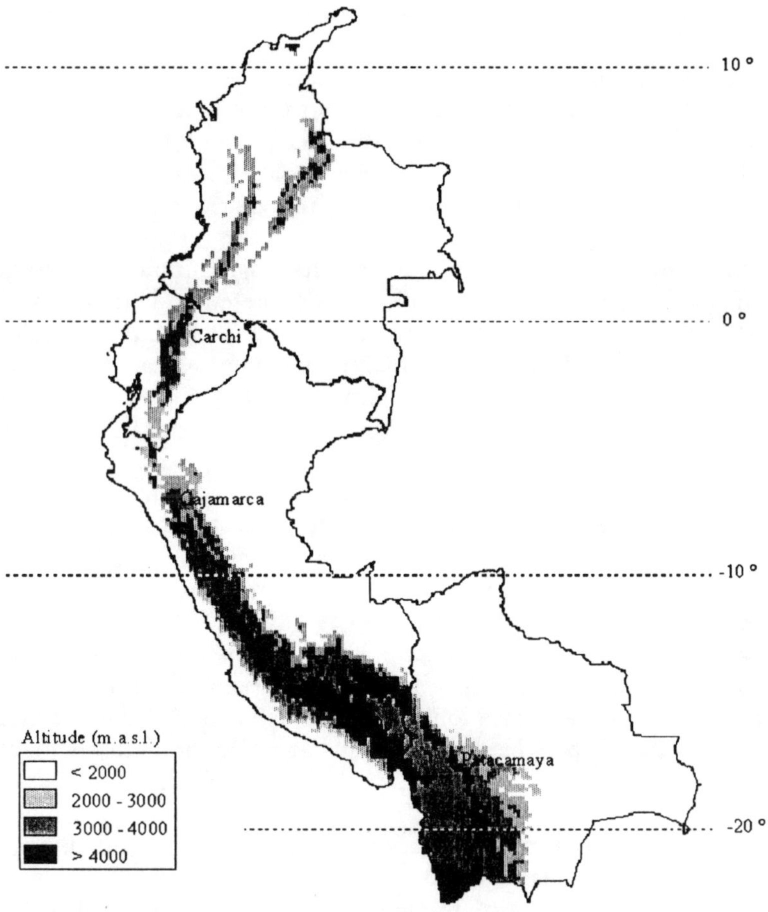

Figure 22.1. Map of the tropical Andean ecoregion above 2000 meters with the four CONDESAN benchmark sites.

there has been little effort to incorporate farmer decision-making into the models. He considers this omission a detriment to serious policy analysis, and recommends that models that aim to provide ex ante policy assessment should link the social and environmental factors with the production elements of farming systems. Antle and Capalbo (1991) note that historically there was little interdisciplinary work between physical sciences and economics and lay blame for that on the lack of incentives for interdisciplinary research, absence of data useful to both groups, and a conceptual model that provides the linkages.

The remainder of this chapter describes a research effort in which this conceptual model is developed and implemented. The implementation was done in a case study in Ecuador which documented the production, environmental, and health impacts of pesticide use in potato production. The research is fully reported in Crissman et al. (1996).

AN INTEGRATED AGRICULTURE/ENVIRONMENTAL MODEL

Policy analysis typically works at national or regional scales. The work frequently uses secondary data that are most often aggregates. However, policy analysis or technology evaluation with sustainable agricultural development objectives must include environmental or other effects that are location specific. In the tropical mountains that are the concern of the global initiative led by CIP, policy analysis in the heterogeneous biophysical setting found there suffers serious deficiencies from the averaging effects of utilizing data that are not location linked. One objective of the pesticide-impacts research is to improve the potential for policy analysis with sustainable agriculture criteria by developing and putting into operation a framework that links macro-type policy to micro-type impacts.

The knowledge base necessary to support informed agricultural or environmental policy involves many disciplines. The conceptual model presented here shows that with coordinated data collection and research design, policy analysis can be conducted on the basis of disciplinary research. In other words, the research process involves interdisciplinary data collection rather than interdisciplinary research.

Typically, policy analysis is conducted at the national or sectoral level using constructs such as the representative producer or consumer, and changes in production or resource utilization are predicted at the national level. This level of aggregation is not useful for the analysis of environmental impacts because the processes that govern environmental impact are location specific. Thus a challenge for including sustainability criteria into policy analysis or technology evaluation is a method to aggregate environmental impact information.

The method used here is to define a common unit of measurement valid to the different disciplines and obtain predictions of impact on those units. In this case the unit is a farm field. By describing the population of these units in statistical terms and estimating impacts on each unit, it is possible to statistically aggregate impacts to a level useful for policy analysis. Then it is possible to define aggregate tradeoffs between economic and environmental outcomes in the form of a transformation frontier, analogous to a production possibilities frontier. This transformation frontier can be used in aggregate analyses of policy or technology impacts.

Figure 22.2 depicts a conceptual framework for policy or technology analysis. The upper part of the figure pertains to the analysis of a unit of land at the farm level. Prevailing prices, policies, technologies, and the physical attributes of the unit of land affect the farmers' management decisions in terms of both land use and input use. These decisions affect agricultural production but also may affect the environment and human health through two distinct but interrelated mechanisms. Farmers first determine which particular hectares of cropland are put into production—the land use decision. Then on the cropland in production, the farmer takes management decisions that determine the application rates of chemicals, water use and land management practices—the input use decision. Physical relationships between the environmental attributes of the land in production and the management practices then jointly determine the agricultural output, and health and environmental impacts associated with the particular unit of land in production. Thus the land use and input use decisions of farmers form the linkage between policy and technology and the environmental and health consequences.

Each unit of land in production has environmental and management characteristics that are functions of prices, policies, technology, and other farm-specific characteristics. As the lower part of the figure illustrates, the distribution of farms and environmental characteristics induces a joint distribution of input use, outputs and environmental impacts. This joint distribution provides the basis for aggregation of the field-specific characteristics. These aggregate values can then be utilized for welfare and policy analysis.

Figure 22.3 illustrates that joint distribution with a hypothetical example of water quality and pesticide use. The matrix represents four production zones, with jointly dis-

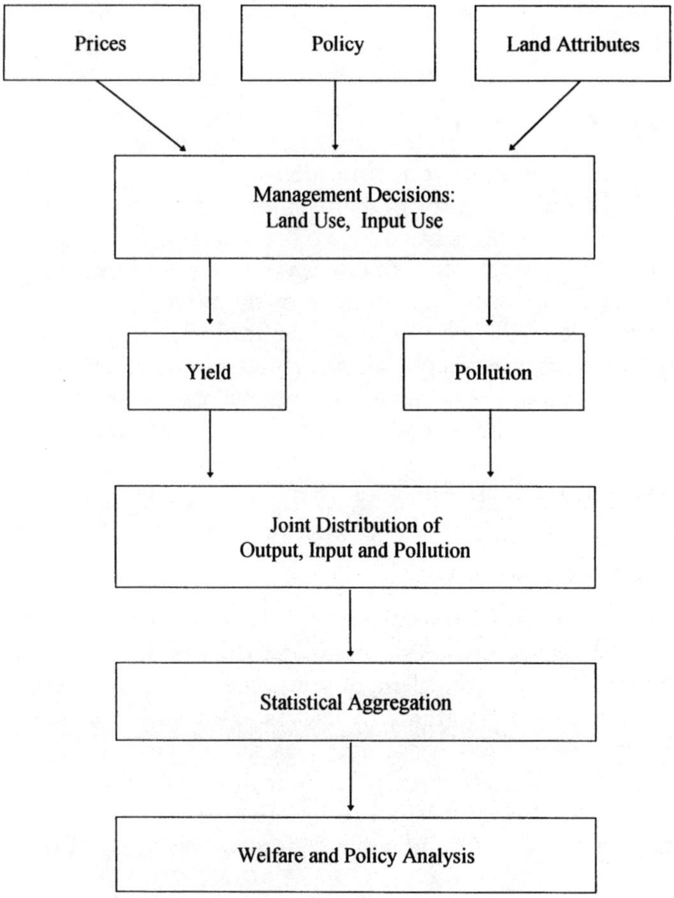

Figure 22.2. A conceptual framework for policy or technology analysis. From Antle and Just, 1991, with permission.

tributed economic and environmental characteristics. The four zones show a simplified high and low representation of agricultural productivity or contamination potential. The cells illustrate the relationship between pesticide use, yield, and water contamination. The example is also useful in calling attention to the heterogeneity of conditions for which policies or technologies must be judged. Extreme heterogeneity of environmental condi-

Area Characteristic	High Contamination	Low Contamination
High Yield	*Area 1*	*Area 2*
	$Y = 2\alpha P$ $C = 2\beta P$	$Y = 2\alpha P$ $C = \beta P$
Low Yield	*Area 3*	*Area 4*
	$Y = \alpha P$ $C = 2\beta P$	$Y = \alpha P$ $C = \beta P$

Figure 22.3. An example of pesticide runoff contamination of water.
Note: Y = yield, C = contamination, and P = pesticide rate.
Adapted from Antle and Just, 1991.

tions and agricultural potential are characteristic of the tropical mountains in which CIP and its collaborators are working.

THE ANDEAN ECOREGION

The Andean ecoregion, stretching from Venezuela to Chile, is home to some of the poorest people in South America. The tropical and subtropical portions of the region are traditional locales for human settlement and have been farmed for millenia. Today, virtually no agricultural frontier remains to be exploited, yet policy makers must call on the region to produce increasing volumes of foods that form the basic food basket of the people of the Andean nations. The challenge of this intensification is that it must occur while the issues of poverty, environmental degradation, and economic growth are also treated. In their study of the possibilities for agricultural intensification in the Andean ecoregion, Walker et al. (1994) review the ecological, agricultural, and socioeconomic setting and conclude that for the main food crops, the forecast is mixed, bright for some commodities and dim for others. The sections below draw heavily on the review.

The Social and Economic Setting

To the casual observer, the high Andes present what is apparently an illogical pattern of land use with large fields of mostly pasture for dairy cows on the valley floors while on the valley walls there are small parcels of field or row crops. This pattern derives from the Spanish colonist practice of land grants to settlers of European origin. Though reduced by programs of agrarian reform, the dominant pattern of land use is still large farms on the valley floor and small farms belonging to indigenous groups on the valley sides. Due largely to population increases, minifundization of agriculture continues in large areas of the Andes. With increasing minifundization, there is anectodal evidence of shortening fallows.

Another factor of the Andean landscape is the lack of forests. There is historic evidence of forests but since the inter-Andean valleys have been occupied for millenia, it is not known when the deforestation occurs. Thus deforestation is not a component of current environmental degradation. Environmental degradation is due to hillside agriculture, the invasion of agriculture into the fragile soils and marginal zones of the paramo, overgrazing in the puna, rainfall-induced soil erosion, and contamination due to the use of agrichemicals.

Population levels in the high Andes reached critical levels in many areas during the 1960s, and migrations began. The migrations have been largely rural-urban but are also directed to new colonization zones, especially toward the Amazon. The result is that rural populations in the sierra are growing at rates only slightly above one percent.

The Biophysical Setting

Based on his work in tropical mountains, Carl Troll (1968) called attention to three factors that define the ecoregion. These factors strongly influence the agricultural systems that are found within the Andean ecoregion. The first is the vertical change from sea level to mountain tops covered in permanent snow. The second is the change from north to south moving from tropical to subtropical latitudes. The third relates the change in rainfall gradients from east to west due to the effects of the prevailing winds on the windward and leeward sides of mountains.

The importance of altitude is reflected in the frequency in which one encounters the phrase *pisos ecologicos* or "ecological floors." Altitude determines temperature and temperature largely dictates which agricultural system will fit into what altitude. Pasture, dairying, maize, and intensive vegetable production with irrigation occupy the valley floor or the neighboring slopes. Moving up slope there is small grain cultivation often with fava beans. Finally at the uppermost levels are tuber crops, particularly potatoes. Frost risk limits crop cultivation at the higher altitudes which convert to high altitude grasslands, where natural pasture provides vegetative cover.

The second factor, latitude, though not as obvious as altitude, is an important indicator of production potential in the Andes. The low latitudes near the equator eliminate seasonal temperature changes. Diurnal temperature change emerges as the dominant factor for determining patterns of climate and plant life, which along with altitude, sharpens the definition of ecological floors. In the higher southern latitudes, seasonal effects become more pronounced, and the second fundamental distinction of the Andean ecoregion is defined. In the northern high Andes, rainfall is possible any month of the year. In the southern Andes, there are distinct dry seasons and a higher incidence of drought and frost risk. These rainfall patterns maintain the two great high altitude grassland systems known as *paramo* and *puna*. These two systems are used to distinguish the northern tropical paramo Andes and the southern subtropical puna Andes. The dividing line for these two systems crosses northern Peru at about the latitude of Cajamarca.

Though they cover only 4% to 5% of the land in the Andes, Andisols, or soils developed on volcanic ash, play an important role in the region. These soils are located just in the areas of dense settlements and are utilized for hillside agricultural production. Cultivation on sloping lands combined with intense rainfall events places degradation through water erosion as the main factor limiting agricultural sustainability (Sentís, 1992).

Despite the obivous heterogenity visible to even the casual observer, farming systems are fundamentally driven by the ecological floors mentioned above. Farmers within an ecological floor frequently face limited options for change. For example in the potato-pasture zone above 3000 m altitude, maize takes over 12 months to mature. Thus with little agricultural frontier remaining, the prospects for intensification rest on improving the productivity of existing systems.

In terms of value of production, milk is the most important commodity produced in the Andes and shows the strongest growth rate of selected important commodities produced in the sierra (Table 22.1). The lack of growth in potato production reflects the diversification of what were formally subsistence diets based on potatoes. The strong growth in Colombia results from the development of the potato processing industry. The processing sector will develop in the other countries and a rebound in potato production will follow there. Wheat production in the mountainous environments has never been able to match the productivity gains of wheat produced elsewhere. Producers in the Andean countries have shifted out of wheat to other more competitive crops. This illustrates

Table 22.1. Annual Rates of Growth (%) in Produciton for Selected Important Commodities in the Andean Ecoregion Between 1961–65 to 1989–93 by Country[a]

	Commodity			
Country	**Milk**	**Potatoes**	**Maize**	**Wheat**
Colombia	2.6	4.2	1.1	-0.1
Ecuador	3.7	0.1	3.9	-0.3
Peru	1.5	0.0	1.8	0.0
Bolivia	5.3	0.1	2.0	2.2

[a]From Walker, Crissman, Darío, Fano, Ortiz, León-Velarde, and Quiroz, 1994.

the role of increasing global integration. Growing trade forces specialization in those areas of comparative advantage, thus the region can expect expansion of some activities while others decline.

THE PESTICIDES IMPACT STUDY

The Case Study

As is obvious from the section above, an application of the conceptual model in the Andes could focus on any of a number of phenomena. We chose to examine the aspects of environmental degradation and health consequences of pesticide use in potato production in northern Ecuador at the CONDESAN Carchi benchmark site (see Fig.22.1). Potatoes in northern Ecuador are part of a potato-dairy farm system. While milk prices are completely controlled, potato prices are allowed to cycle based on supply and demand factors. The potato food system in Ecuador is also affected by numerous sectoral and macro policies that have resulted in, among other things, a subsidy to pesticide prices of nearly 30%. Potato farmers depend heavily on pesticides to maintain crop yields. Production in northern Ecuador is also affected by the sharply increasing demand for potatoes in Colombia and newly liberalized trade laws that ease the exportation of potatoes and milk. Thus an application of the model to pesticide use allows the examination of numerous "what if" questions related to agriculture, environment, health, trade, and macro policies and technology changes in an ecologically fragile production system where an intensification of production is expected.

The case study was conducted in two adjacent watersheds in the principal potato producing zone of Ecuador. Production data were collected in a farm-level survey on 40 farms during 1990–1992. Because crops are planted and harvested continuously throughout the calendar year, data were collected for parcels, where a parcel is defined as a single crop cycle on a farmer's field. A total of 320 potato parcels were used in the estimation sample representing 178 different fields.

Detailed parcel-level data were collected on a monthly basis. Potato production in Ecuador is management intensive, and there are as many as 20 distinct operations during the six-month crop cycle. Post-harvest farmer recall of detailed data on pesticide use is unlikely to be accurate. Thus, the investment in monthly visits was deemed essential to the success of the data collection effort. Sample selection and data collection are described in Crissman and Espinosa (1992).

Late blight fungus (*Phytophthora infestans*) is the principal disease, and the tuber-boring Andean weevil (*Premnotrypes vorax*) and several foliage-damaging insects are the principal pests affecting production. These are controlled with dithiocarbamate fungicides and carbamate and methamidophos insecticides.

Transformation of the Conceptual Model to an Operational One

Cost benefit analysis provides the framework to organize a coherent approach to the problem of incorporating environmental and health costs into policy analysis and technology evaluation. Figure 22.4 shows the structure of the costs and benefits associated with pesticide use. There are several distinct phases in the analysis. First is to determine how the farmer will react to a change in policy or technology. Note that this is the crucial linkage. The model places farmer decision-making in the central role linking policy or technology changes and the eventual agricultural, environmental, and health impacts.

Second is to quantify the changes in costs and benefits. Note that only a selection of

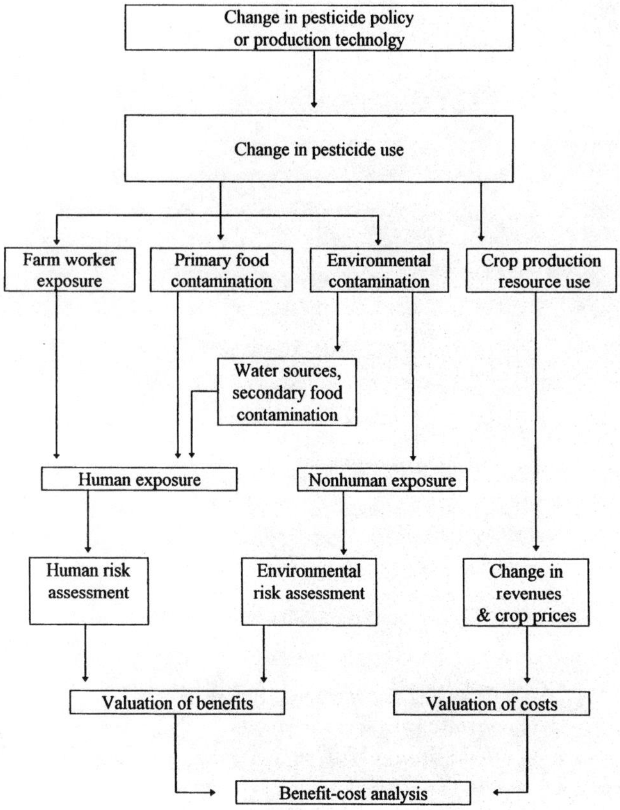

Figure 22.4. Major components of a benefit-cost analysis of a change in pesticide use. Adapted from Antle and Capalbo, 1991.

the possible pathways of pesticides is pursued. Limited by time and budget, and considering local production and environmental conditions, the research team prioritized among the possible pathways and focused on pesticide leaching and applicator exposure as the most important.

The third step is to express the benefits and costs in a common unit that reflects their valuation by the affected individuals. These are commonly monetary units. The final step is to determine the net impact by balancing the costs and benefits.

The Simulation Model

The conceptual model is transformed into an operational one in a simulation model based on an econometric potato production model, a soil pesticide leaching model, and a health effects model. Figure 22.5 illustrates the logical flow of the economics and soil science portions of the simulation model. The economics portion consists of four components; first are the distributions of the economic and physical characteristics of the fields in the study watershed. Next are the net returns distributions of the principal crops in the rotation (potatoes and pasture for milk production). The third component is a system of dynamic factor demand equations for the three categories of pesticides as described in Antle et al. (1994). The final component is a restricted revenue function used to predict the value of production. The third component of the economics portion outputs sequences of quantities and timing of the pesticides used in a particular parcel.

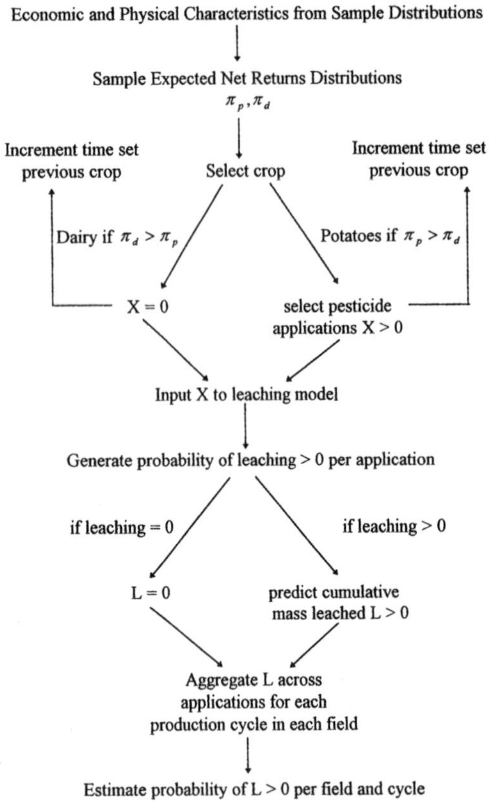

Figure 22.5. Integraged economic/ physical simulation model. From Antle, Crissman, Wagenet and Hutson, 1996, with permission.

The leaching model is a simplified version of the LEACHM model designed by Hutson and Wagenet (1993). The simplified model, LEACHA, replaces various deterministic routines of LEACHM with functional relationships. This detailed process model was parameterized with soils and other physical data from the watersheds, the pesticide data from the economics portion of the model, and simulated with a 30-year series of rainfall and temperature data (Wagenet et al. 1996). The distribution of cumulative leaching beyond the root zone was found to be positively skewed, with many observations at or near zero, but also with a long positive tail (Antle et al., 1996).

The unit of analysis is a parcel of land. Note that the model is stochastic. At several points, the model samples from distributions that are derived from the physical and economic data. Thus any two runs of the model are not expected to be alike. This is advantageous in two ways. First, in real life, economic and physical processes are stochastic. Thus the model can produce those lucky and unlucky outcomes in addition to the expected average outcomes, just like real life. Second, by utilizing distributions constructed from the sample data, you can appeal to rules of aggregation. The statistical aggregation result uses the possibility of the adding up of individual characteristics (in this case, individual parcels and the management decisions made on them) to obtain summary statistics for the group. With reasonable assumptions about the structure of production, this result permits statistically valid extrapolation beyond the data collection site, an important implication for policy analysis (Antle, 1986).

The model also provides for the decision on land use by offering a choice to put the parcel in pasture or potatoes. The model then provides for the selection of management practices, in this case the preoccupation is the quantity and timing of pesticide use. Worth repeating here is that by jointly planning the research, the economics and soil science team obtained economic data valid for and compatible with soil science modeling, and soil science data in turn useful for aggregation to policy analysis.

Presentation of Results

Use of Transformation Frontiers

To facilitate interpretation of the model results, the tradeoffs between agriculture, health, and the environment are presented in a series of pairwise comparisons illustrated with transformation frontiers, a concept similar to the production possibilities frontier familiar to students in introductory economics courses. The hypothetical transformation frontier in Figure 22.6 illustrates the tradeoffs between agricultural output and environmental quality. The transformation frontier represents all the possible pairs of outcomes for a given technology. The slope of the transformation frontier provides information about the opportunity cost of environmental quality in terms of foregone output. The curve represents the changes in outputs that are generated by varying input use for a given technology. Thus a move from point A to B could be induced by changing levels of pesticide use. A change in technology may induce a new frontier. For example, adoption of a new, higher-yielding disease resistant variety may cause a shift from point A to point C, which gives an increase in agricultural output without any change in environmental quality.

Construction of an Empirical Frontier

An empirical frontier is constructed by imposing different policy scenarios on the model. Recall that the model is stochastic; during a loop through the model, five different data distributions are sampled. Thus the outcomes of the model can be plotted as a scatter diagram. Figure 22.7 is a scatter diagram of the comparison of fungicide leaching with value of potato production. The triangles numbered 1 are the base case produced with simulations using the actual data set. Each separate triangle represents the average outcome of 30 different fields, each with five production cycles. This is repeated 30 times. Recall that changing relative prices cause movements along a frontier. The squares 2, 3, and 4 represent a tax of 30, 60, and 90% on the fungicides. This tax causes a reduction in fungicide use, with a consequent reduction in fungicide leaching and value of potato production. The reduced value of potato production comes from two sources. Incomes are reduced because costs increase and because farmers dedicate less area to potatoes. The ovals 5, 6 and 7 represent an increase in potato prices of 30, 60, and 90%. The price increases increase the profitability of production and cause farmers to plant more area and make more profitable extra applications of fungicides, so both fungicide leaching and value of

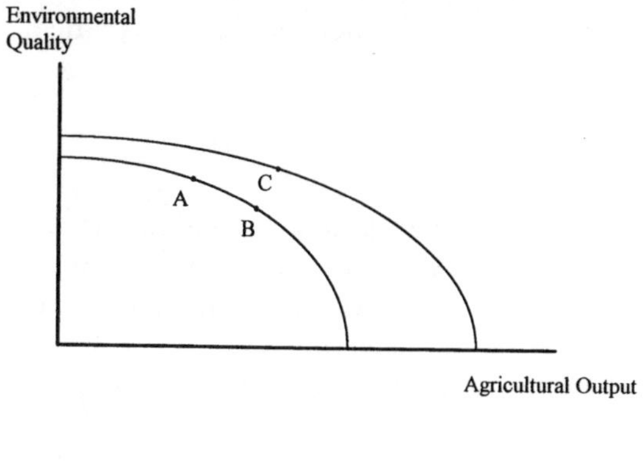

Figure 22.6. Feasible combinations of enviornmental quality and agricultural output.

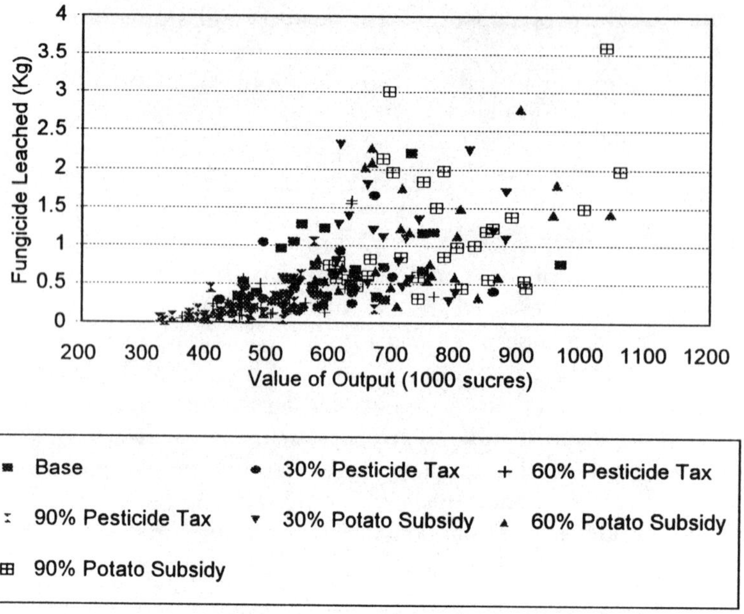

Figure 22.7. Scatter plot of fungicide leaching and value of agricultural output.

production increase. To make interpretation easier, a least squares estimation plots a transition frontier through the scatter diagram. There are 31,500 loops through the model represented in this curve.

IMPLICATIONS FOR POLICY/TECHNOLOGY ANALYSIS

Several issues relevant to policy analysis can be addressed through the use of the transformation frontiers. Politicians implicitly use transformation frontiers every day. By nature of their job, they are concerned with winners and losers resulting from policy decisions. The transformation frontier is simply a concrete expression of what is usually a mental calculation. Transformation frontiers can be constructed with sustainability indicators; indeed, the comparisons made in the study reported here can be considered indicators similar to those defined by Harrington et al. (1994). The analysis of indicators is strenghtened by the sensitivity analysis that is implicit in a transformation frontier. In the example, the politician or analyst can readily see if the sacrifice of a single unit of environmental quality will result in a gain of a single unit or five units of agricultural production. Politically determined weights would then guide the decision as to whether the size of sacrifice is acceptable.

Case Study Results

The empirical results from the case study are of interest to policy makers in Ecuador. First, water contamination from pesticide leaching is not significant; second, applicator and farm family poisoning is much worse than we suspected; and third, farmers are not overspending and applying pesticides irrationally.

These results run counter to the public debate in Ecuador where farmers are depicted as making irrational, excessive use of pesticides and that there is widespread environmental contamination as a result. International environmental activists consider, that within the Latin American region, the environmental lobby in Ecuador is exceptionally well organized. Ecuador just concluded its national and regional elections. Environmental issues received considerable coverage in the national press and are well placed in the political debate of the current political campaigns. However as the study results show, the debate can be uninformed and more polemical than substantive.

There is discussion in Ecuador about proposals to ban certain classes of pesticides, including carbofuran. The proposals are based on the presumption of environmental contamination. However, the research shows a ban on carbofuran would have a major negative effect on potato production and provide minimal improvement in the environment.

The worker safety lobby in Ecuador is not nearly as well organized as the environmental lobby and so there is little public discussion about occupational health issues, and that is where our research shows the debate should be. With pesticide use an essential part of efficient potato production, the policy debate should be on applicator safety, and instead of bans, education programs should be among the policy solutions. Clearly, solutions that derive from a technology policy should emphasize IPM.

Stochastic Results vs. Absolute Limit Policies

In the industrial countries environmental policy is frequently enacted via regulations. Examples are the Environmental Protection Agency (EPA) regulations on groundwater quality. These regulations are typically written in absolute terms; e.g., mancozeb in groundwater may not exceed x parts per million. Thus to comply with the regulation, you need to design a strategy that meets the regulatory values with 100% certainty. As can be seen in the scatter plot in Figure 22.7, even in the outcomes of the base case (the triangles) fungicide leaching can vary from 0.4 to 2.4 kg active ingredient per parcel. Further, as the mean increases, variability increases. Designing policy that provides 100% insurance requires very conservative thinking. More logical and possibly much less costly are regulations written in probabilistic terms. Thus the absolutist language would be revised to read: You should be 95% certain that mancozeb in groundwater does not exceed x PPM. Written in this language, the regulation permits policy strategies that can contemplate that remote unlucky outcome that is in the tail of the distribution.

Heterogeneity of Environment

An important consideration of policy analysis that incorporates environmental impacts is the variability of the impact of the policy. The research watersheds were zoned by the soil scientists into four microregions (Figure 22.8). While three of these have similar leaching characteristics, the fourth zone is many times more prone to leaching. As illustrated in Figure 22.3, there is considerable heterogeneity present even in small areas. By linking environmental data to economic data, it is possible to differentiate such phenomena and it is possible to contemplate something other than uniform policy options. As the regulatory powers of developing countries are improved, it is possible to contemplate cases of environmental policy to improve groundwater that uses agricultural zoning instead of taxes or prohibitions on pesticides. Transformation frontiers can be utilized to identify those zones where, for example, the transfer of a technology can be environmentally sound, as recommended by Silva and Uehara (1985).

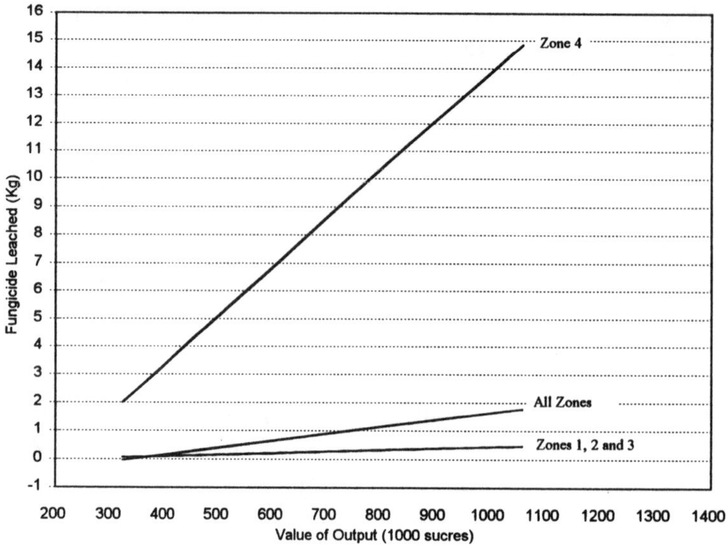

Figure 22.8. Feasible combinations of fungicide leaching and value of agricultural output by microregion.

Technology Impact Analysis

The most important research program in CIP is the development of late blight resistant potato varieties. In those agriculture systems where farmers employ fungicides, the benefits of late blight resistance are not necessarily increased yields but in reduced fungicide expenditures and as the figures show, reduced environmental and health effects. From a private perspective, the reduced expenditures and thus increased revenues are the only benefits balanced against the investment in research. However, with sustainability criteria included in the calculation of costs and benefits, the indirect benefits of environmental and health impacts are legitimately included. In cost benefit analysis, the problem with indirect costs and benefits is assigning values, especially to those factors that do not have values established in markets. Clearly, in the changes in the post-Rio political environment, more serious consideration is given the environment and health and thus their values are increased. These values are important to entities such as CIP where research priorities have to be justified and returns on research investment documented. Figure 22.9 shows the reduction in leaching from two different scenarios of reduced fungicide utilization due to varietal resistance. As ex ante technology analysis, these frontiers can offer powerful support to a decision maker to justify an investment in a screening and variety release program.

CONCLUSIONS

The inclusion of sustainability criteria into policy in both developed and developing countries has spurred substantial changes in the mandates of agricultural research institutions and the manner in which research is conducted. One outcome of these changes is an increased emphasis on systems modeling. However, numerous observers have noted that a weak link in most systems modeling work is the lack of economic criteria in the models.

The research reported in this chapter developed a conceptual model that provides linkages between macro-policy and micro-type impacts. The linkage is provided by the inclusion of a farmer decision-making model. That model shows how farmers react to policy or technology changes through land use and input use decisions. The conceptual model was

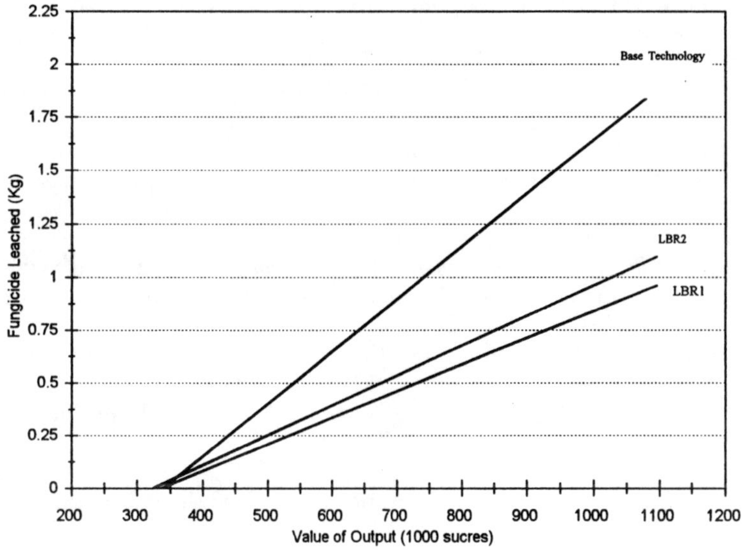

Figure 22.9. Changes in fungicide leaching due to adoption of late blight resistant varieties.

operationalized and applied in a case study of environmental and health impacts of pesticide use in potato production. The simulation model produces results that can be aggregated up for extrapolation beyond the actual case study area, making the model useful for policy analysis. Transformation frontiers are introduced as an analytical tool for their power in summarizing large amounts of data and the intuitive ease of use by nonspecialists.

Although the conceptual model is applied in an economics and soil science study of pesticide impacts, the model is flexible and can be applied to other research questions addressable from different disciplines. Since the model is statistically based, it can utilize data sets generated from various sources, such as the primary data collection effort of the case study or from GIS-based data collection. The model is compatible with other efforts in systems modeling such as those pursued by ICASA, and will eventually be incorporated into them.

REFERENCES

Anonymous. "Tradeoffs in Sustainable Agriculture and the Environment in the Andes: A Decision Support System for Policy Makers." A proposal submitted to the Soil Management CRSP by Montana State University and Collaborating Institutions, Jan. 1996.

Antle, J.M. Aggregation, Expectations, and the Explanation of Technological Change. *J. Econometrics* 33, pp. 213–236, 1986.

Antle, J.M. and S.M. Capalbo. Physical and Economic Model Integration for Measurement of the Environmental Impacts of Agricultural Chemical Use. *Northeastern J. Agric. Resour. Econ.* 20, pp. 68–82, 1991.

Antle, J.M. and R.E. Just. Effects of Commodity Program Structure on Resource Use and the Environment. In *Commodity and Resource Policies in Agricultural Systems*, R.E. Just and N.E. Bockstael, Eds., Springer-Verlag, New York, 1991.

Antle, J.M, S.M. Capalbo, and C.C. Crissman. Econometric Production Models with Endogenous Input Timing: An Application to Ecuadorian Potato Production. *J. Agric. Resour. Econ.* 19, pp. 1–18, 1994.

Antle, J.M., C.C. Crissman, R.J. Wagenet, and J.L. Hutson. Empirical Foundations for Environment-Trade Linkages: Implications of an Andean Study. In *Agriculture, Trade, and the Environment: Discovering and Measuring the Critical Linkages,* M.E. Bredahl, N. Ballenger, J.C. Dunmore and T.L. Roe, Eds., Westview Press, Boulder, CO, 1996.

(CIP) International Potato Center. *Consortium for the Sustainable Development of the Andean Ecoregion.* CIP, Lima, 1995.

Crissman, C.C., J.M. Antle, and S.M. Capalbo, Eds. Getting Pesticides Right: Tradeoffs in Environment, Health, and Sustainable Agricultural Development. Book manuscript in preparation, 1996.

Crissman, C.C. and P. Espinosa. *Agricultural Chemical Use and Sustainability of Andean Potato Production: Project Design and Pesticide Use in Potato Production in Ecuador.* Paper presented at the Rockefeller Foundation Seminar on "Measuring the Health and Environmental Effects of Pesticides," Bellagio, Italy, 30 March–3 April, 1992.

Dent, J.B., G. Edwards-Jones, and M.J. McGregor. Simulation of Ecological, Social and Economic Factors in Agricultural Systems. *Agric. Sys.* 49, pp. 337–351, 1995.

Harrington, L., P. Jones, and M. Winograd. *Operationalizing Sustainability: A Total Factor Productivity Approach.* Paper presented at the "Land Quality Indicators Conference," CIAT, Cali, Colombia, June, 1994.

Hutson, J.L. and R.J. Wagenet. "A Pragmatic Field-Scale Approach for Modeling Pesticides." *J. Environ. Qual.* 22, pp. 494–99, 1993.

Quiroz, R.A., R. Darío Estrada, C.U. León-Velarde, and H.G. Zandstra. Facing the Challenge of the Andean Zone: The Role of Modeling in Developing Sustainable Management of Natural Resources. In *Eco-Regional Approaches for Sustainable Land Use and Food Production,* J. Bouma, A. Kuyvenhoven, B.A.M. Bouman, J.C. Luyten, and H.G. Zandstra, Eds., Kluwer Academic Publishers, Dordrecht, 1995.

Sentís, I.P. "La erodabilidad de los Andisoles en Latino America." In *Suelos Ecuatoriales* 22, pp. 33–43, 1992.

Silva, J.A. and G. Uehara. "Transfer of Agrotechnology." In *Soil-Based Agrotechnology Transfer. Benchmark Soils Project,* J.A. Silva, Ed., Department of Agronomy and Soil Science, Hawaii Institute of Tropical Agriculture and Human Resources, College of Tropical Agriculture and Human Resources, University of Hawaii, 1985.

(TAC) Technical Advisory Committee. *Report on Ecoregional Initiatives.* Mid Term Meeting Jakarta, Indonesia, May 1996.

Troll, C. The Cordilleras of the Tropical Americas. Aspects of Climatic Phyto-Geographical and Agrarian Ecology. In *Geo-Ecology of the Mountainous Regions of the Tropical Americas.* C.Troll, Ed., Dummlers and Verlag, Bonn,1968, pp. 15–56.

Wagenet, R.J., J.L. Hutson and C.H. Ducrot. Describing Pesticide Movment in Potato Production in Carchi Soils. In *Getting Pesticides Right: Tradeoffs in Environment, Health and Sustainable Agricultural Development,* C.C. Crissman, J.M. Antle, and S.M. Capalbo, Eds., book manuscript in preparation, 1996.

Walker, T., C.C. Crissman, R. Darío Estrada, H. Fano, O. Ortiz, C. León Velarde, and R.Quiroz. *Prospects for Agricultural Intensification in the Andean Ecoregion.* Paper presented at the 2020 Ecoregional Worksop, Airlie House, Virginia, Nov. 7–10, 1994.

Six

Future Priorities

TWENTY-THREE

A Decade of Progress at IBSRAM and Future Priorities

Eric T. Craswell

INTRODUCTION

Agenda 21, the IFPRI 2020 Vision study and the recent World Food Summit express confidence in the capacity of the world to feed the burgeoning population into the twenty-first century, but note the urgent need to address the negative impacts of agricultural intensification on land and water resources. Soil constraints to food production were relatively neglected by the international agricultural research centers set up in the wake of the Green Revolution, because the notion that the development of new high-yielding crop varieties could largely solve food production problems was simple and attractive. Although some CGIAR centers and U.S. and European institutions were conducting excellent research on tropical soils, the overall effort was not well coordinated. Furthermore, soil researchers in developing countries were not receiving the high level of support from the CGIAR centers that their colleagues in the crop improvement fields were enjoying.

In the late 1970s, the soil scientists at the CGIAR centers, including Drs. Nyle Brady and Les Swindale, then Directors General of IRRI and ICRISAT, recognized the need to strengthen soils research through the creation of a small coordinating agency that would initiate networks of researchers working on common problems. The plans for such an agency, which became the International Board for Soil Research and Management (IBSRAM), were largely developed at a workshop at IRRI in 1979 that was cosponsored by Cornell University. However it was not until 1983, after an international workshop on soil research in Townsville, Australia, that IBSRAM was officially created. It took another two years before activities really started when, at the invitation of the Royal Thai Government, the headquarters office was established at the Thai Department of Land Development, in Bangkok.

IBSRAM pioneered a decentralized mode of operation, based on cooperation between National Agricultural Research and Extension Systems (NARES) and the international research community working through soil management networks designed to tackle the major problems of land degradation. IBSRAM has also been a catalyst in promoting new ideas and new modes of operation for soil, water, and nutrient management research. The most recent product of this catalysis is an international program of soil, water, and nutrient management research that has been integrated in the CGIAR program. IBSRAM coconvenes this program with CIAT.

This chapter reviews and catalogues activities since 1985, focusing largely on the soil management networks that have been the backbone of IBSRAM's programs. A decentralized approach through a network or consortium of NARES is the only effective way to organize an international research on natural resources problems such as soil constraints to food production. This approach is required not only because the soils are dispersed throughout the farms of the targeted developing countries, but also because the sustainability of the research effort itself depends on the NARES in those countries to undertake the research themselves over the long term. Training and capacity building are therefore a central plank of the IBSRAM approach. Methodological guidelines for the research were the first outputs of the networks. Dissemination of appropriate information was also a major issue, so IBSRAM organized workshops and published Proceedings, as well as *Soil Management Abstracts* and newsletters.

IBSRAM'S SOIL MANAGEMENT NETWORKS

AFRICALAND Management of Acid Soils Network

Objectives

To adapt, test, and validate technologies to manage acid soils for sustainable crop production under reduced fallow periods that would be acceptable to local farmers. The second main objective was to strengthen NARES in undertaking soil management research.

Countries Involved

Burundi, Cameroon, Congo, Cote d'Ivoire, Madagascar, Nigeria, Zambia.

Duration

Initiated in 1986, operating 1988-1994/95.

Training/Capacity Building

Activities were coordinated and backstopped by IBSRAM, workshops, regular annual meetings; training courses were jointly held for the Acid Soils and Land Development networks. Each training seminar for NARES was attended by participants from 16 to 20 African countries. The IBSRAM Quality Assurance Program took care of comparable sample analyses.

Activities

Common-core experiments in all network member countries: Comparison of farmers' practices with improved, high- and low-input technologies (mulch, compost, green manure, liming, mineral fertilizer, herbicides, improved varieties).

- Monitoring of soil chemical and physical parameters, erosion, yields, root growth.
- Farmers' days (station visits) for demonstration and dissemination.
- Training and conferences.

Main Results

1. Effects on soils:

- Site- and crop-specific recommendations for lime application.
- Need of supplementary fertilization to avoid overliming effects on very poor soils.

- High N inputs could be harmful, even in soils of originally moderate acidity, by increasing Al-toxicity.
- A regular lime application according to the concept of Kamprath was in some countries not sufficient to prevent increases in soil acidity.
- After clearing, the level of soil organic matter only can be stabilized by including fallows in the rotation, even under high fertilizer input.

2. Effects on yields:

- Nutrient inputs are needed to achieve and maintain sustainable levels of yield.
- Highest yields and biomass return with high fertilizer input but low to moderate input levels were most profitable. However, without regular maintenance and inputs reacidification occurs.

Training/Capacity Building

IBSRAM organized workshops, regular annual meetings, and training courses held jointly for the Acid Soils and Land Development networks. Each training seminar for NARES was attended by participants from 16 to 20 African countries. The IBSRAM Quality Assurance Program was developed to strengthen the capacity of laboratories to undertake comparable sample analyses.

AFRICALAND Land Development for Sustainable Agriculture Network

Objectives

To test different technologies of land-clearing and postclearing soil management techniques in different environments in order to elucidate the most appropriate techniques for the diverse physical and socioeconomic situations in tropical Africa. The second main objective was to strengthen NARES in undertaking soil management research.

Countries Involved

Cameroon, Cote d'Ivoire, Ghana, Nigeria, Tanzania, Uganda.

Duration

Initiated in 1986, operating 1988-1994 (some long-term experiments are still going on).

Activities

Common-core experiments (comparison of manual, semimechanized, and fully mechanized land-clearing methods; as well as of mechanical plowing, minimum tillage, zero tillage).

- Monitoring of soil chemical and physical parameters, erosion, yields, root growth.
- Farmers' days (station visits) for demonstration and dissemination.
- Training and conferences.

Main Results

- Data on the impact of mechanical clearing on soil physical properties and erosion on a range of soils.
- Manual clearing with minimum or conventional tillage, especially slash-and-burn followed by minimum tillage is the most favorable technique for soil physical parameters, soil biomass, root penetration, maize yield, and erosion.
- Positive impacts on yields appear after 2-3 cropping seasons.
- The effect of fertilization is higher than the effect of conservation farming, independent of the kind of clearing.

- Fertilizer input is required to reduce or stop steady decline of yields under multiseasonal cropping independent of other soil management practices and methods.

Training/Capacity Building

IBSRAM organized workshops, regular annual meetings, and training courses, jointly held for the Acid Soils and Land Development networks. Each training seminar was attended by participants from 16 to over 20 African countries. The IBSRAM Quality Assurance Program was also linked to this network.

AFRICALAND Management of Upland Soils Network

Objectives

To assist NARES in applied and adaptive research with regard to the development and evaluation of sustainable land use systems; i.e., those which are technically sound, environmentally appropriate, economically viable, able to reduce production risks, acceptable to small-scale farmers, and which will lead to sustainable cropping.

Countries Involved

The network started as an amalgamation of the former two *AFRICALAND* networks. After termination of several projects, trials are going on in Cameroon, Cote d'Ivoire and Ghana.

Duration

1995–1997

Activities

Farmer managed on-farm trials accompanied by station experiments.

- Socioeconomic surveys and participatory rural appraisals (PRA), needs assessment, planning, project monitoring and evaluation (PME) with the farming communities in the research areas.
- Realization of conceptual research on promising alternative soil management options, on station and on farm.

First Results:

- NARES-NGO-extension service linkages established.
- NARES trained in participatory on-farm research.

Training/Capacity Building

Annual workshops in 1995, 1996, 1997; training courses in PRA (1995), PME (1996, 1997), training seminars and sample exchange for Quality Assurance Program, student exchange, NARES staff visits at Advanced Research Institutes (ARIs).

AFRICALAND Management of Vertisols Network

Objectives

- To test and adapt improved techniques available for achieving better Vertisol management needs in a range of semiarid zones in order to determine those technologies that can be introduced most easily and effectively into farming.
- To conduct research on tillage, soil management, water conservation, nutrient management, and cropping systems.
- To strengthen the research capabilities of the national cooperating agencies.

Countries Involved
Ethiopia, Ghana, Kenya, Nigeria, Tanzania, Togo, Zimbabwe.

Duration
1986–1996

Activities
Common-core experiments to compare farmers' practices with improved technologies (such as broad beds and furrows; cambered beds, ridges, etc.).

Main Results
Development and testing of improved options for drainage and water conservation under different climatic conditions; e.g., cropped broad beds and furrows for humid Vertisol areas in Ethiopia, cambered beds for humid areas in Ghana, and broad ridges and cropped furrows for semiarid regions in Zimbabwe.

Training/Capacity Building
Conferences 1986, 1987, 1989, training workshops 1988, 1989,1991.

ASIALAND Management of Acid Soils Network

Objectives
To alleviate problems associated with agricultural systems in rainfed infertile acid uplands in Southeast Asia.

Countries Involved
Indonesia, Malaysia, the Philippines, Thailand, Vietnam.

Activities
Core experiments in the first five years of the network on the benefits of organic matter, lime and fertilizers on alleviating the problems of agricultural production on infertile acid soils. Subsequent experiments concentrate on the management of acid soil infertility through the appropriate use of organic and inorganic amendments, with the major focus on phosphorus and particularly the interactions between organic and inorganic phosphorus. The experimental sites cover a wide range of climates, soils, and cropping systems. The annual rainfall ranges from < 1100 mm to > 3000 mm, the latitudes from O to 21°, and the soils include a range of Ultisols.

Main Results
The responsiveness of the sites varied, as does the response to the different treatments, particularly the different organic matter treatments.

- High yields were associated with highest inputs of some types of organic matter (e.g., farmyard manure), fertilizers, and lime.
- In some instances, low inputs of lime, fertilizer, and organic matter, often in combination with improved cultivars, resulted in improved yields after several crops.
- Little or no response to some organic matter amendments (e.g., compost from alang-alang, foliage from rubber trees).
- Addition of lime was of variable value; significant increases in yield were observed where soil was high in exchangeable Al, but there was little effect where soils had low exchangeable Al.
- In some cases, the addition of organic matter significantly reduced Al saturation, but in other cases there was no effect on pH or Al saturation.
- The quality of organic matter was recognized as a significant factor.
- Appropriate P management is required.

Training/Capacity Building

- Training workshop on site selection, characterization and establishment of experiments for the management of acid soils in Asia (1990)—47 participants from seven countries.
- Training workshop on soil analysis and soil quality control (1991)—16 participants from five countries.
- Training workshop on interpretation of data from upland experiments (1992)—24 participants.
- Training workshop on socioeconomic survey methods (1993)—13 participants.
- Nine-week training courses (1994, 1995, 1996) on transfer of land management technologies for profitable and sustainable agriculture. Participants from Cambodia, Laos, Myanmar, Thailand, and Vietnam.
- Soil data for sustainable land use: a training workshop (1995).

ASIALAND Management of Sloping Lands Network

Objectives
To help in conserving soil resources on sloping lands in the region through research and promotion of appropriate land development and management technologies for sustainable agriculture.

Countries Involved
China, Indonesia, Laos, Malaysia, the Philippines, Thailand, Vietnam.

Duration
Phase 1 (1988–1991); Phase 2 (1992–1994); Phase 3 (1995–1997).

Activities

- Site selection and characterization to ensure the relevance of the agroecological environment in which the research is conducted.
- Agronomic research based on common-core experiments on the comparative advantages of different cropping systems and soil-conservation technologies, such as alley cropping, grass strips, and the use of a legume cover crop.
- Supporting field and laboratory research on soil erosion and conservation, soil moisture, and organic-matter management.
- Socioeconomic studies and evaluation of the research outlined above, including baseline surveys, quantification of costs and benefits, and acceptability by farmers.

Main Results
Soil conservation practices and soil loss. The effects of different technologies on soil loss at the various network sites are quite clear. At all sites, soil loss under the farmers' practice was significantly higher than that with the improved practices. The effectiveness of hedgerows in reducing soil loss was substantial in all years at all sites. The reduction in soil loss by using alley cropping ranged from 33% to 100%. It was observed that rill erosion was very serious on the plot where there was no soil conservation measure (the farmers' practice). Under the farmers' practice, most of the surface soil on the upslope was eroded and deposited at the downslope. It was also observed that, due to the filtering out and accumulation of sediment by the hedgerows, miniterraces were formed above the hedgerows in the alley-cropping treatment.

Effect of soil erosion on plant growth and yield. The effects of the different improved technologies on crop yield over time, as compared to those of the farmers' practice, showed that most of the improved technologies gave yields as high as those from the farmers' practice in the first year and better in later years. From years two to four in Indonesia and the Philippines, alley cropping gave yields that were 80% and 68% higher, respectively, than those of the farmers' practice; when fertilizer was used, this figure increased to 159% in the Philippines.

Economic analysis. In general, annual net returns from the improved technologies were less than those from the farmers' practice for the first two years of the trial. In later years, the economic advantage of the improved technologies over the farmers' practice was more obvious, as shown by the example from the Philippines. In the first year of the experiment (1989), all the improved technologies had higher negative net returns than the farmers' practice, due to the additional cost of establishing the hedgerows. The highest negative return was with banana hedgerows (-\$962 ha^{-1}) and this was due to the higher cost of banana, compared to Gliricidia plus napier grass. From the second year onward, the net returns from all the alley-cropping treatments were higher than those of the farmers' practice by \$94 to \$1,117 ha^{-1} averaged over three years (from the second to the fourth year). The highest net returns came from the banana hedgerows, due to the additional income generated.

Impacts on farmers. Although some research on conservation farming carried out in the past was introduced to farmers by various agencies, including NGOs, the extent of adoption of conservation practices by farmers was limited. This may be due to the lack of long-term research planning and follow-up and to insufficient involvement of farmers in the research program. It is the strategy of the network to develop close linkages between farmers, researchers, and extensionists, and to provide follow-up on a long-term basis so that sustained adoption of conservation farming practices can be achieved. A full-scale extension program has not been conducted yet but the adoption by farmers of some of the soil-conservation technologies identified and tested by the network has already occurred in some of the network-participating countries.

On policy-makers. Self-sufficiency in food production is a general policy for most countries and attempts to achieve this have often resulted in the degradation of land resources. The network has made good progress in convincing policy-makers in these countries that soil conservation measures are important. On this basis, the network has been able to stress the value of biological soil-conservation measures.

Training/Capacity Building. One of the objectives of the network is to strengthen the capability and competence of the NARES with regard to their ability to conduct adaptive research. Institution strengthening through the network was achieved through training, workshops, meetings, and national and international interaction. Scientists from the network attended the 9-week training course on transfer of land management technologies for profitable and sustainable agriculture. Publications include eight annual reports and more than 60 technical papers published and presented at various international meetings and workshops.

PACIFICLAND Management of Sloping Lands Network

Objectives

- Develop a collaborative research network on soil and land management of sloping lands.
- Strengthen the capability of national research agencies in undertaking research on soil management on sloping lands, with special regard to promoting sustainable management that is both economic and acceptable to farmers.

- Test and validate existing knowledge on the management of sloping lands.
- Facilitate a system for the exchange of information on soil and land management.
- Assist in the development of appropriate technologies on the management of sloping lands.
- Quality assurance in soil and plant analysis.

Countries Involved

Fiji, Papua New Guinea, Vanuatu, Western Samoa, Tonga, and the Solomon Islands.

Duration

Phase 1 (1991 to 1995); Phase 2 (1996 to 1999).

Activities

Core experiments were designed to measure the impact of land management practices on productivity and soil erosion; a Quality Assurance Program was also developed for laboratories in the region.

Main results

In Fiji:

- Only small amounts of erosion on the different treatments.
- Farmers' practice gives highest rates of soil erosion; there was evidence of rill formation with erosion rates in ginger plots of up to 20 t/ha in 1995 (single erosive event).
- Yields are often higher with farmers' practice than with pineapple and vetiver treatments, even after adjusting yields to take into account land used by the hedgerow.

In Papua New Guinea:

- Generally soil loss and runoff occur when rainfall is above 2 mm/hr (i.e., at Aiyura, this represents the minimum value at which rainfall becomes erosive).
- Vetiver hedgerows appear more promising in controlling soil loss than the other improved practices and the simulated farmers' practice.
- Significant changes in pH, Mg, K, Na, Al, CEC, base saturation and P have occurred in soils after cropping with sweet potato, suggesting that continuous cropping will lead to a decline in soil fertility.

In Vanuatu:

- Highest amounts of erosion (to 23 t/ha) were recorded on plots with traditional practices compared to plots with hedgerows (0.6 t/ha).
- In traditional practices plots, cassava yields were higher than in hedgerow plots (26 c.f. 16 t/ha) due possibly to greater competition for nutrients and light in the hedgerow plots.

In Western Samoa:

- Soil losses were very low—highest losses (to 30 kg/ha) recorded in the fallow period under farmers' practice, whereas hedgerow plots lost even less.
- Farmers' practice plus pineapple and banana hedges, and farmers' practice plus Flemingial/Erythrina hedges, yielded higher than the other treatments—an effect due more to the addition of mulch than to erosion control. Mulch was estimated to contribute about 400 kg N. 40 kg P, and 150 kg of K per ha.

(Solomon Islands and the Kingdom of Tonga have only recently joined the network and data are not available from these countries.)

Training/Capacity Building
Annual network workshops and training courses have involved up to 60 scientists from the region.

CORE ACTIVITIES AND PARADIGM DEVELOPMENT

Soil, Water, and Nutrient Management (SWNM)

In 1993 IBSRAM coordinated the efforts to prepare a global position paper on Soil, Water, and Nutrient Management Research—A New Agenda, (Greenland et al., 1994) that was supported by a group of donors chaired by Ralph Cummings Jr. of USAID. The paper was presented at the midterm CGIAR meeting by lead author, Prof. Dennis Greenland. Other authors included Robert Rhoades, Hari Eswaran, Christian Valentin, and Glyn Bowen. In September 1994 IBSRAM and other international centers, NARES, land users and donors met in Zschortau (Germany) and adopted the Zschortau Plan—an international program and action plan to establish a harmonized global program of research on sustainable land management. It particularly emphasizes that SWNM should be an interdisciplinary, multi-institutional and farmer needs-driven program. The SWNM program has now paved the way for the establishment of a new CGIAR system-wide program on SWNM research. This program operates through consortia created to address land degradation problems or themes of global significance. Activities were designed to focus initially on specific ecoregions: nutrient depletion in sub-Saharan Africa, managing acid soils in the humid tropics of Latin America, optimizing soil water use in semiarid Africa and West Asia, and managing soil erosion in the warm humid tropics and subtropics of Asia. IBSRAM and CIAT, the CGIAR center in Colombia, are coconvertors of the SWNM program. IBSRAM is also coconvertor with PCARRD of the Management of Soil Erosion Consortium (MSEC) which IBSRAM coordinates.

Research Paradigm Evolution

In the Greenland paper, a 'user-participatory approach to soil, water and nutrient management' is advocated along the lines of a 'farmer-back-to-farmer paradigm.' It states that 'we have identified the critical issue as the failure to start the research process at the user level, and to establish a continuing mechanism for interchange of knowledge between the farmers and other practitioners and the researchers.' It speaks in favor of a 'farmer-back-to-farmer approach' that begins with a joint participatory diagnosis of the major land development problems by scientists and farmers. Traditional knowledge is given high priority in diagnosis and research design. The methods proceed through successive phases through which problems are more clearly defined and potential solutions are sought. Each step involves farmers and scientists, culminating in a cycle of joint on-farm testing, evaluation, modification, and retesting of the supposedly improved technology.

The shift toward adaptive on-farm research with the genuine participation of farmers was substantiated by the NARES in IBSRAM's longest-standing network, the *ASIALAND* Sloping Lands Network. They jointly agreed on a set of new guidelines for participatory on-farm research by the end of 1995, outlined by a paper presented by Prof. Patanothai titled 'Guidelines of Methodology for On-Farm Research.' It suggests basing the whole on-farm research process, its priority setting, and design of alternatives on the principles and procedures of farming system research (FSR), i.e., on an understanding of the existing farming system, of its biophysical and socioeconomic environment, on an analysis of farmers' constraints and opportunities, and on comprehension of their goals, motivations, and perceptions which influence their decision-making. The paper recommends:

- Recognizing the importance of an interdisciplinary dimension by assigning social scientists or farming-system research specialists to the research team.
- Recognizing the research process as being dynamic and iterative, with researchers designing the technologies to be tested by farmers; their feedback will result in an ongoing learning process with more fine-tuned technologies being adapted to farmers' specific circumstances.

The research paradigm has evolved further over the last two years, as described in a recent IBSRAM social scientist's paper, 'Moving Further from Consultative to Interactive Participation'. The paper describes the need to move ahead in the attempts to establish a genuine participatory approach for on-farm trials by proceeding from a more passive and extractive participation, based on farmers' consultations, to an interactive form of participation based on partnership and dialogue. Some of the key elements of this strategy include:

- A genuine and effective interdisciplinary research program.
- An attempt to link traditional, indigenous knowledge with scientific knowledge.
- The design of research trials, planned with farmers, and based on the community's inherent potentials and constraints.
- A client-driven agenda, in which farmers organize experimentation, evaluate results, and transmit recommendations.
- An adaptation of RRA/PRA techniques for the particular purpose of on-farm research design and planning, innovation testing, monitoring, and evaluation.
- Identification of the needs for institutional and policy (organizational, managerial) changes to make participation an integral, stable, and long-term component of the research work.

Training courses have been developed on Participatory Monitoring and Evaluation (PME) in On-Farm Research which include:

- Discussion of factors determining the relationship between farmers, researchers, or extensionists (expectations, suspicion, deference, and courtesy notions).
- Identification, comparison, and evaluation of key elements of the conventional and participatory views of the farmer-researcher relationship.
- Investigation of preferable communication styles regarding listening skills, body language, and interviewing.
- Jointly defining the concept of participatory monitoring and evaluation (PME). Discussion of its benefits vis-à-vis conventional ME and its advantages and disadvantages.
- Identification and adaptation of PRA tools and techniques most suitable for PME purposes, such as impact mapping and diagramming, evaluation matrix, farmers' own records, open-ended evaluation, community innovation assessment, etc.

As IBSRAM's various network activities have moved from experimental, station-based research to on-farm research, the inclusion of socioeconomic data and farming system analysis, the recognition of traditional sources of knowledge, and the active involvement or participation of farmers in research design, implementation, and evaluation have become critically important objectives. All networks, and the new consortium on managing soil erosion, will adopt and adapt the new research paradigm.

Framework for Evaluating Sustainable Land Management (FESLM)

For many years, the FAO Framework for Land Evaluation (1976) has proved useful for assessing the suitability of land use alternatives, but it does not allow the sustainability of land management to be determined. In fact, there has been—until recently—no adequate procedure for assessing sustainable land management, and this is seen to be a factor limiting the development of sustainable land management technologies. For that reason, IBSRAM, in partnership with several national and international organizations, in 1991 made a major commitment to the development of a 'Framework for the Evaluation of Sustainable Land Management' (FESLM). The framework aims to assess the sustainability of current land systems of land management in different environments and to establish the probability that improved systems of land management will enhance the likelihood of achieving sustainability.

IBSRAM believes that there is a need to identify how new technologies can be used to increase agricultural production while at the same time preserving the environment, thus contributing significantly to the development of innovative agricultural policies and programs in support of sustainable land use. The intention is to develop biophysical and socioeconomic criteria, indicators, and thresholds which can provide objective and, where possible, quantitative indices of sustainability. The cooperative FESLM thus substantially assisted toward meeting the objectives of sustainable land management outlined in Agenda 21 .

IBSRAM has undertaken, in partnership with other international institutions, national government agencies and academic institutions, the following studies on the evaluation of sustainability of land management on:

- Vertisols in a low-input system in Zimbabwe and in a medium- to high-input system in Queensland, Australia.
- Sloping lands under low-input systems in Mindanao, Philippines, and under a medium-to high-input system in Queensland, Australia.
- Upland rice systems in northern Thailand.
- Rubber cultivation on small farms in Malaysia.

In order to enhance the FESLM (Smyth and Dumanski, 1993) through the development of criteria, indicators, and thresholds, the Methodological Guidelines for IBSRAM's Soil Management Networks (1994, 2d ea.) have also evolved. This publication provides a tentative list of indicators considered appropriate for assessing sustainable land management, particularly at the plot or farm level.

Decision Support System for Sustainable Land Management

In an attempt to operationalize the practical use of an international FESLM, IBSRAM in collaboration with Agriculture Canada and cooperators from Indonesia, Thailand, and Vietnam is presently developing a Decision Support System (DSS) to assist users in diagnosing sustainable land management problems and identifying constraints in achieving sustainability. Major outputs of this project so far include:

- On-farm research sites have been visited and surveyed for socioeconomic and biophysical data in Chiang Rai (Thailand), Jambi (Indonesia), and Hoa Bin (Vietnam) province. During the survey, besides conventional questionnaires for household and village level information, a number of PRA tools were applied. Data have been synthesized and interpreted.
- A protocol as a guideline for conducting case studies within the framework of a DSS for

SLM has been developed, giving particular emphasis for the identification of diagnostic criteria, the establishment of indicators and the definition of threshold levels for sustainable land use. It maintains that the establishment of a partnership of mutual trust and understanding between the farmers and the evaluation team is of paramount importance. Failure to develop this rapport will result in an evaluation which has limited relevance for the farmers and as a consequence, a low level of acceptability and take up.

- A comprehensive Tool Kit for Participatory On-Farm Research has been developed and tested during the field visits. It includes, among others, socioeconomic questionnaires on household level and village level information, a gender information sheet, a framework for innovation assessment, 12 PRA techniques, a form for socioeconomic diagnosis and evaluation as well as background information, and examples and guidelines to each tool. The tool kit will be the basis for the training of NARES cooperators. It will be handed out to them in order to enable our cooperators and their staff members to conduct socioeconomic data collection for on-farm research planning, implementation, monitoring, and evaluation in an effective, efficient, and more participatory way.

A prototype Decision Support System is being developed, based on the knowledge gathered in the field surveys and utilizing the FESLM as a framework. IBSRAM is currently revising its strategy to take account of developments in information science. For example, IBSRAM organized a workshop on the concept of Resource Management Domains (RMDs) that can be used to facilitate the transfer of technologies across regions and improve the efficiency of research on sustainable land management. The RMDs, the FESLM, the DSS, and the new research paradigm are critically important tools for research on sustainable land management. IBSRAM now sees the development of such tools as a central task for the organization into the twenty-first century.

ACKNOWLEDGMENTS

The progress made by IBSRAM over the decade would not have been possible without the enthusiastic commitment of colleagues in the NARES. To them and the many donor agencies that have supported the work, I express our sincere gratitude.

REFERENCES

Greenland, D.J., Bowen, G., Eswaran, H., Rhoades, R. and Valentin, C. 1994. Soil, Water, and Nutrient Management Research—A New Agenda. IBSRAM Position Paper. IBSRAM, Bangkok.

International Board for Soil Research and Management (IBSRAM). 1994. *Methodological Guidelines for IBSRAM's Soil Management Networks*, 2d ed. IBSRAM, Bangkok.

Smyth, A.J. and Dumanski, J. 1993. FESLM: An International Framework for Evaluating Sustainable Land Management. World Soil Resources Report.73. FAO, Rome.

TWENTY-FOUR

The Pilot Linkage Program: An Experiment to Enhance Interinstitutional Collaborations

Charles R. Curtis

INTRODUCTION

Good science requires excellent communications. To advance in today's post-Soviet world, the international scientific community needs to collaborate and communicate more than ever. How might today's U.S. and international scientists maintain these essential contacts?

Many informed sources tell us that agriculture is undergoing major policy shifts that will profoundly change how food is produced, processed, distributed, and marketed in the USA and the world (Council for Agricultural Science and Technology, 1994; Science and Education 2000, 1995). The shift is forcing producers and agribusiness alike to make major changes in order to be successful in the evolving global agricultural industry. Globalization of markets and international trade are expected to increase along with more domestic markets and international competition. More industrialization of agriculture in the developed countries is expected to take place, and new technologies are evolving for precision farming and precision agriculture for food and fiber production. One of the most serious questions is how our community of research scientists will keep up and gain the necessary in-country experience, training, and cultural awareness for effective networked international research. Even more significant questions are "what is the appropriate research agenda for international research, and specifically how will American scientists secure appropriate international awareness, training and research experience in host countries facing the future food and environmental challenges?"

For a large share of the world's population to be malnourished, illiterate, sick, and without resources is not only unethical but wasteful (Pinstrup-Anderson, 1996). Our own national agricultural research base must be supported by a vibrant international agricultural research system. We currently have a grossly inadequate international agricultural research investment. Some believe it is critical that certain types of research, such as modern molecular biotechnology, be focused on agricultural problems of developing countries. Walsh (1990) stated that it is perhaps redundant to speak of "international and agricultural research." Agricultural research means international research which benefits the

U.S. as well as overseas farmers in the global community. This strongly suggests new partnerships and strategies need to be forged among many research-capable organizations and institutes, and in particular the International Agricultural Research Centers (IARCs) and their affiliates.

There is an often-stated expression that "agriculture does not end at the farm gate." It would also be true that modern agricultural research does not end at the borders of the USA. We share with the rest of the world most of the economic threats to a farmer's daily problems (or existence) from plant diseases, insects, and weeds, weather, and markets. There will be future needs for highly informed and culturally aware professional scientists to combat these daily problems.

THE PILOT LINKAGE PROGRAM

The Pilot Linkage Program for international agricultural research has been in effect for about five years. This short-term exchange program allowed on-site opportunties for U.S. scientists to conduct joint research with their counterparts in the IARCs and their affiliates. The program originated from the former USDA/Office of International Cooperation and Development (OICD) and was based on seeking solutions to two concerns: (1) that U.S. research cooperation with the centers was diminishing, and (2) those entering international research needed critical in-country experience and networking with scientific peers to maintain their scientific base.

It is well known that the IARCs contain valuable germ plasm collections to protect the genetic base for agriculture. IARCs conduct many highly relevant international research programs to preserve the natural resource base, production, and sustainability. These research centers provide excellent opportunities and a sound, historic basis for U.S. scientists to initiate and maintain collaboration with scientific colleagues.

As a first step in problem identification, a conference sponsored by the International Science and Education Council Research Committee was held in Washington, D.C. in 1989 titled "International Agricultural Research: Its Relationship to U.S. Agricultural Productivity and Competitiveness." A proceedings of the conference was published (Curtis and Mezainis, 1990). The conference objectives were to:

1. share information on the status of international research and its benefits to U.S. agriculture
2. explore various models for international collaborative research
3. seek advice on priority research issues and explore strategies for support of international research.

The keynote address by Leo Walsh (1990) was titled "International Agricultural Research: Helping Others While We Help Ourselves." Walsh emphasized that international agricultural research is the major hope for sustained economic development in the world's poor countries. He also indicated that international research benefits U. S. farming indirectly by building stronger economies abroad which in turn can purchase U. S. agricultural products. Direct benefits are realized through improved technologies that can be utilized by U. S. farmers.

As a second step in helping U.S. scientists locate research opportunities with the IARCs and their affiliates, a 170-page directory was published in 1993 entitled "Research Opportunities for U.S. Scientists at the International Agricultural Research Centers and Affiliates" (Curtis and Mezainis, 1993). In the directory, a chapter was prepared on each of the 22 Research Centers. Each chapter contains sections on the center's background, programs, specialty areas, types of training programs, staff available, and facilities. In addition, some centers supplied information on existing projects with U.S. institutions and collaborators. The directory was widely distributed and at that time provided a reasonably current description of the research programs and capabilities.

The Pilot Linkage Program played a lead role in supporting 13 international research initiatives (Table 24.1). Requests for proposals were sent out and U. S. scientists were invited to apply. Following proposal evaluation and selection by the OICD, the overall effort was coordinated through The Ohio State University Research Foundation. Table 24.1 illustrates some of the international research initiatives undertaken or supported by the Pilot Linkage Program.

It was recently reported that 800 million people—one out of every six persons in developing countries—do not have access to the food they need for healthy and productive lives (Pinstrup-Anderson, 1996). They are food insecure. We must accelerate efforts to relieve this situation by expanding international agricultural research focused on developing countries. In addition, national agricultural research must be supported or connected to an international agricultural system. Brady (1990) believes these types of networks can

Table 24.1. Pilot Linkage Programs Sponsored by the Research and Science Exchange Division, International Cooperation and Development, Foreign Agricultural Service/USDA

Investigator	Title of Project	Center
Dr. David B. Marx University of Nebraska	Cooperative Development of Statistical Methods Useful for Agricultural Research in the Dry Areas	ICARDA
Dr. Michele Schoeneberger University of Nebraska	Windbreak Technology: Managing the Tree/Crop Interface	ICRAF
Dr. Edwin J. Anderson	Imparting Virus Resistance to Cowpea Varieties Through Applied Biotechnology	IITA
Dr. Lawrence E. Datnoff	An Alternative to Fungicides for Disease Management	CIAT
Dr. Raymond I. Carruthers	Biological Control of Sweet Potato White Fly, *Bemisia tabaci*	AVRDC
Dr. James Jones	A Systems Approach to Characterizing the Sustainability of a Hillside Farming System	CIAT
Dr. Jess D. Reed	Genetic Variation in the Nutritive Value of *Faidherbia Albida* and *Balanites Aegyptiaca*	ICRAF/ICRISAT
Dr. Sieglinde Snapp	Genetic Mapping of Root Architectural Traits Controlling Productivity of Beans in Low-P Soils	CIAT
Dr. Christina Gladwin	Mexico's Structural Adjustment	CIMMYT
Dr. Giles Waines	Population Biology of Wild Wheats and Goat Grasses in Syria	ICARDA
Dr. James Quick	Evaluation of Heat Tolerance in Spring Wheats	CIMMYT
Dr. Don Wallace	Advancing Understanding of Plant Biology Using Applied Yield Trials	ICRISAT
Dr. C. S. Prakash	Developing DNA Amplification Fingerprint System for Peanut	—

serve as a means of getting research work done, not just as a means of communicating research. Research networking increases the quality of the research, and work done within a network will likely be monitored critically by other scientists. Networking provides countries, particularly those in Africa, with contacts and an opportunity they would not be able to afford or utilize individually. An attractive option now is to look at existing or alternative electronic technologies as a means of strengthening and expanding communications for international research networking.

Linkages are needed to enable developed-country laboratories to address priority needs as these countries see them. An important contribution of the IARCs and their affiliates is their participation in international networks which serve to link institutions and U.S programs having mutual interests. One of the current interests is in biotechnology. In 1991 it was stated that relatively little research is being done to develop biotechnologies for crops such as cassava, sorghum, millets, pigeon peas, cowpeas, and plantain which are especially important to the developing countries (Barker and Plucknett, 1991). However, the IARCs' involvement in biotechnology has expanded rapidly, and this research offers the potential for increasing productivity while reducing environmental problems. For the bulk of developing countries the net of IARCs is likely to provide a key linkage to biotechnology research. Since there is a growing concern about the ability of U.S. agriculture to compete successfully in other world markets, our USA biotechnology scientists should be engaged in this type of applied research with extensive linkages to other parts of the world (Barker and Plucknett, 1991). Failure to achieve scientific linkages leads to a less effective position for the global scientific community. In the future the productivity gap between developed and developing countries is likely to widen. It has been suggested that to facilitate technology transfer, direct telecommunication linkages will need to be expanded between researchers in developed country laboratories and counterparts in developing countries. This will greatly facilitate a two-way flow of information because IARCs are likely to be central in transfer of biotechnology in particular. They will need to expand their capacity for assistance in this area (Barker and Plucknett, 1991).

How can networking of researchers benefit the USA? Challenges are constantly made to be more competitive and relevant to today's post-Soviet world. Agriculture has undergone major transformations and does not end at the farm gate or markets. We must focus on the issue of value-added agriculture with the knowledge of world markets, economics, consumption patterns, and demography. At stake for the U.S. scientist are advances in research programs. It will be essential to stay knowledgeable and remain on the forefront of biological sciences. Clearly this will not be possible working alone and only within U.S. borders. To do so could result eventually in a scientific establishment that will be limited and not relevant to modern agriculture.

Who will be guiding and teaching the next generation of international agricultural scientists and professionals? Any meaningful program to improve international research must include support, assistance, or allowances for the specialized in-country experience. Faculty and scientists in industry and government who wish to dedicate their careers to international research now stand at a crossroad of various choices. One choice is to deliberately evolve programs away from an international component. Alternatively, one can hope a crucial nucleus of the nation's pool of scientists becomes interested in and committed to international issues. Participation in overseas research programs engages scientists in becoming more aware of global issues and design projects addressing those issues. Fostering linkages and synergies among international scientists integrates global research and provides a focus for meaningful investments to solve critical world food and fiber problems.

NOTE

The contents of this paper do no necessarily reflect the views or policies of the U.S. Department of Agriculture.

REFERENCES

Barker, R. and D. Plucknett. Agricultural Biotechnology: A Global Perspective. In *Agricultural Biotechnology, Issues and Choices*, B. R. Baumgardt and M. A. Martin, Eds. Purdue University Agricultural Experiment Station, W. Lafayette, IN, 1991.

Brady, M. C. U.S. Foreign Assistance: Its Role in International Agricultural Research. In *International Agricultural Research: Its Relationship to U. S. Agricultural Productivity and Competitiveness*, C. R. Curtis and V. E. Mezainis, Eds., conference held in Washington, D.C. April 10, 1989 as an activity of the International Research Committee, International Science and Education Council, The Ohio State University, 1990, pp. 37–41.

Council for Agricultural Sciences and Technology. Challenges Confronting Agricultural Research at Land Grant Universities. Issue Paper Number 5, November, 1994. Council for Agricultural Science and Technology, Ames, IA, 1994.

Curtis, C. R. and V. E. Mezainis. Research Opportunities for U. S. Scientists at the International Reseach Centers and Affiliates. The Ohio State University, 1993.

Curtis, C. R. and V. E. Mezainis, Eds. *International Agricultural Research: Its Relationship to U. S. Agricultural Productivity and Competitiveness*, Conference held in Washington, D.C. April 10, 1989 as an activity of the International Research Committee, International Science and Education Council, The Ohio State University, 1990.

Pinstrup-Anderson (PER). Paper prepared for "Globalization and Linkages to 20-20: Challenges and Opportunities for OECD Countries," Organization for Economic Cooperation and Development (OECD) Paris, France, June 17–18, 1996.

Science and Education 2000. National Agricultural Research and Extension Users Advisory Board. U. S. Department of Agriculture, Washington, D.C., September 1995.

Walsh, L. M. International Agricultural Research: Helping Others While We Help Ourselves. In *International Agricultural Research: Its Relationship to U. S. Agricultural Productivity and Competitiveness*, Curtis, C. R. and Mezainis, V. A., Eds., conference held in Washington, D.C. April 10, 1989 as an activity of the International Research Committee, International Science and Education Council, The Ohio State University, 1990.

TWENTY-FIVE

Research Priorities for Sustainable Use of Soil and Water Resources

Goro Uehara

INTRODUCTION

The three purposes for undertaking long-term, sustainable soil and water use research are to (1) *understand* ecosystem processes and mechanisms, (2) integrate and synthesize knowledge of processes and mechanisms to *predict* system behavior and performance, and (3) enable research clients to apply this predictive capability to *control* outcomes of policy and management decisions. All three purposes are inextricably linked so that the third purpose, which is what our customers want, cannot be achieved without completing the first and second purposes.

Understanding of processes and mechanisms is what scientists do best, and control of outcomes is what our research clients want most. Very often what scientists do best does not lead to what clients need because the second purpose of research is not fulfilled. This situation is the root cause of one of the great tragedies of science and technology; namely, that the knowledge scientists have and take for granted is not accessible to people who need it most. As a consequence, critical soil and water policy and management decisions that influence outcomes of sustainable land use are made without benefit of this knowledge.

Much of this problem can be traced to the way science is conducted and how researchers are rewarded. Since prediction and control are not possible without understanding of processes and mechanisms, scientist are rewarded and encouraged to concentrate on process-level research. Efforts to channel scientists into prediction research has not been especially successful. Most prediction research has been confined to behavior of components within a system, as for example, predicting water movement in soils. But most clients, including policy makers and farmers face systems problems not solvable by knowledge from a single discipline. To predict the growth and development of a crop in the soil-plant-atmosphere continuum, for example, requires integration of knowledge from at least plant science, soil sciences, meteorology, physiology, hydrology, and systems modeling. Sharing credit of interdisciplinary research is often painfully difficult and many scientists find it easier to return to individual disciplinary research. Research to understand processes and mechanisms should not be an end in itself, but the barriers to predic-

Table 25.1. Customers of Long-Term Soil and Water Research

Intermediate Customers	End-Customers
• Agribusiness	• Bankers
• Banks	• Business executives
• Consultants	• Elected officials
• Extension agents	• Farmers
• Nongovernment organizations	• Home owners
• Regulatory agencies	• Policy makers

tion research virtually make it so. The purpose of this chapter is to describe a research strategy that connects basic research to client needs.

IDENTIFYING CUSTOMERS AND THEIR NEEDS

Sustainable use of soil and water resources involves users and customers. Our customers are farmers, ranchers, policy makers, regulatory agencies, consultants, extension agents, bankers, agribusinesses, and home owners. They need information that will enable them to make economically and environmentally sound decisions. They need answers to "what if" questions related to their occupations and livelihood. Their information needs should be the driving force for research prioritization, but for the most part, that is not yet the case. To alter the situation we must first identify our customers and their needs. Table 25.1 lists potential customers of soil and water research.

LINKING RESEARCH TO CUSTOMERS' NEEDS

Research to understand processes and mechanisms is so remotely related to customers' needs that a major effort in knowledge synthesis is required to transform process level information into prediction models. Not only is building of prediction models a major undertaking requiring specialists from many disciplines, the number of scientists with skills to lead system level modeling is small. In addition, the interdisciplinary requirements render prediction research expensive owing to the large number of scientists involved. All of these factors contribute to the neglect of prediction research, and in the end, prevent customers from benefitting from research that was intended for them. One consequence of this is the low esteem some development officials have of research and researchers. When many years of technical assistance fail to bring about desired change, one often hears the comment "the last thing we need is more research." Much of the antagonism nonscientists have for science and scientists stems from our failure to synthesize what we know about processes and mechanisms to produce the predictive tools customers need and can use to diagnose and solve problems.

Africa is an example where knowledge generated by research remains underutilized. Any number of reasons can be cited for this situation including low education, poor infrastructure, political instability, and lack of credit. For many African nations the hope for change resides in improving agriculture first. This is true for most countries in which the majority of people depend on agriculture for their livelihood.

Theodore Schultz stated in his award-winning book *Transforming Traditional Agriculture* that agricultural development efforts should invest more in research that enables farmers to exercise choice. To choose implies that one must have options. Choices are made on the basis of anticipated outcomes, and many risk-adverse farmers refuse to exercise choice when offered promising options because the options do not come with assurances of success.

It is not enough to indicate the long-term average outcome of adopting an untried option. The average outcome may tower over the farmers' current output, but farmers instinctively understand that the risk of farming resides not in the long-term averages, but in the variances or the tails of probability distributions. What farmers need most is the ability to explore the future with a wide array of locally implementable options. They need help in predicting and visualizing outcomes of their choices. It is in this regard that agricultural science has failed the farmers of the world.

PREDICTION RESEARCH

Knowledge synthesis is the heart of prediction research. It is not enough to know about the biochemistry of photosynthesis or the physics of soil water and nutrient transport, or the microbial ecology of pathogens, or the taxonomy of plants, animals and soils. Scientists are very good at describing their own subject matter specialties in great detail, but spend little time connecting, for example, water and nutrient dynamics in soils to photosynthesis with the eventual aim of predicting the growth and development of a crop in the soil-plant-atmosphere continuum. The synthesis of isolated pieces of knowledge into a connected whole so that the behavior and performance of the integrated whole can be simulated constitutes the field of systems analysis and simulation. By synthesizing and integrating knowledge, one is able to assess, for example, not only the effect of applying fertilizer nitrogen on crop growth, but the effect of the amount and frequency of rainfall on nitrogen leaching and its potential for groundwater contamination, the volatilization of nitrogen and its oxides into the atmosphere, and even the sequestration of carbon in soils. The capacity to simultaneously assess the effect of nitrogen, or any other input, on productivity and environmental sustainability enables customers to evaluate alternative ways to maximize yield while minimizing the unwanted side-effects of human interactions with the environment and maximizing benefits to society. This is what research on sustainable agriculture should enable our customers to do.

THE MINIMUM DATA SET

Knowledge synthesis and integration is necessary but not sufficient to enable customers to simulate systems behavior and performance. To be sufficient, organized knowledge must be fueled by a minimum set of input data. The organized knowledge serves as an inference engine and the minimum data set as the fuel to energize the engine. The process-based knowledge enables the inference engine to operate globally, whereas the minimum set of input data enables the inference engine to simulate outcomes on a site-specific basis. The minimum data set therefore must include such site-specific data as soil properties, weather variables such as daily rainfall, maximum and minimum air temperature, solar radiation, day length, relative humidity, crop information such as maturity type and photoperiod sensitivity, and management information such as plant population, planting date, row spacing, fertilizer and pest control measures, irrigation, and any other factor that significantly affects outcomes (IBSNAT, 1988; 1990).

It should be noted that the minimum data set includes management factors which are under human control, as well as environmental factors such as rainfall, temperature, solar radiation, and day length over which humans have little or no control. Not only are the latter variables not controllable, their magnitudes even for a few days in the future is not forecastable to the degree necessary to reliably predict ecosystem outcomes. How, then, can we explore the future with predictions of what is likely to happen when we are unable to predict future weather?

Instead of future weather, the minimum data set consists of site-specific, long-term, historical weather data. With long-term data, the sustainability of any number of practices can be simulated over many years to see which practice produces outcomes desired by the customer. Because the outcomes are simulated over many years, they can be presented as a probability distribution associated with a mean and a variance. The assumption we make is that historical weather can be used as a surrogate for future weather. If this assumption holds, the probability distribution provides a measure of two key properties of sustainable agroecosystems, namely productivity and stability (Conway, 1985). Productivity is defined as yield or profit per unit input of land, labor and capital, and stability refers to the fluctuations in yield or profit over time. In this case, productivity is represented by the average yield or profit, and stability by the variance. A high variance indicates large fluctuations in yield leading a feast or famine situation. Land use strategies that lead to high means and low variances are what customers look for. In high risk environments a low variance is often more important than high average yields.

VALIDATION

Our predictive capabilities will always be limited by imperfect and incomplete knowledge of how nature works. But our customers want our model of nature to mimic reality to a degree that enables them to use model outputs to support decision making. There are two ways to test a model's integrity. One can test model output against results of experiments especially designed for that purpose. Good agreement between observed and predicted results for several key events provides the bases for using predictive tools for decision support.

Another way to gain customer confidence is to demonstrate the model's capability to simulate historical events remembered by local audiences. The exceptionally good and bad years, or events in the lower and upper tails of the probability distributions should show up as outliers in the history of the region. Not only should the prediction model be able to simulate past events, it should also be able to simulate alternative actions that would have minimized damage to the environment from those events.

CONTROL RESEARCH

The goal of control research is to enable customers to exercise choice. It should allow customers to explore the future with options which they would not otherwise consider. Sustainable use of soil and water resources is achievable if policy makers and users of soil and water resources can look for new options whenever a problem is encountered. Control research is based on the premise that it is faster, cheaper, and more convenient to use process-based simulation models to search for options than to conduct trial and error experiments on the system itself. The demand for more food on less land for more people will place increasing pressures on our soil and water resources. There will not be sufficient time or technicians to assist the millions of customers in an interconnected world who must comply with sustainable soil and water use policies and practices. The tools for diagnosing problems and evaluating alternative solutions to deal with them must be placed in the hands of customers or change agents working closely with customers. These diagnostic and problem solving tools are often referred to as decision support systems or decision aids.

A major task of control research is to develop decision aids that customers will use.

Decision aids developed by scientists rarely meet the user-friendliness customers require. One way to ensure user-friendliness is to involve customers very early in the product development phase. Customers should clearly be involved in the problem identification stage to determine what needs to be predicted and controlled. Research prioritization, therefore, should be first and foremost, customer-driven.

A good time to test customer reaction to the prototype decision aid is during its field validation. Customer suggestions should be carefully noted and incorporated into the product. The process should be iterative, dynamic and self-correcting so that when the product is released to the general public, it will have gone through several cycles of customer evaluation.

RESEARCH PRIORITIZATION

Human curiosity about how nature works has been the basis for doing basic research. This curiosity-driven research continues today, but may need to be supplemented by a *customer-driven* curiosity research. In customer-driven curiosity research, the first research priority is set by customer needs. Customer needs in turn determine the direction of prediction research. Prediction research relies mainly on existing knowledge of processes and mechanism. Libraries and the experience and personal knowledge of scientists are the primary sources of information. Prediction research will almost always identify knowledge gaps that block predictions. Research to fill knowledge gaps should be the engine that drives the prioritization process.

Research needs will also be identified during the validation exercise. The failure of the decision aid to mimic events in the field will require either new knowledge or new ways to find how each part is connected to all other parts of the system. By completing all three purposes of research, it is possible to channel curiosity-driven research to meet customer needs.

CONCLUSION

The connections among understanding (basic) research, prediction research, minimum data sets, validation, control research, and customer needs are shown in Figure 25.1. Feedback loops reach out from the customers to ensure that the research is directed to meet customer needs.

Two examples of international projects that utilize this research approach are the Integrated Nutrient Management Decision Support Systems project currently underway under the Soil Management Collaborative Research Support Program (TropSoils, 1992), and the International Benchmark Sites Network for Agrotechnology Transfer project (IBSNAT, 1993). Both projects funded by the Agency for International Development produce knowledge products designed for use by intermediate customers identified in Table 25.1.

In these schemes, customer needs define the product the interdisciplinary team develops, and the product, in turn, specifies the knowledge to be synthesized and integrated. Since the sustainable use of soil, water, and land resources rests in the hands of customers, they, not researchers, must have the means to find, evaluate, and implement sustainable land use policies and practices. But it is the scientists' responsibility to transform knowledge into decision aids that customers will and can use to make better economic and environmental choices for themselves and society.

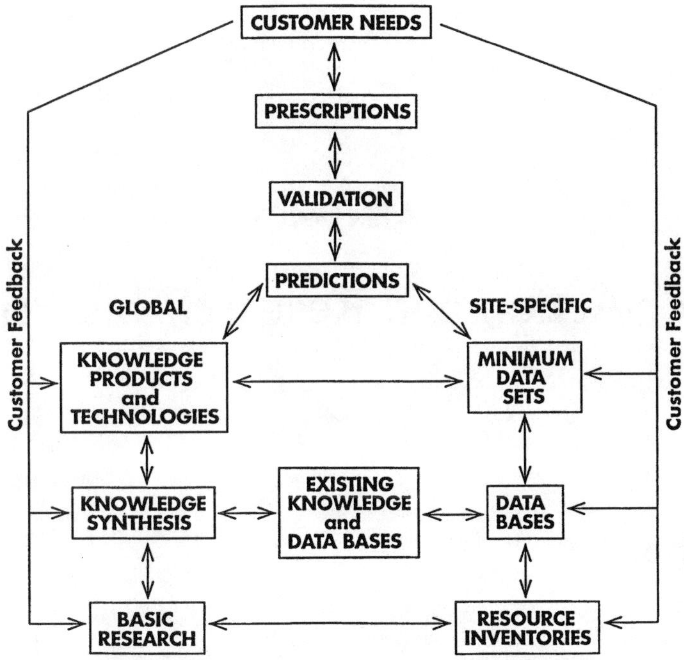

Figure 25.1. Connecting basic research to customer needs.

REFERENCES

Conway, G.R. Agroecosystem Analysis. *Agricultural Admission*. 20:32–55, 1985.

IBSNAT. *Experimental Design and Data Collection Procedure for IBSNAT: Minimum Data Set for Systems Analyses and Crop Simulation*, 3rd edition, Technical Report 1. Department of Agronomy and Soil Science, University of Hawaii, Honolulu, 1988.

IBSNAT. Documentation for IBSNAT Crop Model Input and Output Files: For the Decision Support System for Agrotechnology Transfer (DSSAT V.2.1). Technical Report 5. Department of Agronomy and Soil Science, University of Hawaii, Honolulu,1990.

IBSNAT. The IBSNAT Decade. Department of Agronomy and Soil Science, College of Tropical Agriculture and Human Resources, University of Hawaii, Honolulu, 1993.

TropSoils. Bulletin No. 92-01. Proceedings of the TropSoils Phosphorus Decision Support Systems Workshop. College Station, TX. March 11–12, 1992. Department of Agronomy and Soil Science, College of Tropical Agriculture and Human Resources, University of Hawaii, Honolulu, 1992.

TWENTY-SIX

Need for Action and Researchable Priorities

R. Lal

INTRODUCTION

Soil quality is an important determinant of agricultural sustainability and environmental quality. Decline in soil quality can adversely impact the economic well-being of people living on it. That is why there has been a growing concern about agricultural sustainability and food security since the mid-1980s because of the recognition of the widespread and severe problems of decline in soil quality due to numerous degradative processes. Hence, there is an inevitable question whether soil resources of the world are adequate to meet the demands of the growing human population.

Despite the recognition of the obvious link between soil quality and agricultural sustainability, there is no clear cause-effect relationship between the two, especially for the major hot spots of the world. For example, soil-related constraints to declining productivity of the wheat-rice system are not known for the principal soils and ecoregions of South Asia. Nor are known the soil factors responsible for declining yield of rice in the continuous rice systems in the Philippines and elsewhere in Southeast Asia.

While the visible symptoms of soil degradation (e.g., rill-interrill and gully erosion, sand dune movement, salinization) are easy to recognize, the impact of these degradative processes on agronomic productivity and sustainability is difficult to generalize. Productivity effects of soil degradative processes depend on a multitude of interacting factors including inherent soil properties, climate, and soil and crop management or the farming/cropping system. It is because of this reason that the quantitative effect of global soil degradation on agricultural productivity is not known. Even though the statistics on soil degradation are being compiled on regional and global scales, impact of degradation on productivity is difficult to assess even on the landscape scale.

Therefore, the issues of soil quality and agricultural sustainability that need to be addressed include the following:

1. Agricultural Sustainability

It is important to identify the criteria of a sustainable agricultural system. Some relevant criteria include the following: meeting the demand for food and fiber into an indefinite future, maintaining and enhancing the resource base, improving and maintaining the environmental quality. Once the criteria are identified, which may be different among agroecoregions and socioeconomic environments, methodologies must be established to evaluate sustainability of agricultural systems. Some relevant criteria (e.g., total factor productivity, total natural resource management productivity) must be validated and adopted under local conditions. These criteria for evaluating the sustainability will change over time, as the demand on natural resources changes due to change in socioeconomic conditions.

2. Soil Quality

While the subject of agricultural sustainability has been studied by biophysical and socioeconomic scientists since early 1980s, the term "soil quality" has been in vogue since only the early 1990s. There is a strong need for establishment and standardization of quantitative methods and indices of monitoring soil quality. Developing indices of soil quality involves identification of key soil properties, establishment of critical limits or threshold values of these properties, development of a rating system, and establishment of suitable indices for assessing soil quality. Soil quality indices must be established for principal soils and major crops and land use systems.

It is also necessary to establish the link or functional relationship between soil quality and agricultural sustainability (Fig. 26.1). Such a relationship may differ among agricultural systems and land uses. In some cases, there may be a threshold value of the soil quality (S_{qt}) index before the sustainability index responds. The magnitude of threshold value may differ among production systems and land uses, and needs to be established.

The relationship between agricultural sustainability and soil quality may not be linear, and depends upon soil and the production system. Such a functional relationship may also be dynamic, exhibit large temporal variations, and change with input and management.

3. Soil Quality and Food Security

In view of the ever-increasing demand for food and fiber, the functional relation between soil quality and sustainability of a production system must be extended to evaluate food

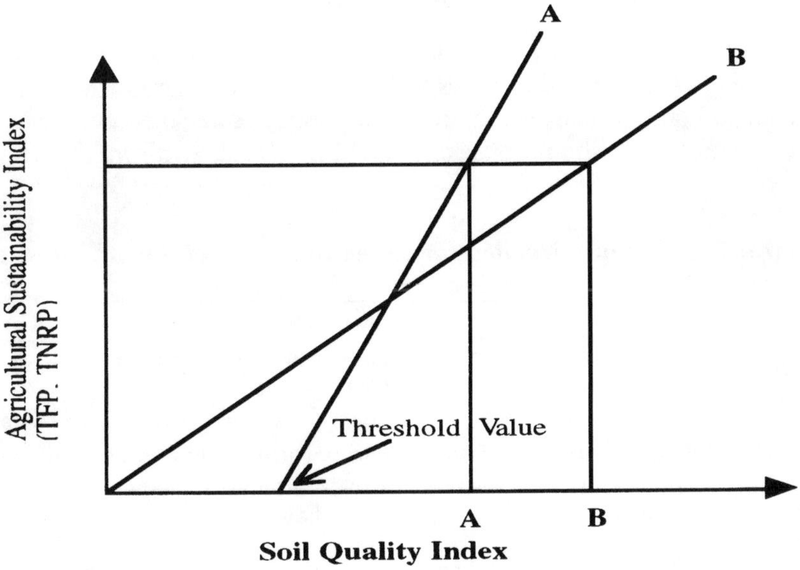

Figure 26.1. Establishing relationship between soil quality and agricultural sustainability.

security at regional, national, or global scale. The question of food security is important because the per capita arable land area is decreasing along with the decline in soil quality due to several degradative processes. The question "Is the world running out of good arable land?" must be answered through: (i) quantitative evaluation of productivity for different soil quality levels or severity of degradative processes at the level of soil order or suborder and for predominant cropping/farming systems, e.g., rice-wheat system on alluvial soils in the Indo-Gangetic plains; (ii) scaling up the soil degradation from the level of soil order to regional and global scale through development of appropriate scaling procedures using modern techniques, e.g., GIS; and (iii) predicting soil quality at different times (50, 100, 200 years) for different land-use scenarios and relating it to potential productivity and future demand.

4. Soil Quality Restoration

In view of the severity of soil degradation, especially in the ecologically-sensitive regions of the tropics and subtropics, it is necessary to adopt a strict code of conduct and define goals for improving soil quality. The production base can be vastly expanded by restoration of the quality of soils degraded by erosion, compaction, salinization, anaerobiosis, and fertility depletion. Adoption of a soil quality restoration program at a global scale would involve development of regional/national "Soil Policy" and its implementation through appropriate incentives. The society/community must pay the price for restoration of degraded soils.

The technical aspect of soil quality restoration involves: (i) establishment of critical limits of key soil properties that determine soil quality, (ii) development of the database to understand temporal changes in these properties in relation to the restorative measures, and (iii) establishing the cause-effect relationship between soil quality and productivity for different input and management systems. The objective is not to preserve soil resources, but to enhance their productivity for human welfare and use by future generations through improvement in soil quality. Methods of soil quality restoration differ among the degradative processes, antecedent soil properties, and the intended land use.

5. Long-Term Soil Quality Management Experiment

There is a dearth of long-term experiments in the tropics which have been specifically designed and are appropriately monitored to assess soil quality. These experiments should be established for major soils and principal land uses in important ecoregions of the tropics (Table 26.1). Such experiments should be established for Oxisols, Ultisols, Alfisols,

Table 26.1. Long-Term Experiments Needed for Assessment of Soil Quality in the Tropics

Ecoregion	Soil	Land Use
1. Humid tropics	Oxisols and Ultisols	Perennials, tree-plantation, pastures, root crops
2. Subhumid tropics	Alfisols, Ultisols	Grain crops (maize, rice, beans), pastures
3. Semiarid tropics	Alfisols, Vertisols, Aridisols	Sorghum, millet, chickpeas, pastoral systems
4. Semiarid subtropics	Inceptisols	Rice-wheat system
5. Tropical highlands	Inceptisols, Andisols	Plantations, rice-wheat system

Vertisols, Aridisols, Inceptisols, Andisols, and other major soils of the tropics. Key soil properties to be monitored should be identified, and methods for their evaluation standardized. These experiments should be established by a central coordinating organization and guaranteed for their continuity for several decades if not centuries. There are some ongoing soil fertility and fertilizer experiments being conducted in the tropics since the early 1900s (e.g., in India). Not only should these experiments be continued but their scope should be broadened by monitoring temporal changes in soil physical, chemical, and biological quality. Establishment of well-designed long-term experiments to establish temporal trends in soil quality under different management systems is a high priority at the international level.

6. Reasons for Lack of Adoption of Proven Technologies

There are numerous soil management technologies that have proved excellent under research conditions but have not been adopted by the farming community. Reasons for the lack of adoption of such technologies must be established. In most cases, such technologies emerged from a narrow disciplinary-based biophysical research, e.g., soil fertility management, erosion control, crop residue management, tillage method, etc. Therefore, the reasons for the lack of adoption may be socioeconomic, e.g., small farm size, labor constraints, lack of credit opportunities, lack of institutional support, gender, and other equity issues. Such reasons may change with time, but their identification is crucial to modification and adaptation of technologies developed.

7. A New Paradigm in Soil Quality Research

The problems of agricultural sustainability, soil quality in relation to productivity and environmental regulatory capacity, and food security are complex. Soil scientists cannot address these complex problems by themselves or in isolation, and must work in close cooperation with other disciplines to address specific soil management issues. The disciplinary composition of such teams may differ depending on specific soil-related constraint (Table 26.2). For example, an effective soil erosion management program can only be developed by a multidisciplinary team comprising soil scientists, hydrologists, climatologists, sedimentologists, geographers, agronomists, agricultural engineers, economists, social scientists, and land use planners.

Table 26.2. A Multidisciplinary and Holistic Approach to Soil Quality Research

Soil-Related Constraints	Multidisciplinary Team Composition
1. Soil erosion management	Soil scientists, hydrologists, climatologists, geographers, sedimentologists, agronomists, agricultural engineers, economists, social scientists, land use planners
2. Soil fertility restoration	Soil scientists, agronomists, plant physiologists, economists, chemists, environmentalists
3. Soil organic carbon sequestration	Soil scientists, climatologists, ecologists, land use planners, economists, political scientists, agronomist
4. Conservation tillage and residue management	Soil scientists, agronomists, agricultural engineers, energy specialists, specialists in IPM, especially weed management
5. Water management (irrigation and drainage)	Soil scientists, hydrologists, agronomists, geologists, agricultural engineers, climatologists, economists

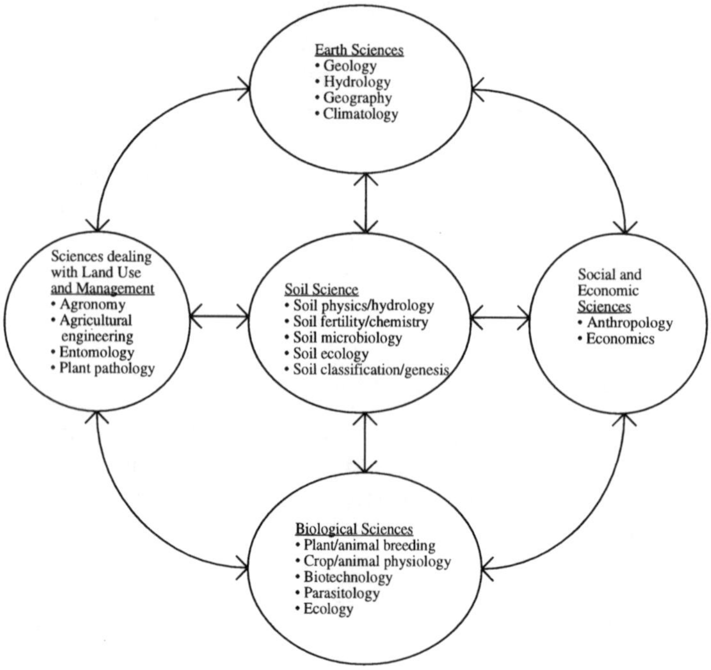

Figure 26.2. An interdisciplinary and a holistic approach to soil science.

One of the principal reasons of modest or even lack of progress by soil science vis-à-vis other disciplines (e.g., ecology, biology, geography, geology) is the narrow and disciplinary-based approach to complex problems that require a multidisciplinary strategy. Soil scientists must develop rapport with and reach out to other disciplines to effectively address these basic issues of food security and environmental quality. Some important disciplines that soil scientists must interact with are outlined in Figure 26.2. These include earth sciences (e.g., geology, geography, hydrology, and climatology) to study basic processes, biological sciences (e.g., genetics, biotechnology, physiology, and parasitology dealing with plants and animals) to understand productivity and its interaction with ecosystems, social and economic sciences (e.g., anthropology, production economics, resource economics, and other issues related to human dimension of the land use) to understand social and economic forces affecting the decision-making processes, and sciences dealing with production and land use (e.g., agronomy, agricultural engineering, plant/animal health) to understand the production dynamics.

Adoption of this multidisciplinary and holistic approach to addressing practical issues of natural resources management does not imply that basic and disciplinary-based research is not important. On the contrary, soil scientists must understand basic processes and properties of soil, their genesis and evolution, and reasons for their spatial and temporal variations. Soil scientists must be strong in their own discipline to be able to effectively interact with other disciplines.

CONCLUSIONS

World soil resources being finite, intensive land use is inevitable to meet the global demands for food and fiber. Intensification of agriculture on existing arable land increases risks of soil and environmental degradation. Therefore, it is important to understand land use and management impacts on soil quality, establish links between soil quality on the one hand and agricultural sustainability, and environmental quality (e.g., water pollution, greenhouse effect) on the other. Establishment of such functional relationships involves identifying criteria for quantitative assessment of sustainability and soil quality.

Restoration of quality of degraded soils is a high global priority. To do so involves establishment of critical limits of key soil properties in relation to specific soil degradative processes, and study of temporal variations in such properties and processes under different restorative measures.

There is a need to adopt a multidisciplinary and a holistic approach to soil science for addressing global issues of food security and environmental quality. In cooperation with anthropologists and economists, soil scientists must identify reasons for the lack of or slow rate of adoption of research-proven technology. Alleviation of soil-specific constraints to an intensive land use must be done in adopting a team approach through establishment of close linkages with other disciplines, e.g., biology, ecology, earth sciences, social and behavioral sciences, and those disciplines involved in production and management.

Soil resources of the world are adequate to meet the needs of the present and future population. Soil scientists can play a key role in achieving food security and enhancing environmental quality by improving/restoring soil quality through development of multidisciplinary and holistic approaches to natural resources management.

INDEX